W0234799

# Morphology Methods

# MORPHOLOGY METHODS

## Cell and Molecular Biology Techniques

Edited by

# Ricardo V. Lloyd, MD, PhD

*Mayo Clinic, Rochester, MN*

Foreword by

## Ronald A. DeLellis, MD

*Weill Medical College of Cornell University and New York Presbyterian Hospital, New York, NY*

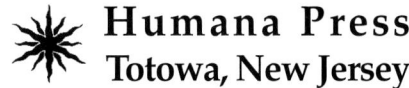 Humana Press
Totowa, New Jersey

© 2001 Humana Press Inc.
999 Riverview Drive, Suite 208
Totowa, New Jersey 07512

**www.humanapress.com**

All rights reserved. No part of this book may be reproduced, stored in a retrieval system, or transmitted in any form or by any means, electronic, mechanical, photocopying, microfilming, recording, or otherwise without written permission from the Publisher.

The content and opinions expressed in this book are the sole work of the authors and editors, who have warranted due diligence in the creation and issuance of their work. The publisher, editors, and authors are not responsible for errors or omissions or for any consequences arising from the information or opinions presented in this book and make no warranty, express or implied, with respect to its contents.

This publication is printed on acid-free paper. ∞

ANSI Z39.48-1984 (American Standards Institute) Permanence of Paper for Printed Library Materials.

Cover illustrations: Backround is figure 18 and inset is figure 5 from Chapter 4, "*In Situ* Detection of Infectious Agents," by Gary W. Procop and Randall Hayden.

Cover design by Patricia F. Cleary.

For additional copies, pricing for bulk purchases, and/or information about other Humana titles, contact Humana at the above address or at any of the following numbers: Tel: 973-256-1699; Fax: 973-256-8341; E-mail: humana@humanapr.com, or visit our Website: http://humanapress.com

**Photocopy Authorization Policy:**

Authorization to photocopy items for internal or personal use, or the internal or personal use of specific clients, is granted by Humana Press Inc., provided that the base fee of US \$10.00 per copy, plus US \$00.25 per page, is paid directly to the Copyright Clearance Center at 222 Rosewood Drive, Danvers, MA 01923. For those organizations that have been granted a photocopy license from the CCC, a separate system of payment has been arranged and is acceptable to Humana Press Inc. The fee code for users of the Transactional Reporting Service is: [0-89603-955-2/01 \$10.00 + \$00.25].

Printed in the United States of America. 10 9 8 7 6 5 4 3 2 1

Library of Congress Cataloging-in-Publication Data

Morphology methods : cell and molecular biology techniques / edited by Ricardo V. Lloyd.
     p. cm.
   Includes bibliographical references and index.
   ISBN 0-89603-955-2 (alk. paper)
   1. Cytology--Technique. 2. Molecular biology--Technique. 3. Fluorescence in situ hybridization. 4. Polymerase chain reaction. 5. Immunohistochemistry. I. Lloyd, Ricardo V.

   QH585 .M67 2001
   571.6'028--dc21

                                                                        2001016959

# Foreword

The past several decades have witnessed an impressive array of conceptual and technological advances in the biomedical sciences. Much of the progress in this area has developed directly as a result of new morphology-based methods that have permitted the assessment of chemical, enzymatic, immunological, and molecular parameters at the cellular and tissue levels. Additional novel approaches including laser capture microdissection have also emerged for the acquisition of homogeneous cell populations for molecular analyses. These methodologies have literally reshaped the approaches to fundamental biological questions and have also had a major impact in the area of diagnostic pathology.

Much of the groundwork for the development of morphological methods was established in the early part of the 19th century by Francois-Vincent Raspail, generally acknowledged as the founder of the science of histochemistry. The earliest work in the field was primarily in the hands of botanists and many of the approaches to the understanding of the chemical composition of cells and tissues involved techniques such as microincineration, which destroyed structural integrity. The development of aniline dyes in the early 20th century served as a major impetus to studies of the structural rather than chemical composition of tissue. Later in the century, however, the focus returned to the identification of chemical constituents in the context of intact cell and tissue structure. Ultimately, it became possible to localize with great precision major classes of nucleic acids, proteins, and individual amino acids, lipids, and carbohydrates on the basis of specific chemical reactions that had been adapted for histological and cytological preparations.

The development of the immunofluorescence technique for the localization of pneumococcal antigens in 1942 by Albert Coons and his colleagues provided the foundation for the subsequent development of enzyme-based immunohistochemical procedures. The advantages of these methods were related to the facts that the reaction products could be visualized directly in the light microscope and that the multistep staining procedures possessed sufficient sensitivity to localize even the small amounts of antigen that survived formalin fixation and paraffin embedding. The subsequent development of increasingly more sensitive immunohistochemical staining sequences and the availability of increasing numbers of monoclonal antibodies together with the development of antigen retrieval technology has made immunohistochemistry clearly one of the most important research and diagnostic tools in the armamentarium of the investigator.

The development of molecular technologies has had an equally profound impact on approaches to the analysis of basic biological questions directly in tissue sections and single cells. *In situ* hybridization techniques, including FISH, have emerged as powerful techniques for the localization of DNA and RNA sequences at the cellular and subcellular levels. Increased sensitivity of *in situ* hybridization has been made possible both by amplification of target sequences through the use of the polymerase chain reaction (PCR) and *in situ* reverse transcriptase (RT)—PCR methods, as well as by approaches to enhance signal amplification, including the use of increasingly more sensitive nonradioactive techniques. Tissue and cDNA microarrays have provided a particularly useful approach for large scale analyses of DNA, RNA, or protein and for studies of gene expression.

The major objective of Dr. Lloyd's book is to present an overview of the major areas relevant to the current practice of molecular morphology. Both he and his distinguished group of contributors have been particularly successful in achieving this goal and in providing succinct overviews of specific methodologies, their major applications and detailed protocols for their performance.

Dr. Lloyd's monograph will undoubtedly find an important place on the bookshelves of basic scientists and clinical investigators who have an interest in correlating morphological and molecular parameters.

*Ronald A. DeLellis,* MD
Professor of Pathology
Weill Medical College of Cornell University
Vice Chair for Anatomic Pathology
New York Presbyterian Hospital
New York, NY

# Preface

Molecular morphologic methods—defined as molecular and cell biologic analyses of tissues in which the architectural integrity and spatial interrelationship of the cells being studied are preserved—have increased rapidly in number and versatility during the past few years. These changes have occurred both in diagnostic pathology and in basic scientific research. Several ongoing developments affecting the pathology and the scientific communities should make this book a valuable resource. First, it is usually difficult for pathologists and investigators interested in molecular morphology to learn rapidly from a single source about methods suitable to specific diagnostic and experimental questions. Second, the completion of the human genome project in the near future will provide the foundation to learn about the functions of myriad of genes with unique roles in specific cells and tissues, so a morphologic basis for the study of human genes and understanding human diseases will be in greater demand. There is no good single source available that discusses in detail the most significant aspects of recent cell biologic techniques by outstanding experts in their fields. Such a book is needed to keep up with scientific research in morphology and recent pathologic diagnostic techniques relevant in the twenty-first century.

Our objective was to produce a book addressing the major areas relevant to molecular morphology today. Many of the chapters include detailed protocols for setting up or performing techniques now in use. Potential pitfalls and anticipated problems are also discussed. Practicing pathologists interested in recent developments and researchers interested in molecular morphology for designing experiments, for teaching undergraduate, graduate, and professional students, or simply for keeping up with the literature detailing molecular morphologic approaches—all will find here the technical and scientific background to accomplish their objectives.

The publication of *Morphology Methods: Cell and Molecular Biology Techniques* would not have been possible without the enthusiastic support and contributions of Mr. Thomas Lanigan, President, and Mr. John Morgan of Humana Press.

*Ricardo V. Lloyd, MD, PhD*

# Contents

# Contributors

- MUHAMMAD AMJAD, PhD • *Laboratories of Transgenic and Recombinant Vaccine, Department of Biology, Lincoln University, Lincoln University, PA*
- OMAR BAGASRA, MD, PhD • *Laboratories of Transgenic and Recombinant Vaccine, Department of Biology, Lincoln University, Lincoln University, PA*
- LISA E. BOBROSKI, MS • *Laboratories of Transgenic and Recombinant Vaccine, Department of Biology, Lincoln University, Lincoln University, PA*
- LISA A. CERILLI, MD • *Fechner Laboratory of Surgical Pathology, University of Virginia Medical Center, Charlottesville, VA*
- JOHN L. FRATER, MD • *Department of Clinical Pathology, Cleveland Clinic Foundation, Cleveland, OH*
- RANDALL HAYDEN, MD • *Clinical and Molecular Microbiology, St. Jude Children's Research Hospital, Cleveland, OH*
- EVA HORVATH, PhD • *Department of Laboratory Medicine, St. Michael's Hospital, University of Toronto, Toronto, Ontario, Canada*
- PHILIP N. HOWLES, PhD• *Department of Pathology and Laboratory Medicine, University of Cincinnati College of Medicine, Cincinnati, OH*
- ERIC D. HSI, MD • *Departments of Anatomic and Clinical Pathology, Cleveland Clinic Foundation, Cleveland, OH*
- JOHBU ITOH, MD, PhD • *Laboratories for Structure and Function Research, Tokai University School of Medicine, Boseidai, Isehara City, Kanagawa, Japan*
- JIN LONG • *Mayo Clinic and Foundation, Rochester, MN*
- NIKIFOROS KAPRANOS, MD, PhD • *Department of Pathology, Molecular Division, Amalia Fleming Hospital, Athens, Greece*
- PAUL KOMMINOTH, MD • *Department of Pathology, Hospital Baden, Baden, Switzerland*
- GEORGE KONTOGEORGOS, MD, PhD • *Department of Pathology, G. Gennimatas General Hospital of Athens, Athens, Greece*
- KALMAN KOVACS, MD, PhD • *Department of Laboratory Medicine, St. Michael's Hospital, University of Toronto, Toronto, Ontario, Canada*
- PAUL J. KURTIN, MD • *Divisions of Anatomic Pathology and Hematopathology, Mayo Clinic, Rochester, MN*
- LARS-INGE LARSSON, DSc • *Division of Cell Biology, Department of Anatomy and Physiology, The Royal Veterinary and Agricultural University, Frederiksberg, Denmark*
- RICARDO V. LLOYD, MD, PhD • *Department of Laboratory Medicine and Pathology, Mayo Clinic and Foundation, Rochester, MN*

- AKIRA MATSUNO, MD, PhD • *Department of Neurosurgery, Teikyo University Ichihara Hospital, Ichihara City, Chiba, Japan*
- TADASHI NAGASHIMA, MD, PhD • *Department of Neurosurgery, Teikyo University Ichihara Hospital, Ichihara City, Chiba, Japan*
- YURI E. NIKIFOROV, MD, PhD • *Department of Pathology and Laboratory Medicine, Univeristy of Cincinnati College of Medicine, Cincinnati, OH*
- R. YOSHIYUKI OSAMURA, MD, PhD • *Department of Pathology, Tokai University School of Medicine, Boseidai, Isehara City, Kanagawa, Japan*
- AUREL PERREN, MD • *Department of Pathology, University Hospital Zürich, Zürich, Switzerland*
- GARY W. PROCOP, MD • *Clinical Microbiology, Cleveland Clinic Foundation, Cleveland, OH*
- XIANG QIAN • *Mayo Clinic and Foundation, Rochester, MN*
- PATRICK C. ROCHE, PhD • *Department of Laboratory Medicine and Pathology, Mayo Clinic, Rochester, MN*
- NAOKO SANNO, MD, PhD • *Department of Neurosurgery, Nippon Medical School, Tokyo, Japan*
- SUSAN SHELDON, PhD • *Department of Pathology, University of Michigan, Ann Arbor, MI*
- SHAN-RONG SHI, MD • *Department of Pathology, Keck School of Medicine, University of Southern California, Los Angeles, CA*
- CLIVE R. TAYLOR, MD, PhD • *Department of Pathology, Keck School of Medicine, University of Southern California, Los Angeles, CA*
- AKIRA TERAMOTO, MD, DMSc • *Department of Neurosurgery, Nippon Medical School, Tokyo, Japan*
- ELENI THODOU, MD, PhD • *Department of Pathology, G. Gennimatas General Hospital of Athens, Athens, Greece*
- RAYMOND R. TUBBS, DO • *Department of Clinical Pathology, Cleveland Clinic Foundation, Cleveland, OH*
- SERGIO VIDAL, DVM, PhD • *Department of Laboratory Medicine, St. Michael's Hospital, University of Toronto, Toronto, Ontario, Canada; Laboratory of Histology, University of Santiago de Compostela, Lugo, Spain*
- KEIICHI WATANABE, MD, PhD • *Department of Pathology, Tokai University School of Medicine, Boseidai, Isehara City, Kanagawa, Japan*
- MARK R. WICK, MD • *Fechner Laboratory of Surgical Pathology, University of Virginia Medical Center, Charlottesville, VA*

# Color Plates

**Color Plates 1–8** follow page 208.

**Color Plate 1** *Fig. 1, Chapter 3.* (**A**) *In situ* hybridization detects EBV in a posttransplant lymphoproliferative disorder. (**B**) *In situ* hybridization to detect EBV in a nasopharyngeal carcinoma metastatic to a cervical lymph node. (**C**) Detection of κ (left) and λ (right) in a plasmacytic lymphoma. (**D**) *In situ* hybridization for albumin to diagnose a hepatocellular carcinoma metastatic to the scapula. (**E**) Chromogranin A and B expression in the adrenal cortex. (**F**) Pro-insulin mRNA expression in the normal pancreatic islets. *See* discussion on pages 36, 39 and full caption on page 37.

**Color Plate 2** *Figs. 1–19, Chapter 4. See* pages 47–55. • *Fig. 1.* Lymph noted with EBV, ISH. • *Fig. 2.* Pharyngeal biopsy with HPV, ISH. • *Fig. 3.* Placenta, involved by CMV, ISH. • *Fig. 4.* Oral cavity (palate) biopsy with HSV, H&E. • *Fig. 5.* Palate biopsy with HSV (*Fig. 4*) confirmed by ISH. • *Fig. 6.* Lung tissue with adenovirus, H&E. • *Fig. 7.* Adenovirus in lung tissue (*Fig. 6*) confirmed by ISH. • *Fig. 8.* Brain biopsy showing JC viral inclusion in oligodendrocytes, H&E. • *Fig. 9.* JC virus in brain biopsy (*Fig. 8*) confirmed by ISH. • *Fig. 10. Cryptococcus neoformans* in an open lung biopsy, GMS. • *Fig. 11.* Identification of the yeast forms in *Fig. 10* as *C. neoformans* using ISH. • *Fig. 12. Histoplasma capsulatum* in an open lung biopsy, GMS. • *Fig. 13.* Confirmation of the yeast forms in *Fig. 12* as *H. capsulatum* using ISH. • *Fig. 14.* Open lung biopsy with *Blastomyces dermatiditis*, GMS. • *Fig. 15.* Confirmation of the yeast forms in *Fig. 14* with ISH. • *Fig. 16.* Endospores and immature spherules of *Coccidiodes immitis* in a lung biopsy, GMS. • *Fig. 17.* Confirmation of *C. immitis* in *Fig. 16* by ISH. • *Fig. 18.* Open lung biopsy showing *Legionella pneumophila*, Warthin-Starry stain. • *Fig. 19.* Identification of *L. pneumophila* in *Fig. 18* by ISH.

**Color Plate 3** *Figs. 5, 6, and 7, Chapter 5. See* pages 74–79 for discussion and additional details. • *Fig. 5.* Two-color microdeletion probe. • *Fig. 6.* A two-color probe for *BCR* and *ABL* genes on a touch preparation from a spleen. • *Fig. 7.* Painting probe for chromosome 9.

**Color Plate 4** *Fig. 3, Chapter 6. See* pages 94–95. (**A**) Examples of numerical chromosome aberrations using an α-satellite, centromere-specific DNA probe. (**B**) Monosomy of chromosome 11 as the dominant abnormality in a mixed somatotrophlactotroph, mostly prolactin-producing pituitary adenoma (FITC/PI). (**C**) Chromosome 1 in ductal carcinoma of the breast. (**D**) Chromosome 2 breast carcinoma of the ductal type. (**E**) Localization of the DiGeorge gene on a chromosome in the 22q11 region in normal metaphase spreads. (**F**) Dual-color FISH in metaphases of normal lymphocytes demonstrates the *bcr* gene on chromosome 22 (green) and the *abl* gene on chromosome 9 (red) (FITC/TR/DAPI). (**G**) Mixture of whole paint (coatsome) DNA probe for chromosome 11 (red) and α-satellite centromere-specific probe or chromosome 2 (green) in normal metaphase spreads (FITC/TR/DAPI). (**H**) All human telomeres demonstrate all chromosomes in metaphase spreads (FITC/PI).

**Color Plate 5** *Figs. 9 and 10,* Chapter 10. • *Fig. 9.* Three-dimensional projection images of ACTH and GH cells in normal rat pituitary. (**A**) ACTH cells are closely associated with the microvessel networks. (**B**) ACTH signals show a granular pattern in the cytoplasm. (**C**) GH cells are clearly visible three-dimensionally by CLSM reflectance mode. (**D**) GH signals appear granular in the cytoplasm three-dimensionally by CLSM reflectance mode. See discussion on page 174 and full caption on page 176. • *Fig. 10.* Three-dimensional reconstructed images of bilaterally adrenalectomized and ACTH administered rat pituitary glands. *See* discussion on pages 174–175 and full caption on page 177.

**Color Plate 6** *Figs. 1–5, Chapter 16.* • *Fig. 1.* Immunoperoxidase stains for κ (**A**) and λ (**B**) immunoglobulin light chains on frozen section of a case of B-cell small lymphocytic lymphoma. *See* discussion and full caption on pages 280, 281. • *Fig. 2.* Immunoperoxidase stains for κ (**A**) and λ (**B**) immunoglobulin light chains on paraffin sections of a case of mantle cell lymphoma. *See* discussion on page 280 and full caption on page 282.• *Fig. 3.* B-cell small lymphocytic lymphoma. H&E section demonstrating the typical cytologic features (**A**). Frozen section immunoperoxidase stains for CD20 (**B**), CD5 (**C**), CD3 (**D**), and CD23 (**E**). *See* discussion on page 283 and full caption on page 284. • *Fig. 4.* Angioimmunoblastic T-cell lymphoma stained in frozen sections for CD2 (**A**) and for CD7 (**B**). *See* discussion on page 286 and full caption on page 287. • *Fig. 5.* Follicular lymphoma stained for bcl-2 paraffin sections. *See* discussion on page 288 and full caption on page 289.

**Color Plate 7** *Figs. 6–13, Chapter 16.* • *Fig. 6.* Mantle cell lymphoma stained for cyclin D1 in paraffin sections. *See* discussion and full caption on pages 289, 290. • *Fig. 7.* CD30-positive anaplastic large cell lymphoma stained for p80 paraffin sections. *See* discussion and full caption on pages 290, 291. • *Fig. 8.* Diffuse large B-cell lymphoma involving a lymph node. H&E section (**A**) and immunoperoxidase stain (**B**). *See* discussion and full caption on pages 292, 293. • *Fig. 9.* Angioimmunoblastic T-cell lymphoma involving a lymph node. H&E section (**A**) and immunoperoxidase stain (**B**) for CD3 performed on a paraffin section of the tumor. *See* discussion and full caption on pages 294, 295. • *Fig. 10.* Immunoperoxidase stain for CD10 performed on a paraffin section of a follicular lymphoma. *See* discussion and full caption on pages 296, 297. • *Fig. 11.* Immunoperoxidase stain for CD23 performed on a paraffin section of an angioimmunoblastic T-cell lymphoma. Same case as *Fig. 9. See* discussion and full caption on pages 298, 299. • *Fig. 12.* Immunoperoxidase stain performed on a paraffin section for the cytolytic granule protein Tia-1 in a case of NK/T-cell lymphoma of the nasal type. *See* discussion and full caption on pages 298, 299. • *Fig. 13.* Immunoperoxidase stain for CD30 performed on a CD30-positive anaplastic large cell lymphoma. *See* full caption and discussion on pages 300, 301.

**Color Plate 8** *Figs. 14–19, Chapter 16* • *Fig. 14.* Hodgkin's lymphoma, nodular sclerosis type involving a lymph node. H&E section demonstrating the large neoplastic cells admixed with small lymphocytes, eosinophils, plasma cells, and macrophages (**A**). Immunoperoxidase stains for CD15 (**B**), CD30 (**C**), and fascin (**D**) performed on paraffin sections. *See* discussion and full caption on pages 301–303. • *Fig. 15.* Hodgkin's lymphoma, nodular lymphocyte predominance type, involving a lymph node. *See* discussion on pages 302, 304, and full caption on page 305. • *Fig. 16.* Acute lymphoblastic leukemia, B-lymphocyte precursor type. *See* discussion and full caption on pages 305, 306. • *Fig. 17.* Acute myelogenous leukemia, FAB subtype M1. *See* discussion and full caption on pages 306, 307. • *Fig. 18.* Langerhans' cell histiocytosis involving the parotid gland. H&E (**A**) exhibits the typical cytologic features of this disorder. Immunoperoxidase stain for CD1a (**B**) performed in paraffin sections. *See* discussion and full caption on pages 307, 308. • *Fig. 19.* Histiocytic sarcoma. Immunoperoxidase stains for CD45 (**A**) and CD68 (**B**), in paraffin sections, and for CD13 on frozen sections of the tumor (**C**). *See* discussion on page 308 and full caption on page 309.

# 1
## Introduction to Molecular Methods

### Ricardo V. Lloyd, MD, PHD

## GENE STRUCTURE

The human genome contains 23 pairs of chromosomes with $3.5 \times 10^9$ nucleotide base pairs (bp) in length. With the near completion of the sequencing of the human gene, the best estimate of functional genes appears to be between 30,000 and 40,000 *(1–3)*. Approximately 90% of the human DNA has no known coding function, but a significant amount of DNA, especially in regions close to functional genes, has regulatory roles in cells. However, most of the functions of much of the DNA remain unknown.

Specific restriction enzymes derived from prokaryotes are used to cleave DNA at specific sequences, which helps in the analysis of the large amount of genomic DNA. For example, EcoRI, which is derived from *Escherichia coli,* cleaves DNA at the G↓ AATTC sequence. Because of the specificity of restriction endonucleases, the DNA can be manipulated by genetic engineering, since DNA from any source cut by EcoRI will produce identical ends that hybridize to any other source of DNA produced by digestion with these restriction enzyme *(4,5)*.

## CHEMICAL COMPOSITION OF NUCLEIC ACIDS

DNA and RNA are made of nucleotides, sugars, and phosphate groups (**Figs. 1–3**). DNA has four bases, the purines adenine and guanine and the pyrimidines cytosine and thymine. In RNA molecules, uracil is substituted for cytosine. The bases are linked to sugar groups (nucleoside), which are in turn linked to phosphates (nucleotides).

DNA and RNA are polynucleotide molecules with alternating series of sugar and phosphate groups (**Fig. 1**). The nucleotide sequences of a polynucleotide is usually written in the 5′ to 3′ order. The nucleotide sequence of DNA contains the genetic information.

The four DNA polynucleotide structures are base-paired in a complementary and parallel manner (**Fig. 2**). One strand runs from 3′ to 5′, and the other strand runs from 5′ to 3′. The DNA structure is twisted in a helix with one complete turn every three base pairs. During hybridization, which is one of the basic tools in molecular analyses, the two strands anneal or come together by base-pairing.

Humans somatic cells contain 46 chromosomes, which are made up of 23 pairs. The 23 pairs represent the haploid number of chromosomes, and the 46 chromosomes

From: *Morphology Methods: Cell and Molecular Biology Techniques*
Edited by: R. V. Lloyd © Humana Press, Totowa, NJ

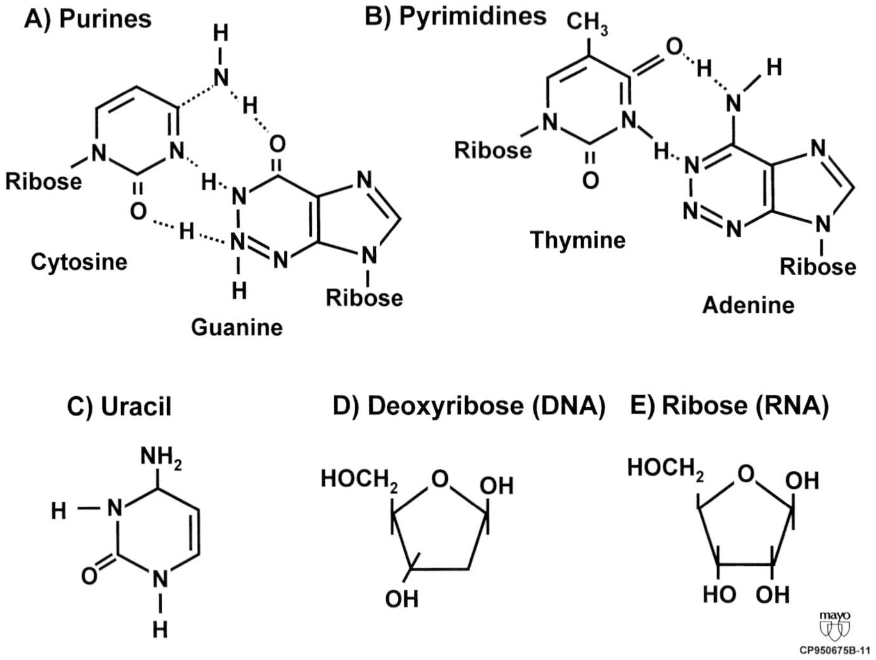

**Fig. 1.** Structure of molecules making up DNA and RNA. The bases (cytosine, guanine, thymine, adenine, and unacil) are linked via hydrogen bonds. (**A**) The purines are linked by three pairs of hydrogen bonds. (**B**) The pyrimidines are linked by two pairs of hydrogen bonds. (**C**) The structure of unacil, present in RNA in place of thymine. (**D**) and (**E**) Structures of the pentose sugars in DNA and RNA.

represent the diploid number, which represents the two homologous sites of each chromosome.

The sex chromosomes are different in males and females; females have two X chromosomes, and males have a Y and an X chromosome. The karyotype represents the complete chromosomal complement of a cell. These karyotypes are usually analyzed by cytogeneticists. Chromosomes are linked by a centromere with two sister chromatid portions. Chromosome maps are based on the position of the banding pattern (easily visualized after staining with specific dyes) and the position of the centromeres. Chromosomes are numbered as 1–22 plus the sex chromosomes. The short arm of the chromosomes is designated as a p and the long arm as q. The p and q arms are divided into various sections. Thus a designation of 5q21 would represent the long arm of chromosome 5, section 2, and band 1 (**Fig. 4**).

## RNA AND PROTEIN SYNTHESIS

Transcription is the process by which RNA is synthesized from DNA (**Fig. 5**). The RNA is complementary to the original DNA strand. RNA is synthesized 5′–3′ by RNA polymerases. There are different forms of RNA. For example, transfer RNA, small nuclear RNA, and ribosomal RNA are end products of transcription. In contrast, messenger RNA (mRNA) constitutes most of the RNA that encodes specific protein products

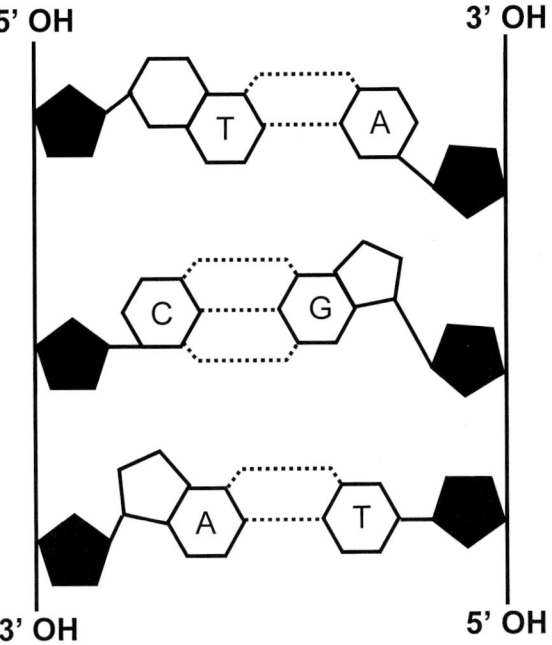

5' OH                    3' OH

3' OH                    5' OH

**Fig. 2.** Structure of double-stranded DNA. The deoxyribonucleotide residues (black areas) of each single strand of DNA are linked by phosphodiester bonds between the 3′ carbon of the first deoxyribose residue and the 5′ carbon of the next residue. The purines and pyrimidines are linked by hydrogen bonds. The two DNA strands are on antiparallel orientations.

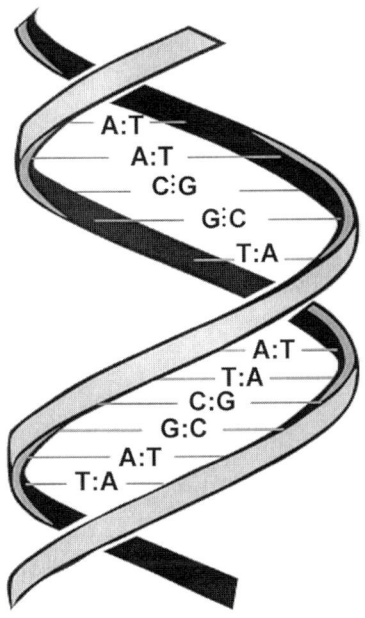

**Fig. 3.** Structure of the double helix, showing the three-dimensional relationship among the bases, sugars, and phosphodiester moieties.

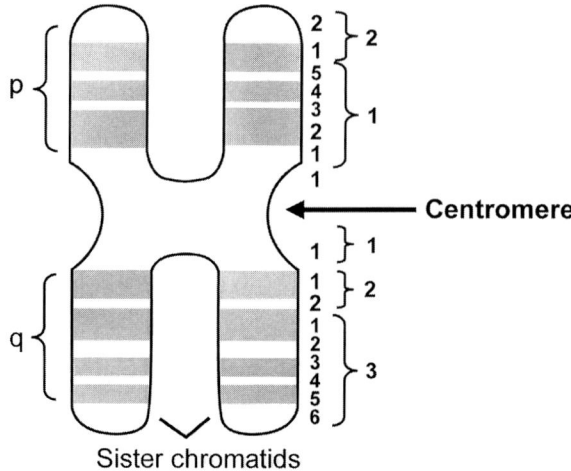

**Fig. 4.** Physical map of chromosome 7 after Giemsa staining and G-banding. Giemsa staining highlights the dark and light bands. The centromere and sister chromatids are shown.

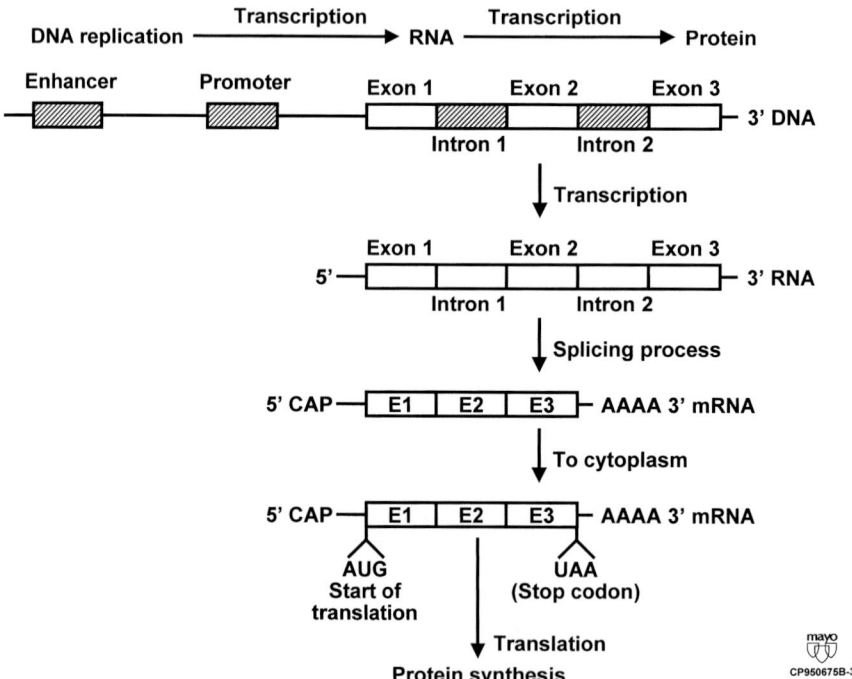

**Fig. 5.** Diagrammatic representation of transcription and translation processes leading to specific protein synthesis. The regulatory sequence with the enhancer and promoter sequences are located in the 5' upstream region. Enhancers bind protein factors and help to determine the rate of transcription. The coding sequences downstream include exons and introns. Introns are spliced out of the mature RNA. AUG is the start codon for protein coding, and there are three stop codons to terminate protein synthesis (UAA, UAG, and UGA). The 5' CAP site is at the 5' end, and the poly(A) tail is present at the 3' end. These sites help to increase the stability of the mature mRNA and optimize protein synthesis.

***Amino acid symbols***

| Amino acid | Symbol | Amino acid | Symbol |
|---|---|---|---|
| Alanine | A | Leucine | L |
| Arginine | R | Lysine | K |
| Asparagine | N | Methionine | M |
| Aspartic acid | D | Phenylalanine | F |
| Cysteine | C | Proline | P |
| Glutamine | Q | Serine | S |
| Glutamic acid | E | Threonine | T |
| Glycine | G | Tryptophan | W |
| Histidine | H | Tyrosine | Y |
| Isoleucine | I | Valine | V |

***The genetic code***

| First position (5'-End) | Second position: U | C | A | G | Third position (3'-End) |
|---|---|---|---|---|---|
| U | UUU F | UCU S | UAU Y | UGU C | U |
| | UUC F | UCC S | UAC Y | UCC C | C |
| | UUA L | UCA S | UAA Stop | UGA Stop | A |
| | UUG L | UCG S | UAG Stop | UGG W | G |
| C | CUU L | CCU P | CAU H | CGU R | U |
| | CUC L | CCC P | CAC H | CGC R | C |
| | CUA L | CCA P | CAA Q | CGA R | A |
| | CUG L | CCG P | CAG Q | CGG R | G |
| A | AUU I | ACU T | AAU N | AGU S | U |
| | AUC I | ACC T | AAC N | AGC S | C |
| | AUA I | ACA T | AAA K | AGA R | A |
| | AUG M | ACG T | AAG K | AGG R | G |
| G | GUU V | GCU A | GAU D | GGU G | U |
| | GUC V | GCC A | GAC D | GGC G | C |
| | GUA V | GCA A | GAA E | GGA G | A |
| | GUG V | GCG A | GAG E | GGG G | G |

**Fig. 6.** The genetic code, with the amino acid symbols at the top and codons with the amino acids they specify at the bottom.

by translation. The RNA in the nucleus is processed by excision of the intron regions and addition of the 5′CAP site and a 3′ poly(A) tail, which increases the stability of the mature mRNA. The introns that do not encode proteins are also removed from the mRNA. The small nuclear RNA assists in the processing of mRNA. The mature mRNA is transported to the cytoplasm, where it is then translated to protein. The nucleotide sequence of the RNA (and parent DNA) determines the amino acid sequence of the newly synthesized proteins.

The genetic code (**Fig. 6**) which consists of specific codons, specifies one or more amino acids that are essential for protein synthesis. Because there are four bases in RNA and $4^3$ (or 64) possible base triplets or codons, the 20 naturally occurring amino acids can be specified by different codons (**Fig. 6**). The methionine codon (AUG) is the start codon for protein synthesis, and UAA, UAG, and UGA are the stop codons.

mRNA translation begins at the 5′ to 3′ end and is carried out on the ribosomes with the assistance of transfer RNA (tRNA). Codons are recognized by tRNA that have an anticodon triplet complement. The tRNA molecules associate with specific amino acids determined by the anticodon sequence. The tRNA then interacts with the complimentary codon on a message that is being translated, and the ribosome catalyzes the addition of the amino acid to the protein during chain elongation.

## HYBRIDIZATION

Hybridization or annealing involves pairing of complementary strands of nucleic acids. Hybrids can be formed between complementary strands of nucleic acids (DNA

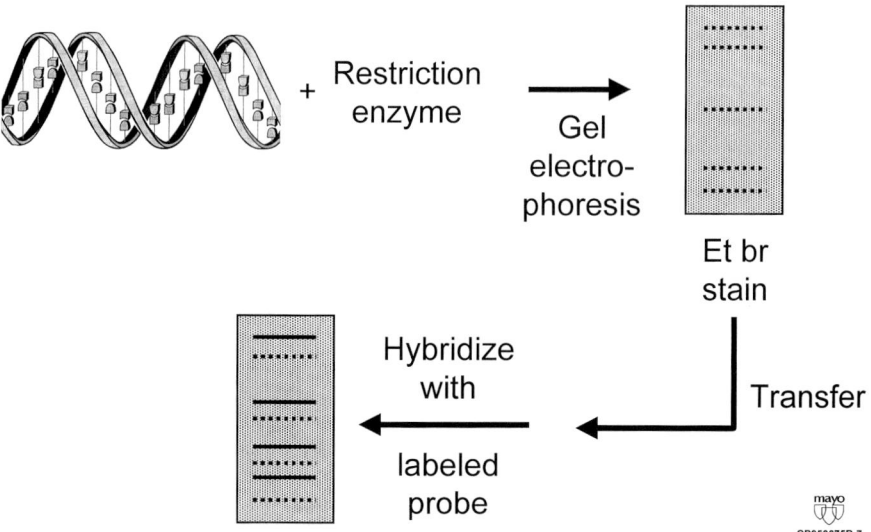

CP950675B-7

**Fig. 7.** Schematic of Southern hybridization. After the genomic DNA is digested by specific restriction endonucleases, the DNA fragments are separated by gel electrophoresis and then transferred to a nylon membrane and denatured with alkali. The single-stranded DNA is hybridized with a labeled probe. Hybrids are formed between the probe and complementary DNA fragments and can be detected with a radioactive or nonradioactive reporter system. Et br, ethidium bromide.

and RNA). These hybrid strands can be reversibly separated by heat treatment or alkaline treatment to disrupt the hydrogen bonds between complementary bases.

Hybridization is one of the more powerful tools in molecular analysis. If a tag or label is placed on one strand of the nucleic acid, it can be used as a probe in the analysis of the nucleic acids. These types of analyses are designated as Southern and Northern hybridization for DNA and RNA analyses, respectively, and *in situ* hybridization for RNA or DNA tissue analysis (**Figs. 7–9**). In the typical Southern or Northern hybridization, the nucleic acids are separated on a gel by electrophoresis. Because nucleic acids are negatively charged, they move to the positive electrode. After denaturation and/or transfer to a solid matrix, they can be hybridized with a known probe to detect the DNA or RNA of interest.

The specificity of the hybridization can be readily controlled by the stringency of the conditions, i.e., pH, temperature, and salt concentration. High stringency can lead to detection of small differences or changes such as a change in a single base, as in a genetic mutation.

Hybridization analyses are used in molecular diagnostics to detect gene mutations or infectious agents in body fluids or tissues or to study gene expression in tumors or other diseased tissues.

## SINGLE-STRAND CONFORMATION POLYMORPHISM ANALYSIS

The single-strand conformation polymorphism (SSCP) technique is a useful method for detecting gene mutations using fresh tissues or paraffin sections *(6–8)*. One can

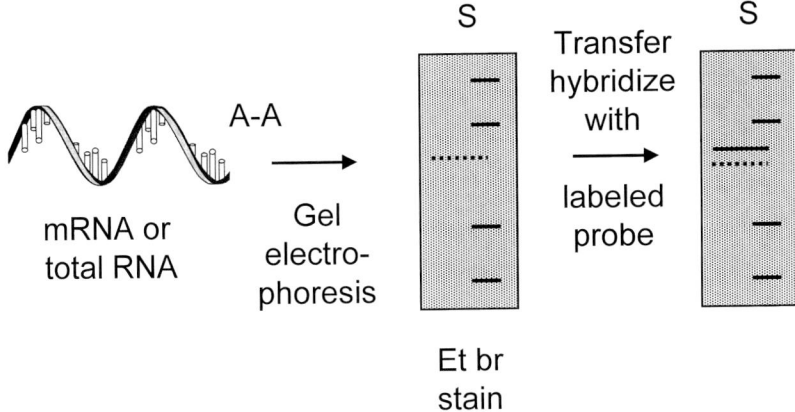

**Fig. 8.** Northern hybridization. Total or poly(A)⁺RNA is separated into fragments of different sizes by gel electrophoresis. Ethidium bromide (Et br) staining allows visualization of the fragments (S) as well as the 28s and 18s ribosomal RNA bands under fluorescence. After it is transferred to a nylon membrane and hybridized with a labeled probe, the mRNA of interest can be detected by autoradiography or with nonradioactive methods. A–A, amino acids.

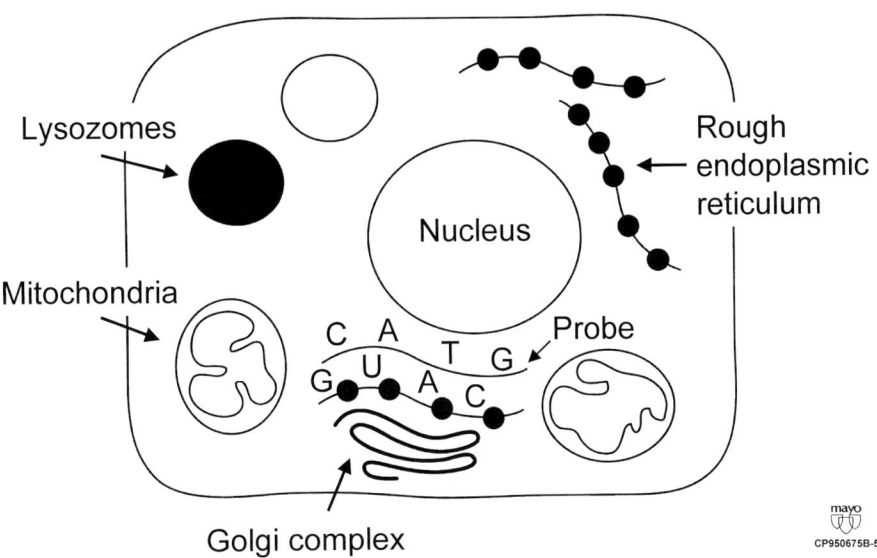

**Fig. 9.** In situ hybridization. The probe (DNA or RNA) is labeled with a radioactive or nonradio-active detection system. The DNA probe is hybridized with the messenger RNA associated with the rough endoplasmic reticulum. After hybridization, the target cells or tissues of interest can be detected by autoradiography or with nonisotopic methods such as with a digoxigenin-labeled probe. Advantages of *in situ* hybridization include more precise intracellular localization of the target and direct visualization of the positive cell relative to the adjacent cells or tissues.

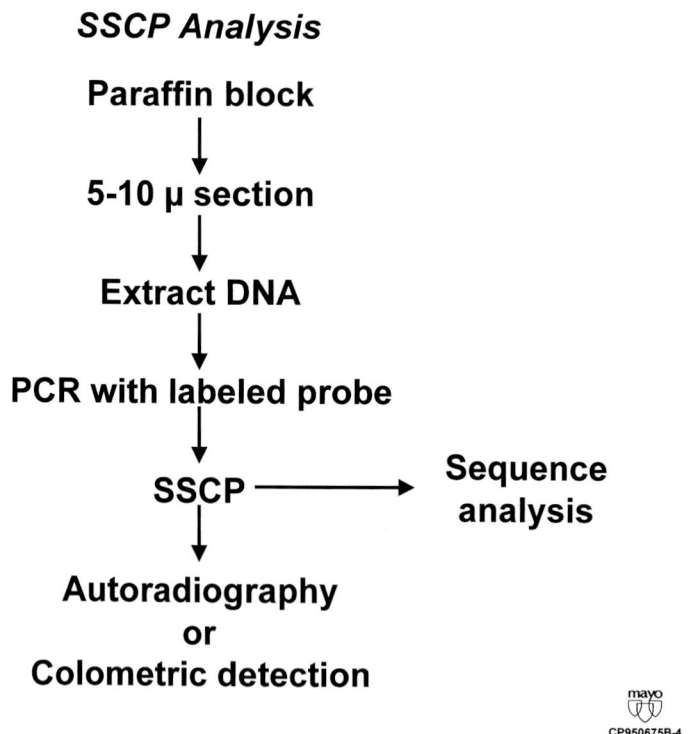

**Fig. 10.** Strategy for performing single-strand conformation polymorphism (SSCP) analysis and sequence analysis from paraffin-embedded tissue sections. The polymerase chain reaction (PCR) products can be visualized on a gel and then used for sequencing. SSCP also can be performed with labeled probes and then analyzed on a polyacrylamide gel to detect differences in migration of the single-stranded DNA compared with a known wild-type DNA. The gene fragments with a shift in the migration pattern can be analyzed by sequencing to detect the gene mutation or other genetic changes.

readily detect sequence changes in a single-stranded DNA fragment. Single-stranded DNA fragments with a unique polyacrylamide gel in a manner dependent on the folded conformation, which is in turn dependent on ionic strength and temperature. The fragments are usually between 50 and 400 bp in length. The SSCP techniques are highly sensitive when combined with polymerase chain reaction. After amplification followed by electrophoresis, mutations in DNA appear as shifts in one or more of the single-stranded DNA bands compared with the wild-type sequence (**Fig. 10**). Samples are usually run at several temperatures with and without glycerol.

## DNA AND TISSUE MICROARRAYS

DNA microarrays consist of thousands of individual gene sequences bound to regions on the surface of a microscopic glass slide or other support surface *(9–13)*. Arrays are usually a few square centimeters in area and may contain 2000–40,000 or more DNA spots. The complete transcription program of an organism or large segments of the transcription program of normal and physiologically altered or abnormal cells can be analyzed rapidly by microarray analysis *(9–11)*. The use of fluorescent labeled cDNA preparations in DNA microarrays allows one to analyze gene expression patterns on a genomic scale. This procedure uses the basic principles of hybridization. A probe hybridizes to multiple defined cDNAs or expressed sequence tags. The probe usually consists of a complex mixture of cDNA fragments derived from mRNA. The probe fragments are usually fluorescent labeled, and they hybridize to the targets on the glass slide. A positive sign is detected by a laser system such as an argon-ion laser and then quantified. Global gene expression can be readily detected with this technique.

Another approach is to use oligonucleotide arrays or DNA chips with synthetic oligonucleotides, which represent thousand of gene sequences synthesized on the surface of small areas of a glass slide.

DNA array technology for gene expression profiling has been applied to many areas of biomedical sciences. A recent dramatic application is the report by Alizadeh et al. *(14),* who used gene expression profiling in B-cell lymphomas to identify two molecularly distinct forms of diffuse large B-cell lymphomas that had gene expression patterns indicative of different stages of B-cell differentiation. Another study of gene-expression profiles in hereditary breast cancers found functional differences between breast tumors with BRCA1 mutations and those with BRCA2 mutations *(15)*.

Tissue microarrays for high-throughput molecular profiling of normal and tumor specimens have been adopted for rapid screening of many tissues *(16,17)*. This new approach to identify clinically relevant molecular changes uses microarrays from different cases of the same type of tissue or tumor, which can range from 50 to a few hundred sections of tissues on one slide, to check for protein or gene expression in these microsamples. In the near future, a combination of tissue microarray technology with pattern recognition software should greatly advance the discovery of new and potentially useful markers with clinical and diagnostic applications.

## REFERENCES

1. International Human Genome Sequencing Consortium (2001) Initial sequencing and analysis of the human genome. *Nature* **409**, 860–921.
2. Venter, J.C., Adams, M.D., Myers, E.W., et al. (2001) The sequence of the human genome. *Science* **291**, 1305–1351.
3. Claverie, J.-M. (2001) Gene number. What if there are only 30,000 human genes? *Science* **291**, 1255–1257.
4. Watson, J. D., Hopkins, N. H., Roberts, J. W., Argetsinger Steitz, J., Weiner, A. M. (1987) *Molecular Biology of the Gene*, 4th ed. Benjamin/Cumming, San Francisco 1987.
5. Lodish, H., Berk, A., Zipinsky, S. L., Matsudaira, P., Baltimore, D., Darnell, J. (2000) *Molecular Cell Biology*. W. H. Freeman, New York.
6. Hensel, C. H., Xiang, R. H., Sakaguchi, A. Y., Naylor, S. L. (1991) Use of single strand conformation polymorphism technique and PCR to detect p53 gene mutations in small cell lung cancer. *Oncogene* **6**, 1067–1071.
7. Orita, M., Iwahana, H., Kanazawa, H., Hayaski, K., Sekiya, T. (1989) Detection of polymorphisms of human DNA by gel electrophoresis as single-strand conformation polymorphisms. *Proc. Natl. Acad. Sci. USA* **86**, 2766–2770.
8. Orita, M., Suzuki, Y., Sekiya, T., Hayashi, K. (1989) Rapid and sensitive detection of point mutations and DNA polymorphisms using the polymerase chain reaction. *Genomics* **5**, 874–879.

9. Schena, M., Shalon, D., Heller, R., Chai, A., Brown, P. O., Davis, R. W. (1996) Parallel human genome analysis: microarray-based expression monitoring of 1000 genes. *Proc. Natl. Acad. Sci. USA* **93**, 10614–10619.

10. De Risi, J. L., Iyer, V. R., Brown, P. O. (1997) Exploring the metabolic and genetic control of gene expression on a genomic scale. *Science* **278**, 680–686.

11. Moch, H., Schraml, P., Bubendorf, L., et al. (1999) High-throughput tissue microarray analysis to evaluate genes uncovered by cDNA microarray screening in renal cell carcinoma. *Am. J. Pathol.* **154**, 981–986.

12. Xu, J., Stolk, J. A., Zhang, X., et al. (2000) Identification of differentially expressed genes in human prostate cancer using subtraction and microarray. *Cancer Res.* **60**, 1677–1682.

13. Kurian, K. M., Watson, C. J., Wyllie, A. H. (1999) DNA chip technology. *J. Pathol.* **187**, 267–271.

14. Alizadeh, A. A., Eisen, M. B., Davis, R. E., et al. (2000) Distinct types of diffuse large B-cell lymphoma identified by gene expression profiling. *Nature* **403**, 503–511.

15. Hedenfalk, I., Duggan, D., Chen, Y., et al. (2001) Gene-expression profiles in hereditary breast cancer. *N. Engl. J. Med.* **344,** 539–548.

16. Kononen, J., Bubendorf, L., Kallioniemi, A., et al. (1998) Tissue microarrays for high-throughput molecular profiling of tumor specimens. *Nature Med.* **4**, 844–847.

17. Moch, H., Schraml, P., Bubendorf, L., et al. (1999) High-throughput tissue microarray analysis to evaluate genes uncovered by cDNA microarray screening in renal cell carcinoma. *Am. J. Pathol.* **154**, 981–986.

# 2

# Laser Capture Microdissection
## *Principles and Applications*

### Ricardo V. Lloyd, MD, PHD

## INTRODUCTION

The acquisition of homogeneous or pure cell populations for cell biologic and molecular analyses has been a difficult challenge for many decades. A variety of approaches have been tried, including macroscopic dissection of tissues from frozen tissue blocks to increase the population of specific cell types *(1,2)*. Some investigators have used irradiation of manually ink-stained sections to destroy cells that were not of interest *(3,4)*. Microdissection with the aid of a microscope and needles has been used by many other investigators *(5,6)*. This latter approach can provide a great deal of precision in collecting homogenous cells, but it is slow and labor-intensive and requires a high degree of manual dexterity. Most of these approaches have not provided the speed, precision, and efficiency needed for research or routine clinical molecular diagnostic use.

The recent development of a laser capture microdissection (LCM) unit by the National Cancer Institute (NCI) group led by Liotta and his colleagues *(7,8)* and by other groups, mainly in Europe *(9,10)*, has provided a rapid and efficient method to capture pure cell populations for molecular and other studies.

### Principles of LCM

Liotta's group at the NCI first reported on the development of a rapid and reliable method of obtaining homogenous population of cells from complex tissues *(7,8)*. The availability of commercial instruments resulted from a joint venture by the NCI and Arcturus Engineering (Mountain View, CA).

With LMC, a complex section of tissues or heterogeneous cell populations on a glass slide is placed on the stage of a specially designed microscope. After the areas or cells of interest are selected, an ethylene vinyl acetate (EVA) transparent film is apposed to the section, and an infrared laser beam is directed at the cell of interest. When the focused laser beam coaxial with the microscope optics is activated, the EVA film above the targeted area melts, surrounds, and holds the cells of interest, which are kept in the film after removal from the glass slide. The film can be placed directly

From: *Morphology Methods: Cell and Molecular Biology Techniques*
Edited by: R. V. Lloyd © Humana Press, Totowa, NJ

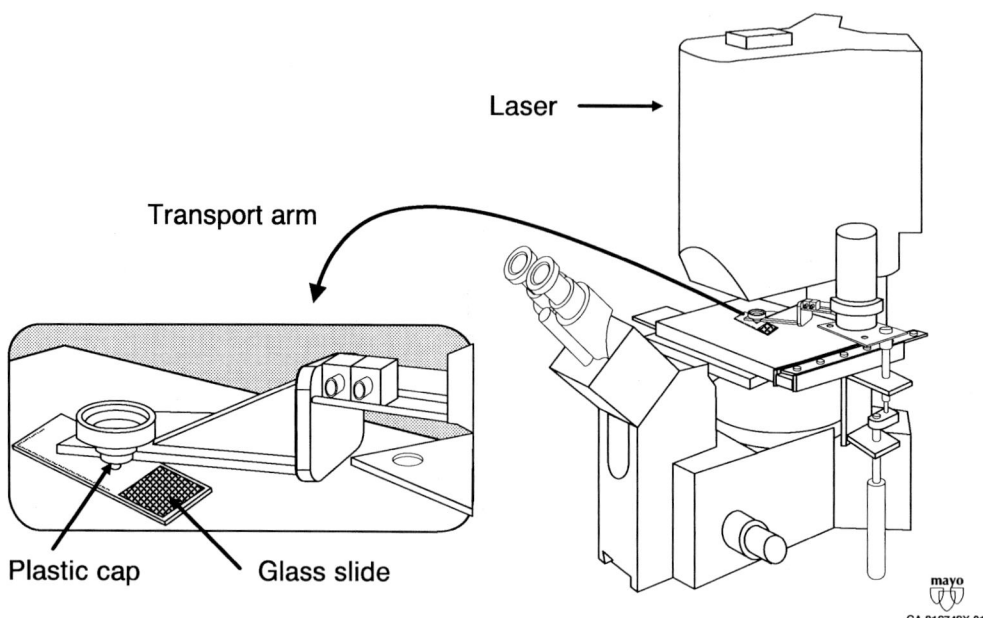

**Fig. 1.** Laser capture microdissection (LCM) with the Pix Cell instrument. The laser beam is focused on the cut section of cells on the glass slide. The transport arm carries the plastic cap over to align it with the specimen and laser beam. The focused laser beam captures the cells of interest, which adhere to the ethylene vinyl acetate (EVA) film that binds to the captured target cells. After collecting the cells of interest, the transport arm carries the cells back to the side.

into DNA or RNA extraction buffers for nucleic acid extraction and analysis or into other buffers for protein analysis. While in the buffer, the cellular material becomes detached from the film and can be used for cellular or molecular analyses after extraction (see Protocols).

Based on numerous reported studies in the literature and our own observations, the LCM procedure does not lead to significant alterations in the morphologic or apparent molecular features of the cells of interest.

### Technical Aspects

The prototypic instrument used for LCM is the Pix Cell™ developed by Arcturus Engineering and the NCI group (**Figs. 1 and 2**). The instrument consists of a laser optics deck and illumination tower with a halogen bulb housing, a microscope, and a slide stage with a vacuum chuck. A joystick with XY control and video camera attachment are attached to the microscope. The XY control is for fine positioning of the sample. The laser can be controlled by amplitude and pulse width adjustments using digital controls on the front panel of the electronic box. When the laser fires, an emission indicator is lit, and the power supply emits a beeping sound. Typical operating parameters for the laser for a 30-μm spot is an amplitude of 30 mW and a pulse width of 5 ms; for a 60-μm spot, an amplitude of 50 mW and a pulse width of 5 ms is used.

When the operator is ready to capture the cells of interest, a plastic cap is transported with the film carrier and placed on the desired position of the tissue or cells. The laser

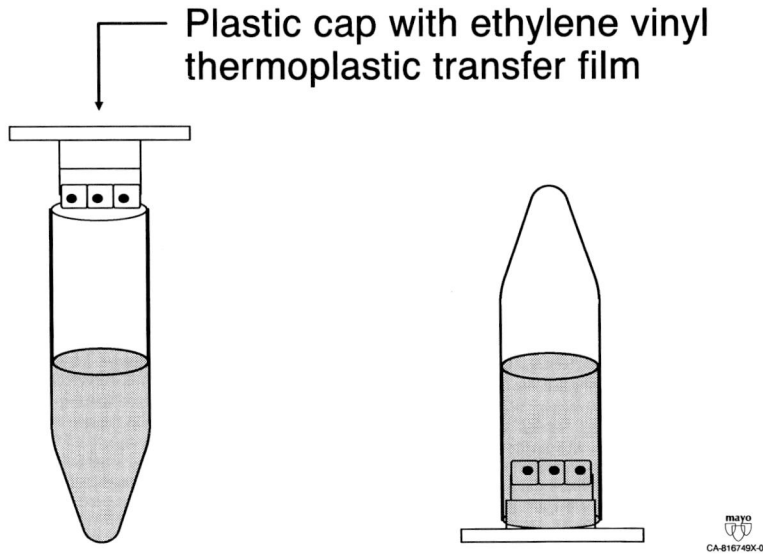

**Fig. 2.** The cells on the EVA film are placed in the microfuge tube and immersed in the appropriate buffer for DNA, RNA, or protein extraction and analysis.

is then activated, and the EVA film binds to the captured target on the tissue. The laser can be activated as often as is needed to capture the desired numbers of cells. Multiple clusters of homogenous cells can be accumulated into the same polymer EVA film, and individual single laser shots can be used to procure specific cell clusters or individual cells **(Fig. 3)**. In addition, multiple shots can be combined to procure complex but homogenous tissue structures. Up to 3000 shots can be captured on one transfer film cap, which may include up to 6000 cells. Since each shot takes less than 1 s to perform, a large number of cells can be captured in a relatively short period.

*Technical Variables*

Technical variables with LCM include the types of analyses that will be done with the tissues relative to fixation and processing, use of frozen versus paraffin tissue sections, and combination of LCM with other techniques such as immunohistochemistry (Immuno-LCM).

Tissues fixed in buffered formalin and embedded in paraffin can be used for routine analysis of DNA and in some cases mRNA as well. Tissues are routinely stained with hematoxylin and eosin for optimum visualization of cellular details during LCM. These stains do not affect the integrity of the DNA or RNA. For optimum RNA analysis, frozen tissue sections are preferred. Surprisingly, the nucleic acids are not adversely affected by exposing the tissues to xylene for a short period. RNAse-free conditions are important for obtaining high-quality RNA samples.

The section for frozen tissue should be cut at 10 µm or less for optimum transfer, because with thicker sections it is more difficult to visualize single cells. Sections have to be completely dry and not cover-slipped for effective LCM transfer so the final xylene rinse facilitates efficient LCM transfer. After the frozen sections are cut, the slides can be used immediately or stained at −80°C until use. We have observed that

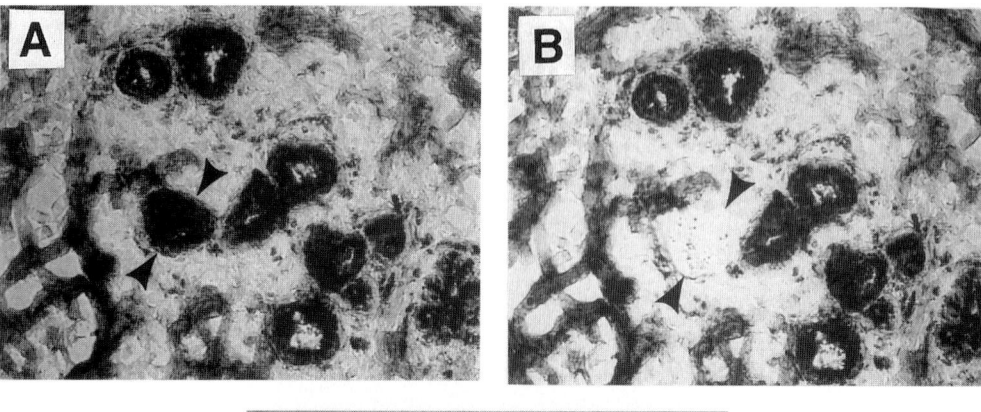

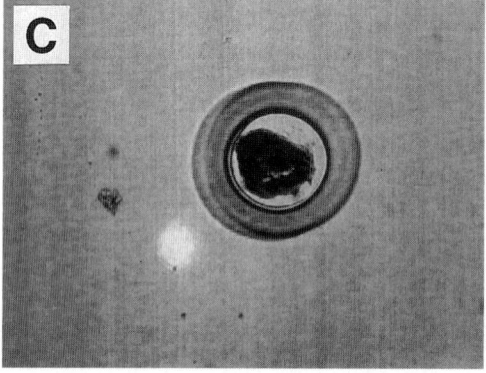

**Fig. 3.** LCM capture of cells from normal breast tissue. (**A**) Groups of normal epithelial cells were selected for LCM (arrowheads). (**B**) Arrowheads indicate the space from which the cells were captured. (**C**) Captured cells on the cap with the transfer EVA film. (Reproduced with permission from ref. *13.*)

starting with fresh alcohols and xylene for each set of experiments avoids many of the technical problems associated with difficult transfers during LCM.

The procedure of combining LCM with immunophenotyping for RNA analysis (immuno-LCM) was first reported from the NCI *(11)* and subsequently by our groups and others for analyzing single cells *(12)* and tissues *(13–15)* (**Fig. 4**). The critical requirements include using RNAse-free condition and using RNAse inhibitors during the immunostaining procedures. A rapid immunostaining procedure minimizes the time of tissue exposure to RNAse.

Various studies have shown that precipitating fixatives such as ethanol and acetone produced better quality reverse transcriptase-polymerase chain reaction (RT-PCR) product amplification compared with crosslinking fixatives such as formaldehyde and glutaraldehyde *(11,12,14)*. However, precipitating fixatives are less effective in some immunostaining procedures compared with crosslinking fixatives *(12)*.

### Other Laser-Based Microdissection Systems

Although LCM is the principal system used, other systems have been developed, mainly in Europe, that are somewhat similar to LCM *(16–21)*. Schutze and Lahr *(16)*

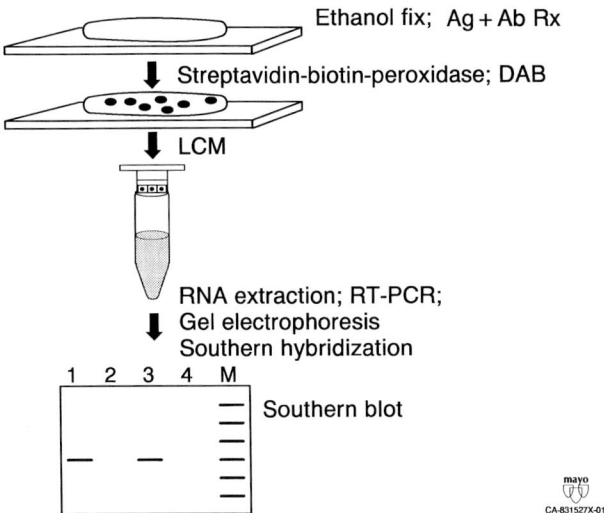

**Fig. 4.** Immuno-laser capture microdissection (LCM). Combined immunohistochemistry and LCM to capture immunophenotyped cells. After ethanol fixing, the tissues are exposed to the specific antibody for about 15 min followed by a streptavidin-biotin-peroxidase reaction and diaminobenzidine (DAB). The immunostained cells (black circles) can be used for LCM and reverse transcriptase-polymerase chain reaction (RT-PCR) and the products analyzed by gel electrophoresis and Southern hybridization.

used a low-power laser to create a gap between the cells of interest and adjacent cells; with an increase in the laser power, they were able to "catapult" the microdissected cells into the cap of a microfuge tube for further analysis without requiring any direct contact with the cells of interest. Using this approach, they could amplify simple cells from archival tissues in the analysis of Ki-ras mutations *(16)*.

Fink and colleagues *(17)* used an ultraviolet light laser to ablate cells that were not of interest and subsequently collected the cells of interest with a stent needle under the control of a micromanipulator *(17)*, which is a modification of the original method of Shibata et al. *(3,4)*. Other investigators have utilized this approach to study microsatellite instability in breast cancer at the single cell level *(18)*. A combination of ultraviolet microbeam microdissection with laser pressure catapulting for the isolation of single chromosomes for development of chromosome-specific paint probes has been reported *(19)*. Becker et al. *(20)* used a combination of ultraviolet laser microbeams with microdissection and molecular characterization of individual cells to detect a novel mutation in the E cadherin gene in single tumor cells *(20)*. A recent study used the PALM Laser-Microbeam System, which allows the contact-free isolation of single cells or groups of cells using the laser pressure catapulting technique and real-time PCR to study the Her-2/Neu gene and topoisomerase II$\alpha$ gene in breast cancer specimens *(21)*.

## Disadvantages of LCM and Related Techniques

Although LCM is faster and easier to perform than manual microdissection, there are a few disadvantages. The principal one is that the tissue is not cover-slipped during LCM, so the refractive indices of the dry sections have a refractile quality that obscures cellular details at higher magnifications. This can be partially overcome with a diffusion filter on the instrument or by using a drop of xylene on the tissues, which provides wetting and refractive-index matching; the xylene usually evaporates rapidly before microdissection *(22)*. Another disadvantage for investigators with a tight budget is the costs of the instrument. Finally, although LCM is faster than manual methods, a great deal of time is still required to collect cells for an experiment compared with biochemical analysis, especially when thousands of cells are collected *(23)*.

## APPLICATIONS OF LCM AND RELATED MICRODISSECTION METHODS

### General Applications

The first reports of the development of LCM illustrated the wide applicability of the technique *(7,8)*. The NCI investigators used this technique to analyze loss of heterozygosity (LOH) of the *BRCA1* gene in familial breast cancer, chromosome 8p in prostate cancer, the *p16* gene in invasive esophageal cancer, and the *MEN1* tumor suppressor gene on chromosome 11q13 in gastrinomas *(7)*. Other reported applications using the original carbon dioxide laser included detection of a single base mutation in exon 2 of the von Hippel-Lindau (VHL) gene in hemangioblastoma RT-PCR amplification of actin, prostate-specific antigen, and matrix metalloproteinase-2 in frozen prostate cancer samples and analysis by gelatin zymography, showing the applicability of the LCM technique for enzyme assays *(7)*.

Suarea-Quian et al. *(24)* used an LCM technique with 1-ms laser pulses focus of 6 μm to demonstrated rapid capture of single cells from different types of tissues including immunostained cells *(24)*.

Glasow and colleagues *(25)* utilized LCM to study the leptin receptor in the adrenal by RT-PCR and were able to identify the cell types producing the leptin receptor. In a subsequent study of prolactin receptor (PRL-R) in the adrenal, LCM was used to demonstrate that PRL-R was produced by cortical, but not medullary, cells in the adrenal gland *(26)*.

In a detailed analysis of B-cell lymphomas, Fend et al. *(27)* used LCM to isolate cell populations with different antigen expression patterns. They combined this with PCR and sequencing of clonal immunoglobulin heavy chain rearrangements and clonal rearrangement to show that low-grade B-cell lymphomas with two distinct morphologic and immunophenotypic patterns in the same anatomic site were frequently biclonal *(27)*. This study demonstrates the power of LCM to detect molecular microheterogeneity in complex neoplasms.

Our laboratory has used LCM from dissociated cells to study gene expression in individual pituitary cells *(12)*. We have also analyzed normal and tumorous breast tissue by LCM and RT-PCR to show that PRL-R was expressed not only in the epithelial component of normal and tumorous breast tissues but also by stromal cells *(13)*, suggesting modulation of mammary stromal cells by PRL **(Figs. 5 and 6)**. In more recent studies, our laboratory has for the first time obtained pure populations of pituitary folliculo-stellate cells and used these for molecular and cell biologic studies that have provided new insights into the role of these cells in pituitary function *(28)*.

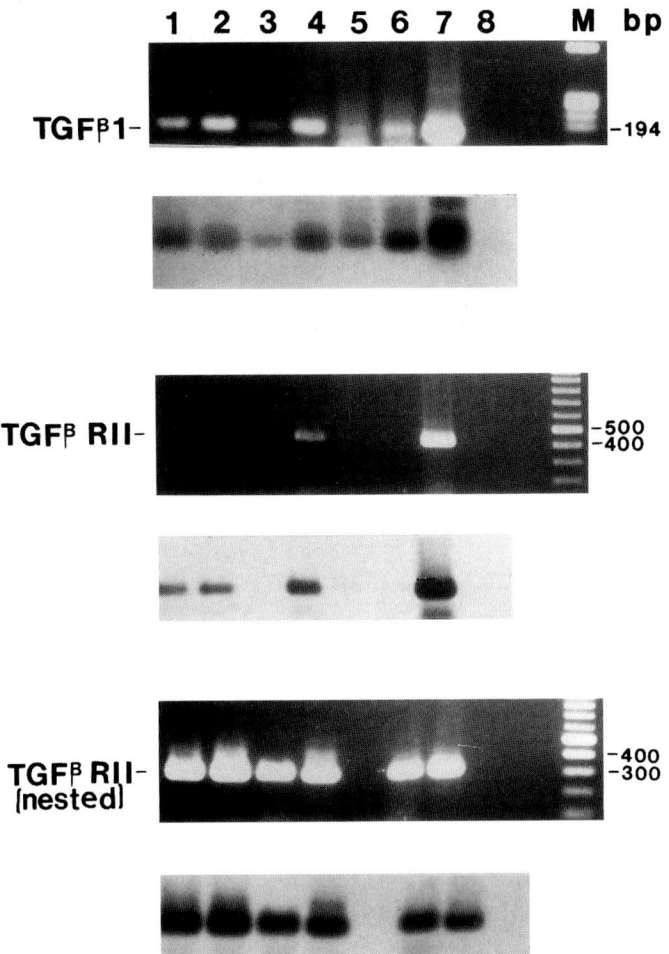

**Fig. 5.** RT-PCR and Southern hybridization examination of transforming growth factor-β1 (TGF-β1) and TGF-β-RII with RNA obtained by LCM for normal and neoplastic breast tissues showing representative examples of the analyses. Lanes 1, 3, and 5 are normal breast tissues. Lanes 2, 4, and 6 are invasive carcinoma. Lane 7 represents a positive control breast tissue, and lane 8 is a negative control without reverse transcriptase. Lanes 1–3 and 4–6 represent matching normal and tumor tissues, respectively. Lane 7 is a positive breast tissue control, and lane 8 is a negative control without reverse transcriptase. M, molecular size marker. The top part of each panel represents the RT-PCR results on the gel after ethidium bromide staining. The bottom part is the Southern hybridization with the internal probes. (Reproduced with permission from ref. *13.*)

In a recent study of synovial sarcoma, Kasai et al. *(29)* used LCM and RT-PCR with paraffin wax-embedded tissues to show that the SYT-SSX fusion transcript was present in both the spindle cell and epithelial areas of biphasic synovial sarcomas. This study also reinforced the idea that RT-PCR can be performed on RNA obtained from formalin-fixed, paraffin-embedded tissue sections. In a related study, Shibutani et al.

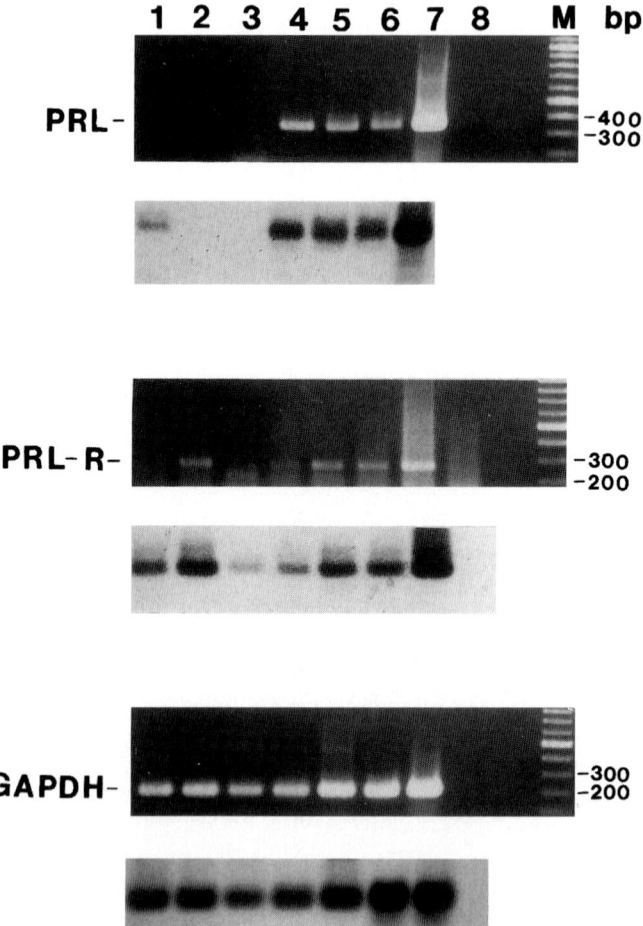

**Fig. 6.** RT-PCR and Southern hybridization examination of prolactin (PRL), prolactin receptor (PRL-R), and GAPDH RNA obtained by LCM showing representative examples of the analyses. Lanes 1, 3, and 5 are normal breast tissues. Lanes 2, 4, and 6 are invasive ductal carcinoma. Lane 7 represents a positive pituitary tissue control, and lane 8 is a negative control without reverse transcriptase. Lanes 1–3 and 4–6 represent matching normal and tumor tissues, respectively. M, molecular size marker. The top part of each panel represents the RT-PCR results on the gel after ethidium bromide staining. The bottom part of each panel is the Southern hybridization with the internal probes. (Reproduced with permission from ref. *13.*)

*(30)* demonstrated that methacarn-fixed, paraffin-embedded tissue could be used for RNA analysis by RT-PCR and for protein analysis by Western blotting after LCM using rat liver as an experimental model.

In a study of renal glomerular cells, Kohda et al. *(31)* used LCM with RT-PCR to detect podoplanin in renal glomerular cells harvested by LCM. They were also able to use LCM to capture pure cell populations from different portions of the tubules, but they could not do this from the collecting ducts.

In a study of papillary and follicular thyroid carcinoma Gillespie et al. *(32)* used LCM and loss of heterozygosity (LOH) analysis of highly polymorphic chromosomes to show that LOH was more common in follicular than papillary carcinoma. In a related study on practical clinical applications of LCM, a group of investigators analyzed a primary duodenal carcinoid as well as tumors in the scalp and cervical lymph nodes from the same patient by LCM with 22 markers. The 3p12 marker showed loss in all three tumors, and the 3pl4.2 marker showed an identical shift in the three tumors indicative of a common microsatellite alteration and supporting the notion that the shared molecular abnormalities indicated a common clonal origin with the duodenal carcinoid as the primary tumor *(33)*.

## Combining LCM with Other Techniques

A combination of other techniques with LCM has been reported by various investigators. Jones et al. *(34)* used LCM and comparative genomic hybridization to study myoepithelial cell carcinomas of the breast and found common alterations such as loss at 16q (3/10 cases) and 17 p (3/10 cases), but they also reported fewer genetic alterations compared with ductal carcinomas of the breast. Shen et al. *(35)* used LCM and genome-wide searching in a study of breast cancers to show that LOH was seen only in ductal carcinoma *in situ* but not in the invasive component of ductal carcinoma *in situ*. LCM was advantageous in reducing the heterogeneity within tumors *(35)*. A combination of LCM with fluorescence *in situ* hybridization (FISH) and flow cytometry was used in a study of normal breast and breast carcinoma from formalin-fixed, paraffin-embedded tissues in which the nuclei were microdissected by LCM *(36)*. Using probes for cyclin-D1 and RB1, this group showed amplification of cyclin-D1 of the two cases studied when the results were compared with touch preparations of nuclei from the same fresh tumor specimens. Unfortunately, the numbers of samples analyzed were too small to evaluate, but the preliminary results are promising.

## DNA Libraries and Arrays

Various studies have used LCM to obtain cells as the starting material to clone cDNAs and to analyze genes by cDNA arrays. Several laboratories have used LCM in the initial phase of generating cDNA libraries. The principal leader in this area is the NCI Cancer Genome Anatomy Project (CGAP) *(37,38)*, under which cDNA libraries from human cancers of prostate, ovary, breast, lung, and the gastrointestinal tract will be produced and sequenced.

cDNA array technology has been combined with LCM in various studies *(39–41)*. Segroi et al. *(40)* used LCM and high-density cDNA arrays to study gene expression in purified normal, invasive, and metastatic breast cell populations from a single patient and combined this approach with real-time quantitative PCR and immunohistochemistry to study tumor progression in this tumor model. Leethanakul et al. *(41)* used LCM and cDNA arrays in studies of squamous cell carcinomas of the head and neck using 5000 cells from normal and tumor tissues. They observed a consistent decrease in cytokeratins and an increase in expression of signal-transducing, cell cycle regulatory proteins, and angiogenic factors, as well as tissue-degrading proteases. Unexpected findings included overexpression of the wnt and notch growth and differentiation regulatory systems by these squamous cell carcinomas *(41)*.

## Proteomic Analysis

The use of LCM for protein analysis had been reported in the original descriptions of LCM *(7)* in which gelatin zymography was used to study gelatinase A (MMP-2) in frozen sections of prostate cancer. In a more recent study, Banks et al. *(42)* used LCM and two-dimensional electrophoresis to examine protein profiles of selected tissue areas. In the cervix they were able to show enrichment of some proteins compared with the whole tissue using these combined techniques. Emmert-Buck et al. *(43)* used normal and neoplastic esophageal squamous mucosa as a model to study proteins captured by LCM followed by two-dimensional polyacrylamide gel electrophoresis (2D-PAGE). They were able to use 50,000 cells to resolve 675 distinct proteins or isoforms with molecular weights between 10 and 200 kDa and isoelectric points between pH 3 and 10. Although 98% of the proteins in normal and tumor samples were identical, 17 of the tumors showed tumor-specific alterations. Seven of these proteins were present only in normal tissues, and 10 of the proteins were present only in the tumors. Cytokeratin I and annexin I were identified as two of the altered proteins.

In a study of LCM-captured prostate tissue analyzed with a quantitative chemiluminescent assay, Simone et al. *(44)* studied prostate-specific antigen (PSA) distribution in normal prostatic intraepithelial neoplasia and carcinoma and found a range of $2 \times 10^4$ to $6.3 \times 10^6$ PSA molecules per microdissected tissue cells. These observations were concordant with the immunohistochemical staining intensity of tissue sections from these different tissue areas *(44)*. Ornstein et al. *(45)* used LCM and PAGE (1D and 2D) to analyze PSA $\alpha_1$-antichymotrypsin (ACT) by Western blotting studies and showed a 30-kDa band that was the expected molecular weight of unbound PSA. Binding studies revealed that PSA recovered from LCM-procured cells had the full ability to bind ACT as in the normal prostate. Electrophoretic studies also showed that the PSA/ACT complex was stable, indicating that the complex was similar in normal and tumorous tissues.

## FUTURE DIRECTIONS

Future advances in LCM will include technologic advances in instrumentation and application of LCM to solve difficult and/or challenging problems in cell biology and biomedical sciences. The further development of laser fields small enough to capture single cells or specific parts of cells such as the nucleus are currently under way. Automation of the LCM system for automatic performance of cell capture is also being developed. A major challenge for the technology is to develop methods to capture live cells that can be used for cell culture and other in vitro studies with living cells.

Major advances can be anticipated in combining LCM with other techniques such as flow cytometry, FISH and comparative genomic hybridization. Significant advances can be anticipated in gene expression studies with microarrays to profile specific human diseases starting from clearly defined histopathologic cell populations. The technique of "molecular fingerprinting" of entire organs such as the prostate has been initiated by one group *(38)*. Cole et al. *(38)* have proposed a model for integrating in three dimensions the data obtained by LCM and microarrays for the in vivo molecular anatomy analysis of normal and neoplastic cells. These types of approaches will rely heavily on LCM for further advances in understanding the molecular aspects of diseases as they relate to tissue structure and function.

## REFERENCES

1. Fearon, E. R., Hamilton, S. R., and Vogelstein, B. (1987) Clonal analysis of human colorectal tumors. *Science* **238**, 193–197.
2. Shibata, D., Hawes, D., Li, Z. H., Hernandez, A. M., Spruck, C. H., and Nichols, P. W. (1992) Specific genetic analysis of microscopic tissue after selective ultraviolet radiation fractionation and the polymerase chain reaction. *Am. J. Pathol.* **141**, 539–543.
3. Shibata, D. (1993) Selective ultraviolet radiation fractionation and polymerase chain reaction analysis of genetic alterations. *Am. J. Pathol.* **143**, 1523–1526.
4. Emmert-Buck, M. R., Roth, M. J., Zhuang, Z., et al. (1994) Increased gelatinase A (MMP-2) and cathepsin B activity in invasive tumor regions of human colon cancer samples. *Am. J. Pathol.* **145**, 1285–1290.
5. Zhuang, Z., Bertheau, P., Emmert-Buck, M. R., et al. (1995) A microdissection technique for archival DNA analysis of specific cell populations in lesions <1 mm in size. *Am. J. Pathol.* **146**, 620–625.
6. Noguchi, S., Motomura, K., Inaji, H., Imaoka, S., and Koyama, H. (1994) Clonal analysis of predominantly intraductal carcinoma and precancerous lesions of the breast by means of polymerase chain reaction. *Cancer Res.* **54**, 1849–1853.
7. Emmert-Buck, M. R., Bonner, R. F., Smith, P. D., et al. (1996) Laser capture microdissection. *Science* **274**, 998–1001.
8. Bonner, R. F., Emmert-Buck, M., Cole, K. et al. (1997) Laser capture microdissection: molecular analysis of tissue. *Science* **278**, 1481–1483.
9. Schütze, K., Lahr, G. (1998) Identification of expressed genes by laser mediated manipulation of single cells. *Nat. Biotechnol.* **16**, 737–742.
10. Fink, L., Seeger, W., Ermert, L., et al. (1998) Real-time quantitative RT-PCR after laser-assisted cell picking. *Nat. Med.* **4**, 1329–1333.
11. Fend, F., Emmert-Buck, M. R., Chuaqui, R., et al. (1999) Immuno-LCM: laser capture microdissection of immunostained frozen sections for mRNA analysis. *Am. J. Pathol.* **154**, 61–66.
12. Jin, L., Thompson, C. A., Qian, X., Kuecker, S. J., Kulig, E., Lloyd, R. V. (1999) Analysis of anterior pituitary hormone mRNA expression in immunophenotypically characterized single cells after laser capture microdissection. *Lab. Invest.* **79**, 511–512.
13. Kuecker, S. J., Jin, L., Kulig, E., Ondraogo, G. L., Roche, P. C., Lloyd, R. V. (1999) Analysis of PRL, PRL-R, TGFβ1 and TGFβ-RII gene expression in normal and neoplastic breast tissues after laser capture microdissection. *Appl. Immunohistochem. Mol. Morphol.* **7**, 193–200.
14. Goldsworthy, S. M., Stockton, P. S., Trempus, C. S., Foley, J. F., Maronpot, R. R. (1999) Effects of fixation on RNA extraction and amplification from laser capture microdissected tissue. *Mol. Carcinog.* **25**, 86–91.
15. Fink, L., Kinfe, T., Stein, M. M., et al. (2000) Immunostaining and laser-assisted cell picking for mRNA analysis. *Lab. Invest.* **80**, 327–333.
16. Schutze, K., Lahr G. (1998) Identification of expressed genes by laser-mediated manipulation of single cells. *Nat. Biotech.* **16**, 737–742.
17. Fink, L., Seeger, W., Ermert, L., et al. (1998) Real-time quantitative RT-PCR after laser-assisted cell picking. *Nat. Med.* **4**, 1329–1333.

18. Dietmaier, W., Hartmann, A., Wallinger, S., et al. (1999) Multiple mutation analyses in single tumor cells with improved whole genome amplification. *Am. J. Pathol.* **154**, 83–95.

19. Schermellch, L., Tyhalhammer, S., Iteckl, W., et al. (1999) Laser microdissection and laser pressure catapulting for the generation of chromosome-specific paint probes. *Biotechniques* **27**, 362–367.

20. Becker, I., Becker, K. F., Rohrl, M. H., Minkus, G., Schütze, K., Höfler, H. (1996) Single-cell mutation analysis of tumors from stained histologic slides. *Lab. Invest.* **75**, 801–807.

21. Lehmann, V., Glöckner, S., Kleeberger, W., von Wasielewski, H. F., Kreipe, H. (2000) Detection of gene amplification in archival breast cancer specimens by laser-assisted micro-dissection and quantitative real-time polymerase chain reaction. *Am. J. Pathol.* **156**, 1855–1864.

22. Simone, N. L., Bonner, R. F., Gillespie, J. W., Emmert-Buck, M. R., Liotta, L. A. (1998) Laser-capture microdissection: opening the microscopic frontier to molecular analysis. *Trends Genet.* **14**, 272–276.

23. Curran, S., McKay, J. A., McLeod, H. L., Murray, G. I. (2000) Laser capture microscopy. *Mol. Pathol.* **53**, 64–68.

24. Suarez-Quian, C. A., Goldstein, S. R., Pohida, T., et al. (1999) Laser capture microdissection of single cells from complex tissues. *Biotechniques* **26**, 328–335.

25. Glasow, A., Haidan, A., Hilbers, U., et al. (1998) Expression of Ob receptor in normal human adrenals: differential regulation of adrenocortical and adrenomedullary function by leptin. *J. Clin. Endocrinol. Metab.* **83**, 4459–4466.

26. Glasow, A., Haidan, P., Gillespie, J., Kelly, P. A., Chrousos, G. P., Bornstein, S. R. (1998) Differential expression of prolactin receptor (PRLR) in normal and tumorous adrenal tissues: separation of cellular endocrine compartments by laser capture microdissection (LCM). *Endocr. Res.* **24**, 857–862.

27. Fend, F., Quintanilla-Martinez, L., Kumar, S., et al. (1999) Composite low grade B-cell lymphomas with two immunophenotypically distinct cell populations are true biclonal lymphomas. A molecular analysis using laser capture microdissection. *Am. J. Pathol.* **154**, 1857–1866.

28. Jin, L., Tsumanuma, I., Ruebel, K. H., Bayliss, J. M., Lloyd, R. V. (2001) Analysis of homogeneous populations of anterior pituitary folliculostellate cells by laser capture microdissection and RT-PCR. *Endocrinology*, in press.

29. Kasai, T., Shimajiri, S., Hashimoto, H. (2000) Detection of SYT-SSX fusion transcripts in both epithelial and spindle cell areas of biphasic synovial sarcoma using laser capture microdissection. *Mol. Pathol.* **53**, 107–110.

30. Shibutani, M., Uneyama, C., Miyazaki, K., Toyoda, K., Hirose, M. (2000) Methacarn fixation: a novel tool for analysis of gene expressions in paraffin-embedded tissue specimens. *Lab. Invest.* **80**, 199–208.

31. Kohda, Y., Murakami, H., Moe, O. W., Star, R. A. (2000) Analysis of segmental renal gene expression by laser capture microdissection. *Kidney Int.* **57**, 321–331.

32. Gillespie, J. W., Nasir, A., Kaiser, H. E. (2000) Loss of heterozygosity in papillary and follicular thyroid carcinoma: a mini review. *In Vivo* **14**, 139–140.

33. Milchgrub, S., Wistuba, II, Kim, B. K., et al. (2000) Molecular identification of metastatic cancer to the skin using laser capture microdissection: a case report. *Cancer* **88**, 749–754.

34. Jones, C., Foschini, M. P., Chaggar, R., et al. (2000) Comparative genomic hybridization analysis of myoepithelial carcinoma of the breast. *Lab. Invest.* **80**, 831–836.

35. Shen, C. Y., Yu, J. C., Lo, Y. L., et al. (2000) Genome-wide search for loss of heterozygosity using laser capture microdissected tissue of breast carcinoma: an implication for mutator phenotype and breast cancer pathogenesis. *Cancer Res.* **60**, 3884–3892.

36. DiFrancesco, L. M., Murthy, S. K., Luider, J., Demetrick, D. J. (2000) Laser capture

microdissection-guided fluorescence in situ hybridization and flow cytometric cell cycle analysis of purified nuclei from paraffin sections. *Mod. Pathol.* **13**, 705–711.

37. Strausberg, R. L., Dahl, C. A., Klausner, R. D. (1997) New opportunities for uncovering the molecular basis of cancer. *Nat. Genet.* **15**, 415–416.
38. Cole, K. A., Krizman, D. B., Emmert-Buck, M. R. (1999) The genetics of cancer—a 3D model. *Nat. Genet. Suppl.* **21**, 38–41.
39. Bornstein, S. R., Willenberg, H. S., Scherbaum, W. A. (1998) Progress in molecular medicine: "Laser capture microdissection." *Med. Klin.* **93**, 739–743.
40. Sgroi, D. C., Teng, S., Robinson, G., LaVangie, R., Hudson, J. R. Jr., Elkahloun, A. G. (1999) In vivo gene expression profile analysis of human breast cancer progression. *Cancer Res.* **59**, 5656–5661.
41. Leethanakul, C., Patel, V., Gillespie, J., et al. (2000) Distinct pattern of expression of differentiation and growth-related genes in squamous cell carcinomas of the head and neck revealed by the use of laser capture microdissection and cDNA arrays. *Oncogene* **19**, 3220–3224.
42. Banks, R. E., Dunn, M. J., Forbes, M. A., et al. (1999) The potential use of laser capture microdissection to selectively obtain distinct populations of cells for proteomic analysis—preliminary findings. *Electrophoresis* **20**, 689–700.
43. Emmert-Buck, M. R., Gillespie, J. W., Paweletz, C. P., et al. (2000) An approach to proteomic analysis of human tumors. *Mol. Carcinog.* **27**, 158–165.
44. Simone, N. L., Remaley, A. T., Chorboneau, L., et al. (2000) Sensitive immunoassay of tissue cell proteins procured by laser capture microdissection. *Am. J. Pathol.* **156**, 445–452.
45. Ornstein, D. K., Englert, C., Gillespie, J. W., et al. (2000) Characterization of intracellular prostate-specific antigen from laser capture microdissected benign and malignant prostatic epithelium. *Clin. Cancer Res.* **6**, 353–356.

# PROTOCOLS

## H&E STAINING

For paraffin sections, start with **step 1.** For frozen sections that have been previously fixed, start with **step 5.**

1. Xylenes to deparaffinize the slides 2 × 5 min each.
2. 100% ethanol 30 s.
3. 100% ethanol 30 s.
4. 95% ethanol 30 s.
5. Purified water rinse.
6. Mayer's hematoxylin 30 s.
7. Purified water rinse.
8. Eosin Y 30 s.
9. 95% ethanol 30 s.
10. 95% ethanol 30 s.
11. 100% ethanol 30 s.
12. Xylene 2 × 3 min.
13. Shake off excess, wipe the slide carefully, and air-dry for at least 2 min.

*NOTE:* Chain reagents (xylenes and alcohols) each day that LCM staining is done and filter hematoxylin before use.

## OPERATING THE PIX CELL SYSTEM

1. If entire instrument is shut off, turn on power strip located behind the computer monitor. Also, turn on power for the TV monitor.
2. On computer screen, click mouse on "shortcut to Arc 100."
3. When instrument serial # is given, click "continue."
4. Select name or enter new name and click "acquire data."
5. Highlight study or enter new study and click "select."
6. Enter slide #, spot size, cap lot, and thickness (usually 10 μm, but this may vary depending on slides cut), and click "continue."
7. Enter laser power of approx. 45 and pulse of approx. 55 (to start) length needed (see procedures for RNA and DNA extractions for this information).
8. Place slide on microscope, and, using TV monitor, find an appropriate starting area that will be easy to find after finishing the microdissection.
9. Click on "before" to obtain a picture of the area selected.
10. Insert a row of arcturus caps with transfer film into slot on the right side of the microscope.
11. If the laser power chosen is below 60, place the optic beam adjust piece without the filter in the indentation on the end of the placement "arm." If laser power chosen is above 60, place the beam adjust piece with the filter on the arm.
12. Using the placement arm, pick up a cap, move the arm all the way over to the left (this will put it directly above the slide), and gently release it so that the cap slowly drops onto the slide.
13. Pick up the white cord with the red button on the end and press the button to get a laser pulse. A dark circle will appear around the area if it "melted." (However, just because an area melts does not necessarily mean that it transferred to the cap.)
14. Use the joystick located to the bottom left of the microscope to move the slide around and get pulses in different areas.
15. After finishing the microdissection, use the arm to pick up the cap very gently (so as not to pick up any other tissue), move it over to the right, and place it on a sterile 0.5-mL microcentrifuge tube.
16. Find the original starting area and clock on "after" to get an image of the completed microdissection. Then look at the slide under a microscope to make sure that most of the areas transferred.
17. Click "done" and then "exit" or "continue."

## DNA EXTRACTION FROM COLLECTED CELLS

1. Strain slides with hematoxylin and eosin according to the written procedure for staining.
2. Aliquot 100 μl 0.05 *M* Trizma buffer, pH 8.3, and 4 μL proteinase K (10 mg/mL) into sterile 0.5-mL Eppendorf microfuge tubes (see **Digestion Buffer** below).
3. Use LCM to capture the desired number of cells from the slide.
4. Place the film cap containing the captured cells on the top of the microfuge tube and use the black cap tool to snap the cap in. (It is *critical* to place the bottom of the film cap just

inside the lip of the microfuge tube, leaving about a 1-cm gap. Using the black cap tool ensures the correct spacing needed to prevent leakage during incubation.)

5. Invert the tubes and incubate in a 55°C water bath for 48 h.
6. Remove tubes from water bath and spin down in microcentrifuge. Discard the film cap and transfer the solution to a new sterile 0.5-mL tube.
7. Boil samples for 8 min at 95°C in a hot block and place on ice. Store DNA at 4°C until ready to use.

### Digestion Buffer

1. 3.03 g Trizma base.
2. 500 mL water.
3. Adjust pH to 8.3.

# RNA EXTRACTION FROM COLLECTED CELLS

1. Obtain samples in cap with transfer film using LCM.
2. Aliquot 200 μL TRIzol reagent into a *sterile* 0.5-mL microcentrifuge tube, then place cap with film on top, and invert tube.
3. Leave samples with TRIzol inverted at room temperature for more than 1 h. (At this point, samples may be stored at −70°C for up to 1 month).
4. Take off cap, add 1 μL glycogen and 40 μL chloroform to each tube, and shake vigorously for 15 s.
5. Incubate samples for 3 min at room temperature.
6. Using Eppendorf centrifuge in cold room, spin for 15 min at 10,000 rpm.
7. Transfer aqueous phase to a new 0.5-mL tube, add 100 μL isopropanol, and vortex.
8. Incubate at room temperature for 10 min.
9. Place in a −70°C freezer for 1 h or longer. Take out and allow to thaw.
10. Centrifuge for 10 min at 10,000 rpm in Eppendorf centrifuge.
11. Very carefully discard the supernatant, as you will probably not see a pellet.
12. Add 200 μL 75% ethanol and vortex.
13. Centrifuge at 8,000 rpm for 5 min in Eppendorf centrifuge.
14. Carefully pour off the supernatant, invert the tube, and air-dry for approximately 5 min.
15. Resuspend the pellet in 10 μL diethyl pyrocarbonate (DEPC) water and use this directly for the RT reaction.

# 3

# *In Situ* Hybridization:
## *Detection of DNA and RNA*

## Long Jin, Xiang Qian, and Ricardo V. Lloyd

## INTRODUCTION

Recent developments in molecular biology are rapidly expanding the technologic advances used to study disease processes. *In situ* hybridization (ISH) represents a unique technique in which molecular biologic and histochemical techniques are combined to study gene expression in tissue sections and cytologic preparations. DNA and RNA can be readily localized in specific cells with this method. The method involves a hybridization reaction between a labeled nucleotide probe and complementary target DNA or RNA. Those hybrids can be detected either by autoradiographic emulsion for radioactively labeled probes or by histochemical chromogen development for nonisotopically labeled probes. ISH localizes gene sequences *in situ* and visualizes the product of gene expression while preserving cell integrity within the heterogeneous tissue, permitting anatomically meaningful interpretations.

ISH analysis of nucleic acids was first described in 1969 *(1,2)*. Since the mid-1980s, ISH techniques have developed rapidly and have become important tools in basic scientific research and clinical diagnosis *(3–8)*. The introduction of nonradioactive probe labeling and detection systems in the late 1970s has made ISH analysis feasible in the diagnostic pathology laboratory as a molecular diagnostic tool and in microbiology for the tissue localization of infectious agents. Other recent developments in the applications of ISH techniques involve localization of multiple target DNA and RNA sequences and increasing the assay sensitivity by *in situ* amplification for both the target nucleic acid and signal detection system.

This chapter discusses the basic principles and approaches of the ISH technique, new advances in the ISH method, and some application of ISH methods for the study of specific diseases.

## *IN SITU* HYBRIDIZATION METHODS

### *Principles*

DNA is a double-stranded nucleic acid chain consisting of two complementary nucleic acid strands made up of four basic deoxynucleotides linked to one another by

From: *Morphology Methods: Cell and Molecular Biology Techniques*
Edited by: R. V. Lloyd © Humana Press, Totowa, NJ

phosphodiester linkage. Each of the four nucleotides has a base, including adenine (A), cytosine (C), guanine (G), and thymidine (T). A is bound to T with two hydrogen bonds, and C is bound to G with three hydrogen bonds. RNA is a single-stranded nucleotide with the base uridine (U) substituted for T. ISH is the process by which there is specific annealing of a labeled nucleic acid probe to a target complementary sequence in fixed tissue or cells. This is followed by visualization of the hybridized signals with isotopic or colorimetric detection methods. One can have DNA-DNA, DNA-RNA, and RNA-RNA ISH, depending on various types of probes and targets concerned.

The main advantages of the ISH method are in its specificity for individual cells in a heterogeneous tissue or cell population, as well as in its sensitivity in detecting low-copy gene expression in cells or chromosomal gene mapping. ISH consists of multiple steps including probe preparation and labeling, tissue preparation, hybridization, and signal detection. Each of these steps must be systematically optimized for each probe and tissue preparation. When applied appropriately, the methods can also be semiquantitative.

## Probe Synthesis and Labeling

Three principal types of probes can be used for ISH, including double-stranded complementary DNA probes, single-stranded antisense RNA probes, and synthetic oligonucleotide probes. Like other hybridization techniques such as Northern or Southern hybridization, probe design, synthesis, and labeling play a key role for ISH with respect to specificity and sensitivity. Probes that work for Northern and Southern hybridization may work as well for tissue hybridization (such as ISH). The choice of probe for ISH must take into account specificity and sensitivity, ease of tissue penetration, stability of hybrids, and reproducibility of the technique. DNA probes can be labeled by nick translation or random primer methods, RNA probes are usually synthesized and labeled by in vitro transcription method, and the 3′ or 5′ end labeling methods are mainly used for oligonucleotide probes. For each method, the labeled nucleotides with reporter can be incorporated into DNA or RNA probe sequences and visualized with the appropriate detection system.

### Double-Stranded DNA Probes

Double-stranded DNA probes can be generated by reverse transcription of mRNA and by amplification of specific sequences of DNA or from cDNA using cloning vectors *(9)*. If probes are labeled using a random primer, the template DNA is linearized, denatured, and annealed to a primer. Starting from the 3′-OH end of the annealed primer, Klenow enzyme synthesizes new DNA along the single-stranded substrate. The size of the probe is between 200 and 1000 bp. In the nick translation labeling reaction, however, the DNA template can be supercoiled or linear. After the DNA is nicked with DNAse I, the 5′-3′ exonuclease activity of DNA polymerase I extends the nick to gaps; then the polymerase replaces the excised nucleotides with labeled ones. The size of the probe after reaction should be about 200–500 bp for optimal hybridization. DNA probes can also be prepared by polymerase chain reaction (PCR) using a DNA template and appropriate primers *(10)*. The probes can be directly labeled during synthesis by the incorporation of nucleotides conjugated to a reporter molecule. Large amounts of labeled probe can be produced by PCR from small amounts (10–100 pg) of linearized plasmid or even from nanogram amounts of genomic DNA. The length of the amplified probe is precisely defined by the 5′ ends of the PCR primers. All three DNA labeling methods allow great flexibility with regard to length of the labeled fragments. In the nick translation reaction, changing the DNAse concentration alters the fragment length. In the random primed labeling reaction, changing the primer concentration affects fragment length. In the PCR procedure, the sequence of the PCR primers controls fragment length. The probes prepared by these three labeling methods require denaturation to produce single-stranded DNA before hybridization.

*Antisense RNA Riboprobes*

Antisense RNA riboprobes are generated by in vitro transcription from a linearized DNA template, incorporating labeled nucleotides. A promoter for RNA polymerases must be present on the vector DNA containing the template. SP6, T3, or T7 RNA polymerase is commonly used to synthesize RNA complementary to the DNA substrate *(11)*. The synthesized transcripts are an exact copy of the sequence (about 300–800 bp) from the promoter site to the restriction site used for linearization. The size of the probe can be adjusted by selection of the linearizing restriction enzyme so that the antisense and sense probes both have the same length. RNA riboprobes are single-stranded and are susceptible to degradation by RNAses. However, cRNA probes have the advantage that RNA-RNA hybrids are more stable than DNA-DNA or DNA-RNA hybrids.

*Oligonucleotide (Oligo) Probes*

Oligonucleotide (oligo) probes consisting of 30–50 bases can be generated with an automated DNA synthesizer. The oligo-probes probably penetrate cells more readily and can produce excellent hybridization signals. With the increasing numbers of cloned genes, oligonucleotide probes can be generated from published cDNA sequences or gene banks and synthesized rapidly and inexpensively. Labeling oligo-probes for ISH is commonly done with 3′ end labeling, or "tailing" with relatively few labeled nucleotides incorporated into each probe. The oligo-probes are relatively less sensitive than longer cDNA or cRNA probes *(4,12)*, so they are generally considered the most suitable for detection of relatively abundant gene expression such as hormone mRNAs *(13–15)*. This disadvantage can be overcome by using a "cocktail" of multiple oligonucleotide probes that are complementary to different regions of the target molecules *(8,16)*. Careful selection of oligo-probes with low or no homology with other nucleotide sequences is most important to ascertain the specificity of ISH *(17)*. By using 3′ end labeling with deoxydinucleotides, about 10–20 labeled nucleotides are incorporated into end of probe sequence with extension after incubation at 37°C for 30–45 min. Oligo-probes may also be generated in the same way as standard riboprobes using a short DNA template or by combining single-strand oligonucleotides with a bacteriophage promoter.

**Tissue Preparation**

Tissue processing including storage and fixation should be optimized to detect intracellular nucleic acids. ISH has been applied to cells (smears, cytospins, or cell pellets), tissue sections (frozen, paraffin, semithin, and ultrathin plastic section), or whole-mount embryos. Nucleic acids are better preserved in frozen tissue than in paraffin-embedded tissues. Intact cells are ideal for ISH, as they have less damaged nucleotide sequences. Both fresh frozen tissues (stored at −70°C) and formalin-fixed archival tissues can be used for ISH after storage for several years. Ideal fixation for ISH should preserve both RNA/DNA and tissue morphology but still allow penetration of probes. Crosslinking fixatives such as 4% paraformaldehyde, 4% formaldehyde, and 1% glutaraldehyde, are most commonly used. Generally, fixation for 20 min to 1 h is sufficient for frozen sections on glass slides. After a brief wash in 2× standard saline citrate (SSC) to remove residual fixative and dehydration of the slides through a graded ethanol series, the slides can be stored at −70°C until hybridized. The slides used for ISH should be pretreated with a suitable coating solution, such as 3-aminoprpyltrime-thoxysilane or poly-L-lysine.

A series of pretreatment steps before hybridization increases the efficiency of hybridization and reduce nonspecific background staining. Some steps are routinely used in most ISH protocols.

Protease treatment (e.g., proteinase K) is considered one of the most important steps to increase the accessibility of the target nucleic acid, especially for paraffin sections with nonisotopic probes. The concentrations of proteinase K (1–2 µg/mL for frozen sections and 20–50 µg/mL for paraffin sections) and the length of treatment (5–30 min) depend on tissue type and length of fixation. Prolonged incubation results in overdigestion, resulting in loss of signal and morphologic integrity *(18)*. A brief post-fixation in paraformaldehyde is commonly recommended.

Pretreatment of slides with microwaving to increase ISH sensitivity, especially for old paraffin-embedded tissues, has been reported *(18)*. For microwaving, sections are immersed in 10 m$M$ citrate buffer, pH 6.0, and microwaved at full power in an 800-W microwave for 5–10-min periods. It should be noted that if microwaving treatment is added, the length of proteinase K treatment should be reduced.

Acetylation of sections using 0.25% acetic anhydride/0.1 $M$ triethanolamine can reduce charged probe binding to tissues and nonspecific binding. Some investigators have noted that acetic anhydride prevents the nonspecific binding of unrelated digoxi-genin-labeled probes to neuroendocrine cells *(15)*. Adding fresh dithiothreitol (DTT) to the hybridization solution containing [35]S-labeled probes can protect the sulfur from oxidation. The background artifacts with [35]S-labeled probes can be dramatically reduced by increasing the DTT concentration *(19)*. Incubation of sections with prehybridization buffer, normally containing 50% formamide, is generally considered useful for mainte-nance of the hybridization stringency, even though some investigators prefer to perform ISH in a formamide-free condition *(20,21)*. The presence of endogenous biotin or alkaline phosphatase should be anticipated in some tissues when nonisotopic probes are used. Endogenous alkaline phosphatase can be inhibited by treatment of sections with 0.2 $N$ HCL and levamisole, and endogenous biotin can be inhibited by biotin-blocking agents.

## Hybridization

Hybridization between the labeled probe and target DNA or RNA is defined by hydrogen bonding and the hydrophobic interactions in equilibrium. Annealing and separation of the two hybrid strands depend on various factors, including temperature, salt concentrations, pH, the nature of the probes and target molecules, and the composi-tion of the hybridization and washing solution. The "melting" temperature ($T_m$) of hybrids is the point at which 50% of the double-stranded nucleic acid chains are separated. The optimal temperature for hybridization is 15–25°C below the $T_m$. There are various formulas for calculating $T_m$, depending on probe length and type of hybrids *(5)*. For a DNA-DNA hybrid with probes longer than 22 bases, the following formula can be used:

$$T_m = 81.5 + 16.5 \log(Na) + 0.41 \ (\%GC) - 0.62 \ (\% \text{ formamide})$$
$$- 500/\text{length of base pairs of probe}$$

where Na = sodium concentration and GC = guanine and cytosine content of the probe.

RNA-RNA hybrids are generally 10–15°C more stable than DNA-DNA or DNA-RNA and therefore require more stringent conditions for hybridization and posthybridization washing.

Hybridization buffer contain reagents to maximize nucleic acid duplex and inhibit nonspecific binding of probe to tissues. Generally, 20–30 µL of hybridization buffer containing labeled probe is enough for each slide. The concentration of probe must be optimized for each probe and for each tissue. Hybridization is done on a cover-slipped slide in a humidified chamber in an oven at 42–45°C for oligo-probes or 50–55°C for cDNA or RNA riboprobes.

After an overnight incubation, careful removal of the cover slips using 2X SSC is followed by posthybridization washing with 2X, 1X, and 0.5X SSC for 10–30 min each. The final washing can be done at 42–45°C (for oligo-probes) or 50–55°C (for DNA or RNA probes) to increase the stringency. If riboprobes are used, treatment with ribonuclease A (RNAse) after hybridization (generally 0.1–1 ng/mL for paraffin sections or 0.1–1 µg/mL for frozen sections) will reduce nonspecific background signal, since the enzyme digests single-stranded, but not double-stranded, RNA hybrids. After additional washing in phosphate-buffered saline (PBS) for 15–30 min, the slides are ready for signal detection.

### Reporter Molecules and Signal Detection

Two major types of reporters for signal detection are used: 1) radioisotope labeling, such as [3]H, [35]S, [125]I, [32]P, and recently [33]P, which is detected with X-ray film and/or emulsion autoradiography; and 2) nonisotope labeling, such as biotin, digoxigenin, fluorescein, hapten, alkaline phosphatase, or bromodeoxyuridine (BrdU), which is visualized by histochemistry or immunohistochemistry (IHC) detection systems.

Radioisotope labeling is the more traditional way to perform ISH. This is still regarded as the most sensitive method of labeling, although under certain conditions nonisotopic methods can be equally sensitive *(22)*. The choice of isotope usually reflects a compromise between the signal strength and resolution. The [35]S-labeled probe is most commonly used, because it gives a reasonable resolution and exposure time (typical exposure about 1 week) compared with other isotopic labeling. The half-life of [35]S is 87 days, and labeled probes should be used within 1 month or so for optimum results. Other isotopes are less commonly used. After counterstaining with hematoxylin and eosin and coverslipping, the silver grains, which represent the hybridization signal, can be easily visualized with a bright-field or a dark-field microscope. Radioactive labeling may be used to detect low-copy sequences, to quantify the ISH signal, and to detect multiple nucleotide sequences for combined studies with immunocytochemistry. These applications are uncommon in diagnostic practice. The disadvantages of isotopic probes include biohazards and a short half-life, as well as the fact that it takes a long time to obtain the final results.

Nonisotopic ISH methods are being used with increasing frequency. The advantages of nonisotopic labeling include greater stability of labeled probes, rapid results, and better resolution. These advantages make nonisotopic labeling probes very attractive in the clinical laboratory. Biotin and digoxigenin are the most commonly used reporters at present because they are readily detectable in tissues.

Biotinylated probes were first used a few decades ago and have been widely used for detection of viral DNA or relative abundance of mRNA at both the light (LM) and electron microscopic (EM) levels. The hybridization signals with biotinylated probes can be detected by anti-biotin antibodies. However, streptavidin or avidin conjugated to alkaline phosphatase (AP), horseradish peroxidase (HPO), fluorochromes, or colloidal gold particles are more frequently used because these molecules have a high binding capacity for biotin. The disadvantages of biotinylated probes include the widespread presence of endogenous biotin in tissues such as liver and kidney, which can result in false-positive results, and the limited success of blocking endogenous biotin in some tissues.

Digoxigenin-labeled probes were introduced a few years ago and are rapidly becoming more widely used compared to biotinylated probes. Digoxigenin probes have higher sensitivity and less background staining than biotinylated probes *(12)*. The fact that digoxigenin is not present in mammalian cells and that anti-digoxigenin antibody does not bind to other biologic material is a particular advantage when studying tissues such as liver or kidney, which may contain endogenous biotin, but not digoxigenin. Digoxigenin is a derivative of the cardiac glycoside digoxin, and the digoxigenin-labeled nucleotides may be incorporated, at a defined density, into nucleic acid probes by DNA polymerases, RNA polymerases, or terminal transferase. Since the digoxigenin label may be added by random primed labeling, nick translation, PCR, 3'-end labeling/tailing, or in vitro transcription, it has been widely used in labeling cDNA, cRNA, or oligonucleotide DNA probes. The signals can be visualized by using an anti-digoxigenin antibody fragment conjugated to AP or HPO with respective substrates that yield insoluble-colored products. Alternatively, unconjugated anti-digoxigenin antibodies and conjugated secondary antibodies may be used. Nonisotopic probes are generally considered less sensitive than the corresponding radioactive probes, and the hybridization results are difficult to quantify.

Fluorescein is a newer nonradioactive labeling alternative. Fluorescein-labeled probes can be used for direct as well as indirect *in situ* hybridization experiments. Fluorescein-dUTP/UTP/ddUTP can be incorporated enzymatically into nucleic acids by different labeling methods. Fluorescein-labeled probes can be visualized directly with a fluorescent microscope after hybridization, and the background is usually low. Alternatively, fluorescein-labeled probes can be detected with an anti-fluorescein antibody-enzyme conjugate or with an unconjugated antibody and a fluorescein-labeled secondary antibody.

## Evaluation

It is important to use various controls to verify the specificity of the ISH signal. A variety of controls can be used: 1) pretreatment of tissues with RNAse or DNAse, depending on the target being tested; 2) omission of the specific probes in hybridization reaction; 3) using an unrelated or sense probe; 4) competitive studies with unlabeled probes before adding labeled probes for hybridization; 5) Northern or Southern hybridization to characterize the nuclei acid species of the hybrids; and 6) combining ISH with immunostaining to localize the translated protein product in the same cells.

It is important to assess the integrity of target RNA during ISH. Loss of RNA or failure of probes to detect target RNA may result in false-negative results, particularly for retrospective studies of paraffin-embedded tissues. To evaluate the preservation of mRNA, control probes including b-actin *(23)*, poly(dT) *(24),* and ribosomal RNA *(25)* can be used.

Most RNA ISH protocols emphasize that the procedure must be performed under RNAse-free conditions, especially for fresh tissue sections. Riboprobe or target mRNA degradation by residual endogenous RNAse activity or by accidental RNAse contamination is considered one of the major potential causes of experimental failure. RNA degradation may occur mainly during pretreatment, since the degradation can be inhibited by formamide, generally at 50% volume in the hybridization buffer during the hybridization step *(26)*. The endogenous RNAse activity might be readily neutralized through paraformaldehyde fixation *(26)*. It is recommended that the labeled probes be in aliquots and stored at −70°C in the presence of RNAse inhibitor. There is usually a detectable degradation of the riboprobes after 1 week *(4,7)*.

The experimental results of ISH with radioactive probes can be readily quantitated or semiquantitated by densitometry counting on film or by silver grain counting, compared with nonisotopic ISH methods. The reliability and accuracy of this type of mRNA quantification are still to be determined, and comparable data can only be gained under very strict conditions, such as different cases in the same set of experiments, or different tissue areas on the same section. Recently, improvements in ISH quantitation have been achieved by using densitometric computer-assisted image analysis *(27)*. Using these new methods, ISH quantification may soon be comparable to that of Northern blot *(27)*.

## Recent Advances

A variety of new ISH techniques have been developed over the past few years. These advances can be divided two categories: 1) combined use of various ISH approaches for simultaneous detection of multiple target nucleic acids at the light microscope (LM) and electron microscope (EM) levels; and 2) increasing the sensitivity of ISH signals by *in situ* amplification for target DNA or RNA or by signal detecting systems.

## Multiple-Target Localization

Double-labeling ISH methods involve the simultaneous detection of multiple-target nucleic acids such as two mRNAs, DNA, and mRNA, or nucleic acids and proteins for combination ISH with IHC techniques *(28–32)*. In performing double ISH, the two probes may be hybridized simultaneously or sequentially. For combined ISH/IHC, ISH is generally performed before IHC because it reduces the chances of RNAse contamination. A variety of combinations of ISH methods have been reported, as discussed below.

A combination of nonradioactive and radioactive ISH methods is mainly used for detection of two mRNAs in the same tissue sections *(19,29,30)*. After histochemical detection for nonisotopic signals, the slides are subject to an autoradiographic approach for radioactive signals. Some investigators have reported that the use of Ilford emulsion, rather than Kodak NTB$_2$ emulsion, can reduce silver grain background by chemographic artifact when digoxigenin and $^{35}$S-labeled oligonucleotide probes for double ISH detection are used simultaneously *(31)*.

Several investigators have successfully used a combination of two nonisotopic labeled probes, mainly biotin and digoxigenin, conjugated to different enzymes (AP or HPO) or fluorescence followed by respective detective systems for a simultaneous localization of multiple mRNA and genomic DNA in the same tissue section, or even in the same cells *(3,32–34)*. Using probes labeled with different fluorochrome molecules and with different excitation and emission characteristics allows simultaneous analysis of different probe signals on chromosomal and cytogenic preparations *(5,33)*. An alternate method is the combination of nonisotopic ISH with fluorescence (F)ISH: the two signals can be visualized by using brightfield and darkfield microscopy *(32)*.

Ultrastructural ISH is used mainly for investigations attempting to correlate molecular function and ultrastructural morphology. Nonradioactive ISH methods with colloidal gold are most commonly used for pre- and/or postembedding EM techniques *(35–37)*. Colloidal gold with different nonoverlapping particle sizes (5–30 nm) greatly facilitates multiple labeling for ISH analysis at the EM level. Bienz and Egger *(35)* reported the simultaneous localization of P$_1$ and P$_2$ virus genome regions in poliovirus-infected cells by double-EM ISH: a digoxenin labeled P$_1$ probe and a biotin-labeled P$_2$ probe were detected through simultaneous direct immunodetection with 10-nm gold-labeled anti-digoxigenin and 6-nm gold-labeled antibiotin antibodies. A combination of ISH with the terminal deoxynucleotidyl transferase dUTP nick end-labeling technique on pre-embedded free-floating sections has been used to analyze specific gene expression and apoptosis at both LM and EM levels *(38)*.

### In Situ Amplification

An example of *in situ* target amplification is *in situ* PCR, a relatively new molecular technique that combines the high sensitivity of PCR with anatomic localization by ISH *(39–41)*. ISH localizes gene sequences at the cellular level, but it generally requires at least 10–20 or more copies of the mRNA of interest in a cell to make the signals visible. A combination of PCR with ISH can amplify DNA or RNA sequences inside single cells and increase the copy numbers to levels readily detectable by ISH methods. There are two main approaches: direct and indirect. In the direct approach a labeled nucleotide is incorporated into the PCR products *(42)*; in the indirect method ISH is performed after *in situ* amplification using a labeled oligonucleotide probe.

Over the past few years, *in situ* PCR has been mainly applied to detect DNA sequences that are not easily detected by conventional ISH. *In situ* PCR to detect low copy numbers of viral genes, especially the human immunodeficiency virus and hepatitis C, has led to significant discoveries about viral diseases *(43)*. *In situ RT-PCR* is the process of detecting mRNA in which mRNA is reverse transcribed to single strands of DNA followed by PCR amplification. Application of *in situ* RT-PCR to detect gene expression is still limited primarily to cell preparations and frozen sections. Only a few applications of this procedure in paraffin sections have been reported *(44)*. Successful amplification of mRNA by *in situ* RT-PCR includes hormone, receptor, and oncogenes

*(41,45–47).* Direct comparison of *in situ* RT-PCR with radioactive or nonradioactive conventional ISH for mRNA analysis shows 2–20-fold increases in hybridization signals *(45,46).* *In situ* RT-PCR has many potential applications, since many genes are known to be expressed in low abundance, and their functions have not been elucidated. *In situ* PCR in EM and a combination of *in situ* PCR with IHC have also been described *(45,48).*

*In situ* self-sustained sequence replication (3SR) is based on the use of primers with attached RNA polymerase initiation sites and the combination of three different enzymes in the same reaction mixture (DNA polymerase, RNAse H, and RNA polymerase), resulting in accumulation of target mRNA through the combination of RT, DNA synthesis, and *in vitro* transcription *(49).* This method has been applied to cytocentrifuge preparation, and much less amplification efficiency compared with liquid-phase 3SR assays.

Primed *in situ* labeling (PRINS) is a rapid, one-step target amplification technique based on the Taq DNA polymerase-mediated incorporation of labeled nucleotides into newly synthesized DNA by a single primer elongation. This newly described technique is used primarily for chromosomal analysis on both metaphase spreads and interphase nuclei, as well as detection of karyotype changes in diagnostic pathology *(50).* The potential problems with the PRINS method include false-positive results due to mispriming and artifacts related to the incorporation of labeled nucleotide into damaged DNA.

Another strategy to improve ISH sensitivity is to use *in situ* amplification with a signal detection system. The tyramide signal amplification method with catalyzed reported deposition (CARD) has been used most frequently in recent years *(51).* This CARD amplification technique allows a 500–1000-fold increase in sensitivity of immunohistochemical signals compared with the conventional avidin-biotin complex method without production of increased background *(52).* This approach to signal amplification has been adapted to ISH in cytocentrifuged material and tissue sections. Wiedorn et al. *(53)* found that CARD ISH can be used as an alternative to *in situ* PCR on routinely processed tissue specimens and that it was superior to *in situ* PCR in reproducibility.

## APPLICATIONS OF *IN SITU* HYBRIDIZATION

ISH methods have found many applications in clinical research and diagnostic pathology. The technique is used in diagnostic practice in the areas of gene expression, infection, and interphase cytogenetics. The ISH technique can be applied to routinely fixed and processed tissues such as biopsy and autopsy material.

### Infectious Diseases

In the field of infectious diseases, ISH has been used mainly to detect viral infections, such as human papillomavirus (HPV) and Epstein-Barr virus (EBV) by the presence of small nuclear RNAs (EBERs). (*See also* Chapter 4.)

#### Viral Infections

Viral infections have been more widely investigated, and the technique has proved useful both in the identification of specific viral infection and in defining the extent of systemic infection. The combined use of broad-spectrum probes and those identifying specific virus subtypes has confirmed a strong positive association between HPV and high-grade cervical intraepithelial neoplasia or progressive disease, with evidence that types 16 and 18 are more likely to be associated with malignant progression *(54,55)*. The test has given some insight into the epidemiology, in that similar patterns of infections were identified in two cohorts separated by 25 years *(56)*. ISH may be a useful adjunct in the investigation of HPV infection of the male genitalia *(57)*. However, the role played by HPV in other lesions, including Bowen's disease and squamous cell papilloma of the bronchus and larynx, is unclear. EBV has been implicated in the pathogenesis of a wide range of human lymphoid and epithelial tumors *(58)* **(Figs. 1A, B)**. It is difficult to detect viral DNA by ISH in latent infection because of the low copy number per cell. However, EBERs encoded by the virus are highly expressed and detectable by nonradioactive ISH. EBV has been detected in some cases of Hodgkin's disease, in a variety of lymphomas, including T-cell lymphomas, and in oral hairy leukoplakia *(59–61)*. The presence of EBV in undifferentiated nasopharyngeal carcinoma has been studied extensively *(62)*. ISH is useful in confirmation of infectious mononucleosis in atypical cases and of renal disease associated with EBV patients who have a negative serology *(63,64)*.

Hybridization to viral DNA and mRNAs in the liver has aided our understanding of the complexity of hepatitis B infection *(65)*. However, these studies have provided conflicting data about the correlation between viral replication and disease activity. More recently, ISH has been used to detect hepatitis C virus (HCV) RNA. However, negative results were found in a significant proportion of HCV-seropositive cases *(66)*. This could be due to the low level of replication of the virus, and amplification techniques may increase the sensitivity. Most studies applied these techniques to formalin-fixed, paraffin-embedded tissue due to its availability and the good preservation of morphology. Very few studies have applied this technique to frozen tissue. However, most groups showed that HCV-positive cells were usually isolated, with occasional clustering, and that usually <20% of cells were positive *(67,68)*. In a few reports, the proportion of HCV RNA-positive cells was found to be very high *(69)*. Most studies showed the localization of HCV in hepatocytes, but some studies showed ISH signals in mononuclear cells, bile duct epithelium, and sinusoidal cells *(70,71)*. HCV may be directly involved in the biliary epithelial damage seen in a proportion of patients with chronic HCV infection *(70,71)*. The level of HCV positivity appeared to correlate with serum aminotransferase, suggesting that HCV might cause damage to the liver cells directly in the absence of overt morphologic changes.

ISH has been used for the identification of cytomegalovirus (CMV) in viral encephalitis associated with the acquired immunodeficiency syndrome (AIDS) and chronic encephalitis associated with AIDS or epilepsy. Pulmonary involvement has been detected in cytologic specimens and the extent of systemic disease documented when the presentation was that of isolated oophoritis *(72)*. CMV infection has been detected in gastrointestinal biopsies in cardiac transplant recipients *(73)*. However, with CMV, there may not be a significantly greater sensitivity with ISH than with histology and immunohistochemistry.

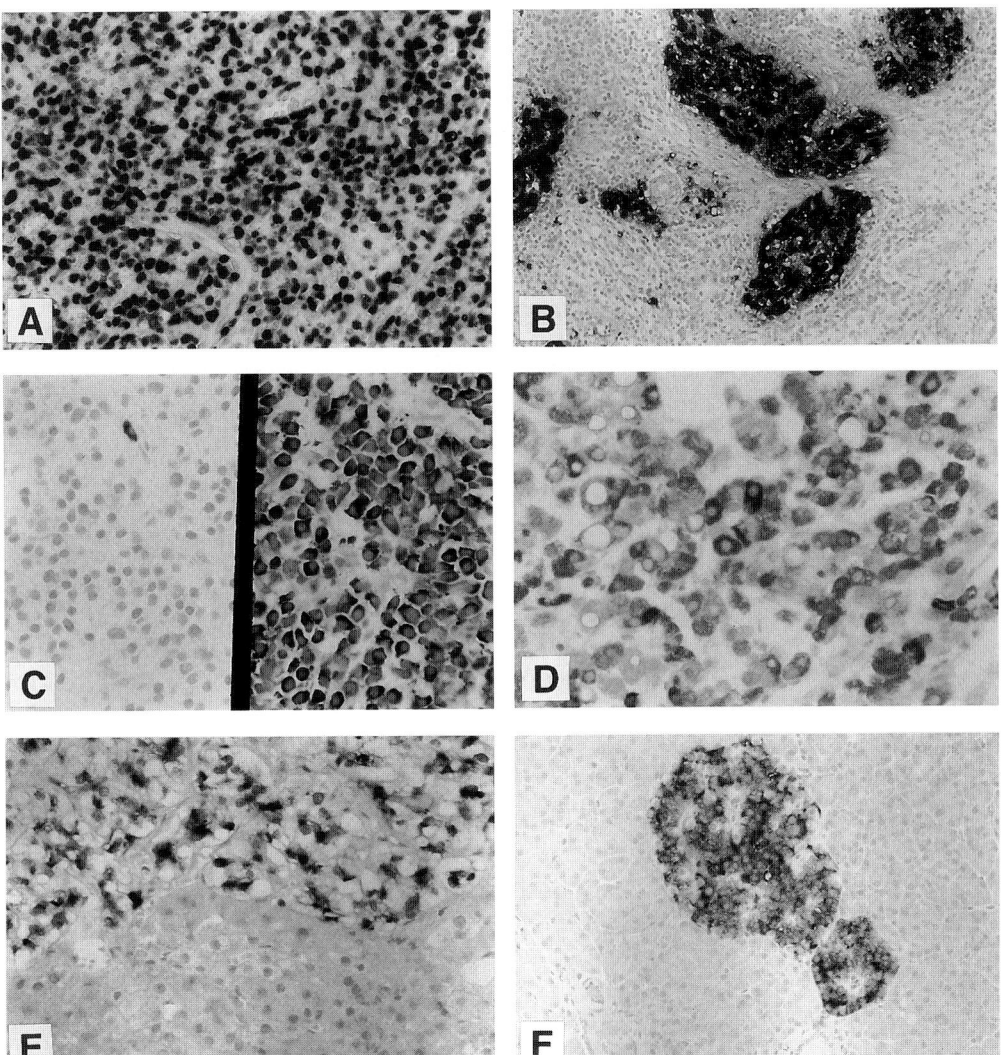

**Fig. 1. (A)** *In situ* hybridization detects EBV in a posttransplant lymphoproliferative disorder. There is diffuse staining of all lymphoid cells for the EBV probe. *In situ* hybridization with digoxigenin-labeled probes and detection by alkaline phosphatase/nitroblue tetrazolium-bromochloroindolyl phosphate (NBT-BCIP) for all figures. **(B)** *In situ* hybridization to detect EBV in a nasopharyngeal carcinoma metastatic to a cervical lymph node. The presence of EBV in this clinical setting is diagnostic of a nasopharyngeal carcinoma. **(C)** Detection of κ (left) and λ (right) light chains in a plasmacytic lymphoma. The tumor shows λ light chain restriction with all the tumor cells expressing this light chain. A few residual normal plasma cells stained positively for κ light chain. **(D)** *In situ* hybridization for albumin to diagnose a hepatocellular carcinoma metastatic to the scapula. Albumin expression is relatively specific for normal and neoplastic liver cells. **(E)** Chromogranin A and B expression in the adrenal cortex. The adrenal cortex in the lower portion of the figure is negative for chromogranin A and B cocktail probe. **(F)** Pro-insulin mRNA expression in the normal pancreatic islets. Most of the islet cells (60–80%) express the pro-insulin mRNA (original magnification ×250 for A, B, D, and E; ×200 for F). (*See* **Color Plate 1**, following page 208.)

Herpesviruses (HHVs) may also be detectable, including herpes simplex in lymphadenitis and endometritis and HHV-6 in erythroderma associated with an infectious mono-
nucleosis-like syndrome and lymphoproliferative states *(74–76)*. The presence of HHV-
8 in Kaposi's sarcoma has been demonstrated using PCR-ISH *(77)*. There is some
evidence to suggest that ISH may not be as sensitive as IHC in the detection of HSV,
but proper validation should be performed with the newer amplification techniques. It
may have an advantage over IHC in early infection.

*Identification of Bacteria and Fungi*

ISH can provide a useful tool for identification of fungi and bacteria *(78)*. Yeast
and yeast-like organism identification in tissue sections can be very difficult, with only
occasional organisms seen. Several common species have overlapping morphologic features. Aspergillosis results in significant mortality in immunosuppressed
patients. Rapid diagnosis is often required to initiate appropriate therapy. The histology
of *Aspergillus* species may overlap with a variety of fungi, so diagnosis often relies
on fungal cultures that can take weeks to complete. ISH targeting *Aspergillus* 5S
ribosomal RNA (rRNA) identified 41 cases of localized aspergillomas in the lung,
brain, sinonasal tract, and ear, as well as 2 cases of invasive aspergillosis involving
pleura and soft tissue of the scapular region *(79–81)*. Allergic fungal sinusitis (AFS)
is a serious form of sinonasal disease that is commonly associated with *Aspergillus* or
*dematiacious* fungi. ISH has identified *Aspergillus/Penicillium* organisms in many
AFS patients.

The diagnosis of *Pneumocystis carinii* by ISH can be performed on paraffin sections.
The reactions were positive in all 12 cases of *P. carinii* pneumonia, but in none of the
infections with other pathogenic agents, including virus, mycobacteria, protozoa, and
fungi. The reactivity and specificity of this method was comparable to that of immunohistochemistry using a monoclonal anti-human *P. carinii* antibody *(82,83)*.

*Staphylococcus epidermidis* and *Staphylococcus aureus* are the most common causes
of medical device-associated infections. The microbiologic diagnosis of these infections
remains ambiguous. The detection and identification of *S. aureus* and *S. epidermidis*
by an ISH method with fluorescence-labeled oligonucleotide probes specific for staphy-
lococcal 16S rRNA have been reported *(84)*. *Helicobacter pylori* is the causative
agent of chronic gastritis and peptic ulcers and is also associated with gastric cancer.
Eradication of *H. pylori* infection has proved to be difficult to confirm. ISH results
were compared with those of culture and conventional histology, with agreement in
most cases *(85)*.

DNA probes have been used to detect both the 18S and 28S ribosomal RNA sequences
of various fungal organisms with a high degree of specificity for each fungus, including
*Blastomyces dermatitis, Coccidiodes immitis, Cryptococcus neoformans, Histoplasma
capsulatum,* and *Sporothrix schenckii (86)*. Probes were tested against 98 paraffin-
embedded tissue specimens, each of which had culture-proved involvement by one of
these organisms. ISH with oligonucleotide DNA probes was uniformly present with
all species-specific probes yielding 100% specificity, including four cases of Grocott
methenamine-silver stain (GMS)-negative staining *(86)*. These results show that ISH,
directed against ribosomal RNA, provides a rapid and accurate technique for the identification of yeast-like organisms in histologic tissue sections.

## Cell Proliferation Analysis

The cell cycle consists of four distinct phases, G1, S, G2, and M. The expression of histone genes is an important step in the process of cell proliferation. Histone H3 mRNA accumulates in the cytoplasm during the S-phase, and then decreases as cells approach the G2-phase. H3 transcription rates increase 10-fold at the onset of the S-phase and are downregulated at the cessation of cell proliferation and during quiescence *(87)*. The mRNA can be detected by ISH. The principal advantage of H3 mRNA determination is that the results are tightly coupled with *de novo* DNA synthesis. Moreover, its short half-life (20 min to 6 h) results in a more accurate reflection of S-phase activity. The proliferative activity of 71 cases of astrocytomas was studied by histone ISH and by immunohistochemistry with an antibody to proliferating cell nuclear antigen, Ki67, and mitoses *(88)*. A significant correlation (with reproducible results) was found among the labeling indices of histone mRNA, proliferation marker labeling indices, and mitotic indices *(88)*.

## Gene Expression

ISH methods are powerful tools for the analysis of gene expression and regulation in normal and pathologic tissues *(89)*. A major advantage of ISH is its ability to localize mRNA at the cellular level in heterogeneous tissues, thus expanding the results of other molecular techniques, such as Northern blot hybridization, for specific gene analysis.

ISH has been used to detect mRNA as a marker of gene expression when levels of protein storage are low, for example, to confirm an endocrine tumor as the source of excess hormone production *(13,90)*. ISH methods have been widely used for the study of mRNA-encoding oncogenes, growth factors and their receptors, hormones and hormone receptors, cytokines, structural proteins, enzymes, collagenase, and others.

There are many practical applications of ISH methods in tumor pathology (**Fig. 1C–F**). The correlation of oncogene expression with prognosis is being investigated in neuroblastomas and epithelial neoplasms such as colon, lung, prostate, and breast carcinomas. Detection of genes encoding cell structural proteins, including tumor-associated markers, represents a potential area of application of ISH methods in pathologic diagnosis. For example, nonisotopic ISH methods for localization of immunoglobulin light chain in mRNAs in hyperplastic and neoplastic lymphoproliferative disorders *(16)* (**Fig. 1C**), albumin mRNA to distinguish between hepatocellular and metastatic carcinomas in the liver (**Fig. 1D**), and chromogranin/secretogranin mRNAs in the classification of neuroendocrine tumors *(8)* (**Fig. 1E**) are used in some diagnostic pathology laboratories. ISH methods used to identify cells or tumors on the basis of their specific mRNA content are different from IHC, which is dependent on the protein content of cells. Thus, ISH identifies the gene products from *de novo* synthesis, rather than nonspecific uptake of proteins by cells, which may result in false-positive immunostaining results. ISH analysis has also been used extensively in studies of endocrine cells and tumors (**Fig. 1F**). For example, a significant number of small cell lung carcinomas with few secretory granules are commonly negative for chromogranin proteins, but the mRNAs may be detected by ISH methods *(8,14)*. ISH studies of gene expression in endocrine tumors (including parathyroid hormones and calcitonin gene-related peptide) have contributed to our understanding of the biology and pathophysiology of various endocrine disorders.

## REFERENCES

1. Gall, J. G. and Pardue, M. L. (1969) Formation and detection of RNA-DNA hybrid molecules in cytological preparation. *Proc. Natl. Acad. Sci. USA* **63**, 378–383.
2. John, H. A., Birnstiel, M. L., and Jones, K. W. (1969) RNA-DNA hybrids at the cytological level. *Nature* **223**, 582–587.
3. Egger, D., Troxler, M., and Bienz, K. (1994) Light and electron microscopic in situ hybridization: non-radioactive labeling and detection, double hybridization, and combined hybridization-immunocytochemistry. *J. Histochem. Cytochem.* **42**, 815–822.
4. Wilcox J. N. (1993) Fundamental principles of in situ hybridization. *J. Histochem. Cytochem.* **41**, 1725–1733.
5. Wilkinson, D. G., ed. (1993) *In Situ Hybridization: A Practical Approach.* IRL Press-Oxford University Press, New York.
6. Panoskaltsis-Mortari, A. and Bucy, R. P. (1995) In situ hybridization with digoxigenin-labeled RNA probes. Facts and artifacts. *Biotechniques* **18**, 300–307.
7. Poulsom, R., Longcroft, J. M., Jeffery, R. E., Rogers, L. A., and Steel, J. H. (1998) A robust method for isotopic riboprobe in situ hybridization to localize mRNAs in routine pathology specimens. *Eur. J. Histochem.* **42**, 121–132.
8. Lloyd, R. V. and Jin, L. (1995) In situ hybridization analysis of chromogranin A and B mRNAs in neuroendocrine tumors with digoxigenin-labeled oligonucleotide probe cocktails. *Diagn. Mol. Pathol.* **4**, 143–151.
9. Temsamani, J. and Agrawal, S. (1996) Enzymatic labeling of nucleic acids (Review). *Mol. Biotechnol.* **5**, 223–232.
10. Mertz, L. M. and Rashtchian, A. (1994) Nucleotide imbalance and polymerase chain reaction: effects on DNA amplification and synthesis of high specific activity radiolabeled DNA probes. *Anal. Biochem.* **221**, 160–165.
11. Witkiewicz, H., Bolander, M. E., and Edwards, D. R. (1993) Improved design of riboprobes from pBluescript and related vectors for in situ hybridization. *Biotechniques* **14**, 458–463.
12. Komminoth, P., Merk, F. B., Leav, I., Wolfe, H. J., and Roth, J. (1992) Comparison of 35S and digoxigenin-labeled RNA and oligonucleotide probes for in situ hybridization. Expression of mRNA of the seminal vesicle secretion protein II and androgen receptor genes in the rat prostate. *Histochemistry* **98**, 217–228.
13. Lloyd, R. V., Cano, M., Chandler, W. F., Barkan, A. L., Horvath, E., and Kovacs, K. (1989) Human growth hormone and prolactin secreting pituitary adenomas analyzed by in situ hybridization. *Am. J. Pathol.* **134**, 605–613.
14. Lloyd, R. V., Jin, L., Kulig, E., and Fields, K. (1992) Molecular approaches for the analysis of chromogranins and secretogranins. *Diagn. Mol. Pathol.* **1**, 2–15.
15. Pagani, A., Cerrato, and Bussolati, G. (1993) Nonspecific in situ hybridization reaction in neuroendocrine cells and tumors of the gastrointestinal tract using oligonucleotide probes. *Diagn. Mol. Pathol.* **2**, 125–130.
16. Weiss, L. M., Movahed, L. A., Chen, Y. Y., et al. (1990) Detection of immunoglobulin light-chain mRNA in lymphoid tissues using a practical in situ hybridization method. *Am. J. Pathol.* **137**, 979–988.
17. Stahl, W. L., Eakin, T. J., and Baskin D. G. (1993) Selection of oligonucleotide probes for detection of mRNA isoforms. *J. Histochem. Cytochem.* **41**, 1735–1740.
18. Oliver, K. R., Heavens, R. P., and Sirinathsinghji, D. J. S. (1997) Quantitative comparison of pretreatment regimens used to sensitize in situ hybridization using oligonucleotide probes on paraffin-embedded brain tissue. *J. Histochem. Cytochem.* **45**, 1707–1713.

19. Miller, M. A., Kolb, P. E., and Raskind, M. A. (1993) A method for simultaneous detection of multiple mRNAs using digoxigenin and radioisotopic cRNA probes. *J. Histochem. Cytochem.* **41**, 1741–1750.

20. Farquharson, M. A., Harvie, R., Kennedy, A., and McNicol, A. M. (1992) Detection of mRNA by in situ hybridization and in Northern blot analysis using oligodeoxynucleotide probes labeled alkaline phosphatase. *J. Clin. Pathol.* **45**, 999–1002.

21. Thomas, G. A., Davies, H. G., and Williams, E. D. (1993) Demonstration of mRNA using digoxigenin labeled oligonucleotide probes for in situ hybridization in formamide free conditions. *J. Clin. Pathol.* **46**, 171–174.

22. Steel, J. H., Jeffery, R. E., Longcroft, J. M., Rogers, L. A., and Poulsom, R. (1998) Comparison of isotopic and non-isotopic labelling for in situ hybridisation of various mRNA targets with cRNA probes. *Eur. J. Histochem.* **42**, 143–150.

23. Singer, R. H. and Ward, D. C. (1982) Actin gene expression visualized in chicken muscle tissue culture by using in situ hybridization with biotinated nucleotide analogue. *Proc. Natl. Acad. Sci. USA* **79**, 7331–7335.

24. Szakacs, J. G. and Livingston, S. K. (1994) mRNA in-situ hybridization using biotinylated oligonucleotide probes: implications for the diagnostic laboratory. *Ann. Clin. Lab. Sci.* **24**, 324–338.

25. Yoshii, A., Koji, T., Ohsawa, N., and Nakome, P. (1995) In situ hybridization of ribosomal RNAs is a reliable reference for hybridizable RNA in tissue sections. *J. Histochem. Cytochem.* **43**, 321–327.

26. Tongiorgi, E., Righi, M., and Cattaneo, A. (1998) A non-radioactive in situ hybridization method that does not require RNAse-free conditions. *J. Neurosci. Methods.* **85**, 129–139.

27. Guiot, Y. and Rahier, J. (1997) Validation of nonradioactive in situ hybridization as a quantitative approach of messenger ribonucleic acid variations. A comparison with northern blot. *Diagn. Mol. Pathol.* **6**, 261–266.

28. Kriegsmann, J., Keyszer, G., Geiler, T., Gay, R. E., and Gay, S. (1994) A new double labeling technique for combined in situ hybridization and immunohistochemical analysis. *Lab. Invest.* **71**, 911–917.

29. Trembleau, A., Roche, D., and Calas, A. (1993) Combination of non-radioactive and radioactive in situ hybridization with immunohistochemistry: a new method allowing the simultaneous detection of two mRNAs and one antigen in the same brain tissue section. *J. Histochem. Cytochem.* **41**, 489–498.

30. Dagerlind, A., Friberg, K., Bean, A. J., and Hokfelt, T. (1992) Sensitive detection using unfixed tissues: combined radioactive and nonradioactive in situ hybridization histochemistry. *Histochemistry* **98**, 39–43.

31. Young, W. S. III and Hsu, A. C. (1991) Observations on the simultaneous use of digoxigenin and radiolabeled oligodeoxyribonucleotide probes for hybridization histochemistry. *Neuropeptides* **18**, 75.

32. Van Wijk, I. J., Van Vugt, J. M. G., Konst, A. A. M., Mulders, M. A. M., Nieuwint, A. W. M., and Oudejans, C. B. M. (1995) Multiparameter in situ hybridization of trophoblast cells in mixed cell populations by combined DNA and RNA in situ hybridization. *J. Histochem. Cytochem.* **43**, 709–714.

33. Gray, J. W., Pinkel, D., and Brown, J. M. (1994) Fluorescence in situ hybridization in cancer and radiation biology (Review). *Radiat. Res.* **137**, 275–289.

34. Speel, E. J. M. (1999) Detection and amplification systems for sensitive, multiple-target DNA and RNA in situ hybridization: looking inside cells with a spectrum of colors. *Histochem. Cell. Biol.* **112**, 89–113.

35. Bienz, K. and Egger, D. (1995) Immunocytochemistry and in situ hybridization in the electron microscope: combined application in the study of virus-infected cells. *Histochemistry* **103**, 328–338.

36. Lloyd, R. V., Jin, L., and Song, J. (1990) Ultrastructural localization of prolactin and chromogranin B messenger ribonucleic acids with biotinylated oligonucleotide probes in cultured pituitary cells. *Lab. Invest.* **63**, 413–419.

37. Sibon, O. C. M., Cremers, F. F. M, Boonstra, H. J., and Verkeij, A. J. (1995) Localization of nuclear RNA by pre- and post-embedding in situ hybridization using different gold probes. *Histochem. J.* **27**, 35–45.

38. Bessert, D. A. and Skoff, R. P. (1999) High-resolution in situ hybridization and TUNEL staining with free-floating brain sections. *J. Histochem. Cytochem.* **47**, 693–701.

39. Long, A. A. (1998) In situ polymerase chain reaction: foundation of the technology and today's options. *Eur. J. Histochem.* **42**, 101–109.

40. Uhlmann, V., Silva, I., Luttich, K., Picton, S., and O'Leary, J. J. (1998) In cell amplification. *J. Clin. Pathol. Mol. Pathol.* **51**, 119–130.

41. Komminoth, P. and Long, A. A. (1995) In situ polymerase chain reaction and its applications to the study of endocrine diseases. *Endocr. Pathol.* **6**, 167–171.

42. Nuovo, G. J., Gallery, F., MaConnell, P., Becker, J., and Bloch, W. (1991) An improved technique for the in situ detection of DNA after polymerase chain reaction amplification. *Am. J. Pathol.* **139**, 1239–1244.

43. Nuovo, G. J., Lidonnici, K., MacConnel, P., and Lane, B. (1993) Intracellular localization of polymerase chain reaction (PCR)-amplified hepatitis C cDNA. *Am. J. Surg. Pathol.* **17**, 683–690.

44. Martinez, A., Miller, M. J., Quinn, K., Unsworth, E. J., Ebina, M., and Cuttitta, F. (1995) Non-radioactive localization of nucleic acids by direct in situ PCR and in situ RT-PCR in paraffin-embedded sections. *J. Histochem. Cytochem.* **43**, 739–747.

45. Jin, L., Qian, X., and Lloyd, R. V. (1995) Comparison of mRNA expression detected by in situ polymerase chain reaction and in situ hybridization in endocrine cells. *Cell Vision* **2**, 314–321.

46. Chen, R. H. and Fuggle, S. V. (1993) In situ cDNA polymerase chain reaction. A novel technique for detecting mRNA expression. *Am. J. Pathol.* **143**, 1527–1534.

47. Patel, V. G., Shum-Siu, A., Heniford, B. W., Wieman, T. J., and Hendler, F. J. (1994) Detection of epidermal growth factor receptor mRNA in tissue sections from biopsy specimens using in situ polymerase chain reaction. *Am. J. Pathol.* **144**, 7–14.

48. Bagasra, O., Sheshamma, T., Hansen, J., Bobroski, L., Saikumari, P., and Pomerantz, R. J. (1995) Application of in situ PCR methods in molecular biology. II. Special applications in electron microscopy. Cytogenics and immunohistochemistry. *Cell Vision* **2**, 61–70.

49. Hofler, H., Putz, B., Mueller, J. D., Neubert, W., Sutter, G., and Gais, P. (1995) In situ amplification of measles virus RNA by the self-sustained sequence replication reaction. *Lab. Invest.* **73**, 577–585.

50. Wilkens, L., Komminoth, P., Nasarek, A., Wasielewski, R. van, and Werner, M. (1997) Rapid detection of karyotype changes in interphase bone marrow cells by oligonucleotide primed in situ labeling (PRINS). *J. Pathol.* **181**, 368–373.

51. Speel, E. J. M., Hopman, A. H. N., and Komminoth, P. (1999) Amplification methods to increase the sensitivity of in situ hybridization: play CARD(S). *J. Histochem. Cytochem.* **47**, 281–288.

52. Merz, H., Malisius, R., Mannweiler, S, et al. (1995) Immunomax. A maximized immunohistochemical method for the retrieval and enhancement of hidden antigens. *Lab. Invest.* **73**, 149–156.

53. Wiedorn, K. H., Kuhl, H., Galle, J., Caselitz, J., and Vollmer, E. (1999) Comparison of in situ hybridization, direct and indirect in situ PCR as well as tyramide signal amplification for the detection of HPV. *Histochem. Cell Biol.* **111**, 89–95.

54. Pollanen, R., Vuopala, S., and Lehto, V. P. (1993) Detection of human papillomavirus

infection by nonisotopic in situ hybridization in condylomatous and CIN lesions. *J. Clin. Pathol.* **46**, 936–939.

55. Herrington, C. S., Evans, M. F., Charnock, F. M., Gray, W., and McGee, J. O'D. (1996) HPV testing in patients with low-grade cervical cytological abnormalities—a flow-up study. *J. Clin. Pathol.* **49**, 493–496.

56. Anderson, S. M., Brooke, P. K., van Eyck, S. L., Noell, H., and Frable, W. J. (1993) Distribution of human papillomavirus types in genital lesions from two temporarily distinct populations determined by in situ hybridization. *Hum. Pathol.* **24**, 547–553.

57. Hippelainen, M. I., Syrjanen, S., Hippelainen, M. J., Saarikoski, S., and Syrjanen, K. (1993) Diagnosis of genital human papillomavirus (HPV) lesions in the male—correlation of peniscopy, histology and in situ hybridization. *Genitourin. Med.* **69**, 346–351.

58. Anagnostopoulos, L. and Hummel, M. (1996) Epstein-Bar virus in tumors. *Histopathology* **29**, 297–315.

59. Chang, K. L., Chen, Y. Y., Shibata, D., and Weiss, L. M. (1992) Description of an in situ hybridization methodology for detection of Epstein-Bar virus RNA in paraffin embedded tissue, with a survey of normal and neoplastic tissue. *Diagn. Mol. Pathol.* **1**, 246–255.

60. Niedobitek, G. (1995) Patterns of Epstein-Barr virus infection in non-Hodgkin's lymphoma. *J. Pathol.* **175**, 259–261.

61. Felix, D. H., Jalal, H., Cubie, H. A., Southam, J. C., Wray, D., and Maitland, N. J. (1993) Detection of Epstein-Barr virus and human papilloma virus type-16 DNA in hairy leukoplakia by in situ hybridization and the polymerase chain reaction. *J. Oral Pathol. Med.* **22**, 277–281.

62. Gan, Y. Y., Fones-Tan, A., and Chan, S. H. (1996) Molecular diagnosis of nasopharyngeal carcinoma: a review. *Ann. Acad. Med. Singapore*, **25**, 71–74.

63. Shin, S. S., Berry, G. J., and Weiss, L. M. (1991) Infectious mononucleosis—diagnosis by in situ hybridization in two cases with atypical features. *Am. J. Surg. Pathol.* **15**, 625–631.

64. Nadasdy, T., Park, C. S., Peiper, S. C., Wenzl, J. E., Oates, J., and Silva, F. G. (1992) Epstein-Barr virus infection associated renal disease—diagnostic use of molecular hybridization technology in patients with negative serology. *J. Am. Soc. Nephrol.* **2**, 1734–1742.

65. Choi, Y. J. (1990) In situ hybridization using a biotinylated DNA probe on formalin-fixed liver biopsies with hepatitis B-virus infections: in situ hybridization superior to immunohisto-chemistry. *Mod. Pathol.* **3**, 343–347.

66. Yamada, S., Koji, T., Nozawa, M., Kiyosawa, K., and Nakane, P. K. (1992) Detection of hepatitis C virus (HCV) RNA in paraffin embedded tissue sections of human liver of non-A, non-B hepatitis patients by in situ hybridization. *J. Clin. Lab. Anal.* **6**, 40–46.

67. Lau, J. Y., Krawczynski, K., Negro, F., and Gonzalez-Peralta, R. P. (1996) In situ detection of hepatitis C virus—a critical appraisal [Review]. *J. Hepatol.* **24(2 suppl.)**, 43–51.

68. Negro, F., Pacchioni, D., Shimizu, Y., et al. (1992) Detection of intrahepatic replication of hepatitis C virus RNA by in situ hybridization and comparison with histopathology. *Proc. Natl. Acad. Sci. USA* **89**, 2247–2251.

69. Negro, F. (1994) Detection of hepatitis C virus RNA by in situ hybridization: a critical appraisal [Review]. *Int. J. Clin. Lab. Res.* **24**, 198–202.

70. Nouri Aria, K. T., Sallie, R., Sangar, D., et al. (1993) Detection of genomic and intermediate replicative strands of hepatitis C virus in liver tissue by in situ hybridization. *J. Clin. Invest.* **91**, 2226–2234.

71. Blight, K., Trowbridge, R., Rowland, R., and Gowans, E. (1992) Detection of hepatitis C virus RNA by in situ hybridization. *Liver* **12**, 286–289.

72. Sharma, T. M., Nadasdy, T., Leech, R. W., Kingma, D. W., Johnson, L. D., and Hanson-Painton, O. (1994) In situ DNA hybridization study of "primary' cytomegalovirus (CMV) oophoritis. *Acta Obstet. Gynecol. Scand.* **73**, 429–431.

73. Murray, J., Farquharson, M., Wheatley, D. J., and McPhaden, A. R. (1996) Cytomegalovirus detection by in situ hybridization in upper gastrointestinal biopsies from cardiac transplant patients. *J. Pathol.* **179 (suppl.)**, 34A.

74. Gaffey, M. J., Benezra, J. M., and Weiss, L. M. (1991) Herpes simplex lymphadenitis. *Am. J. Clin. Pathol.* **95**, 709–714.

75. Remadi, S., Finci, V., Ismail, A., Zacharie, S., and Vassilakos, P. (1995) Herpetic endometritis after pregnancy. *Pathol. Res. Pract.* **191**, 31–34.

76. Sumiyoshi, Y., Akashi, K., and Kikuuchi, M. (1994) Detection of human herpes virus-6 (HHV-6) in the skin of a patient with primary HHV-6 infection and erythroderma. *J. Clin. Pathol.* **47**, 762–763.

77. Boshoff, C., Schulz, T. F., Kennedy, M. M., et al. (1995) Kaposi's sarcoma-associated herpesvirus infects endothelial and spindle cells. *Nat. Med.* **1**, 1274–1278.

78. Reiss, E., Tanaka, K., Bruker, G., et al. (1998) Molecular diagnosis and epidemiology of fungal infections [Review]. *Med. Mycol.* **36, (suppl. 1)** 249–57.

79. Perez-Jaffe, L. A., Lanza, D. C., Loevner, L. A., Kennedy, D. W., and Montone, K. T. (1997) In situ hybridization for *Aspergillus* and *Penicillium* in allergic fungal sinusitis: a rapid means of speciating fungal pathogens in tissues. *Laryngoscope* **107**, 233–240.

80. Hanazawa, R., Murayama, S. Y., and Yamaguchi, H. (2000) In-situ detection of *Aspergillus fumigatus*. *J. Med. Microbiol.* **49**, 285–290.

81. Park, C. S., Kim, J., and Montone, K. T. (1997) Detection of *Aspergillus* ribosomal RNA using biotinylated oligonucleotide probes. *Diagn. Mol. Pathol.* **6**, 255–260.

82. Hayashi, Y., Watanabe, J., Nakata, K., Fukayama, M., and Ikeda, H. (1990) A novel diagnostic method of *Pneumocystis carinii*. In situ hybridization of ribosomal ribonucleic acid with biotinylated oligonucleotide probes. *Lab. Invest.* **63**, 576–580.

83. Kim, J., Yu, J. R., Hong, S. T., and Park, C. S. (1996) Detection of *Pneumocystis carinii* by in situ hybridization in the lungs of immunosuppressed rats. *Korean J. Parasitol.* **34**, 177–84.

84. Krimmer, V., Merkert, H., von Eiff, C., et al. (1999) Detection of *Staphylococcus aureus* and *Staphylococcus epidermidis* in clinical samples by 16S rRNA-directed in situ hybridization. *J. Clin. Microbiol.* **37**, 2667–2673.

85. Karttunen, T. J., Genta, R. M., Yoffe, B., Hachem, C. Y., Graham, D. Y., and el-Zaatari, F. A. (1996) Detection of *Helicobacter pylori* in paraffin-embedded gastric biopsy specimens by in situ hybridization. *Am. J. Clin. Pathol.* **106**, 305–311.

86. Hayden, R. T., Qian, X., Roberts, G. D., and Lloyd, R. V. (2001) In situ hybridization for the identification of yeast-like organisms in tissue section. *Diagn. Molec. Pathol.* **10**, 15–23.

87. Assy, N. and Minuk, G. Y. (1997) Liver regeneration: methods for monitoring and their applications [Review]. *J. Hepatol.* **26**, 945–952.

88. Rautiainen, E., Haapasalo, H., Sallinen, P., Rantala, I., Helen, P., and Helin, H. (1998) Histone mRNA in-situ hybridization in astrocytomas: a comparison with PCNA, MIB-1 and mitoses in paraffin-embedded material. *Histopathology* **32**, 43–50.

89. DeLellis, R. A. (1994) In situ hybridization techniques for the analysis of gene expression: applications in tumor pathology. *Hum. Pathol.* **25**, 580–585.

90. Lloyd, R. V., Jin, L., and Chandler, W. F. (1996) In situ hybridization studies in human pituitaries, *Pituitary Adenomas* (Landolt, A. M., Vance, M. L., and Reilly, P. L., eds.), Churchill Livingstone, New York, pp. 47–58.

## IN SITU HYBRIDIZATION

### IN *SITU* RNA LOCALIZATION USING DIGOXIGENIN-LABELED RIBOPROBES FOR PARAFFIN SECTION

#### Day 1

1. Deparaffinize with xylene for 5 min, ×2.
2. 100% ethyl alcohol (ETOH) for 3 min, ×2.
3. 95% ETOH for 3 min, ×2.
4. Wash in phosphate-buffered saline (PBS) for 3 min.
5. Incubated in 10 m$M$ citric acid buffer (pH 6.0) in plastic jars for 3 min, then treat with microwave for 3 min, ×3 (2 jars each time), and add diethyl pyrocarbonate (DEPC)-$H_2O$ after each boiling.
6. Keep in citric acid buffer at room temperature for 20 min.
7. Wash in PBS for 3 min.
8. Incubate in proteinase K (25 µg/mL) at 37°C for 10 min.
9. Wash in PBS for 3 min, ×2.
10. Incubate in 0.2 $N$ HCl for 10 min.
11. Wash in PBS for 3 min, ×2.
12. Incubate in 0.25% acetic anhydride in 0.1 M triethanolamine for 10 min.
13. Wash in 2× standard saline citrate (SSC) for 3 min.
14. Incubate in prehybridization buffer for 30 min.
15. Prepare riboprobe mix: dilute probe in prehybridization buffer, heat it at 85°C for 5 min, and then keep in ice.
16. Hybridization: add 20–50 µL riboprobe mix to slide, cover-slip and incubate in a moisture chamber at 50–55°C for 16 h.

#### Day 2

1. Dip slides in 2× SSC to remove cover slips for 5 min.
2. Wash in PBS for 3 min.
3. Incubate in RNAse A solution (0.1–1 ng/mL in PBS) at 37°C for 30 min.
4. Wash in PBS for 5 min.
5. Wash in 2× SSC at 50°C for 10 min.
6. Wash in 1× SSC at 50°C for 10 min.
7. Wash in 0.5× SSC at 50°C for 10 min.
8. Incubate in buffer A for 1 min.
9. Incubate in buffer A with 1% sheep serum and 0.03% Triton X-100 for 20 min.
10. Add anti-Dig antibody (1/200 in above buffer, from Boehringer Mannheim) and incubate for 3 h.
11. Wash in buffer A for 5 min, ×2.
12. Wash in buffer B for 5 min, ×2.
13. Develop in nitroblue tetrazolium/bromochloroindolyl phosplate (NBT/BCIP) in the dark for 30–60 min, monitoring under the microscope.
14. Stop the reaction in buffer B and wash briefly in $H_2O$.
15. Counterstain nucleus with fast red, dehydrate, and coverslip.

### For Frozen Section

1. Omit **steps 1** and **2** of day 1.
2. Treat with proteinase K at 1–2 µg/mL for 10 min.
3. Digest with RNAse A at 0.1–1 µg/mL at 37°C for 30 min.

## *IN SITU* RNA LOCALIZATION USING DIGOXIGENIN-LABELED OLIGONUCLEOTIDE PROBES FOR PARAFFIN SECTION

### Day 1

1. Deparaffinize with xylene for 5 min, ×2.
2. 100% ETOH for 3 min, ×2.
3. 95% ETOH for 3 min, ×2.
4. Wash in PBS for 3 min.
5. Incubate in 10 m$M$ citric acid buffer (pH 6.0) in plastic jars for 3 min, then treat with microwave for 3 min, ×3 (2 jars each time), and add DEPC-H$_2$O after each boiling.
6. Keep in citric acid buffer at room temperature for 20 min.
7. Wash in PBS for 3 min.
8. Incubate in proteinase K (25 µg/mL) at 37°C for 10 min.
9. Wash in PBS for 3 min, ×2.
10. Incubate in 0.2 $N$ HCl for 10 min.
11. Wash in PBS for 3 min, ×2.
12. Incubate in 0.25% acetic anhydride in 0.1 $M$ triethanolamine for 10 min.
13. Wash in 2× SSC for 3 min.
14. Incubate in prehybridization buffer for 30 min.
15. Prepare oligo-probe mix: dilute probe in prehybridization buffer (0.5–2 ng/µL).
16. Hybridization: add 20–50 µL oligo-probe mix to slide, cover-slip, and incubate in a moisture chamber at 42–44°C for 16 h.

### Day 2

1. Dip slides in 2× SSC to remove cover slips.
2. Wash in 2× SSC at 42–44°C for 10 min.
3. Wash in 1× SSC at 42–44°C for 10 min.
4. Wash in 0.5 × SSC at 42–44°C for 10 min.
5. Incubate in buffer A for 1 min.
6. Incubate in buffer A with 1% sheep serum and 0.03% Triton X-100 for 20 min.
7. Add anti-Dig antibody (1/200 in above buffer, from Boehringer Mannheim) and incubate for 3 h.
8. Wash in buffer A (1 $M$ NaCl, 0.1 $M$ Tris-HCl, 2 m$M$ MgCl$_2$, pH 7.5) for 5 min, ×2.
9. Wash in buffer B (0.1 $M$ NaCl, 0.1 $M$ Tris base, 5 m$M$ MgCl$_2$, pH 9.5) for 5 min, ×2.
10. Develop in NBT/BCIP in the dark for 30–60 min, monitoring under the microscope.
11. Stop the reaction in buffer B and wash briefly in H$_2$O.
12. Counterstain nucleus with fast red, dehydrate, and cover-slip.

### For Frozen Section

1. Omit **steps 1** and **2** of day 1.
2. Treat with proteinase K at 1–2 µg/mL for 10 min.

# 4

# *In Situ* Detection of Infectious Agents

Gary W. Procop, MD and Randall Hayden, MD

## INTRODUCTION

*In situ* hybridization (ISH) has a wide variety of uses in diagnostic pathology *(1–4)*. ISH for the detection and differentiation of microorganisms is a science that in many respects is still in its infancy. Early in the development of these technologies, viruses in tissue sections were the natural targets for identification. These methods have proved useful for such identification, especially for viruses that are difficult to culture or when infection was not suspected and tissues were not submitted for culture *(5–7)*. Unfortunately, many of the early ISH methods were complicated, required a high level of technical expertise, and may have been viewed as a detection method for the anatomic pathology laboratories alone (i.e., direct competition for clinical virology). These factors may have discouraged some from using this technology. Furthermore, microorganisms, which may be detected by ISH, have been successfully found through alternate technologies. The greatest competing technologies for ISH are immunohistochemistry (IHC), solution hybridization, and nucleic acid amplification (NAA). In many instances, these alternate technologies have proved equally effective, possibly more sensitive, and sometimes more user friendly and/or cost effective than ISH *(8)*. ISH, however, remains a useful and important tool for the identification of microorganisms in histologic sections, and its uses are expanding to include the direct identification of microorganisms in clinical specimens and rapid culture confirmation.

There has been a dramatic evolution of the molecular biology industry over the past 10 years, which has provided a variety of tools to the molecular pathologist and microbiologist. The relative ease of nucleic acid synthesis to produce high-quality ISH probes has made this method of microbial detection highly versatile. Nucleic acid probes directed against ribosomal RNA targets can provide a definitive genus and species identification for a variety of organisms in a fraction of the time of culture. These probes may be labeled with fluorescent reporter molecules, or secondary signal amplification methods may be employed to enhance assay sensitivity. Incredibly, with ISH and advanced signal amplification methods, it is possible to detect genes present in only a single copy *(9–11)*. This method holds great potential for detection of the genes responsible for antimicrobial resistance in bacteria and fungi and could conceivably replace conventional susceptibility testing.

From: *Morphology Methods: Cell and Molecular Biology Techniques*
Edited by: R. V. Lloyd © Humana Press, Totowa, NJ

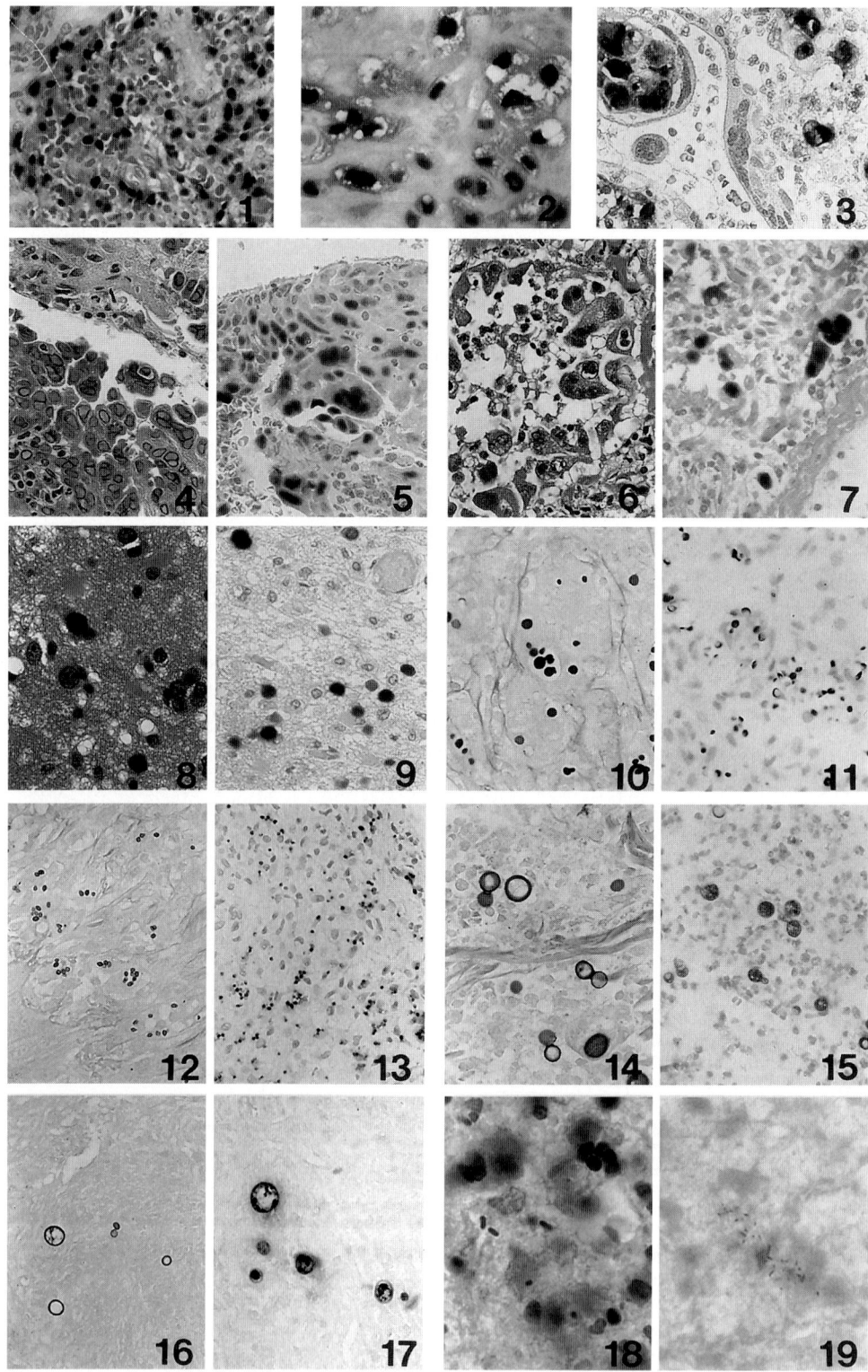

With the passage of time, the concept of ISH has become less foreign to the clinical microbiologist, probably because of the influx of other diagnostic molecular methods such as polymerase chain reaction (PCR) with Southern blot or solution hybridization. ISH applications for microorganisms have also expanded from viruses to include bacteria, fungi, and parasites. These methods have recently been shown to be particularly useful for the detection of fastidious microbes that are difficult to culture and for rapid culture confirmation. *(5–7,12,13)*. For example, ISH methods have been used to identify *Legionella* and *Histoplasma* definitely in histologic sections and to identify bacteria in blood culture rapidly, without the delay of subculture and subsequent biochemical testing *(12,13)* (**Figs. 12, 13, 18** and **19**). Furthermore, some have employed ISH directly on the patient specimen on a rapid basis and substantially improved the time to identification for *Aspergillus* species when only hyphae are seen *(14)*. The maintenance of tissue morphology afforded by ISH may provide an additional benefit, by helping to discriminate contaminants from true infectious pathogens. This could be of particular import in the cases of organisms such as *Aspergillus* that may be either saprophytic or pathogenic. These expanded applications of ISH, as well as the standardization of more user-friendly methods and the use of automation, are likely to establish ISH as a rapid and effective means of organism identification in the modern microbiology and pathology laboratory *(3,15)*.

## VIRUS DETECTION

Early in the development ISH techniques, viruses were the most common infectious agents detected. The presence of these viruses was often suspected because of the cytopathic effect seen in histologic sections or cytologic preparations. The most common human pathogenic viruses that ISH is used to detect are cytomegalovirus (CMV), herpes

◄ **Fig.** **1.** Lymph node with EBV, ISH.
**Fig.** **2.** Pharyngeal biopsy with HPV, ISH.
**Fig.** **3.** Placenta, involved by CMV, ISH.
**Fig.** **4.** Oral cavity (palate) biopsy with HSV, H&E.
**Fig.** **5.** Palate biopsy with HSV (**Fig. 4**) confirmed by ISH.
**Fig.** **6.** Lung tissue with adenovirus, H&E.
**Fig.** **7.** Adenovirus in lung tissue (**Fig. 6**) confirmed by ISH.
**Fig.** **8.** Brain biopsy showing JC viral inclusion in oligodendrocytes, H&E.
**Fig.** **9.** JC virus in brain biopsy (**Fig. 8**) confirmed by ISH.
**Fig. 10.** *Cryptococcus neoformans* in an open lung biopsy, GMS.
**Fig. 11.** Identification of the yeast forms in **Fig. 10** as *C. neoformans* using ISH.
**Fig. 12.** *Histoplasma capsulatum* in an open lung biopsy, GMS.
**Fig. 13.** Confirmation of the yeast forms in **Fig. 12** as *H. capsulatum* using ISH.
**Fig. 14.** Open lung biopsy with *Blastomyces dermatidis*, GMS.
**Fig. 15.** Confirmation of the yeast forms in **Fig. 14** with ISH.
**Fig. 16.** Endospores and immature spherules of *Coccidiodes immitis* in a lung biopsy, GMS.
**Fig. 17.** Confirmation of *C. immitis* in **Fig. 16** by ISH.
**Fig. 18.** Open lung biopsy showing *Legionella pneumophila*, Warthin-Starry stain.
**Fig. 19.** Identification of *L. pneumophila* in **Fig. 18** by ISH.
(Original magnification ×200 for Figs. 1–17; ×400 for Figs. 18 and 19. *See* **Color Plate 2** following page 208.)

simplex virus (HSV), human papillomavirus (HPV), Epstein-Barr virus (EBV), adenovirus, and the JC polyoma virus **(Figs. 1–9;** see **Protocols 8, 16–28)**. In addition to the detection of viruses and the confirmation of cause of a cytopathic effect, ISH has also been used differentiate the high- and low-risk subtypes of HPV, to type herpes simplex viruses, and to assess the effects of treatment *(29–33,17)*. The presence of a virus in tissue culture is often confirmed by using specific antibody stains, but ISH may also be used *(34)*. ISH has also helped to demonstrate associations and possible associations between particular viruses and specific neoplastic conditions *(35–37)*. HPV, which is associated with dysplasia/squamous cell carcinoma of the uterine cervix, EBV, which is associated with posttransplant lymphoproliferative disorder and other malignancies, and human herpes virus 8 (HHV8), which is associated with Kaposi's sarcoma, have all been successfully detected in the respective neoplastic conditions by ISH *(35,36,38–48)*.

This technology has come into direct competition with immunohistochemistry and other molecular techniques. In many institutions, immunohistochemistry, which has a wide variety of applications in anatomic pathology, has replaced ISH or prevented the institution of ISH for the detection of these agents. This may, in part, be secondary to efforts to reduce duplication of like technologies and streamline laboratory operations to lower operating costs. Additionally, the high-quality commercial availability of antibodies directed against many of these viruses, the use of semiautomated techniques, and the rapid turn-around time for IHC stains also contributed to the use of IHC rather than ISH for detection of these agents *(26–28,48)*. More recently, ISH has had to compete with the accurate, rapid and cost-effective solution hybridization technology (i.e., Hybrid-Capture®, Digene, Beltsville, MD) for the detection and differentiation of HPV subtypes. Although these issues may have limited the widespread routine use of ISH techniques, the recent availability of semiautomated platforms and the increasing number of commercially available probes promise to put these methods within the reach of many more laboratories.

Interestingly, the use of ISH for the detection of viruses in histologic sections may come full circle. Many pathologists, particularly those skilled in infectious disease pathology, may choose not to order additional confirmatory tests such as immunohistochemistry or ISH, particularly if viral cytopathic effects and the histopathologic features of the disease are characteristic. Furthermore, the rapid direct immunofluorescent testing now performed in many clinical virology laboratories provides same-day confirmation of many viral infections, making ISH and IHC unnecessary. For these reasons and possibly others, some laboratories have experienced a decline in requests for IHC or ISH testing for viruses. A recent review of IHC antibody use for the detection of viruses at the Cleveland Clinic Foundation disclosed that orders for HSV, CMV, and HPV stains were among the most underused. These expensive commercial antibody stains were expiring prior to use and had to be discarded, resulting in a monetary loss for that section. In this regard, ISH offers an attractive alternative to IHC. Probes for ISH generally offer an equal or even greater degree of specificity when compared with IHC, but they may be ordered in small volumes, frozen, and validated in-house, to help avoid expiration and monetary loss. Additionally, since any nucleic acid sequence may be synthesized, ISH probes may be designed for a wide variety of viruses, including those for which IHC stains are unavailable. Perhaps it is time once again to examine which method is best for the detection of viruses in histologic sections.

Although ISH has been shown to be useful for the diagnosis and study of viral diseases in histologic sections, its utility in the routine clinical virology laboratory remains unclear. ISH has been used to confirm positive viral cultures, but, as with IHC in histologic sections, in clinical virology ISH has to compete with high-quality commercially available antibody stains that are used for culture confirmation. The detection of other viruses, such as CMV and HPV, has been successfully accomplished using solution hybridization, which is less labor intensive and more cost effective than ISH. Finally, ISH has to compete with NAA, including quantitative methods, for the detection of viruses such as HIV, HCV, and EBV. With the variety of molecular methods currently being used in clinical virology and numerous new applications on the horizon, the role of ISH in the clinical virology laboratory remains unclear.

## BACTERIA DETECTION

Fluorescently labeled oligonucleotide probes have been used for the rapid detection of bacteria, and the technical parameters of hybridization have been explored *(49)* (Braun-Howland). These techniques have been used by microbiologists to study bacterial physiology, enumerate viable bacteria, and examine biofilms *(50–54)*. Such methods have been used to study *Legionella* and *Escherichia coli* in water *(55,56)* and have been employed by the food industry to study the bacterium that causes necrotizing hepatopancreatitis of farm-raised shrimp *(57)*. Applications also exist in veterinary medicine. ISH has been used to study *Sepulina* and *Mycoplasma hyopneumoniae* infections in pigs, which cause enterocolitis and respiratory infections, respectively *(58,59)*. It has been used to investigate the role of *Haemobartonella felis* in infectious feline anemia and to study *Chlamydia trachomatis* infections in swine and mice *(60–62)*.

These techniques have also been used to examine a variety of agents of human infections. In formalin-fixed, paraffin-embedded specimens ISH has been used to detect fastidious bacteria. Several researchers have examined ISH for the detection of *Helicobacter pylori* in gastric biopsies and concur that this method is sensitive, specific and useful *(63–65)*. Recently, ISH has been shown to be comparable to rapid PCR (i.e., Light-Cycler™, Roche Diagnostics, Indianapolis, IN) for the detection of *Legionella pneumophila* and was found to be superior to the traditional Warthin-Starry stain (**Figs. 18 and 19**). The use of ISH has been explored for the diagnosis of *Chlamydia,* given the difficulty of identifying inclusions during cytologic or histopathologic examination. The role of *Chlamydia trachomatis* in chronic prostatitis, coronary artery atherosclerosis, reactive arthritis, and stillbirth has been examined using ISH *(66–69)*. The presence of bacteria, undetectable by conventional Gram stain but identifiable by ISH, has been demonstrated in the sinus-nasal mucosa and mucus of a few patients with chronic sinusitis *(70)*. The distribution of bacteria in the intestinal mucus layer in patients with inflammatory bowel disease (IBD) has also been studied by ISH and compared with patients without IBD, in an effort to describe differences in the protective capacity of the intestinal mucus in this population *(71)*.

Although it is clear that these techniques are capable of accurately identifying bacteria, they have not become commonplace in the microbiology laboratory. In the clinical microbiology laboratory, ISH, like Gram stain, but more highly specific, may be useful for the direct identification of bacteria in clinical specimens *(72,73)*. Such technologies are clearly important if the agent under investigation is uncultivable. In this regard, ISH has been used to study the epidemiology of oral treponemes associated with periodontal disease *(74)*. A careful choice of specific ISH probes may be used to rapidly identify particular organisms that are likely to be associated with particular diseases. For example, Matsuhisa et al. *(72)* used ISH for the rapid identification of bacteria in phagocyte smears from patients with septicemia. This methodology, however, has not been confirmed in a large cohort of patients with septicemia. Furthermore, considering that many patients with bacteremia often have few circulating bacteria per milliliter of blood, the sensitivity of such methods would have to compete with the highly sensitive automated blood culture instruments used today. Similarly, Krimmer et al. *(73)* used ISH to detect *Staphylococcus* species in specimens associated with prosthetic orthopedic devices. Although useful for the direct identification of bacteria in clinical specimens, another, possibly more practical use of ISH in the microbiology laboratory may be the rapid identification and differentiation of common microorganisms in cultures.

ISH has been used in the clinical microbiology laboratory for the rapid identification of bacteria in positive blood cultures. Jansen et al. *(12)* and Kempf et al. *(13)* simultaneously published articles examining the ability of ISH to detect and differentiate bacteria in positive blood culture bottles using probes directed against specific target sequences present in bacterial rRNA. Both genus-inclusive probes (i.e., all staphylococci or all Enterobacteriaceae) and species-specific probes (i.e., *Staphylococcus aureus* or *E. coli*) were used. For the identification of *S. aureus* in blood cultures, Jansen et al. *(12)* showed a sensitivity of 66.7%, whereas Kempf et al. *(13)* demonstrated a sensitivity of 100%, even though overlapping, nearly identical probe sequences were used. In both studies, the specificity for the identification of *S. aureus* was 100%. This ability to provide rapid speciation of staphylococci in blood cultures would probably have a significant clinical impact by rapidly identifying *S. aureus,* the organism more likely to be associated with serious disease. Furthermore, the rapid identification of *S. aureus* by ISH allows for the judicial use of other rapid molecular methods for detecting the *mecA* gene, the gene responsible for methicillin resistance. In these blood culture/ISH studies, the accurate and rapid identification of other Gram-positive cocci (streptococci and enterococci) and a variety of Gram-negative bacilli *(12,13)* was also achieved. Possibly the most attractive feature of this technology is the greatly enhanced time to identification for ISH compared with conventional methods. Kempf et al. *(13)* demonstrated that with a limited number of ISH probes, the cause of 96.5% of positive blood cultures could be identified by ISH within 2.5 h, whereas conventional culture took between 1 and 3 days.

This construction of ISH batteries directed against common pathogens of particular diseases may prove useful in areas other than blood culture. This approach has been studied for the simultaneous detection and identification of common respiratory pathogens in cultures from patients with cystic fibrosis *(75)*. Although it is slightly less sensitive than culture, Hogardt et al. *(75)* used ISH to detect and identify *Pseudomonas aeruginosa, Burkholderia cepacia, Stenotrophomonas maltophilia, Haemophilus influenzae,* and *S. aureus* and demonstrated 90% sensitivity and 100% specificity of ISH compared with culture.

Mycobacteriology is another area in which ISH may be useful for organism identification in direct specimens and in culture. The utility of the direct identification of mycobacteria has obvious clinical importance, considering the slow growth of these organisms and the fastidious nature of some mycobacteria, such as *M. leprae, M. genevense,* and *M. haemophilum.* ISH with peptide nucleic acid probes has been used to distinguish tuberculous and nontuberculous mycobacteria rapidly in smears from mycobacterial cultures *(76,77).* This technology has also been used for rapid identification of *M. tuberculosis* and nontuberculous mycobacteria species directly in sputa that contained acid-fast bacilli *(78).* ISH applications have also been used in histologic sections to identify definitely *M. leprae,* an organism that cannot be cultured in vitro *(79).*

## FUNGUS DETECTION

The identification of fungi in histologic sections and cytologic preparations relies primarily on the morphologic features of the organism present *(6).* Unfortunately, in daily practice yeast forms are often routinely characterized as "consistent with *Candida,*" and hyaline septate hyphae are often subsumed under the diagnosis of "consistent with *Aspergillus.*" In most instances these diagnoses are correct, given the prevalence of disease caused by these organisms. Important differences, however, do exist in the severity and treatment of fungal disease based on the type of infecting organism. For example, some *Candida* species, such as *C. glabrata* and *C. krusei,* are more likely to be resistant to fluconazole, one of the most common drugs used to treat *C. albicans* infections. Similarly, there are differences in antifungal fungal susceptibility profiles between *Aspergillus* species and other hyaline septate molds, such as *Fusarium* and *Pseudallesheria boydii.* Advances in drug development for the treatment of systemic fungal disease have broadened the armamentarium of the infectious disease clinician, which in the past was essentially limited to amphotericin B; now a more accurate description of the fungal pathogen present is required for optimal therapy.

Although certain morphologic features may suggest that a hyaline septate mold belongs to a genus other than *Aspergillus,* these differences are subtle, require mycologic expertise, and may call for more time than a busy surgical pathologist may be willing to allot *(80,81).* Furthermore, these morphologic differences are not infallible *(81).* Fungi have not traditionally been considered a target for identification by ISH. However, ISH methods have been developed for fungi and used both in the research setting *(82,83)* and for the histopathologic and microbiologic diagnosis of fungal infections *(13,84).* There is now substantial evidence that ISH can be used to differentiate morphologically similar fungi accurately. Several researchers have shown that ISH methods may be used to identify *Aspergillus* species in histologic sections *(85–87).* Two groups accomplished this by targeting the fungal rRNA for hybridization while the third group targetted the proteinase gene *(85–87).* Recently, an ISH method has been adapted for the rapid identification of *Aspergillus* species in pulmonary cytology specimens *(14).* Such methods afford specific organism identification, which is usually not available from routine direct examination, and provides useful information that may guide antifungal therapy. If, in the future, pressures are brought to bear on surgical pathologists to identify fungi more accurately in histologic sections, the tools now available to accomplish this task include IHC, ISH, and NAA.

As with other organism groups, ISH is particularly useful for uncultivable agents. Several researchers have used ISH to detect the uncultivable fungus, *Pneumocystis carinii (88–90)*. Unlike most conventional histochemical stains, but like IHC, ISH detects the trophozoite form of the organism, as well as the cyst forms *(88)*. Immunohistochemical or ISH staining methods for *P. carinii* may be particularly useful when few organisms are present or when confusion exists between *Pneumocystis* and small yeast forms wherein buds cannot be identified.

ISH has also been applied to the difficult histologic differential diagnosis of morphologically similar yeast and yeast-like forms in human tissues. The nature of disease and therapeutic options also differ for this group of organisms, which bespeaks the need for accurate organism identification. ISH methods have been used successfully to identify accurately *Cryptococcus neoformans* and the parasitic forms of the dimorphic fungal pathogens *Histoplasma capsulatum, Blastomyces dermatitidis,* and *Coccidioides immitis* in human tissue sections *(84)* (**Figs. 10–17**). The yeast cells of *Candida albicans, Candida tropicalis,* and the agent of South American blastomycosis, *Paracoccidioides brasiliensis,* have been identified using ISH methods *(91–93)*. The laboratory diagnosis of systemic fungal disease is often first made by the histopathologist. Therefore, accurate ISH methods for detection of fungi in histologic sections would represent a significant advance in diagnosis, especially for organisms like *H. capsulatum* that produce suggestive, but not diagnostic, forms in tissue and require extended culture incubation for growth.

As with bacteria, ISH may prove useful for the identification of fungi in the clinical mycology laboratory. These methods have been used for the direct identification of common yeast in blood cultures with 100% specificity *(13)*. Additionally, because saprophytic, nonpathogenic molds exist that demonstrate conidiation in culture similar to that of the systemic dimorphic pathogens, molecular probes or other techniques are necessary for confirmation of these fungi. It follows then that the same ISH probes used to differentiate fungi in tissue sections may be useful for culture confirmation. In this manner, *in situ* methods may compete with traditional assimilation and morphologic methods of yeast and mold identification, especially as automation of ISH methods becomes more widely available.

## PARASITE DETECTION

Many parasites are efficiently and accurately identified in histologic sections or microbiologic preparations of specimens. Therefore, ISH methods have not been used extensively in clinical or anatomic pathology for the identification of parasites. There are some parasites, however, such as the microsporidia, that are difficult to detect in both histologic sections and in the stool preparations in the microbiology laboratory. If detected, electron microscopy is needed for speciation. ISH has been used for the diagnosis of the most common microsporidia, *Enterocytozoon bieneusi,* which causes gastroenteritis in patients with the acquired immunodeficiency syndrome *(94)*. Apart from the detection of parasites in clinical specimens, ISH has been used as a research tool to study the pathophysiology of a wide variety of parasitic diseases including trypanosomes, plasmodia, helminths, and trematodes *(95–98)*.

## CONCLUSIONS

The ISH detection of microorganisms in clinical specimens began as a method of detecting and differentiating viruses in histologic sections. Although it is still useful in this regard, alternative technologies such as IHC, PCR, and solution hybridization can also serve as primary diagnostic modalities in this setting. Given the availability of a wide variety of genetic sequences, and therefore a wide variety of targets, ISH is a useful method for detecting emerging viral pathogens and possibly genes that confer resistance. This technology, however, has been shown to be useful for the detection of more than just viruses. ISH has been used to identify or study bacteria, fungi, and parasites. The use of these probes in histologic sections affords pathologists the opportunity to render a more highly specific organism identification than has ever before been possible, while at the same time viewing infective agents in a maintained morphologic background (i.e., *in situ*). This is possible without the delays of culture and is particularly useful for organisms that cannot be cultured in vitro. The recent identification of bacteria, mycobacteria, and fungi in direct specimens and culture has revealed new uses for ISH. The application of ISH to positive blood cultures could dramatically decrease the time to definitive identification of infectious agents and obviate the need for subculture and biochemical testing. It appears that the Golden Age of ISH for the detection of microorganisms is on the horizon.

## REFERENCES

1. Sklar, J. (1985) DNA hybridization in diagnostic pathology. *Hum. Pathol.* **16**, 654–658.
2. Speel, E. J. (1999) Robert Feulgen Prize Lecture. Detection and amplification systems for sensitive, multiple-target DNA and RNA in situ hybridization: looking inside cells with a spectrum of colors. *Histochem. Cell Biol.* **112**, 89–113.
3. McNicol, A. M. and Farquharson, M. A. (1997) In situ hybridization and its diagnostic applications in pathology. *J. Pathol.* **182**, 250–261.
4. Jin, L. and Lloyd, R. V. (1997) In situ hybridization: methods and applications. *J. Clin. Lab. Anal.* **11**, 2–9.
5. Chandler, F. W. (1997) Approaches to the pathologic diagnosis of infectious diseases, in *Pathology of Infectious Diseases* (Connor, D. H. and Chandler, F. W., eds.), Appleton & Lange, Stamford, CT, pp. 3–7.
6. Chandler, F. W. and Watts, J. C. (1987) *Pathologic Diagnosis of Fungal Infections.* ASCP Press. Chicago, IL.
7. Azumi, N. (1997) Immunohistochemistry and In situ hybridization techniques in the detection of infectious organisms, in *Pathology of Infectious Diseases*, (Connor, D. H. and Chandler, F. W., eds.), Appleton & Lange, Stamford, CT, pp. 35–44.
8. Trofatter, K. F. Jr. (1997) Diagnosis of human papillomavirus genital tract infections. *Am. J. Med.* **102**, 21–27.
9. Zehbe, I., Hacker, G. W., Su, H., Hauser-Kronberger, C., Hainfeld, J. F., and Tubbs, R. (1997) Sensitive in situ hybridization with catalyzed reporter deposition, streptavidin-Nanogold, and silver acetate autometallography: detection of single-copy human papillomavirus. *Am. J. Pathol.* **150**, 1553–1561.
10. Cheung, A. L., Graf, A. H., Hauser-Kronberger, C., Dietze, O., Tubbs, R. R., and Hacker, G. W. (1999) Detection of human papillomavirus in cervical carcinoma: comparison of

peroxidase, Nanogold, and catalyzed reporter deposition (CARD)-Nanogold in situ hybridization. *Mod. Pathol.* **12**, 689–696.

11. Hacker, G. W. (1998) High performance Nanogold-silver in situ hybridisation. *Eur. J. Histochem.* **42**, 111–120.

12. Jansen, G. J., Mooibroek, M., Idema, J., Harmsen, H. J. M., Welling, G. W., and Degener, J. E. (2000) Rapid identification of bacteria in blood cultures by using fluorescently labeled oligonucleotide probes. *J. Clin. Microbiol.* **38**, 814–817.

13. Kempf, V. A., Trebesius, K., and Autenrieth, I. B. (2000) Fluorescent in situ hybridization allows rapid identification of microorganisms in blood cultures. *J. Clin. Microbiol.* **38**, 830–838.

14. Zimmerman, R. L., Montone, K. T., Fogt, F., and Norris, A. H. (2000) Ultra fast identification of *Aspergillus* species in pulmonary cytology specimens by in situ hybridization. *Int. J. Mol. Med.* **5**, 427–429.

15. Takahashi, T. and Ishiguro, K. (1991) Development of an automatic machine for in situ hybridization and immunohistochemistry. *Anal. Biochem.* **196**, 390–402.

16. Delvenne, P., Arrese, J. E., Thiry, A., Borlee-Hermans, G., Pierard, G. E., and Boniver, J. (1993) Detection of cytomegalovirus, *Pneumocystis carinii*, and *Aspergillus* species in bronchoalveolar lavage fluid. A comparison of techniques. *Am. J. Clin. Pathol.* **100**, 414–418.

17. Botma, H. J., Dekker, H., van Amstel, P., Cairo, I., and van den Berg, F. M. (1995) Differential in situ hybridization for herpes simplex virus typing in routine skin biopsies. *J. Virol. Methods* **53**, 37–45.

18. Forman, M. S., Merz, C. S., and Charache, P. (1992) Detection of herpes simplex virus by a nonradiometric spin-amplified in situ hybridization assay. *J. Clin. Microbiol.* **30**, 581–584.

19. Heggie, A. D. and Huang, Y. T. (1993) Rapid detection of herpes simplex virus in culture by in situ hybridization. *J. Virol. Methods* **41**, 1–7.

20. Hukkanen, V., Haarala, M., Nurmi, M., Klemi, P., and Kiiholma, P. (1996) Viruses and interstitial cystitis: adenovirus genomes cannot be demonstrated in urinary bladder biopsies. *Urol. Res.* **24**, 235–238.

21. Wu, T. C., Kanayama, M. D., Hruban, R. H., Au, W. C., Askin, F. B., and Hutchins, G. M. (1992) Virus-associated RNAs (VA-I and VA-II). An efficient target for the detection of adenovirus infections by in situ hybridization. *Am. J. Pathol.* **140**, 991–998.

22. Itoh, S., Irie, K., Nakamura, Y., Ohta, Y., Haratake, A., and Morimatsu, M. (1998) Cytologic and genetic study of polyomavirus-infected or polyomavirus-activated cells in human urine. *Arch. Pathol. Lab. Med.* **122**, 333–337.

23. Keh, W. C. and Gerber, M. A. (1988) In situ hybridization for cytomegalovirus DNA in AIDS patients. *Am. J. Pathol.* **131**, 490–496.

24. Randhawa, P. S., S. Finkelstein, Scantlebury, V., et al. (1999) Human polyoma virus-associated interstitial nephritis in the allograft kidney. *Transplant* **67**, 103–109.

25. Sando, Z., Taban, F., Mathez-Loic, F., Anagnostopoulou, I. D., and Remadi, S. (1996) Risks of diagnostic errors in pathology research of post-abortion herpetic endometritis: limitations of immunohistochemistry in situ hybridization. *Ann. Pathol.* **16**, 279–281.

26. Strickler, J. G., Manivel, J. C., Copenhaver, C. M., and Kubic, V. L. (1990) Comparison of in situ hybridization and immunohistochemistry for detection of cytomegalovirus and herpes simplex virus. *Hum. Pathol.* **21**, 443–448.

27. Strickler, J. G., Rooney, M. T., d'Amore, E. S., Copenhaver, C. M., and Roche, P. C. (1993) Detection of Epstein-Barr virus by in situ hybridization with a commercially available biotinylated oligonucleotide probe. *Mol. Pathol.* **6**, 208–211.

28. Strickler, J. G., Singleton, T. P., Copenhaver, C. M., Erice, A., and Snover, D. C. (1992) Adenovirus in the gastrointestinal tracts of immunosuppressed patients. *Am. J. Clin. Pathol.* **97**, 555–558.

29. Multhaupt, H., Bruder, E., Elit, L., Rothblat, I., and Warhol, M. (1993) Combined analysis of cervical smears. Cytopathology, image cytometry and in situ hybridization. *Acta Cytol.* **37**, 373–378.

30. Tweddel, G., Heller, P., Cunnane, M., Multhaupt, and Roth, K. (1994) The correlation between HIV seropositivity, cervical dysplasia and HPV subtypes 6/11, 16/18, 31/33/35. *Gyn. Oncol.* **52**, 161–164.

31. Tomita, T., Garcia, F., and Mowry, M. (1992) Herpes simplex hepatitis before and after acyclovir treatment. Immunohistochemical and in situ hybridization study. *Arch. Pathol. Lab. Med.* **116**, 173–177.

32. Tomita, T., Chiga, M., Lenahan, M., and Balachandran, N. (1991) Identification of herpes simplex virus infection by immunoperoxidase and in situ hybridization methods. *Virchows Arch. Pathol. Anat. Histopathol.* **419**, 99–105.

34. McClintock, J. T., Mosher, M., Thaker, S. R., et al. (1991) Culture confirmation of cytomegalovirus and herpes simplex virus by direct enzyme-labeled DNA probes and in situ hybridization. *J. Virol. Methods* **35**, 81–91.

35. Li, J. J., Huang, Y. Q., Cockerell, C. J., and Friedman-Kien, A. E. (1996) Localization of human herpes-like virus type 8 in vascular endothelial cells and perivascular spindle-shaped cells of Kaposi's sarcoma lesions by in situ hybridization. *Am. J. Pathol.* **148**, 1741–1748.

36. Ohshima, K., Kikuchi, M., Kobari, S., Masuda, Y., Yoneda, S., and Takeshita, M. (1995) Demonstration of Epstein-Barr virus genomes, using polymerase chain reaction in situ hybridization in paraffin-embedded lymphoid tissues. *Pathol. Res. Pract.* **191**, 139–147.

37. Flaegstad, T., Andresen, P. A., Johnsen, J. I., et al. (1999) A possible contributory role of BK virus infection in neuroblastoma development. *Cancer Res.* **59**, 1160–1163.

38. Papadaki, H. A., Stefanaki, K., Kanavaros, P., et al. (2000) Epstein-Barr virus-associated high-grade anaplastic plasmacytoma in a renal transplant patient. *Leuk. Lymph.* **36**, 411–415.

39. van de Rijn, M., Bhargava, V., Molina-Kirsch, H., et al. (1997) Extranodal head and neck lymphomas in Guatemala: high frequency of Epstein-Barr virus-associated sinonasal lymphomas. *Hum. Pathol.* **28**, 834–839.

40. Pingel, S., Hannig, H., Matz-Rensing, K., Kaup, F. J., Hunsmann, G., and Bodemer, W. (1997) Detection of Epstein-Barr virus small RNAs EBER-1 and EBER-2 in lymphomas of SIV-infected rhesus monkeys by in-situ hybridization. *Int. J. Cancer* **72**, 160–165.

41. Braun-Howland, E. B., Danielsen, S. A., and Nierzwicki-Bauer, S. A. (1992) Development of a rapid method for detecting bacterial cells in situ using 16S rRNA-targeted probes. *Biotechniques* **13**, 928–934.

42. Vasef, M. A., Ferlito, A., and Weiss, L. M. (1997) Nasopharyngeal carcinoma with emphasis on its relationship to Epstein-Barr virus. *Ann. Otol. Rhinol. Laryngol.* **106**, 348–356.

43. Chan, J. K., Tsang, W. Y., Ng, C. S., Wong, C. S., and Lo, E. S. (1995) A study of the association of Epstein-Barr virus with Burkitt's lymphoma occurring in a Chinese population. *Histopathology* **26**, 239–245.

44. Choi, Y. J., Kang, C. S., Shin, W. S., et al. (1994) Epstein-Barr virus associated posttransplant malignant lymphoma in renal allograft recipients. *J. Korean Med. Sci.* **9**, 162–168.

45. Uner, A. H., Hutchinson, R. E., and Davey, F. R. (1994) Applications of in situ hybridization in the study of hematologic malignancies. *Hematol. Oncol. Clin. North Am.* **8**, 771–784.

46. Hennig, E. M., Di Lonardo, A., Venuti, A., Holm, R., Marcante, M. L., and Nesland, J. M. (1999) HPV 16 in multiple neoplastic lesions in women with CIN III. *J. Exp. Clin. Cancer Res.* **18**, 369–377.

47. Southern, S. A., Graham, D. A., and Herrington, C. S. (1998) Discrimination of human papillomavirus types in low and high grade cervical squamous neoplasia by in situ hybridization. *Diagn. Mol. Pathol.* **7**, 114–121.

48. Cho, N. H., Joo, H. J., Ahn, H. J., Jung, W. H., and Lee, K. G. (1998) Detection of human

papillomavirus in warty carcinoma of the uterine cervix: comparison of immunohistochemistry, in situ hybridization and in situ polymerase chain reaction methods. *Pathol. Res. Pract.*
**194**, 713–720.

49. Braun-Howland, E. B., Danielsen, S., and Nierzwicki-Bauer, S. A. (1992) Development
of a rapid method for detecting bacterial cells in situ using 16S rRNA-targeted probes.
*Biotechniques* **13**, 928–934.

50. DeLong, E. F., Taylor, L. T., Marsh, T. L., and Preston, C. M. (1999) Visualization and
enumeration of marine planktonic archaea and bacteria by using polyribonucleotide probes
and fluorescent in situ hybridization. *Appl. Environ. Microbiol.* **65**, 5554–5563.

51. Harmsen, H. J., Gibson, G. R., Elfferich, P., et al. (2000) Comparison of viable cell
counts and fluorescence in situ hybridization using specific rRNA-based probes for the
quantification of human fecal bacteria. *FEMS Microbiol. Lett.* **183**, 125–129.

52. Harmsen, H. J., Kengen, H. M., Akkermans, A. D., Stams, A. J., and de Vos, W. M. (1996)
Detection and localization of syntrophic propionate-oxidizing bacteria in granular sludge
by in situ hybridization using 16S rRNA-based oligonucleotide probes. *Appl. Environ.
Microbiol.* **62**, 1656–1663.

53. Laramee, L., Lawrence, J. R., and Greer, C. W. (2000) Molecular analysis and development
of 16S rRNA oligonucleotide probes to characterize a diclofop-methyl-degrading biofilm
consortium. *Can. J. Microbiol.* **46**, 133–142.

54. Manz, W., Szewzyk, U., Ericsson, P., Amann, R., Schleifer, K. H., and Stenstrom, T. A.
(1993) In situ identification of bacteria in drinking water and adjoining biofilms by hybridization with 16S and 23S rRNA-directed fluorescent oligonucleotide probes. *Appl. Environ.
Microbiol.* **59**, 2293–2298.

55. Manz, W., Amann, R., Szewzyk, R., et al. (1995) In situ identification of Legionellaceae
using 16S rRNA-targeted oligonucleotide probes and confocal laser scanning microscopy.
*Microbiology* **141**, 29–39.

56. Prescott, A. M. and Fricker, C. R. (1999) Use of PNA oligonucleotides for the in situ
detection of *Escherichia coli* in water. *Mol. Cell Probes* **13**, 261–268.

57. Loy, J. K., Dewhirst, F. E., Weber, W., et al. (1996) Molecular phylogeny and in situ
detection of the etiologic agent of necrotizing hepatopancreatitis in shrimp. *Appl. Environ.
Microbiol.* **62**, 3439–3445.

58. Boye, M., Jensen, T. K., Moeller, K., Leser, T. D., and Jorsal, S. E. (1998) Specific detection
of the genus *Serpulina*, *S. hyodysenteriae* and *S. pilosicoliin* porcine intestines by fluorescent
rRNA in situ hybridization. *Mol. Cell Prob.* **12**, 323–330.

59. Kwon, D. and Chae, C. (1999) Detection and localization of *Mycoplasma hyopneumoniae*
DNA in lungs from naturally infected pigs in situ hybridization using a digoxigenin-labeled
probe. *Vet. Pathol.* **36**, 308–313.

60. Berent, L. M., Messick, J. B., Cooper, S. K., and Cusick, P. K. (2000) Specific in situ
hybridization of *Haemobartonella felis* with a DNA probe and tyramide signal amplification.
*Vet. Pathol.* **37**, 47–53.

61. Alakarppa, H., Surcel, H. M., Laitinen, K., Juvonen, T., Saikku, P., and Laurila, A. (1999)
Detection of Chlamydia pneumoniae by colorimetric in situ hybridization. *APMIS* **107**, 451–
454.

62. Chae, C., Cheon, D. S., Kwon, D., et al. (1999) In situ hybridization for the detection and
localization of swine *Chlamydia trachomatis*. *Vet. Pathol.* **36**, 133–137.

63. Bashir, M. S., Lewis, F. A., Quirke, P., Lee, A, and Dixon, M. F. (1994) In situ hybridization
for the identification of *Helicobacter pylori* in paraffin was embedded tissue. *J. Clin. Pathol.*
**47**, 862–864.

64. Park, C. S. and Kim, J. (1999) Rapid and easy detection of *Helicobacter pylori* by in situ
hybridization. *J. Korean Med. Sci.* **14**, 15–20.

65. Barrett, D. M., Faigel, D. O., Metz, D. C., Montone, K., and Furth, E. E. (1997) In situ

hybridization for *Helicobacter pylori* in gastric mucosal biopsy specimens: quantitative evaluation of test performance in comparison with the CLOtest and thiazine stain. *J. Clin. Lab. Anal.* **11**, 374–379.

66. Gumus, B., Sengil, A. Z., Solak, M., et al. (1997) Evaluation of non-invasive clinical samples in chronic chlamydial prostatitis by using in situ hybridization. *Scand. J. Urol. Nephrol.* **31**, 449–451.

67. Jantos, C. A., Nesseler, A., Waas, W., Baumartner, W., Tillmanns, H., and Haberbosch, W. (1999) Low prevalence of *Chlamydia pneumoniae* in atherectomy specimens from patients with coronary heart disease. *Clin. Infect. Dis.* **28**, 988–992.

68. Berlau, J., Junker, U., Groh, A., and Straube, E. (1998) In situ hybridization and direct fluorescence antibodies for the detection of *Chlamydia trachomatis* in synovial tissue from patients with reactive arthritis. *J. Clin. Pathol.* **51**, 803–806.

69. Gencay, M., Puolakkainen, M., Wahlstrom, T., et al. *Chlamydia trachomatis* detected in human placenta. *J. Clin. Pathol.* **50**, 852–855.

70. Hwang, P. H., Montone, K. T., Gannon, F. H., Senior, B. A., Lanza, D. C., and Kennedy, D. W. (1999) Application of in situ hybridization techniques in the diagnosis of chronic sinusitis. *Am. J. Rhinol.* **13**, 335–338.

71. Schultsz, C., Van Den Berg, F. M., Ten Kate, F. W., Tytgat, G. N., and Dankert, J. (1999) The intestinal mucus layer from patients with inflammatory bowel disease harbors high numbers of bacteria compared with controls. *Gastroenterology* **117**, 1089–1089.

72. Matsuhisa, A., Saito, Y., Sakamoto, Y., et al. (1994) Detection of bacteria in phagocyte-smears from septicemia-suspected blood by in situ hybridization using biotinylated probes. *Microbiol. Immunol.* **38**, 511–517.

73. Krimmer, V., Merkert, H., von Eiff, C., et al. (1997) Detection of *Staphylococcus aureus* and *Staphylococcus epidermidis* in clinical sample by 16S rRNA-directed in situ hybridization. *J. Clin. Microbiol.* **37**, 2667–2673.

74. Moter, A., Hoenig, C., Choi, B. K., Riep, B., and Gobel, U. B. (1998) Molecular epidemiology of oral treponemes associated with periodontal disease. *J. Clin. Microbiol.* **36**, 1399–1403.

75. Hogardt, M., Trebesius, K., Geiger, A. M., Hornef, M., Rosenecker, J., and Heesemann, J. (2000) Specific and rapid detection by fluorescent in situ hybridization of bacteria in clinical samples obtained from cystic fibrosis patients. *J. Clin. Microbiol.* **38**, 818–825.

76. Padilla, E., Manterola, J. M., Rasmussen, O. F., et al. (2000) Evaluation of a fluorescence hybridisation assay using peptide nucleic acid probes for identification and differentiation of tuberculous and non-tuberculous mycobacteria in liquid cultures. *Eur. J. Clin. Microbiol. Infect. Dis.* **19**, 140–145.

77. Stender, H., Lund, K., Petersen, K. H., et al. (1999) Fluorescence in situ hybridization assay using peptide nucleic acid probes for differentiation between tuberculous and nontuberculous mycobacterium species in smears of mycobacterium cultures. *J. Clin. Microbiol.* **37**, 2760–2765.

78. Stender, H., Mollerup, T. A., Lund, K., Petersen, K. H., Hongmanee, P., and Godtfredsen, S. E. (1999) Direct detection and identification of *Mycobacterium tuberculosis* in smear-positive sputum samples by fluorescence in situ hybrization (FISH) using peptide nucleic acid (PNA) probes. *Int. J. Tubercul. Lung Dis.* **3**, 830–837.

79. Arnoldi, J., Schluter, C., Duchrow, M., et al. (1992) Species-specific assessment of *Mycobacterium leprae* in skin biopsies by in situ hybridization and polymerase chain reaction. *Lab. Invest.* **66**, 618–623.

80. Liu, K., Howell, D. N., Perfect, J. R., and Schell, W. A. (1998) Morphologic criteria for the preliminary identification of *Fusarium, Paecilomyces*, and *Acremonium* species by histopathology. *Am. J. Clin. Pathol.* **109**, 45–54.

81. Watts, J. C. and Chandler, F. W. (1998) Morphologic identification of mycelial pathogens in tissue sections. A caveat. *Am. J. Clin. Pathol.* **109**, 1–2.

82. Taga, M. and Murata, M. (1994) Visualization of mitotic chromosomes in filamentous fungi by fluorescence staining and fluorescence in situ hybrization. *Chromosoma* **103**, 408–413.

83. Tenberge, K. B., Stellamanns, P., Plenz, G., and Robenek, H. (1998) Nonradioactive in situ hybridization for detection of hydrophobin mRNA in the phytopathogenic fungus *Claviceps purpurea* during infection of rye. *Eur. J. Cell. Biol.* **75**, 265–272.

84. Hayden, R. T., Qian, X., Roberts, G. D., and Lloyd, R. V. (2000) In-situ hybridization for the identification of yeast and yeast-like organisms in tissue sections. Abstract #992, United States and Canadian Academy of Pathology, 89th Annual Meeting, New Orleans, LA. *Mod. Pathol.* **13**, 169A.

85. Kobayashi, M., Sonobe, H., Ikezoe, T., Hakoda, E., Ohtsuki, Y., and Taguchi, H. (1999) In situ detection of *Aspergillus* 18S ribosomal RNA in invasive pulmonary aspergillosis. *Intern. Med.* **38**, 563–569.

86. Park, C. S., Kim, J., and Montone, K. T. (1997) Detection of *Aspergillus* ribosomal RNA using biotinylated oligonucleotide probes. *Diagn. Mol. Pathol.* **6**, 255–260.

87. Hanazawa, R., Murayama, S. Y., and Yamaguchi, H. (2000) In-situ detection of *Aspergillus fumigatus*. *J. Med. Microbiol.* **49**, 285–290.

88. Kim, J., Yu, J. R., Hong, S. T., and Park, C. S. (1996) Detection of *Pneumocystis carinii* by in situ hybridization in the lungs of immunosuppressed rats. *Korean J. Parasitol.* **34**, 177–184.

89. Hayashi, Y., Watanabe, J., Nakata, K., Fukayama, M., and Ikeda, H. (1990) A novel diagnostic method of *Pneumocystis carinii*. In situ hybridization of ribosomal ribonucleic acid with biotinylated oligonucleotide probes. *Lab. Invest.* **63**, 576–580.

90. Kobayashi, M., Urata, T., Ikezoe, T., et al. (1996) Simple detection of the 5S ribosomal RNA of *Pneumocystis carinii* using in situ hybridisation. *J. Clin. Pathol.* **49**, 712–716.

91. Lischewski, A., Amann, R. I., Harmsen, D., Merkert, H., Hacker, J., and Morschhauser, J. (1996) Specific detection of *Candida albicans* and *Candida tropicalis* by fluorescent in situ hybridization with an 18S rRNA-targeted oligonucleotide probe. *Microbiology* **142**, 2731–2740.

92. Lischewski, A., Kretschmar, M., Hof, H., Amann, R., Hacker, J., and Morschhauser, J. (1997) Detection and identification of *Candida* species in experimentally infected tissue and human blood by rRNA-specific fluorescent in situ hybridization. *J. Clin. Microbiol.* **35**, 2943–2948.

93. De Brito, T., Sandhu, G. S., Kline, B. C., et al. (1999) In situ hybridization in paracoccidioidomycosis. *Med. Mycol.* **37**, 207–211.

94. Velasquez, J. N., Carnevale, S., Labbe, J. H., Chertcoff, A., Cabrera, M. G., and Oelemann, W. (1999) In situ hybridization: a molecular approach for the diagnosis of the microsporidian parasite *Enterocytozoon bieneusi*. *Hum. Pathol.* **30**, 54–58.

95. Pereira, M. C., Singer, R. H., and de Meirellas, M. N. (2000) Ultrastructural distribution of poly (A)+ RNA during *Trypanosoma cruzi*-cardiomyocyte interaction in vitro: a quantitative analysis of the total mRNA content by in situ hybridization. *J. Eukaryot. Microbiol.* **47**, 264–270.

96. Jambou, R., Hatin, I., and Jaureguiberry, G. (1995) Evidence by in situ hybridization for stage-specific expression of the ATP/ADP translocator mRNA in *Plasmodium falciparum*. *Exp. Parasitol.* **80**, 568–571.

97. Unnasch, T. R., Bradley, J., Beauchamp, J., Tuan, R., and Kennedy, M. W. (1999) Characterization of a putative nuclear receptor from *Onchocerca volvulus*. *Mol. Biochem. Parasitol.* **104**, 259–269.

98. Zurita, M., Bieber, D., and Mansour, T. E. (1989) Identification, expression and in situ hybridization of an eggshell protein gene from *Fasciola hepatica*. *Mol. Biochem. Parasitol.* **37**, 11–17.

# PROTOCOLS

## DETECTING ADENOVIRUS

ISH with *biotinylated probes* is used to detect adenovirus DNA in paraffin sections. Slides required for the above test include patient slides labeled for positive probe, negative probe, and DNA probe, (use DNA procedure) and a known positive quality control slide.

### DEPARAFFINIZATION (RNASE-FREE)

| | |
|---|---|
| 1. Dry paraffin slides in 70°C oven | 30 min |
| 2. Deparaffinize in xylene, two changes. | 7 min each |
| 3. 100% ethanol, two changes. | 2 min or 15 dips each |
| 4. 95% ethanol, two changes. | 2 min or 15 dips each |
| 5. 2× standard saline citrate (SSC) rinse, two changes. | 2 min each |
| 6. 0.2 *N* HCl (0.2 *N* hydrochloric acid). | 20 min |
| 7. 2× SSC rinse, two changes. | 2 min each |
| Delete **steps 8–11** if tissue is not B-5 fixed. | |
| 8. Weigert's iodine (remove mercury deposits). | 5 min |
| 9. 0.02% diethyl pyrocarbonate (DEPC)-treated water rinse. | 1 min |
| 10. 5% sodium thiosulfate. | 1 min |
| 11. 2× SSC rinse, two changes. | 2 min each |

### MICROWAVE PRE-TREATMENT (RNASE-FREE)

| | |
|---|---|
| 1. Preheat microwave oven (*high power*) with deionized water in a beaker for 1 min. Use 400 mL of water in beaker for 1 Coplin jar of slides. Use 50 mL less of water for each additional Coplin jar. Example: 300 mL water for 3 Coplin jars. | |
| 2. Place slides into white plastic (9 count) RNAse-free Coplin jar(s) and cover with 10 m*M* citric acid (pH 6.0). | |
| 3. Microwave (*high power*) stopping every 3–5 mɪn, to bring solution volume up to original level by adding 0.02% DEPC-treated water to Coplin jar(s). | 10 min total |
| 4. Remove Coplin jars from microwave oven, keeping slides in citric acid solution. | 15 min |
| 5. 2× SSC rinse, two changes. | 2 min each |

### TISSUE DIGESTION (RNASE-FREE)

| | |
|---|---|
| 1. Apply 300 µL (0.3 mL) proteinase K, diluted with phosphate-buffered saline (PBS) 1/20 dilution of 500 µg/mL stock diluted to 25 µg/mL) to each slide and incubate within humid tray in 50°C oven. | 5 min |
| 2. Rinse with 2× SSC, two changes. | 3 min each |
| 3. Mix 50 mL of 0.1 *M* triethanolamine and 0.3 mL of acetic anhydride (add just before use) and place slides in solution. | 15 min |
| 4. 2× SSC rinse, two changes. | 5 min total |

## HYBRIDIZATION: *DNA BIOTINYLATED PROBES (RNASE-FREE)*

1. Apply
   a. 30–40 µL Hepatitis A virus (HAV) biotin probe (Enzo, original concentration of 20 µL diluted to 0.2 ng/µl, using prehybridization buffer) to each appropriately labeled *negative* probe slide.
   b. ADV-biotin probe (Enzo, original concentration of 20 ng/µL diluted to 2.0 ng/µL, using prehybridization buffer) to each appropriately labeled *positive* probe slide.
   c. DNA probe (Enzo Blur 8 Human alu repeat, original concentration of 20 ng/µL diluted to 2.0 ng/µL, using prehybridization buffer) to each *DNA* probe slide.
      Working dilution may vary with different lot numbers.
2. Cover-slip, after application of each probe type, with glass RNAse-free Sigmacote-treated cover slip.
3. Place slides on metal tray and put in 95°C oven (denature).                    5 min
   **Caution:** Use appropriate protection when removing tray and slides from oven—very hot!
4. Place slides into humid tray and hybridize in 50°C oven.                        2 h

## STRINGENCY WASHING

Stringency washes are performed in shaking waterbath using prewarmed solutions. Place each probe type into a separate rinse container for stringency washing.

1. Wash in 2× SSC (room temperature), two changes.                           3 min each
2. Wash in *prewarmed* 0.5× SSC (37°C) rinse, and place in 37°C shaking waterbath. (DNA probe slide requires (37°C) 0.5× SSC rinse and 37°C shaking waterbath.)                                                       20 min

## PROBE DETECTION

1. Buffer A (pH 7.5) rinse on orbital shaker.                                      3 min
2. Apply 300 µL of streptavidin-AP (BRL, 1/100; dilute with buffer A + 1% normal swine serum and 0.3% Triton X-100) to each slide and incubate within humid tray in 37°C oven.                        20 min
3. Buffer A rinse on orbital shaker, two changes.                            3 min each
4. Buffer C (pH 9.5) rinse on orbital shaker.                                      3 min
5. Prepare *NBT/BCIP + levamisole solution* by mixing the following in order (300 µL per slide):

| Buffer C (pH 9.5) | 1.0 mL | 2.0 mL | 3.0 mL | 4.0 mL | 5.0 mL |
|---|---|---|---|---|---|
| Add: nitroblue tetrazolium (NBT; BRL) (invert to mix) | 4.4 µL | 8.8 µL | 13.2 µL | 17.6 µL | 22.0 µL |
| Add: bromochloroindolyl phosphate (BCIP; BRL) (invert to mix) | 3.3 µL | 6.6 µL | 9.9 µL | 13.2 µL | 16.5 µL |
| Add: levamisole (Vector) (invert to mix) | 8.0 µL | 16.0 µL | 24.0 µL | 32.0 µL | 40.0 µL |
| Total | 1.0 mL | 2.0 mL | 3.0 mL | 4.0 mL | 5.0 mL |

6. Wipe around section and apply NBT/BCIP + levamisole solution to slides. Wrap tray with aluminum foil and develop the reaction for up to 1 h. *Monitor slides closely the first 10 min* and then every 5–10 min until a strong signal free of background (nonspecific staining) is reached (approximately *60 min*). Keep slides wrapped in aluminum foil during development. Less time may be needed for some cases to prevent background staining.                                                       30–60 min
7. Buffer C rinse, using some agitation.                               1 min

## COUNTERSTAINING, DEHYDRATION, AND COVER-SLIPPING

1. 0.1% Nuclear Fast Red.                                              3 min
2. Buffer C rinse, using agitation.                                    1 min
3. 95% ethanol, two changes.                                   15 dips each
4. 100% ethanol, two changes.                                  15 dips each
5. Xylene, three changes.                                      15 dips each
6. Cover-slip using a xylene-based synthetic mountant
   (i.e., Cytoseal, Stephens Scientific).

# EPSTEIN-BARR VIRUS (EBER I/II)

ISH with *digoxigenin probes* is used to detect Epstein-Barr virus in paraffin sections. Slides required for the above test are: patient slides labeled for positive probe, negative probe, and RNA probe (use RNA procedure) and a known positive quality control slide.

## DEPARAFFINIZATION (RNASE-FREE)

1. Dry paraffin slides in 70°C oven (minimum 30 min).                 30 min
2. Deparaffinize in xylene, two changes.                         7 min each
3. 100% ethanol, two changes.                         2 min or 15 dips each
4. 95% ethanol, two changes.                          2 min or 15 dips each
5. 0.02% DEPC-treated water rinse, two changes.                  2 min each
6. Weigert's iodine (remove mercury deposits).                        5 min
7. 0.02% DEPC-treated water rinse.                                    1 min
8. 5% sodium thiosulfate.                                             1 min
9. 0.02% DEPC-treated water rinse, two changes.                  2 min each

## TISSUE DIGESTION (RNASE-FREE)

1. Apply 300 µL (0.3 mL) of proteinase K, diluted with PBS (1/20 dilution of 500 µg/mL) stock diluted to 25 µg/mL) to each slide and incubate within humid tray in 50°C oven.       10 min
2. 0.02% DEPC-treated water rinse, two changes.                  3 min each
3. Mix 50 mL of 0.1 *M* triethanolamine and 0.3 mL of acetic anhydride (add just before use) and place slides in solution.       15 min
4. 0.02% DEPC-treated water rinse, two changes.                5 min total

## HYBRIDIZATION: *OLIGONUCLEOTIDE PROBES (RNASE-FREE)*

1. Apply 30–40 µL calcitonin digoxigenin (Mayo source, original
   concentration of 8 ng/µL diluted to 0.5 ng/µL (working dilution),
   using prehybridization buffer) to each appropriately labeled *negative* probe slide.
   EBER I/II-digoxigenin probe (Mayo source, original concentration
   of 8 ng/µL diluted to *working dilution* using prehybridization buffer)
   to each appropriately labeled *positive* probe slide. *Working dilution
   may vary with different lot numbers.*
   Cover-slip, after each probe type, with glass RNAse-free Sigmacote-
   treated cover slip.
   Place slides on metal tray and put in 95°C oven (denature).                 5 min
   **Caution:** Use appropriate protection when removing tray and slides
   from oven—very hot!
2. Place slides into humid tray and hybridize in 50°C oven.                    2 h

## STRINGENCY WASHING

Stringency washes are performed in a shaking waterbath using *prewarmed* solutions.
Place each probe type in a separate rinse container for stringency washing.

1. Wash in 2× SSC (room temperature), two changes.                        3 min each
2. Wash in *prewarmed* 0.5× SSC (*37°C*) rinse and place in *37°C*
   shaking waterbath.                                                         20 min

## PROBE DETECTION

1. Buffer A (pH 7.5) rinse on orbital shaker.                                 3 min
2. Apply 300 µL of sheep anti-digoxigenin-AP (BM, 1/100; dilute
   with buffer A + 1% normal swine serum and 0.3% Triton X-100)
   to each slide and incubate within humid tray in 37°C oven.                20 min
3. Buffer A rinse on orbital shaker, two changes.                         3 min each
4. Buffer C (pH 9.5) rinse on orbital shaker.                                 3 min
5. Prepare *NBT/BCIP + levamisole solution* by mixing the following
   in order (300 µL per slide):

| Buffer C (pH 9.5) | 1.0 mL | 2.0 mL | 3.0 mL | 4.0 mL | 5.0 mL |
|---|---|---|---|---|---|
| Add: NBT (BRL) (invert to mix) | 4.4 µL | 8.8 µL | 13.2 µL | 17.6 µL | 22.0 µL |
| Add: BCIP (BRL) (invert to mix) | 3.3 µL | 6.6 µL | 9.9 µL | 13.2 µL | 16.5 µL |
| Add: Levamisole (Vector) (invert to mix) | 8.0 µL | 16.0 µL | 24.0 µL | 32.0 µL | 40.0 µL |
| Total: | 1.0 mL | 2.0 mL | 3.0 mL | 4.0 mL | 5.0 mL |

6. Wipe around section and apply NBT/BCIP + levamisole solution to slides.
   Wrap tray with aluminum foil and develop the reaction. *Monitor slides
   closely the first 5–10 min* and then every 5–10 min until a strong signal
   free of background (nonspecific staining) is reached. Keep slides wrapped
   in aluminum foil during development. Less time may be needed for some
   patient cases to prevent background staining.                         20–40 min
7. Buffer C rinse, using some agitation.                                      1 min

## COUNTERSTAINING, DEHYDRATION, AND COVER-SLIPPING

| | |
|---|---:|
| 1. 0.1% Nuclear Fast Red. | 3 min |
| 2. Buffer C rinse, using agitation. | 1 min |
| 3. 95% ethanol, two changes. | 15 dips each |
| 4. 100% ethanol, two changes. | 15 dips each |
| 5. Xylene, three changes. | 15 dips each |
| 6. Cover-slip using a xylene-based synthetic mountant (i.e., Cytoseal, Stephens Scientific). | |

# Fluorescent *In Situ* Hybridization

## Susan Sheldon, PhD

## INTRODUCTION

Fluorescence in situ hybridization (FISH) arose from a marriage of classical DNA hybridization in solution to modern molecular biologic techniques, most notably the use of restriction endonucleases and, later, the polymerase chain reaction (PCR). The former has made identification of both genes and relevant interspersed sequences possible, whereas without the latter, many widely used probes would not be available. FISH allows one to localize a specific DNA sequence to a specific chromosome, region of a chromosome, or cell type. Generally, the morphology of the chromosome, cell, or tissue of interest is preserved to permit unambiguous identification of the target *(1–3)*. Consequently, the technique has been widely used in a variety of venues. Gene mapping *(4)*, chromosome identification *(5)*, demonstration of gene amplification in some solid tumors *(6,7)*, identification of chimeric populations, minimal residual disease, and tumor cells admixed in normal tissue *(8,9)* are just a few examples. The advantages of modern FISH techniques include:

1. Simple, straightforward, standardized techniques
2. Wide assortment of probes available commercially
3. Double labeling and use of multiple probes on a single cell
4. Rapid turnaround times (anywhere from 4 h to a maximum of 48 h).

Comparing modern FISH technology with the original experiments described by Pardue and Gall *(10)*, which entailed the generation of tritium-labeled DNA or RNA probes by providing cell cultures with either tritiated thymidine or labeled nucleotide triphosphates, it is clear that FISH is simpler. The relative ease is emphasized as additional complex procedures follow labeling of the nucleic acids, i.e., the so-called satellite sequences (so-called because they floated above the main band DNA in a sucrose or CsCl gradient) were isolated by density gradient centrifugation and annealed to chromosome spreads on slides. The radioactive signal was localized using autoradiography, which involves coating the slide carrying the specimen with a photographic emulsion (literally gelatin containing silver halide crystals); the emulsion must be handled in the dark. Once dry, the coated slides are exposed in the dark, usually at 4°C for days to weeks, before development, drying, and counterstaining. Double-labeling experiments were possible *(11)* using isotopes with different path lengths, such

From: *Morphology Methods: Cell and Molecular Biology Techniques*
Edited by: R. V. Lloyd © Humana Press, Totowa, NJ

as tritium and $P^{32}$ and detecting them by using two coatings of emulsion separated by a layer of cellophane.

FISH was developed in part due to the technical difficulties and limitations imposed by autoradiography, in part because of pressure from various regulatory agencies to reduce the use of radioisotopes, and in part because molecular biology was evolving. Investigators seeking ways of avoiding labor-intensive classical cytogenetic analysis by automation propelled the field of molecular cytogenetics. For example, various combinations of fluorescent dyes with affinity for nucleotides and/or DNA were used to stain suspensions of isolated chromosomes. The chromosome suspensions were then analyzed by flow cytometry or fluorescence-activated cell sorting *(12)*. Although flow cytometry had limited success in completing a karyotype, some of the fluorochromes and dyes were later used for FISH.

## NONISOTOPIC LABELING PRIOR TO FISH

Prior to the advent of modern molecular biology, numerous studies were undertaken to insert substituents into nucleotides or their precursors. The main obstacles were uptake of the modified nucleotide through both the cell and nuclear membrane, appropriate incorporation into DNA or RNA, and visualization of the labeled product.

One of the earliest solutions to the isotope problem was to use mercurated, rather than tritiated nucleotides *(13,14)*. Pyrimidines were mercurated either chemically or enzymatically and incorporated into DNA or RNA. Following a hybridization reaction, the signal was localized by its electron density, or by indirect immunofluorescence using a hapten-sulfhydral conjugate, in which the sulfur ligand reacted with the mercury. In other studies, fluorescent derivatives of nucleotides or polynucleotides of interest were produced *(15–17)* with varying degrees of success; the adducts could then be visualized. Later, the thymidine analog bromodeoxyuridine (BrdU), which was widely used by cytogeneticists during the 1970s to demonstrate sister chromatid exchanges, was incorporated into some probes; its presence is visualized with an antibody directed against the BrdU.

*In situ* hybridization changed dramatically when, in 1981, Langer et al. *(18)* revolutionized probe technology by synthesizing biotin-labeled polynucleotides that could be detected with avidin conjugated to either a fluorochrome or peroxidase. DNA technology became accessible to any reasonably sophisticated immunohistochemistry laboratory.

Early FISH publications demonstrated the localization of unique-sequence human genes *(19–21)*, but the first probes for human genes to become commercially available were the α-satellite, or centromere probes. Three types of FISH probes are commonly used and commercially available: α-satellites, which bind to the pericentromeric region of a chromosome, unique-sequence probes, which recognize DNA sequences associated with disease genes, and painting probes, which bind to a series of sequences on a specific chromosome (summarized in **Table 1**). The availability of some probes was markedly curtailed following the patenting of "blocking DNA" and subsequent legal action by the patent holder, a scenario becoming all too common in modern molecular biology. The remainder of this discussion describes in more detail the three types of probes (their use and interpretation of the results), followed by a methods section, using examples of commonly used probes.

**Table 1**
**Common Types of FISH Probes**

| Class | Size (Kb) | Chromosomal location |
|---|---|---|
| Repeated sequence | | |
| —α-Satellite | 150–900 | Centromeres of each chromosome |
| —β-Satellite | 150–900 | Centromeres of chromosomes 1, 9, and 16 |
| —Classic satellite | 150–900 | Centromeres of chromosomes 1, 9, 16 and Yq |
| —Telomere | 10 | Telomere of each chromosome |
| | | |
| Unique-sequence | | |
| —Painting | 80–500 | Covers entire chromosome |
| —Gene-specific | 80–500 | Specific locus on chromosome[a] |
| —Subtelomere | 20–100 | Chromosome-specific unique sequence[b] |

[a] Most of these probes are a solution with a marker to identify the specific chromosome.

[b] These sequences are located just proximal to the telomere and may be associated with specific syndromes.

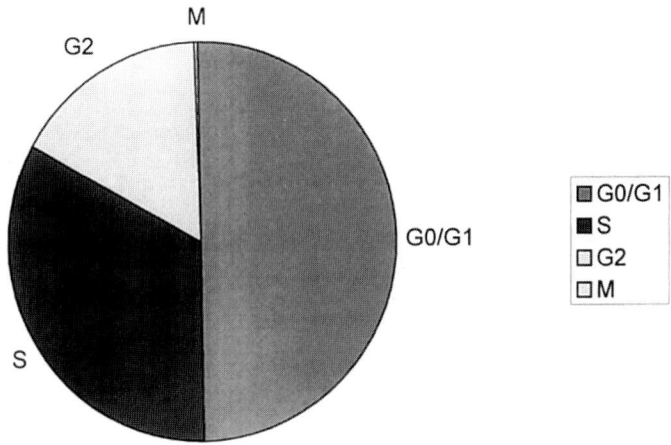

**Fig. 1.** Diagram of the cell cycle. In cycling or dividing cells, the time between mitoses is about 24 h. The initial phase, G1, is approximately 12 h. DNA synthesis takes 6–8 h, followed by a second gap, or G2, which lasts about 4 h. Mitosis requires about 20 min.

## CHROMOSOME STRUCTURE AND FUNCTION

To make some sense of FISH signals, a brief review of chromosome function and structure is useful. This discussion is relevant to the interpretation of results.

Conventional cytogenetic analysis (sometimes referred to as chromosomal cytology) can only be performed on cycling cells, as chromosomes can only be visualized during mitosis. The phases of the cell cycle are summarized in **Fig. 1**, with the caveat that most cells in adults are not cycling and will not divide under ordinary circumstances; these cells are in a phase called *G0*, the G standing for gap. Normal cellular functions such as protein synthesis are occurring. This generalization applies to most neoplastic entities as well, which often precludes their cytogenetic evaluation. When cells enter

**A**

- two DNA double helices
- four strands of DNA
- two chromatids
- two copies of any DNA probe will bind

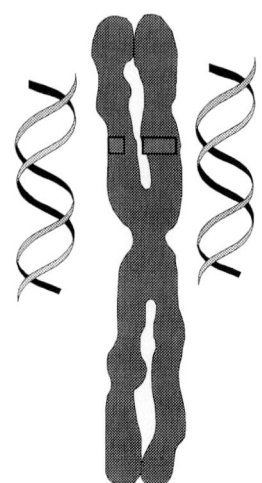

**B**

telomere

centromere

chromatids

telomere

**Fig. 2. (A)** Chromosome structure. Chromosomes condense sufficiently for analysis after completion of DNA synthesis. The chromosomes are composed of four strands of DNA or two double helices. Two copies of any DNA probe, shown as small boxes, will bind to each chromosome. **(B)** Each sister chromatid (arrows) is composed of a DNA double helix. The sister chromatids are attached at the centromere, which appears as a constriction at metaphase. The terminal (distal from the centromere) ends of the chromosomes are called telomeres.

into a proliferative cycle, which in normal cells lasts on average 24 h, they move from G0 to G1; G1 lasts about 12 h. The DNA synthetic or S-phase lasts for 6–8 h. A second gap, or G2, phase takes about 4 h, although it may last from 2 to 6 h in normal cells. This is followed by mitosis, which is completed in 20 min. *Interphase cytogenetic analysis* is actually an *in situ* hybridization analysis, as it is performed on cells in interphase, in other words, on cells that are not dividing.

Once a given chromosome has completed DNA synthesis, that chromosome is composed of four DNA strands or two DNA double helices, as seen in **Fig. 2A.** At metaphase, each double helix forms a sister chromatid; sister chromatids remain associated at the centromere **(Fig. 2B)**. Whereas one chromosome may have completed DNA synthesis, the other chromosome of that pair may not complete DNA synthesis at the same time, e.g., the maternally derived chromosome 2 may go through DNA synthesis before the paternally derived chromosome 2 does. The former fact explains why in FISH on chromosome spreads, two closely juxtaposed signals are seen on a single chromosome; the latter fact explains why in FISH on nuclei, two, three, or four signals are seen in cells that are diploid. Although chromosome banding techniques are a great advance in the accurate identification of chromosomes and in gene mapping, it should be noted that a single chromosome band may encompass 5–10 million bp, or 10–50 genes.

Mitosis itself is composed of four phases. Prophase is when the chromosomes condense and the nuclear membrane and the nucleoli dissolve. So-called high-resolution chromosome analysis is done during mid to late prophase; although it may seem that only a single chromatid is present, the chromosomes are extremely elongated, making resolution of the chromatids, as well as the centromere, difficult. During metaphase the chromosomes line up, forming the metaphase plate in preparation for the sister chromatids to segregate to the two poles during anaphase. Conventional cytogenetic analysis is done during metaphase, when the condensing chromosomes are shorter and less likely to overlap than they are at prophase. The sister chromatids and centromere are easily resolved as the cell moves towards anaphase. During anaphase the sister chromatids separate and begin to move to opposite poles; the cell membrane begins to invaginate in order to form two daughter cells. Telophase is signaled by the condensation of the chromosomes into nuclei, reformation of the nuclear membrane and nuclei, and pulling apart of the daughter cells.

## GENERAL CLASSES OF DNA PROBES

### α-Satellite Probes

The work of Pardue and Gall *(10)* was with mouse satellite DNA, which appears uniformly in the centromeric region of each chromosome; the structure of the human chromosome is different. In humans, the degree of repetitiveness of the centromeric sequences and the concomitant heterochromatization is not as pronounced, with some exceptions that are now termed *classical satellites*. This is relevant to FISH because a repeated sequence called an α-satellite has been described in human chromosomes. α-Satellite sequences are composed of 171-bp A-T-rich "alphoid monomers" arranged in tandem repeats that range from 150 to 9000 kb; with few exceptions, each human chromosome has a specific α-satellite sequence, which is localized to the centromere of the chromosome *(22,23)*. Consequently, α-satellite probes can identify a given chromosome, provided that its centromere is present. As the number of repeats varies between chromosomes, so will the size of the FISH signal; in other words, the relative size of the signal from a given chromosome, chromosome 3, for instance, will generally be larger than the signal from chromosome 5 and may be smaller than that from chromosome 7 in any given individual. As suggested earlier, in addition to α-satellite DNA, there are β and classic satellite DNA, with probes available for each. The latter binds to constituitive (A-T-rich) heterochromatin present around the centromeres of chromosomes 1, 9, and 16 and on distal Yq. The classic satellite is further broken down into satellites I, II, and III, based on buoyant density. β-Satellites are also located in the pericentromeric region *(24)*.

The chromosomes that do not have specific α-satellite probes are the so-called D- and G-group chromosomes, chromosomes 13–15, 21, and 22, which are acrocentric and carry the ribosomal RNA genes on a "stalk" just above their short arms. To confuse the issue further, the D and G chromosomes are topped by a structure referred to as a "satellite," which is unrelated to satellite sequences in either mouse or human. The α-satellite sequences are located near the centromere, and not near the structure called the satellite in these acrocentric chromosomes. There is sufficient sequence homology between the α-satellite region of chromosomes 13 and 21, and 14 and 22 *(25,26)* to preclude the use of conventional α-satellite probes to distinguish these chromosomes. This becomes relevant when selecting probes for evaluation of cells for trisomy or confirming that the chromosome of interest is present when using FISH to evaluate chromosomes for structural abnormalities. For instance, an α-satellite probe is not a good choice to distinguish trisomy 13 from trisomy 21, an issue that often arises in prenatal diagnosis. Similarly, if one were evaluating a highly aneuploid cell line for the presence of translocations involving genes near the retinoblastoma locus on chromosome 13, it would be prudent to use a probe for another gene on chromosome 13, but not an α-satellite probe, to confirm that the cell had two copies of chromosome 13.

α-Satellites are often thought of as chromosome enumeration probes. They are easy to use, and in general will give a large, bright signal with little background. They are used in a variety of settings including demonstration of changes of ploidy in tumors, distinguishing diploid from triploid populations in hydropic pregnancies, identifying numerical abnormalities in leukemia and lymphoma, sex determination, and for prenatal diagnosis of trisomy and of sex chromosome anomalies. In **Fig. 3**, an α-satellite probe for the X chromosome was used; a single signal from the fluorescein is noted in both a single chromosome and the interphase nuclei. The result is clear: one X centromere is present; the interpretation, however, could be a female with Turner's syndrome, a normal male, or a person with an intact X centromere, in an otherwise structurally abnormal X chromosome. α-Satellite probes are not informative for demonstrating most structural abnormalities, such as a translocations or deletions, unless they involve the centromere.

*Nomenclature*

As FISH techniques came into increased use for both investigations and research, it became obvious that standard cytogenetic nomenclature was not adequate to report a result. Interphase cytogenetic analyses were particularly problematic. Consequently, the International Standing Committee on Human Cytogenetic Nomenclature (ISCN) met and issued a revision of the standard nomenclature in 1995 *(27)*. To summarize for α-satellite or chromosome enumeration probes, there are three general ways to report a result. If both a standard karyotype and *in situ* hybridization have been done, the results of the karyotype are reported, followed by a period and "ish," followed by the *in situ* hybridization result and the probe used. For example, if there were some question about the identity of an extra chromosome 22 that was resolved using FISH, the result would be reported as 47,XX,+22.ish 22(D22Z1×3). If, on the other hand, no karyotype was performed, and three signals were noted with a probe for chromosome

---

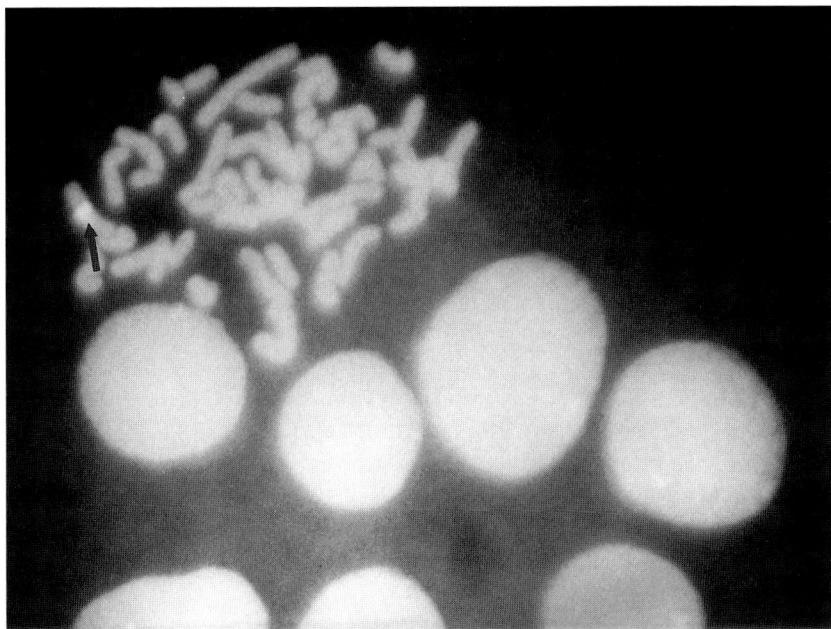

**Fig. 3.** FISH performed on cultured human lymphocytes, prepared for conventional cytogenetic analysis. The probe used is DXZ1, an alpha satellite directed against the X chromosome's centromere. A single bright signal (arrow) is seen on a chromosome and in the interphase nuclei. Counterstain is ethidium bromide.

18, the result would be written as nuc ish 18cen (D18Z1×3). Although the shorthand appears nightmarishly complex, clinical cytogenetics reports must still include an interpretation or explanation of the nomenclature.

*Telomere Probes*

Before leaving the topic of repeated sequences, another type of repeat probe should be noted, the TTAGGG$_n$ repeat at the terminus of the chromosome arm or telomere **(Fig. 2B)**; this is the signal for DNA polymerase that it has reached the end of the chromosome. This sequence is repeated over and over to a total size of about 10 kb *(28)*. Although this is a repeated sequence, the signal generated by a FISH study is generally quite small. Some of the more interesting work done with these probes is to utilize them to evaluate whether an apparently terminal translocation is actually reciprocal *(29)*, as cytogenetic dogma maintains, or is indeed terminal.

**Unique-Sequence Probes**

Unique-sequence probes recognize DNA sequences or portions of genes that are associated with a phenotype of interest. They are associated with either constitutional disorders or acquired disorders. In the former an abnormal sequence is present or a normal sequence is absent in all cells in the affected individual. With acquired disorders, such as malignancies, only those cells involved in the malignant process either will have an abnormal sequence, generally arising from a translocation, or will have lost a normal sequence.

**Table 2**
**Some Common FISH Probes for Genetic Disorders and Microdeletion Syndromes**

| Disorder | Chromosomal locus | Gene or genes involved |
|---|---|---|
| Cornelia de Lange syndrome | 3q26.3 | CDL[a] |
| Wolf-Hirschorn syndrome | 4p16 | WHSCR |
| Cri du chat syndrome | 5p15.2 | Two genes: cry and facies |
| Williams syndrome | 7q11.2 | Elastin |
| Retinoblastoma syndrome | 13q14 | RB1 |
| Angelman's syndrome | 15q11.2 | GABRB3 |
| Prader-Willi syndrome | 15q11.2 | SNRPN |
| Rubinstein-Taybi syndrome | 16p13.3 | CREB binding protein[a] |
| Smith-Magenis syndrome | 17p11.2 | FL11 |
| Miller-Diecker syndrome | 17p13.3 | LIS |
| DiGeorge syndrome (velocardiofacial) | 22q11.2 | D22S75 or Tuple1 |
| Kallmann syndrome | Xp22 | |
| Steroid sulfatase deficiency | Xp22 | Steroid sulfatase |

[a] Although FISH is diagnostic, these probes are not commercially available.

*Constitutional Disorders and Microdeletion Syndromes*

The term *microdeletion syndromes* refers to disorders in which a gene is disrupted, but the aberration is too small to be seen by conventional cytogenetic analysis. Although they are occasionally seen in high-resolution analysis, these syndromes are most often diagnosed by molecular means. Examples of constitutional microdeletions include DiGeorge, Williams, Prader-Willi, and Angelman syndromes **(Table 2)**. In genetic evaluation of these patients, such as one suspected of having DiGeorge syndrome, one might begin by reviewing a karyotype, concentrating on the DiGeorge critical region, which is the region proximal to the centromere on chromosome 22 at band q11.2. This may or may not demonstrate a cytogenetically apparent deletion. Further evaluation is done using a FISH probe called D22S75 that binds to a portion of the region that is often deleted in patients with the syndrome; the probe is visualized by indirect immunofluorescence. The probe, like most microdeletion probes, consists of a cocktail containing the sequence of interest as well as a probe for the chromosome on which the gene is located, in this case distal 22q **(Fig. 4)**. If the gene is deleted, the signal proximal to the centromere will be absent on one chromosome 22, whereas both the proximal and distal signals will be present on the other chromosome 22. Closely paired signals at each locus, corresponding to each of the sister chromatids, are seen in this figure. Interpretation of the interphase nuclei in the figure is not as straightforward; if four signals are present, there is no deletion, whereas three signals suggest a deletion. Depending on the cell cycle, however, it may become difficult to distinguish neighboring signals on different chromosomes from paired signals on sister chromatids. Propidium iodide, a red fluorescent DNA dye, is used as a counterstain, so the nuclei or chromosomes appear red.

Two-color fluorescence has made interpretation of microdeletion studies easier. Directly conjugated probe cocktails are available in which the gene of interest and the chromosome identifier are labeled with fluorochromes of different colors, such as

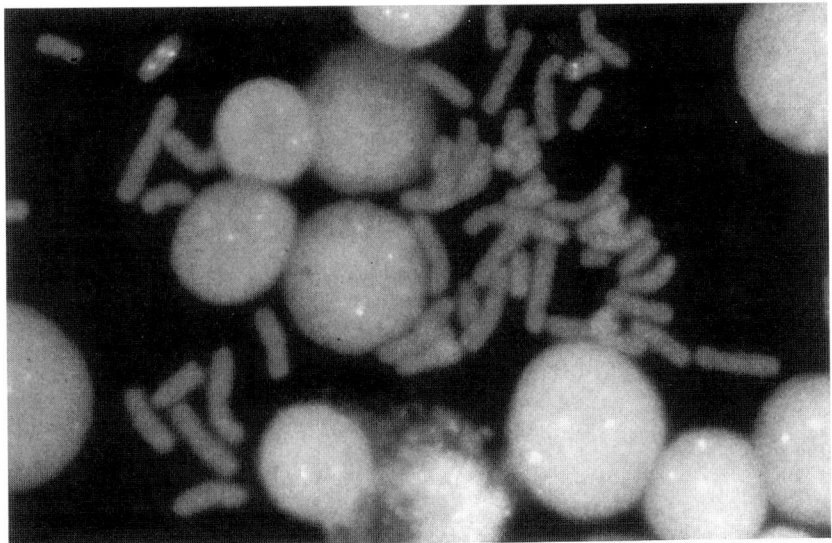

**Fig. 4.** Microdeletion associated with DiGeorge or velocardiofacial syndrome demonstrated by FISH. Note that the chromosomes 22 have a probe binding distal from the centromere that identifies the chromosome as well as a probe for TUPLE1, which is often deleted in this collection of syndromes.

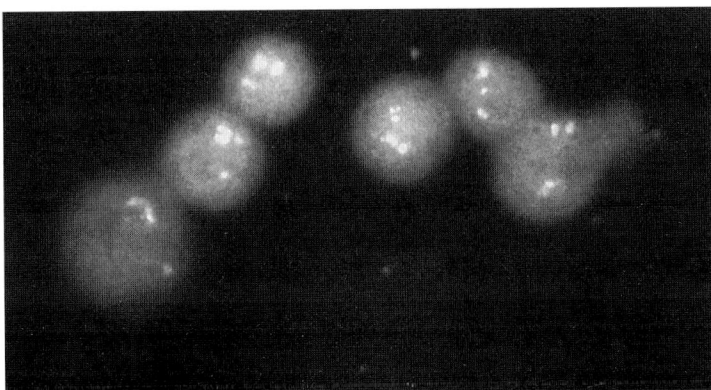

**Fig. 5.** A two-color microdeletion probe. The green signal is a marker for chromosome 15, in this case the PML gene; the red signal shows the Prader-Willi region on proximal 15q.

fluorescein and Texas Red or rhodamine. In this illustration, chromosome 15 is identified by a fluorescein-labeled probe for the PML gene; SNRPN, often deleted in patients with Prader-Willi syndrome, is labeled with rhodamine (**Fig. 5** and **Color Plate 3**, following page 208). With a two-color system, there is generally no question of whether one is seeing loss of the gene of interest or loss of a whole chromosome, as the chromosome label is one color and the gene is another color.

*Acquired Disorders and Malignancies*

Ever increasing numbers of probes for evaluation of acquired disorders such as solid tumors and leukemia are available. The simplest approach is to examine cells for gains

**Table 3**
**Common FISH Probes for Malignancies**

| Disorder[a] | Chromosomal locus | Gene or genes involved |
|---|---|---|
| Neuroblastoma | 2p23 | N-myc |
| Prostate | 8p22 | Lipoprotein lipase |
| Carcinoma or lymphoma | 8q24 | C-myc |
| Breast (prostate, other) | 17q12 | Her2/neu |
| Carcinoma | 17p13 | p53 |
| Breast | 20q13 | Amplified in breast cancer |
| Ki-1 lymphoma | 2p23;5q35 | ALK;NPM |
| MDS/AML | 5q33-q34 | 5q- critical region |
| MDS/AML | 7q31 | 7q- critical region |
| AML-M2 | 8q22;21q22.1 | AML1;ETO |
| CML/AML/ALL | 9q34;22q11.2 | BCR;ABL |
| ALL/AML | 11q23 | MLL |
| ALL | 12p13;22q22.3 | ETV6 (TEL);AML1 |
| CLL | 13q14 | D13S25 |
| Lymphoma (follicular) | 14q32;18q21 | IgH;BCL2 |
| AML-M4eo | 16p13q22 | CBFP |
| Promyelocytic leukemia | 15q2217q21 | PML;RARα |
| MDS or Polycythemia vera | 20q12 | 20q critical region |

[a] MDS, myelodysplastic syndrome; AML, acute myelogenous leukemia; CML, chronic myelogenous leukemia; ALL, acute lymphocytic leukemia; CLL, chronic lymphocytic leukemia.

or losses of entire chromosomes, such as trisomy 8 in myeloid disorders, trisomy 12 in chronic lymphocytic leukemia, or trisomy 1 in melanoma using α-satellite probes. Unique sequence probes for genes such as *N-myc* in neuroblastoma or *HER-2/neu* in breast and other malignancies can be used to demonstrate the degree of amplification of the normal cellular genes. Neoplastic tissues can also be examined for relevant translocations using specific unique-sequence probes (**Table 3**). The most widely used example of this is a BCR/ABL probe for evaluation of chromosomes or cells for a *BCR/ABL* fusion gene. This probe is available in a variety of configurations, although generally probes for a portion of the *BCR* sequence on chromosome 22 and for a portion of the *ABL* sequence on chromosome 9 are contained in a single solution.

The probes either have biotinylated and digoxigenin-labeled nucleotides, which can then be detected separately by fluorochrome-labeled antibodies, or are directly labeled with different fluorochromes. These sequences will be identified by their respective fluorochromes: a normal cell will have two signals from chromosome 9 and two from chromosome 22. Cells in which a translocation has occurred will display either two closely juxtaposed signals of different colors, or a fusion signal of an intermediate color, i.e., if *BCR* is labeled with fluorescein, and *ABL* is labeled with rhodamine, a yellow signal is generated by a fusion signal (**Fig. 6** and **Color Plate 3** following page 208). A blue fluorescent DNA dye, 4,6-diamino-2-phenyl-indole (DAPI), is used as a counterstain so that the nucleus or chromosomes can be seen as well. A recent advance is D-FISH, in which both the Philadelphia chromosome and the reciprocal translocation

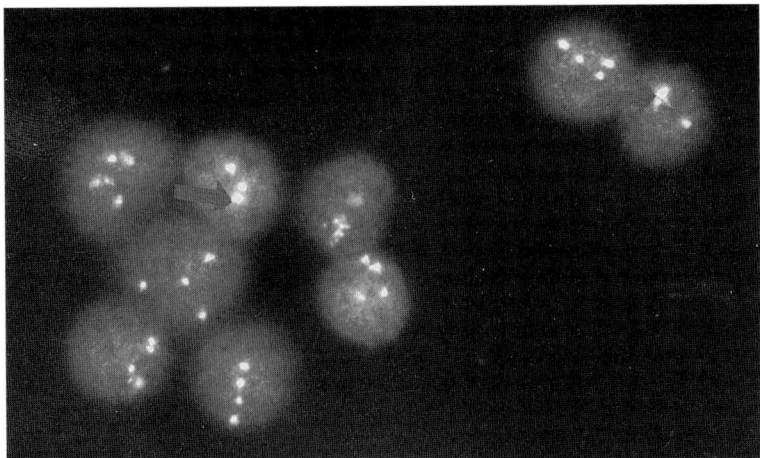

**Fig. 6.** A two-color probe for *BCR* and *ABL* genes on a touch preparation from a spleen. Note occasional fusion signals (arrows), which are suggestive of a translocation.

on chromosome 9 appear as fusion signals, eliminating some of the ambiguity generated by chromosomes that are merely juxtaposed and not actually translocated *(30)*.

*Nomenclature*

Reporting results for unique-sequence probes (ISCN) again makes the distinction between observations from a classical cytogenetic analysis and for the probe. As with α-satellites, if a karyotype is performed, that result is reported first, i.e., 46,XY, followed by ".ish". If a deletion at a given locus is noted, it is reported as .ish del(7), followed by the breakpoints of the deletion and the name of the probe used, such as .ish del(7)(q11.2q11.2)(ELN−); in this case, only one copy of the elastin probe is seen, ELN−. Because this is an interstitial deletion, albeit a microdeletion, it has by definition starting and stopping points, hence the naming of two breakpoints. If the probe were used, but a deletion was not seen, it would be reported as .ish 7(q11.2)(ELN×2), i.e., two copies of the probe were seen at an appropriate locus. As with α-satellite probes, if only an interphase analysis is performed, then the result is reported as nuc ish(q11.2)(ELN×2).

Reporting results for probes that yield a fusion signal is different, because the term *con* is used. For example, if one were doing *in situ* hybridization for a translocation between chromosomes 9 and 22 on a smear from a patient with chronic myeloid leukemia, it might be reported as nuc ish 9q34(ABL×2),22q11.2(BCR×2)(ABL con BCR×1). This means that there are two signals from chromosome 9 and two from chromosome 22, but one of the signals from each has been fused.

## Painting Probes

Painting probes, or whole chromosome paints (wcp), are constructed from a collection of unique DNA sequences that cover or are specific for a single chromosome. A good painting probe covers the entire chromosome, with little or no crossreactivity with the other chromosomes, with the following caveat. Because the painting probes are constructed to bind to unique-sequence DNA, the more highly repetitive DNA sequences

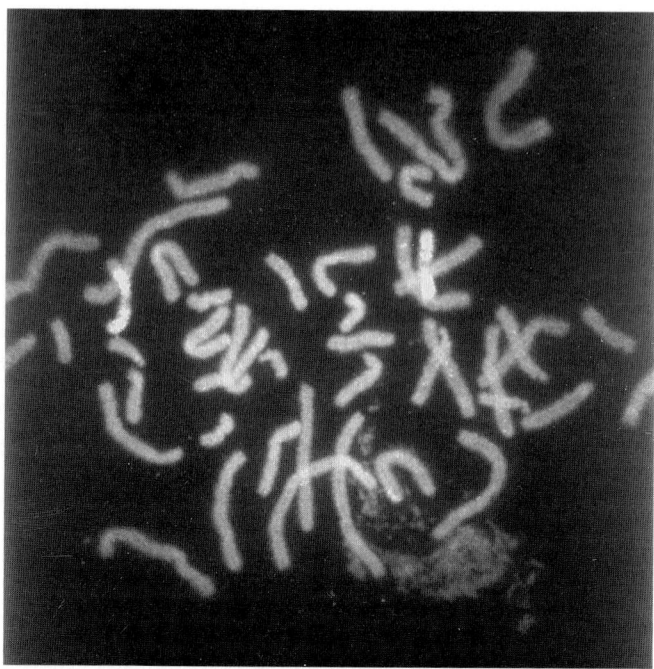

**Fig. 7.** Painting probe for chromosome 9. Note that the centromeres, which contain repeated sequences, do not paint. Often a faint banding pattern can be seen with paints because the distribution of unique sequences is not even. (*See* **Color Plate 3** following page 208.)

at the centromeres and telomeres may be spared **(Fig. 7)**. The original painting probes were made by lysing mitotic cells and sorting out individual chromosomes using a flow cytometer *(31–33)*. One would wind up with, for example, a tube of intact chromosome 17. The chromosomes would then be cut with a restriction endonuclease and the unique sequences on chromosome 17 amplified by so-called Alu PCR or, more correctly, interspersed repetitive sequence (IRS) PCR *(34,35)*. Alu and other intermediate repetitive sequences are interspersed through the human genome; consequently, these sequences can be used to create primers for PCR amplification of the unique sequences between the Alu sites. Because the exact nature of the unique sequences is not known, this is considered an anonymous DNA library: the genes for p53, BRCA-1, and the retinoic acid receptor may be included in this mixture, but it would be both impossible and unnecessary to identify them as individual bands. Then, either the PCR products are cloned into plasmids or cosmids, or further PCR reactions are run, this time in the presence of a compound such as biotin or digoxigenin attached to one or more of the bases in order to label the resulting probe.

Some of the commercial painting probes are available directly conjugated with a fluorochrome, whereas others use an indirect method and require incubation with an antibody or binding protein, such as an avidin-fluorochrome or anti-dogoxigenin-fluorochrome for their detection. Each method has its advantages. The indirect method allows for amplification of a weak signal; the direct method is faster and easier to use when more than one chromosome is being probed on the same slide. The specificity of the binding of a probe to its target sequence is in part obtained by adjusting the stringency of the hybridization conditions.

Painting probes are useful for identification of chromosomes carrying complex translocations that do not include the centromere, i.e., pieces of chromosomes attached to the ends of or inserted into other chromosomes. They are also useful for interphase analysis of translocations for which a specific probe is not available. For instance, if a patient's tumor carried an unusual translocation between chromosomes 9 and 22, not involving the usual breakpoints in a Philadelphia chromosome, using a painting probe for chromosome 22, the analysis would proceed as follows.

Normal cells will have two copies of chromosome 22; mitotic cells will have two brightly staining signals. Cells with a translocation, however, will have three signals: a small chromosome 22, a normal chromosome 22, and a chromosome 9 with a small piece of chromosome 22 translocated on to it. Only the end of chromosome 9 with the fluorescein-tagged probe for chromosome 22 would be seen. Chromosomes other than 22 will be stained red with a propidium iodide counterstain, as will the nuclei of cells that are not in mitosis. A more sophisticated approach might involve a probe for chromosome 9 labeled with a second fluorochrome, such as rhodamine, as well as the probe for chromosome 22. This type of analysis has been taken one step further with the marketing of slides impregnated with combinations of three different painting probes in a series of eight squares, making it possible to analyze a cell suspension for the presence of common tumor translocations with a single hybridization reaction.

*Spectral Karyotyping and M-FISH*

M-FISH and spectral karyotyping (SKY) are specialized methods for using painting probes so that each pair of chromosomes is visualized as a different color (reviewed in ref. *36*). Each uses a set of five fluorochromes in combination to generate the colored probes; the former uses combinations of filter sets to visualize the fluorescent chromosomes *(37,38)*, whereas the latter uses interference spectroscopy or interferometry *(39)*. These methods are useful for analysis of complex cytogenetic changes.

*Nomenclature for Whole Chromosome Paints*

The most common settings in which these probes are used are identification of a supernumerary unidentified marker chromosome (mar) or when additional chromosomal material is present on the terminus of a chromosome. In the former case, a result might be reported as 47,XX,+mar.ish der(15)(wcp15+). In the case of additional material present on a chromosome 17, which is demonstrated to be part of chromosome 18 by FISH, the result might be written as 46,XX,add(17)(p13).ish der(17)t(17;18)(p13;p or q)(wcp18+). The ISCN text *(27)* should be referred to for more detailed information.

## IN *SITU* HYBRIDIZATION PROCEDURES

FISH procedures rely on the fact that nucleic acid sequences will bind to one another if sufficient sequence homology is present. All FISH procedures involve denaturation of the probe, denaturation of the target sequence, placing the probe and target in contact and allowing them to reanneal, all of which may be done simultaneously. As outlined above, the probe and the target must be visualized, either by direct or indirect fluorescence of the probe. Nonspecific binding of the probe to the cellular DNA is reduced or eliminated by a wash following the reannealing or hybridization step.

Unlike Southern and PCR-based analyses, much of the specificity of FISH probes is derived from the posthybridization wash. The concept of *stringency* has been mentioned. This refers to the ability of DNA sequences to bind to one another when the sequence homology is not exact, for instance, sequences that are rich in adenine and thymine (A-T) and highly repeated will often hybridize to one another. To eliminate this nonspecific binding, one would use a high-stringency wash. Factors that increase the stringency, and make it more difficult for inexactly matched hybrids to remain bound, include increasing the temperature (from 37°C to 43°C), increasing the formamide concentration (from 50 to 70%), decreasing the salt concentration (from 0.4X standard saline citrate [SSC] to 0.25X SSC). α-satellite probes contain the most repeated sequences and will require a higher stringency posthybridization wash than will either unique-sequence probes or whole chromosome paints.

### Some Comments on Selection of Probes and Controls

The American College of Medical Genetics (ACMG) has developed standards for use of FISH probes in clinical practice *(40)*. Even in the absence of the intent to use FISH clinically, it is probably wise to follow some of the guidelines to validate new probes and new lots of probes on cells that are known to be normal as well as those known to be carrying the deletion that the probe is supposed to demonstrate. Appropriate positive and negative controls should be run with a FISH study; often the normal homologous chromosome will serve as a control. However, in interphase studies, particularly those looking for numerical abnormalities, it is useful to evaluate results from a FISH probe not thought to be involved in the process. This allows one to confirm that the cells in question have a trisomy 8, for example, and are not triploid.

Probes generated in the user's laboratory ("home brews") should also be validated and used with both positive and negative controls. Guidelines for validation and use are outlined by the ACMG (section G) in its Guidelines for Molecular Genetics Laboratories.

Methodologies for unique sequence probes and whole chromosome paints are similar, with the caveat that paints are difficult to interpret on sections. This is because in the average 4-μm tissue section the DNA sequence of interest may not be in the plane of the section, or a single chromosome may be snaking in and out of the section, generating multiple signals.

### Examples

The examples given in the Protocols section will serve to illustrate the specifics of using FISH probes (*see* pp. 84–88). In the first example, the specimen is a lymph node from a patient suspected of having involvement by a tumor that has a trisomy for chromosome 3. The lymph node was processed for routine histology. An α-satellite probe can be used on a paraffin section of the lymph node to evaluate it for metastatic disease. The second example is a touch preparation from the spleen of a patient who had been treated for chronic myelogenous leukemia, but who is experiencing a new leukocytosis but no basophilia. Massive splenomegaly was noted, prompting splenectomy. A BCR/ABL probe is used on the alcohol-fixed smear to determine whether the leukocytes represent recurrent disease or a reactive process.

Formalin-fixed, paraffin-embedded tissue can be used in FISH studies; a few steps are added to the general procedure to remove the paraffin and the protein-DNA crosslinks *(41–45)*. FISH can be done on standard sections, allowing retention of tissue morphology, although some authors disaggregate thick sections to make a cell suspension *(46,47)*. FISH can also be done on specimens prepared by routine methods in the cytopathology laboratory, although an acid-alcohol postfix is useful on air-dried and hair-sprayed slides.

## ACKNOWLEDGMENTS

The author would like to thank Nanci Lefebvre and Beth Cox at the University of Michigan Cytogenetics Laboratory, Ann Arbor, MI, for assistance with the FISH studies. The staff of Diagnostic Cytogenetics, Seattle, WA, contributed some of the slides for two-color FISH.

## REFERENCES

1. Haferlach, T., Winkermann, M., Loffler, H., et al. (1996) The abnormal eosinophils are part of the leukemic cell population in acute myelomonocytic leukemia with abnormal eosinophils. *Blood* **87**, 2459–2463.
2. Bentz, M., Schroder, M., Herz, M., Stilgenbauer, S., Lichter, P., and Dohner, H. (1993) Detection of trisomy 8 on blood smears using fluorescence in situ hybridization. *Leukemia* **7**, 752–757.
3. Thompson, C. T., LeBoit, P. E., Nederlof, P. M., and Gray, J. W. (1994) Thick section fluorescence in situ hybridization on formalin-fixed, paraffin-embedded archival tissue provides a histogenetic profile. *Am. J. Pathol.* **144**, 237–243.
4. Price, P. M. and Hirschorn, K. (1975) In situ hybridization for gene mapping. *Cytogenet. Cell Genet.* **14**, 395–401.
5. Popp, S., Jauch, A., Schindler, D., et al. (1993) A strategy for the characterization of minute chromosome rearrangements using multiple color fluorescence in situ hybridization with chromosome-specific DNA libraries and YAC clones. *Hum. Genet.* **96**, 527–532.
6. Taylor, C. P., McGuckin, G., Bown, N. P., et al. (1994) Rapid detection of prognostic genetic factors in neuoblastoma using fluorescence in situ hybridization on tumour imprints and bone marrow smears. *Br. J. Cancer* **69**, 445–451.
7. Press, M. F., Bernstein, L., Thomas, A., et al. (1997) HER-2/neu gene amplification characterized by fluorescence in situ hybridization. *J. Clin. Oncol.* **15**, 2894–2904.
8. Zhan, L., Chang, K-S., Estey, E. H., Hayes, K., Deisseroth, A. B., and Liang, J. C. (1995) Detection of residual leukemic cells in patients with acute promyelocytic leukemia by the fluorescence in situ hybridization method. *Blood* **85**, 495–499.
9. Buno, I., Wyatt, W. A., Zinsmeister, A. R., Dietz-Band, J., Silver, R. T., and Dewald, G. (1998) A special fluorescent in situ hybridization technique to study peripheral blood and assess the effectiveness of interferon therapy in chronic myeloid leukemia. *Blood* **92**, 2315–2321.
10. Pardue, M. L. and Gall, J. G. (1970) Chromosomal localization of mouse satellite DNA. *Proc. Natl. Acad. Sci. USA* **168**, 1356–1358.

11. Prescott, D. M. and Bender, M. (1962) RNA and protein synthesis during mitosis in mammalian tissue culture cells. *Exp. Cell Res.* **26**, 260–268.

12. Cremer, C., Gray, J. W., and Ropers, H-H. (1982) Flow cytometric characterization of a Chinese hamster × man hybrid cell line retaining the human Y chromosome. *Hum. Genet.* **60**, 262–266.

13. Dale, R. M. and Ward, D. C. (1975) Mercurated polynucleotides: new probes for hybridization and selective polymer fractionation. *Biochemistry* **14**, 2458–2469.

14. Hopman, A. H. N., Wiegant, J., and VanDuijn, P. (1987) Mercurated nucleic acid probes, a new principle for non-radioactive in situ hybridization. *Exp. Cell Res.* **169**, 357–368.

15. Ward, D. C., Reich, E., and Stryer, L. (1969) Fluorescence studies of nucleotides and polynucleotides. I. Formycin, 2-aminopurine riboside, 2,6-diaminopurine riboside, and their derivatives. *J. Biol. Chem.* **244**, 1228–1237.

16. Ward, D. C. and Reich, E. (1972) Fluorescence studies of nucleotides and polynucleotides. II. 7-Deazanebularin: coding ambiguity in transcription with base pairs containing fewer than two hydrogen bonds. *J. Biol. Chem.* **247**, 705–719.

17. Ward, D. C., Horn, T., and Reich, E. (1972) Fluorescence studies of nucleotides and polynucleotides. 3. Diphosphopyridine nucleotide analogues which contain fluorescent purines. *J. Biol. Chem.* **247**, 4014–4020.

18. Langer, P. R., Waldrop, A. A., and Ward, D. C. (1981) Enzymatic synthesis of biotin-labeled poly nucleotides: novel nucleic acid affinity probes. *Proc. Natl. Acad. Sci. USA* **78**, 6633–6637.

19. Cremer, T., Landegent, J., Bruckner, A., et al. (1986) Detection of chromosome aberrations in the human interphase nucleus by visualization of specific target DNAs with radioactive and non-radioactive in situ hybridization techniques; diagnosis of trisomy 18 with probe LI.84. *Hum. Genet.* **74**, 346–352.

20. Julien, C., Bazin, A., Guyot, B., Forstier, F., and Defos, F. (1986) Rapid prenatal diagnosis of Down's syndrome with in situ hybridization of fluorescent DNA probes. *Lancet* **2**, 863–864.

21. Lichter, P., Cremer, T., Tang, C-J., Watkins, P. C., Manuelidis, L., and Ward, D. C. (1988) Rapid detection of human chromosome 21 aberrations by in situ hybridization. *Proc. Natl. Acad. Sci USA* **85**, 9664–9668.

22. Manuelidis, L. (1978) Chromosomal location of complex and simple repeated human DNAs. *Chromosoma* **66**, 23–32.

23. Willard, H. F. (1985) Chromosome-specific organization of human alpha satellite DNA. *Am. J. Hum. Genet.* **37**, 524–532.

24. Waye, J. S. and Willard, H. F. (1989) Human beta satellite DNA: genomic organization and sequence definition of a class of highly repetitive tandem DNA. *Proc. Natl. Acad. Sci. USA* **86**, 6250–6254.

25. Choo, K. H., Vissel, B., Brown, R., Filby, R. G., and Earle, E. (1988) Homologous alpha satellite sequences on human acrocentric chromosomes with selectivity for chromosomes 13, 14 and 21. *Nucleic Acids Res.* **16**, 1273–1284.

26. Jorgensen, A. L., Kolvraa, S., Jones, C., and Bak, A. L. (1988) A subfamily of alphoid repetitive DNA shared by the NOR-bearing human chromosomes 14 and 22. *Genomics* **3**, 100–109.

27. ISCN, Mitelman, F., ed. (1995) *An International System for Human Cytogenetic Nomenclature.* S. Karger, Basel.

28. Moyzis, R. K., Buchingham, J. M., and Cram, L. S. (1988) A highly conserved repetitive DNA sequence, $TTAGGG_n$, present at the telomeres of human chromosomes. *Proc. Natl. Acad. Sci. USA* **85**, 6622–6625.

29. Park, V. M., Gutashaw, K. M., and Wathen, T. M. (1992) The presence of interstitial telomeric sequences in constitutional chromosome abnormalities. *Am. J. Hum. Genet.* **50**, 914–923.

30. Dewald, G., Stallard, R., Alsaadi, A., et al. (2000) A multicenter investigation with D-FISH BCR/ABL1 probes. *Cancer Genet. Cytogenet.* **116**, 97.

31. Lichter, P., Cremer, T., Border, J., Manuelidis, L., and Ward, D. (1988) Delineation of individual human chromosomes in metaphase and interphase cells by in situ suppression hybridization using recombinant DNA libraries. *Hum. Genet.* **80**, 224.

32. Pinkel, D., Landegent, J., Collins, C., et al. (1988) Fluorescence in situ hybridization with human chromosome specific libraries. *Proc. Natl. Acad. Sci. USA* **85**, 9138–9142.

33. Van Dilla, M., Deaven, L., Albright, K., et al. (1986) Human chromosome-specific DNA libraries: construction and availability. *Biotechnology* **4**, 537–552.

34. Nelson, D., Ledbetter, S., Corbo, L. et al. (1989) Alu polymerase chain reaction: a method for rapid isolation of human-specific sequences from complex DNA sources. *Proc. Natl. Acad. Sci. USA* **86**, 6686–6690.

35. Ledbetter, S., Nelson, D., Warren, S., and Ledbetter, D. (1990) Rapid isolation of DNA probes within specific chromosome regions by interspersed repeated sequence polymerase chain reaction. *Genomics* **6**, 475–481.

36. Jalal, S. M. and Law, M. E. (1999) Utility of multicolor fluorescent in situ hybridization in clinical cytogenetics. *Genet. Med.* **1**, 181–186.

37. Spiecher, M. R., Ballard, S. G., and Ward, D. C. (1996) Karyotyping human chromosomes by combinatorial multi-fluor FISH. *Nat. Genet.* **12**, 368–375.

38. Eils, R., Uhrig, S., Saracoglu, K., et al. (1998) An optimized, fully automated system for fast and accurate identification of chromosome rearrangements by multiplex-FISH (M-FISH). *Cytogenet. Cell Genet.* **82**, 160–171.

39. Schrock, E., duManoir, S., Veldman, T., et al. (1996) Multicolor spectral karyotyping of human chromosomes. *Science* **273**, 494–497.

40. American College of Medical Genetics. (1999) *Standards and Guidelines for Clinical Genetics Laboratories*, 2nd ed., Rockville, MD.

41. Wolfe, K. and Herrington, C. (1997) Interphase cytogenetics and pathology: a tool for diagnosis and research. *J. Pathol.* **181**, 359–361.

42. van de Kaa, C., Nelson, K., Ramaekers, F., Vooijs, P., and Hopman, A. (1991) Interphase cytogenetics in paraffin sections of routinely processed hydatidiform moles and hydropic abortions. *J. Pathol.* **165**, 281–287.

43. Poddighe, P., Ramaekers, F., and Hopman, A. (1992) Interphase cytogenetics of tumors. *J. Pathol.* **166**, 215–224.

44. Hopman, A., van Hooren, E., van de Kaa, C., Vooijs, P., and Ramaekers, F. (1991) Detection of numerical chromosome aberrations using in situ hybridization in paraffin sections of routinely processed bladder cancers. *Mod. Pathol.* **4**, 503–513.

45. Long, A., Mueller, J., Schwartz, J., Barrett, K., Schwartz, R., and Wolfe, H. (1992) High specificity in situ hybridization. *Diagn. Mol. Pathol.* **1**, 45–57.

46. Arnoldus, E. P. J., Dreef, E., Noodermeer, I. A., et al. (1991) Feasibility of in situ hybridization with chromosome specific DNA probes on paraffin wax embedded tissue. *J. Clin. Pathol.* **44**, 900–904.

47. Seto, E. and Yen, TSB. (1987) Detection of cytomegalovirus infection by means of DNA isolated from paraffin-embedded tissues and dot hybridization. *Am. J. Pathol.* **127**, 409–413.

# PROTOCOLS

## FISH

The sample for the lymph node is a routine 4-μm paraffin section cut on positively charged slides (PLUS or silane-coated) and then heated in a 60°C oven for no more than 15 min. The control, both positive and negative, is blood cells present in the specimen, which should have a diploid number of signals. For the evaluation of BCR/ABL, a series of touch or imprint preparations were made of a spleen removed from a patient with possible chronic myelogenous leukemia. One slide was fixed and stained with Wright-Giemsa to confirm the presence of tumor, and the rest were processed as described below. Internal controls include the presence of two signals in the majority of cells (corresponding to the normal cellular genes) and the absence of multiple signals in stroma and blood cells.

### MATERIALS

#### Fixing Smears and Imprints
1. Methanol.
2. Glacial acetic acid.

A mixture of three parts methanol to one part acetic acid is prepared within 1 h of use. This fixative is referred to as acid-alcohol, or sometimes as modified Carnoy's.

#### Dewaxing Paraffin Sections
1. Xylene
2. Absolute ethanol, 95% ethanol, 80% ethanol, 70% ethanol.
3. Deionized water.

#### Thiocyanate Pretreatment of Paraffin Sections
1. 1 $M$ sodium thiocyanate solution; store protected from light.

#### Enzymatic Digestion of Paraffin Sections
1. Pepsin 4 mg/mL in 0.2 $M$ HCl (store stock at −20°C)
   or
2. Proteinase K 25 μg/mL in PBS (store stock at −20°C).

Prepare these solutions using sterile reagents and containers; thaw at 37°C and use immediately.

#### Denaturation
Prepare all the following solutions using sterile water and containers.

1. 20× SSC: 3 $M$ NaCl (175.32 g/L), 0.3 $M$ NaCitrate (88.23 g/L), 1 L deionized water. Dissolve the two salts separately and mix. This is a stock solution. The pH will need to be adjusted to 6.8–7.0 upon dilution.
2. 70% formamide, pH 7.0, in 2× SSC (see *Note 1*).

## Hybridization

The probes used are commercially available and are provided with a suitable hybridization buffer. The exact ingredients of the hybridization buffers are proprietary, but they generally contain 50% formamide, 10% dextran sulfate, 0.01% sheared salmon sperm DNA in 2× SSC. The BCR/ABL (Oncor, Gaithersburg, MD) does not require dilution. The α-satellite or chromosome enumeration probe (CytoCell, Oxfordshire, UK) is supplied affixed to a plastic cover slip and comes with hybridization buffer, as described above.

1. Hybridization chamber: these are available commercially, *or* use a plastic box with a lid, *or* a glass baking dish covered with plastic wrap; place one or two damp paper towels in the bottom to maintain the humidity.
2. Rubber cement.
3. 37°C incubator, without $CO_2$, or 37°C oven, or slide warmer.

## Posthybridization Wash

1. 50% formamide in 2× SSC, pH 7.0, or 0.4× SSC, pH 7.0.
2. Phosphate-buffered detergent (PBD): phosphate-buffered saline, pH 6.8 (1 L), 130 m$M$ sodium chloride (FW = 58.4) 7.59 g, 7 m$M$ dibasic sodium phosphate (FW = 142) 0.99 g, 3 m$M$ monobasic sodium phosphate (FW = 138) 0.41 g. Dissolve in deionized water in order given and then add Triton X-100, 0.05% (0.05 mL/L).

## Localization of Probe

Anti-digoxigenin antibody labeled with appropriate fluorochrome or avidin labeled with appropriate fluorochrome; counterstains include DAPI or propidium iodide in Antifade (Sigma, St. Louis, MO).

## METHODS

### Preparing the Target

In working with paraffin-embedded tissue, one must remove 1) the paraffin (deparaffinization or dewaxing), 2) the protein that has been crosslinked to the DNA, and 3) DNA crosslinks. Many manufacturers of DNA probes recommend that the sections not be baked at 60°C; this depends on the probe. It may be helpful to bake for 15 min or so to improve adhesion of the section to the slide. Slides should be positively charged, such as silane-treated slides, to improve adhesion.

Fix smears and imprints, such as the spleen in this example, by laying the slides on a horizontal surface; cover the surface of the specimen with acid-alcohol for 2 min. The excess fix is tilted off, and the slide is allowed to air-dry for at least 1 h. Skip to the **Denaturation** section below. Paraffin sections, such as the section of lymph node in this example, require dewaxing and several additional steps **(steps 1–3)** to permit binding of the probe.

1. Dewaxing. Slides are incubated in a series of Coplin jars at room temperature as follows: xylene, two changes, 10 min each; xylene/ethanol (50:50) 5 min; ethanol, two changes, 5 min each; air-dry at least 5 min.
2. Thiocyanate pretreatment. 1 $M$ sodium thiocyanate in deionized water; heat to 80°C and incubate slides for 10 min. This can be done on a slide warmer. Flood the slide with about

2 mL of the reagent to avoid evaporation. Following incubation, rinse the slide with deionized water, two changes, 2 min. Briefly drain slides to remove excess water.

3. Enzymatic pretreatment. Several methods are described. This laboratory uses pepsin; some of the commercial kits use proteinase K. Pepsin 4 mg/mL in 0.2 $M$ HCl (store stock at –20°C). Incubate at 37°C for 5–15 min (may require up to 45 min; see *Note 2*). The lymph node was treated for 20 min. Rinse with deionized water, two changes, 2 min. Dehydrate through graded alcohols, 70%, 80%, 95%, 100%, and allow to air-dry. This is critical to avoid diluting the formamide in the following, denaturation step.

## Denaturation

Generally, both the probe and the target DNA will require denaturation, but this depends on the manufacturer's instructions. Some probes require no denaturation, and some will tolerate co-denaturation, in which the probe solution is placed on the target area of the slide, a cover slip is sealed in place with rubber cement, and then the slide is heated to between 70° and 100°C for 2 : 10 min. To denature the cells or sections on a slide in 70% formamide.

1. 70% formamide in 2× SSC, pH 7.0 (the pH is critical; see *Note 1*) is heated in a water bath in a Coplin jar, with times and temperatures as follows (see *Note 3*):
   a. Paraffin-embedded sections: 8–12 min, 85°C; the section of lymph node was denatured for 10 min, although this step is not in the probe manufacturer's directions.
   b. Touch preps, smears, chromosome spreads: 2 min, 72°C; the touch prep of spleen was denatured for 2 min.
2. Immediately place the slide in cold (–20°C) 70% ethanol to stop the denaturation and prevent renaturation; after 2 min, (see *Note 4*), transfer to 90, 95, and 100% ethanol, 2 min each, and allow to air-dry. Serial dehydration is crucial for paraffin sections; 70% followed by absolute ethanol is adequate for thin smears, chromosome preparations, and so on.

## Hybridization

Both the paraffin section of lymph node and the touch prep of spleen were denatured and dehydrated as outlined above. For the BCR/ABL probe, skip to **step 3**; for the chromosome enumeration probe, skip to **step 4**.

1. Some probes require dilution into hybridization buffer. This buffer contains 70% formamide, some kind of carrier DNA, buffers, and dextran sulfate (to keep the solution in place on the slide). The last ingredient makes it difficult to pipet when cold. Bring the stock solution of the probe to room temperature, briefly vortex, and centrifuge in a microcentrifuge for 2–3 s to collect the contents at the bottom of the tube. If the probe is labeled with a fluorochrome, it may be useful to work in subdued light at this point (see *Note 5*). If the probe has passed its expiration date, altering the dilution is often useful (see *Note 6*).
2. Some probes require pre-denaturation and/or preannealing; the manufacturer will state this in the package insert. An aliquot is removed to a microcentrifuge tube and heated to 72°C for 5–10 min. It is then placed on ice for a few minutes (for α-satellites, to prevent reannealing) or at 37°C for 30 min (for paints, to allow the repeated sequences to preanneal and prevent their nonspecific binding to all chromosomes).
3. The BCR/ABL probe does not require dilution but must be denatured. Using a micropipet, approximately 10 µL of probe per slide is pipetted into a microcentrifuge tube and denatured in a water bath at 72°C for 5 min. Place the probe on the surface of the section or smear and cover with a 22-mm² cover slip. Seal the edges of the cover slip with rubber cement. Do not try to be neat: you have to peel the cement off again.

4. The α-satellite probe used on the section of lymph node is usually denatured by a co-denaturation step, per the manufacturer's instructions. Co-denaturation is a procedure by which the probe is placed on the slide, and it and the target DNA are denatured simultaneously and then allowed to hybridize. This technique can be employed for many probes but must be used with a very nice collection of probes supplied affixed to cover slips (CytoCell, UK). *Notes 5–7* are applicable. For paraffin sections, place 10 μL of hybridization buffer on the section, place the cover slip with the attached probe on the slide, and seal with rubber cement. Allow the rubber cement to dry for 5 min, and then place the slide on a 75°C hotplate for 5 min to denature the section.

5. Place the slides in the hybridization chamber, and incubate at 37°C for 30 min to overnight. For paraffin sections and new probes, it is useful to start with the longer incubation and decrease the time on subsequent trials. In these examples, the probe is incubated for 16 h.

## Posthybridization Wash

1. Peel the rubber cement from the cover slip; often the cover slip comes off, too. If it does not, "swish" the slide in a beaker of PBD to float the cover slip off.
2. Immediately place slide in one of the following solutions:
   a. 50% formamide, 37°C, 15 min (paraffin sections, irrespective of probe, paints, unique-sequence probes).
   b. 65% formamide, 43°C, 15 minutes (alpha and beta satellite probes) or
   c. 0.4× SSC, 72°C, 2 minutes, for directly labeled probes; this is the rapid wash technique, see *Note 8*.

   For both BCR/ABL and the α-satellite used in this example, the rapid wash was used.

3. Following a formamide wash, place slides in 2× SSC, 37°C, 5 min; for the rapid SSC wash, follow with 30 s in PBD. Hybridized slides can be held in PBD at 4°C overnight prior to the next step.

## Localization of the Probe

1. Directly conjugated probes require only a rinse with PBD and mounting with a glass coverslip (**step 5** below).
2. Rinse the slides with PBD to remove the SSC. At this point, the placenta section is counter-stained (**step 5**).
3. Apply 15–20 μL of anti-digoxigenin/fluorochrome conjugate or avidin/fluorochrome conjugate, cover with a plastic cover slip (see *Note 9*), and incubate at 37°C for 15–30 min. The BCR/ABL probe required a fluorescein-conjugated avidin incubation for 30 min followed by rhodamine-conjugated anti-digoxigenin, but it is now available directly conjugated (see *Note 10*).
4. Working in subdued light, remove the cover slip, and rinse the slide with PBD, three times for 2 min each rinse.
5. Apply a counterstain, such as propidium iodide for fluorescein-labeled probes, or DAPI for probes labeled with a red fluorochrome, in "antifade"; apply a glass cover slip (22 × 40 or 24 × 50 mm).

## Visualizing the Probe

1. A 100-W mercury vapor light source is recommended, particularly for unique sequence probes (see *Note 11*).
2. Locate the sections or cells under low power; air bubbles are in the same focal plane as the cells.
3. The presence of label is scored under high power, either a 100× or 60× oil objective. For some larger probes, such as some α-satellites, a 40× or 60× water objective may be adequate.

4. Score a minimum of 250–500 nuclei for so-called interphase cytogenetics. This is particularly necessary for tissue sections, as the entire nucleus may not be in the section.
5. Cells labeled with a two-color probe should have at least one signal present in the nucleus in each color to demonstrate that the probes worked and contained the appropriate labels.
6. Slides may be stored either before or after viewing (see *Note 11*).
7. If results are unsatisfactory, many specimens can successfully be destained (see *Notes 12 and 13*) and/or rehybridized (see *Note 14*).

## NOTES

1. Formamide becomes basic as it degrades; adjust the pH of the solution with HCl. Deionizing the formamide and freezing it in polypropylene tubes will increase the stability. Ultra-pure formamide (BRL, Bethesda, MD), stored frozen, will maintain its pH for years.
2. To determine whether the digestion is adequate, cover-slip the wet slide and view with a fluorescent microscope with a fluorescein isothiocyanate (FITC) filter. If there is green fluorescence, more digestion time is needed. If distinct nuclei are not seen, the tissue is overdigested.
3. For specimens to be denatured in formamide, it is useful to prewarm the slides on a slide warmer prior to placing them in the formamide. Denature no more than four slides at a time to maintain the temperature of the formamide, and recheck the temperature between batches of slides.
4. For most specimens, there is a tradeoff between denaturation sufficient to allow the probe to bind and over-denaturation so that morphology is lost. In general, do not exceed 2 min unless the specimen is paraffin embedded. The exception is older specimens, which can be denatured for 2½–3 min.
5. For most commercial probes, subdued light is more than adequate. When you are working with only a few slides, and the fluorochrome is *not* directly conjugated, the relatively brief exposure to room light does not appear to be a problem.
6. Probes that are 6 months to a year or more past their stated expiration date have been used successfully by using a higher ratio of probe to diluent (twice the recommended amount of probe works well). If the probe does not require dilution, a larger volume (20 µL rather than 10 for a 22-mm² area) is often successful.
7. At least one vendor (CytoCell, UK) supplies their probes bound to cover slips. These can be used quite successfully. After placing the cover slip over the hybridization solution on the target area of the slide and sealing with rubber cement, it is important to wait at least 5 min before doing the denaturation step, as the probe requires this time to dissolve off the cover slip.
8. A variety of posthybridization washes are described. For probes that are conjugated to a fluorochrome, a "quick wash" is preferable.
9. Some manufacturers of fluorochrome conjugates sell kits that include a variety of "blocking reagents," designed to reduce nonspecific binding of avidin to the tissue, buffers for washing, and polypropylene cover slips for use during incubation of the conjugate. "Plastic" cover slips can be cut from Parafilm, which will reduce the scratching of the specimen during this incubation. A 22 × 50-mm² surface area is preferable for this step.
10. Two-color indirect immunofluorescence can be completed by adding both fluorochrome-labeled antibodies simultaneously. In some cases, adding separate aliquots of the fluorochromes effectively dilutes each in half; in other cases it may be necessary to do separate incubations, add marked excess of antibody solution, and increase the incubation time or increase the concentration of each one.
11. Most commercial probes are designed for viewing with a 100-W rather than a 50-W mercury vapor lamp or equivalent. Signals may not be visible at 50 W. A 100-W bulb cannot be put in a 50-W lamp socket.

12. Slides can be stored in the dark in the refrigerator for 3–7 days. Some slides may be usable for several months if stored in a freezer. In either case, wipe the oil off the cover slip before storing, as it may be necessary to remount the cover slip prior to further viewing.

13. If the signal from a probe is faint, sometimes it can be improved by removing some of the counterstain. Propidium iodide is readily removed by a brief rinse in PBD. DAPI is much more difficult to remove; the slides will have to be processed through two or three 5-min washes in PBD, preferably in the dark.

14. Most specimens can be rehybridized or reprobed at least once. Wipe any oil from the cover slip, and then remove the cover slip, either by gently prying it off with a razor blade or by soaking in PBD. Rinse in fresh PBD and dehydrate. Any probe bound to the slide will be removed during denaturation.

# 6

# Practical Applications of the FISH Technique

## George Kontogeorgos, MD, PhD, Nikiforos Kapranos, MD, PhD, and Eleni Thodou, MD, PhD

## INTRODUCTION

Fluorescence *in situ* hybridization (FISH), often referred to as molecular cytogenetics, is currently recognized as a reliable technique and an alternative to classic cytogenetics (karyotyping), with many applications for detecting chromosomal abnormalities. The technique can analyze interphase nuclei of resting cells or metaphase spreads of mitotic cells. In addition, specific protocols for karyotyping analysis combine FISH with classic cytogenetics.

FISH was introduced in 1988 by Pinkel et al. *(1)*, who utilized a nonisotopic, fluorescence technique for cytogenetics. However, the credit for initial application of interphase cytogenetics is given to Lichter et al. *(2)*, who in 1988 applied a radioactive DNA probe to demonstrate chromosomal aberrations in nonmitotic human nuclei.

## FISH TECHNIQUE

The main principle of FISH is the formation of the hybridization product of a DNA probe with the target chromosomal DNA **(Fig. 1)**. Briefly, a denatured fluorescent-labeled DNA probe is added to the denatured specimen DNA and used to perform hybridization. Posthybridization washings, under specific stringency conditions, remove the unbound probe. After nuclear counterstaining, the specimen is ready for study with a fluorescent microscope equipped with appropriate excitation filters and a photographic camera. Digital image analysis systems equipped with appropriate software can enhance the fluorescent signals.

Metaphase FISH is ideal for detecting locus-specific gene rearrangements and for generating high-resolution physical genetic maps *(3)*. However, it is often difficult to obtain metaphase spreads from solid tumors. Selective cell growth of nontumorous cells (lymphocytes, fibroblasts), which may result in loss of overall genetic information and lead to false results, represents the main disadvantage of this technique *(4,5)*. In addition, metaphase FISH requires specific equipment and expertise in tissue cultures.

Interphase FISH is useful in detecting normal and defective genes such as deletions or gains. The technique is ideal for analyzing solid tumors, particularly those with a low proliferation rate. Nuclei from touch imprints, specimens of fine needle aspiration

From: *Morphology Methods: Cell and Molecular Biology Techniques*
Edited by: R. V. Lloyd © Humana Press, Totowa, NJ

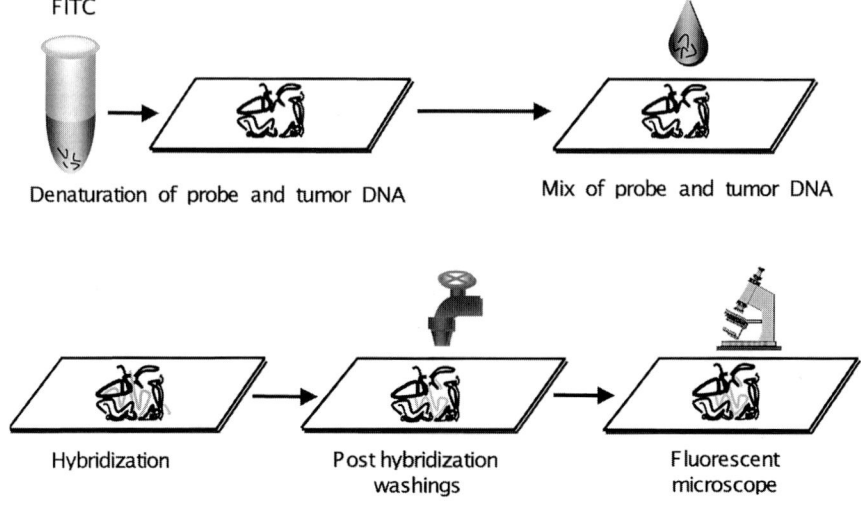

**Fig. 1.** Principles of the FISH technique.

biopsies, preparations of centrifuged biologic fluids, and nuclei isolated from frozen or paraffin-embedded tissues are appropriate for interphase FISH analysis *(6–11)*. The technique can also be applied directly to frozen or paraffin sections; however, cautious interpretation of the results is required due to partial loss of nuclear DNA mass during tissue sectioning *(7–14)*.

## Tissue Preparation and Fixation

Several fixatives including formalin are suitable; however, in our hands, fixation in methanol/acetic acid or chilled acetone gives excellent results. Careful digestion with proteinase K before application of the technique is often crucial, particularly for tissue sections *(15)*. Overdigested tissues show loss of nuclear borders; the cells appear "ghostly" or lost. These preparations should be discarded, and the experiment should start again with less proteinase K digestion time, depending on the degree of overdigestion *(16)*. By examining tissue sections under the fluorescence microscope, underdigested tissue produces persistent autofluorescence of the background, often associated with poor propidium iodide staining. In that case, additional digestion, depending on the amount of background fluorescence, is required.

## Utility of Fluorescent Microscope

FISH analysis requires high-quality standards of fluorescent microscopy for optimal visualization of the results. An epiilluminated microscope is preferable. A 100-W high-pressure mercury lamp is required, particularly for evaluation of dual- or triple-color analysis *(17)*. It is optimal for detecting the weak fluorescent signals of sequence-specific probes. Evaluation and recording of fluorescent signals requires high-power, dry, or oil objective lenses. Only nonfluorescing immersion oil should be used. Filters are specifically designed for certain fluorochromes used for probe labeling and counterstain. A wide range of filters and combinations, which allow only certain wavelengths

**Table 1**
**Selection of Filters for Optimal Fluorescent Microscopy, According to Detection Systems, Fluorochromes, and Counterstains**

| Probe label | Detection reagent | Counterstain | Filter needed to visualize counter-stain excitation/ emission (nm) | Filter needed to visualize probe signal excitation/ emission |
|---|---|---|---|---|
| Digoxigenin | FITC/anti-digoxigenin | Propidium iodide (PI) | 520/610 | 490/525 (long-pass) |
| | Rhodamine/ anti-digoxigenin | DAPI | 365/480 | 540/550 |
| Biotin | FITC-avidin | PI | 520/610 | 490/525 (long pass) |
| | Texas Red-avidin | DAPI | 365/480 | 540/525 |
| Digoxigenin and biotin | Dual-color detection | DAPI | 365/480 | Triple-bandpass |

From: *Workshop Manual. Introduction to Molecular Cytogenetics: Tissue in situ hybridization.* Oncor, Gaithersburg, MD, pp. 18, 1995.

of light to pass through, are used for single or simultaneous dual-signal fluorescence detection (**Table 1**). Optimal use of filters can subject the fluorescent signal to the lowest amount of excitation light to retain fluorescence as much as possible. The microscope should be equipped with a photographic camera. A 400-ASA film is appropriate, although 100-ASA films provide higher resolution. Alternatively, a digital, preferably cool, camera allows direct transfer of the digitized signals to a computer for storage and further processing using proper software *(16)*.

## COMPARATIVE GENOMIC HYBRIDIZATION

Comparative genomic hybridization (CGH) is a sophisticated molecular technique that provides overall genomic information in a single experiment. This technique was introduced by Kallioniemi *et al.* in 1992 *(18)*, and it is considered today as the state of the art of fluorescence technology in the analysis of human chromosomes. Given that CGH is applied in uncultured cells, it is ideal for screening of solid tumors, particularly those that have a low growth rate for tumor-specific, nonbalanced numerical and structural aberrations *(18,19)*.

Briefly, DNA extracts from tumor and blood lymphocytes are separately labeled with two-color different fluorochromes (fluorescein isothiocyanate [FITC] and Texas Red) (**Fig. 2**). Equal quantities of the two DNAs, mixed with unlabeled Cot-1 DNA, to suppress cross-hybridization of repetitive sequences, are denatured and hybridized to already prepared denatured metaphase spreads from normal lymphocytes. Posthybridization washings follow to remove the excessive, unbound probe, and then air-dried slides are counterstained. Digitized three-color images taken from hybridized metaphase spreads with a fluorescent microscope are then processed with an image analysis system using specific software for karyotyping analysis. The system (based on estimation of the fluorescence intensity ratio between green and red signals, comparing them with that of simultaneously hybridized normal DNA in the same metaphase spreads) automatically reports losses and gains. The system displays all colorized chromosomes and generates ideograms with vertical bars indicating losses and gains of the corresponding chromosomal regions for each individual chromosome *(20,21)*.

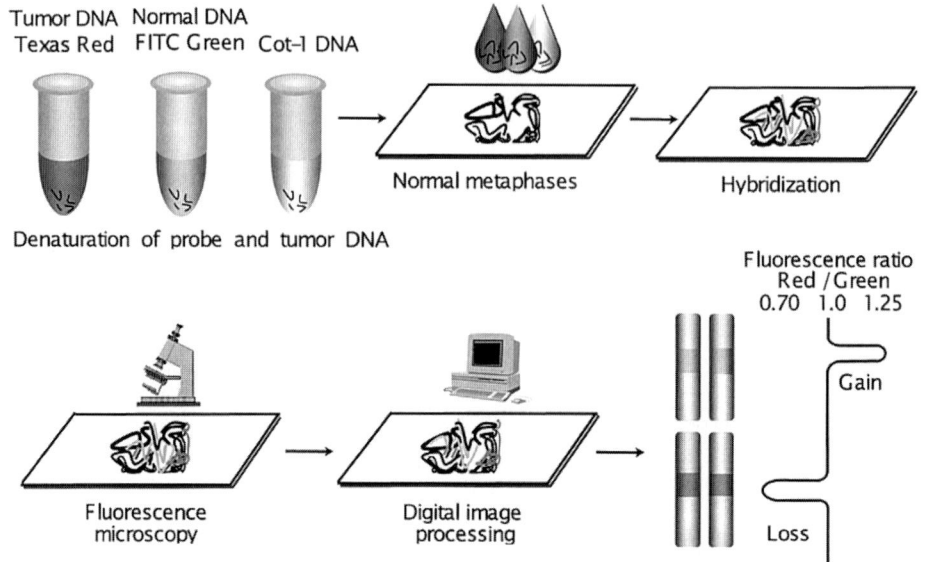

**Fig. 2.** Principles of the CGH technique.

## PRACTICAL APPLICATIONS

### Detection of Numerical Aberrations

α-Satellite centromeric, β-satellite pericentromeric, or classical satellite probes, specific for tandem repeats of DNA sequences of individual chromosomes, are widely used, mostly with interphase FISH, to detect copy number aberrations of chromosomes *(3,22)* (**Fig. 3**). Alternatively, paint (coatsome) probes, complementary to hundreds of

---

(*See* **Color Plate 4** following page 208.)                                              ▶

Fig. 3. (**A**) Examples of numerical chromosome aberrations using an α-satellite, centromere-specific DNA probe. The nucleus in the left panel shows two fluorescent signals corresponding to a pair of normal chromosome 11, whereas more than two signals, as shown in the nucleus of the right panel are characteristic for polysomies (FITC/PI, 400×). (**B**) Monosomy of chromosome 11 as the dominant abnormality in a mixed somatotroph-lactotroph, mostly prolactin-producing pituitary adenoma (FITC/PI). (**C**) Chromosome 1 in ductal carcinoma of the breast. Most cell exhibit three or more fluorescent signals (FITC/PI). (**D**) Chromosome 1 breast carcinoma of the ductal type. All cells display aberrant fluorescent signals (FITC/PI). (**E**) Localization of the DiGeorge gene on a chromosome in the 22q11 region in normal metaphase spreads. A pair of fluorescent signals corresponds to the alleles of the DiGeorge gene. The third signal marks the centromeric region of chromosome 22 (FITC/PI). (**F**) Dual-color FISH in metaphases of normal lymphocytes demonstrates the *bcr* gene on chromosome 22 (green) and the *abl* gene on chromosome 9 (red) (FITC/TR/DAPI). (**G**) Mixture of whole paint (coatsome) DNA probe for chromosome 11 (red) and α-satellite centromere-specific probe for chromosome 2 (green) in normal metaphase spreads (FITC/TR/DAPI). (**H**) All human telomeres demonstrate all chromosomes in metaphase spreads (FITC/PI, original magnification ×400 for A and B; ×160 for C and D; ×600 for E–G; ×720 for H.) E–H: *from personal experiments (G.K., H.T.) in the Appligene Oncor Labs, Illkirch, France.*

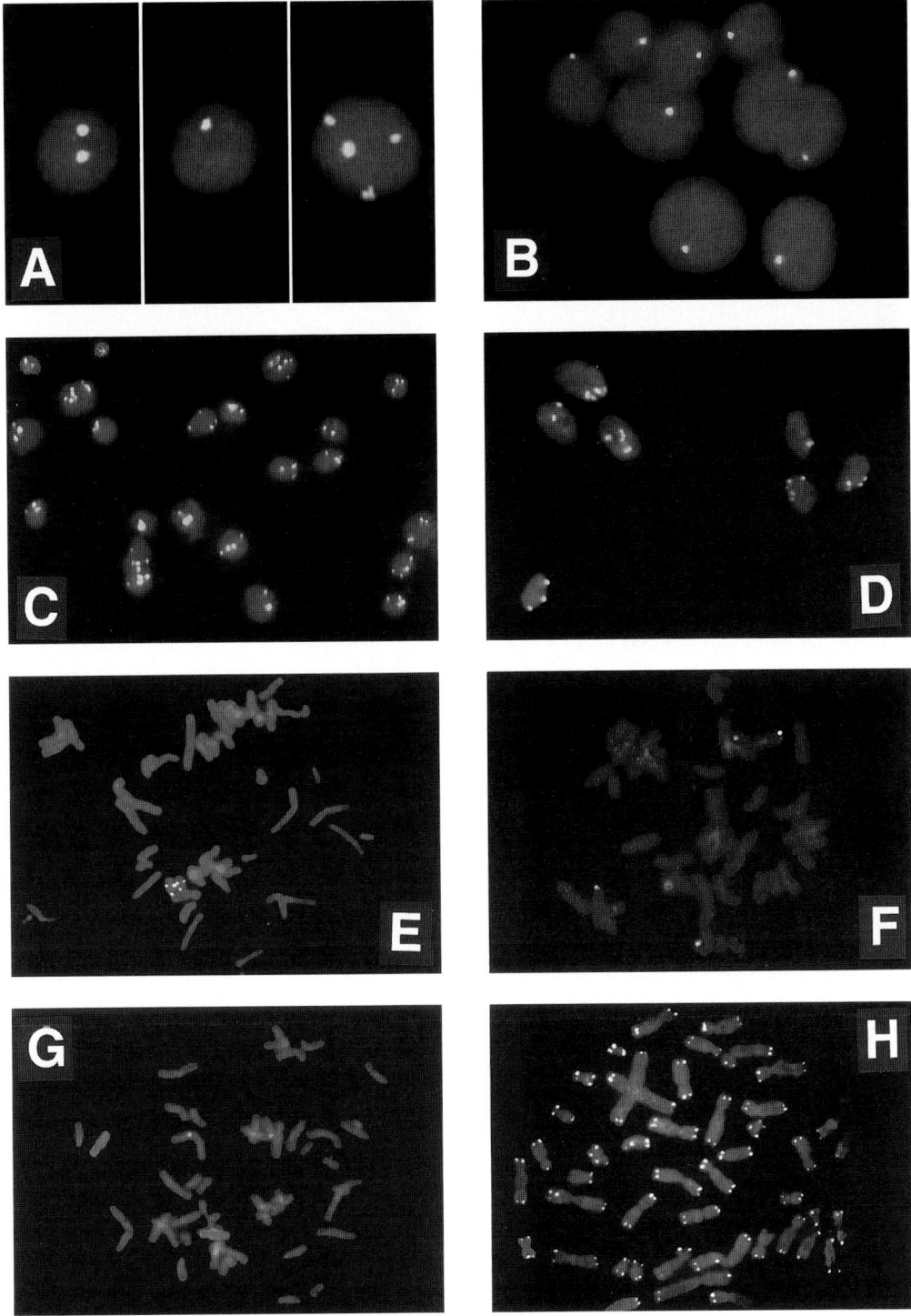

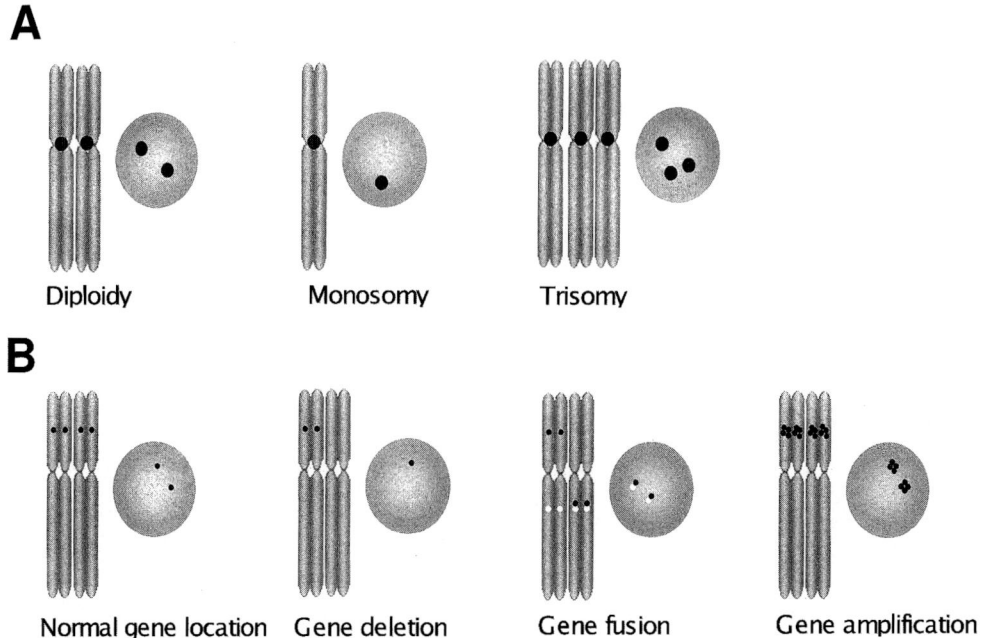

**Fig. 4. (A)** Uses of centromeric and locus-specific probes. **(B)** Use of locus-specific probes to detect structural aberrations.

unique DNA sequences specific to part or the entire length of the chromosome, can be used for metaphase FISH analysis *(3,5)* **(Fig. 4)**. Probes to the total human genome can hybridize the entire chromosome and distinguish it from somatic cell hybrids *(18)*. Lastly, probes complementary to the highly conserved repetitive telomeric DNA sequence (TTAGGG)$_n$ can also demonstrate all human chromosomes in metaphase spreads and differentiate them from other species *(23)* **(Fig. 4)**.

*Aneuploidy in Solid Tumors*

Interphase FISH utilizing centromere-specific DNA probes discloses numerical aberrations and thus provides information on tumor DNA ploidy. Normal diploid cells display two centromeres for each pair of chromosome. The presence of one centromere, corresponding to loss of one chromosome copy, is characteristic for monosomy, whereas aneuploid cells with polysomies display more than two centromeres, corresponding to extra chromosome copies **(Fig. 3)**.

Numerical aberrations in some solid tumors (such as of chromosomes 1 and 17 in breast cancer, 7 and 12 in follicular thyroid tumors, and of chromosome 8 in prostate carcinomas and 11 in pituitary adenomas) are frequent and in most cases are correlated with the DNA index. Thus, FISH is useful to estimate aneuploidy in these tumors *(14,24–27)*.

*Aneuploidy in Prenatal Diagnosis*

FISH has many applications for prenatal, and postnatal diagnosis and for the detection of preimplantation genetic abnormalities. Specific DNA probes can rapidly detect numerical aberrations, such as trisomies of chromosomes 13, 18, and 21 in Patau's, Edwards', and Down's syndromes, respectively, and other aberrations of X and Y sex chromosomes in other syndromes such as Klinefelter's and Turner's *(28)*. Aneuploidy of these chromosomes includes the most common chromosomal aberrations leading to birth defects. Interphase FISH in uncultured fetal cells can rapidly detect with high accuracy approximately 95% of aneuploidies in high-risk pregnancies *(29)*. However, prenatal diagnosis of chorionic villous sampling or amniotic fluid by interphase FISH analysis should be considered together with classic karyotyping. It has therefore been suggested that interphase FISH can be used for diagnosis as an alternative to classic cytogenetics, when abnormalities are found on ultrasound scan or by maternal serum biochemical screening *(30)*.

### Detection of Structural Abnormalities

Sequence-specific cosmid probes can detect a single gene copy in both interphase and metaphase preparations. They can localize a single gene on its chromosome region in metaphase spreads and thus are used for generating high-resolution genetic maps **(Fig. 4)**. These specific unique-sequence DNA probes can identify marker chromosomes, other structural aberrations, balanced or unbalanced translocations, and complex rearrangements **(Fig. 4)**. They are also useful in detecting gene amplifications, gains, or deletions. However, interphase FISH is not always appropriate in identifying translocations, unless the defective gene is translocated to the close proximity of another known reference gene or if there is a gene fusion. In both instances, the utility of two probes labeled with different fluorescent dyes is necessary **(Fig. 4)**. A typical example of gene fusion is the *bcr/abl* gene rearrangement in leukemias *(31,32)*.

Locus- or sequence-specific probes can be used in combination with centromeric or paint chromosome-specific probes for simultaneous demonstration of numerical and structural abnormalities on a defective chromosome.

### bcr-abl *Rearrangement*

*bcr-abl* rearrangement, also known as Philadelphia chromosome, is the result of the reciprocal translocation t(9;22)(q34;q11), which juxtaposes the *abl* oncogene on chromosome 9 with the breakpoint cluster region *(bcr)* of the gene on chromosome 22. This rearrangement is typical of chronic myeloid leukemia, serving as a diagnostic marker and for estimation of an abnormal clone, which is of major importance in monitoring the patient's response to treatment with interferon or bone marrow transplantation *(33)*. In addition, this defect occurs in some cases of acute myeloid and lymphoblastic leukemia. The FISH technique requires two- or three-color analysis to detect with accuracy the bicolor fluorescent signal of the fused *bcr/abl* genes *(17)* **(Fig. 4)**.

### c-erbB-2 *(Her2-neu)*

The *c-erbB-2* gene is a transmembrane receptor located on chromosome 17q11. Gene amplification results in overexpression of the encoded p185 protein in approximately one-third of breast cancers *(34,35)*. The presence of the functionally activated *c-erbB-2* receptor seems to be responsible for reduced patient survival and for alterations to chemotherapy response *(36)*. Recently, herceptin, a monoclonal antibody against *c-erbB-2*, was introduced for clinical use, particularly for metastatic breast cancer refractory to chemotherapy *(37)*. Therefore, *c-erbB-2* analysis became of great importance for prognosis and treatment. Interphase FISH is ideal for detection and enumeration of gene amplification *(35)*. A commercially provided kit for FISH analysis approved by the Food and Drug Administration is now available. A centromeric probe specific for chromosome 17 and a sequence-specific probe for *c-erbB-2* are necessary for FISH analysis. A scoring system utilized with FISH is based on the ratio of gene copy to centromeric copy number. Controls from cell lines showing a normal and amplified *c-erbB-2* gene should be used for testing the technique and for counting the fluorescent signals. The expected normal ratio is 1.0; when the ratio grater than 2.0, the gene is considered amplified. Although overexpression of c-erbB-2 protein can be detected by immunohistochemistry, FISH analysis is preferable, particularly for tumors showing a moderate degree of typical membranous immunostaining *(38)*.

## Androgen Receptor Gene Amplification

Amplification of the androgen receptor (AR) gene located on chromosome Xq11-q13 may occur during androgen deprivation therapy of recurrent carcinomas from patients who failed androgen endocrine treatment *(39)*. Untreated prostate cancers show no AR gene amplification *(40)*. High levels of AR amplification are demonstrated in specimens from recurrent tumors, but in none of those taken from the same patients before therapy *(41)*. Clinicopathologic studies indicate that AR amplification occurs mostly in tumors that initially showed good response for more than 12 months following endocrine therapy *(39)*.

## DiGeorge's Syndrome

DiGeorge's syndrome represents a developmental defect of the third and fourth pharyngeal pouches associated with and characteristic of facial dysmorphism, congenital heart defect, and hypoplasia or aplasia of the thymus and parathyroids. The DiGeorge gene is located on chromosome 22q11. FISH and DNA dosage analysis showed deletion of the DiGeorge gene region in all 16 cases studied *(42)*. In another investigation of 23 patients with DiGeorge's syndrome by prometaphase chromosome analysis and/or by FISH, four patients displayed an interstitial deletion in band 22q11.2, whereas in 18 patients, the deletion was disclosed only by the FISH technique *(43)*. Detection of the DiGeorge defect gene is very important in diagnosis, prognosis, and genetic counseling.

## Kallmann's Syndrome

Kallmann's syndrome is an inherent disorder characterized by hypogonadotropic eunuchoidism, anosmia, and hyposomia. The affected individuals may also have borderline normal intelligence, cleft lip and palate, deafness, and renal and cardiac anomalies. Males are predominately affected. The patients show microdeletion of the KAL gene located in the Xp22.3 region. FISH is a reliable technique for detecting the KAL-1 deletion in people with normal karyotypes but features consistent with Kallmann's syndrome *(44)*.

## Ewing's Sarcoma/Primitive Neuroectodermal Tumors

The balanced translocation t(11;22)(q24;q12) is specific for Ewing's sarcoma/primitive neuroectodermal tumors. This translocation results in the fusion of the EWS gene on chromosome 22q12 to the 3′ portion of the FLI-1 gene on chromosome 11q24. Recent reports have shown detection of EWS/FLI-1 fusion using dual-color interphase FISH analysis in frozen and paraffin sections *(45,46).*

## ACKNOWLEDGMENTS

This work was supported in part by the E-52/96 Award of the Medical Council of Greece (to Dr. Kontogeorgos).

## REFERENCES

1. Pinkel, D., Landegent, J., Collins, C., et al. (1988) Fluorescence in situ hybridization with human chromosome-specific libraries: detection of trisomy 21 and translocations of chromosome 4. *Proc. Natl. Acad. Sci. USA* **85**, 9138–9142.
2. Lichter, P., Cremer, T., Borden, J., Manuelidis, L., and Ward, D. C. (1988) Delineation of individual human chromosomes in metaphase and interphase cells in situ suppression hybridization using recombinant DNA libraries. *Hum. Genet.* **8**, 235–246.
3. Meyne, J. and Moyzis, R. K. (1994) In situ hybridization protocols, in *Methods in Molecular Biology*, vol. 33 (Choo, K. H. A., ed.), Humana, Totowa, NJ, pp. 63–74.
4. Wolman, S. R. (1997) Applications of fluorescence situ hybridization techniques in cytopathology. *Cancer* **81**, 193–197.
5. Cowan, J. M. (1994) Fishing for chromosomes. The art and its applications. *Diagn. Mol. Pathol.* **3**, 224–226.
6. Cajulis, R. S. and Frias-Hidvegi, D. (1993) Detection of numerical chromosomal abnormalities in malignant cells in fine needle aspirates by fluorescence in situ hybridization of interphase cell nuclei with chromosome-specific probes. *Acta Cytol.* **37**, 391–396.
7. Kim, S. Y., Lee, J. S., Ro, J. Y., Gay, M. L., Hong, W. K., and Hittelman, W. N. (1993) Interphase cytogenetics in paraffin sections of lung tumors by non-isotopic in situ hybridization. Mapping genotype/phenotype heterogeneity. *Am. J. Pathol.* **142**, 307–317.
8. Zitzelsberger, H., Szucs, S., Weier, H. U., et al. (1994) Numerical abnormalities of chromosome 7 in human prostate cancer detected by fluorescence in situ hybridization (FISH) on paraffin-embedded tissue sections with centromere-specific DNA probes. *J. Pathol.* **172**, 325–335.
9. Xiao, S., Renshaw, A., Cibas, E. S., Hudson, T. J., and Fletcher, J. A. (1995) Novel fluorescence in situ hybridization approaches in solid tumors. Characterization of frozen specimens, touch preparations, and cytological preparations. *Am. J. Pathol.* **147**, 896–904.
10. McManus, D. T., Patterson, A. H., Maxwell, P., et al. (1999) Interphase cytogenetics of chromosomes 11 and 17 in fine needle aspirates of breast cancer. *Hum. Pathol.* **30**, 137–144.
11. Kontogeorgos, G. (2000) The art and applications of FISHing in endocrine pathology. *Endocr. Pathol.* **11**, 123–136.
12. Hopman, A. H. N., van Hooren, E., van de Kaa, C. A., Vooijs, P. G. P., and Ramaekers, F. C. S. (1991) Detection of numerical chromosome aberrations using in situ hybridization in paraffin sections of routinely processed bladder cancers. *Mod. Pathol.* **4**, 503–513.
13. Blanco, R., Lyda, M., Davis, B., Kraus, M., and Fenoglio-Preiser, C. (1999) Trisomy 3 in gastric lymphomas of extranodal marginal zone B-cell (mucosa-associated lymphoid tissue) origin demonstrated by FISH in intact paraffin tissue sections. *Hum. Pathol.* **30**, 706–7112.
14. Kontogeorgos, G., Kapranos, N., Kokka, E., Orphanidis, G., and Rologis, D. (1999) Molecular cytogenetics of chromosome 11 in pituitary adenomas: a comparison of fluorescence in situ hybridization and DNA ploidy study. *Hum. Pathol.* **30**, 1377–1382.
15. Kapranos, N., Kontogeorgos, G., Frangia, K., and Kokka, E. (1997) Effect of fixation on

interphase cytogenetic analysis by direct fluorescence in situ hybridization on cell imprints. *Biotech. Histochem.* **72**, 148–151.

16. Workshop Manual. (1995) Tissue in situ hybridization, in *Introduction to Molecular Cytogenetics*. Oncor, Gaithersburg, MD, pp. 1–23.

17. Sinclair, P. B., Green, A. R., Grace, C., and Nacheva, E. R. (1997) Improved sensitivity of BCR-ABL detection: a triple-probe, three-color fluorescence in situ hybridization system. *Blood* **90**, 1395–1402.

18. Kallioniemi, A., Kallioniemi, O. P., Sudar, D., et al. (1992a) Comparative genomic hybridization for molecular cytogenetic analysis of solid tumors. *Science* **258**, 818–821.

19. Klein, C. A., Schmidt-Kittler, O., Schardt, J. A., Pantel, K., Speicher, M. R., and Riethmuller, G. Comparative genomic hybridization, loss of heterozygosity, and DNA sequence analysis of single cells. *Prod. Natl. Acad. Sci. USA* **96**, 4494–4499.

20. Kallioniemi, O., Kallioniemi, A., Piper, J., et al. (1994) Optimizing comparative genomic hybridization for analysis of DNA sequence copy number of changes in solid tumors. *Genes Chromosomes Cancer* **10**, 231–243.

21. Kjellman, M., Kallioniemi, O. P., Karhu, R., et al. (1996) Genetic alterations in adrenocortical tumors detected using comparative genomic hybridization correlate with tumor size and malignancy. *Cancer Res.* **56**, 419–423.

22. Kiechle-Schwarz, M., Decker, H. J., Berger, C. S., Fiebig, H. H., and Sandberg, A. A. (1991) Detection of monosomy in interphase nuclei and identification of marker chromosomes using biotinylated alpha-satellite DNA probes. *Cancer Cell Genet. Cytogenet.* **51**, 23–33.

23. Kontogeorgos, G. and Kovacs, K. (1997) FISHing chromosomes in endocrinology. *Endocrine* **5**, 235–240.

24. Criado, B., Barros, A., Suijkerbuijk, R. F., et al. (1995) Detection of numerical alterations for chromosomes 7 and 12 in benign thyroid lesions by in situ hybridization. Histological implications. *Am. J. Pathol.* **147**, 136–144.

25. Bova, G. S. and Isaacs, W. B. (1996) Review of allelic loss and gain in prostate cancer *World J. Urol.* **14**, 338–346.

26. Botti, C., Pescatore, B., Mottolese, M., et al. (2000) Incidence of chromosomes 1 and 17 aneusomy in breast cancer and adjacent tissue: an interphase cytogenetic study. *J. Am. Coll. Surg.* **190**, 530–539.

27. Marinho, A. F., Botelho, M., and Schmitt, F. C. (2000) Evaluation of numerical abnormalities of chromosomes 1 and 17 in proliferative epithelial breast lesions using fluorescence in situ hybridization. *Pathol. Res. Pract.* **196**, 227–233.

28. Moore, G. E., Ruangvutilert, P., Chadzimeletiou, K., et al. (2000) Examination of trisomy 13, 18 and 21 foetal tissues at different gestational ages using FISH. *Eur. J. Hum. Genet.* **8**, 223–228.

29. Feldman, B., Ebrahim, S. A., Hazan, S. L., Gyi, K., Johnson, A., and Evans, M. I. (2000) Routine prenatal diagnosis of aneuploidy by FISH studies in high-risk pregnancies. *Am. J. Med. Genet.* **90**, 233–238.

30. Thein, A. T., Abdel-Fattach, S. A., Kyle, P. M., and Soothill, P. W. (2000) An assessment of the use of interphase FISH with chromosome specific probes as an alternative to cytogenetics in prenatal diagnosis. *Prenat. Diagn.* **20**, 275–28.

31. Acar, H., Stewart, J., and Connor, M. J. (1997) Philadelphia chromosome in chronic myelogenous leukemia: confirmation of cytogenetic diagnosis in Ph positive and negative cases by fluorescence in situ hybridization. *Cancer Genet. Cytogenet.* **94**, 75–78.

32. Garcia-Isidoro, M., Tabernero, M. D., Garcia, J. L., et al. (1997) Detection of the bcr/abl translocation by fluorescence in situ hybridization: comparison with conventional cytogenetics and implications for minimal residual disease detection. *Hum. Pathol.* **28**, 154–159.

33. Yanagi, M., Shinjo, K., Takeshita, A., et al. (1999) Simple and reliably sensitive diagnosis

and monitoring of Philadelphia chromosome-positive cells in chronic myeloid leukemia by interphase fluorescence in situ hybridization of peripheral blood cells. *Leukemia* **13**, 542–552.

34. Liu, E., Thor, A., He, M., Barcos, M., Ljung, B. M., and Benz, C. (1992) The HER2 (*c-erbB-2*) oncogene is frequently amplified in in situ carcinomas of the breast. *Oncogene* **7**, 1027–1032.

35. Kallioniemi, A., Kallioniemi, O. P., Kurisu, W., et al. (1992b) ErbB-2 amplification in breast cancer analyzed by fluorescence in situ hybridization. *Proc. Natl. Acad. Sci. USA.* **89**, 5321–5325.

36. Fitzgibbons, P. I., Page, D. I., Weaver, D., et al. (2000) Prognostic factors in breast cancer. College of American Consensus Statement 1999. *Arch. Pathol. Lab. Med.* **124**, 966–978.

37. Pegram, M. D., Lipton, A., Hayes, D. F., et al. (1998) Phase II study of receptor-enhanced chemosensitivity using recombinant humanized anti-p185 HER2/neu monoclonal antibody plus cisplatin in patients with HER2/neu-overexpressing metastatic breast cancer refractory to chemotherapy treatment. *J. Clin. Oncol.* **16**, 2659–2671.

38. Hoang, M. P., Sahin, A. A., Ordòñez, N., and Sneige, N. (2000) HER-2/*neu* gene application compared with Her-2/*neu* protein overexpression and interobserver reproducibility in invasive breast carcinoma. *Am. J. Clin. Pathol.* **113**, 852–859.

39. Koivisto, P., Kononen, J., Palmberg, C., et al. (1997) Androgen receptor gene amplification: a possible molecular mechanism for androgen deprivation therapy failure in prostate cancer. *Cancer Res.* **57**, 314–319.

40. Miyoshi, Y., Uemura, H., Fujinami, K., et al. (2000) Fluorescence in situ hybridization evaluation of c-myc and androgen receptor gene amplification and chromosomal anomalies in prostate cancer in Japanese patients. *Prostate* **43**, 225–232.

41. Visakorpi, T., Hyytinen, E., Koivisto, P., et al. (1995) In vivo amplification of the androgen receptor gene and progression of human prostate cancer. *Nat. Genet.* **9**, 401–406.

42. Levy-Mozziconacci, A., Wernert, F., Scambler, P., et al. (1994) Clinical and molecular study of DiGeorge sequence. *Eur. J. Pediatr.* **153**, 813.

43. Demczuk, S., Desmaze, C., Aikem, M., et al. (1994) Molecular cytogenetic analysis of a series of 23 DiGeorge syndrome patients by fluorescence in situ hybridization. *Ann. Genet.* **37**, 60–65.

44. Hou, J. W., Tsai, W. Y., and Wang, T. R. (1998) Detection of KAL-1 gene deletion with fluorescence in situ hybridization. *J. Formos. Med. Assoc.* **98**, 448–451.

45. Kumar, S., Pack, S., Kumar, D., et al. (1999) Detection of EWS-FLI-1 fusion in Ewing's sarcoma/primitive neuroectodermal tumor by fluorescence in situ hybridization using formalin-fixed paraffin-embedded tissue. *Hum. Pathol.* **30**, 324–330.

46. Monforte-Muñoz, H., Terrada-Lopez, D., Affendie, H., Rowland, J. M., and Triche, T. J. (1999) Documentation of EWS gene rearrangement by fluorescence in situ hybridization (FISH) in frozen sections of Ewing's sarcoma-peripheral primitive neuroectodermal tumor. *Am. J. Surg. Pathol.* **23**, 309–315.

47. Workshop Manual (1995) Metaphase and interphase chromosomes, in *Introduction to Molecular Cytogenetics*. Gaithersburg, MD, pp. 1–70.

48. Gibas, M., Gibas, Z., and Sandberg, A. A. (1984) Technical aspects of cytogenetic analysis of human solid tumors. *Karyogram* **10**, 25–27.

49. Thodou, E., Ramyar, L., Cohen, A., Singer, W., and Asa, S. L. (1995) A serum free system for primary cultures of human pituitary adenomas. *Endocr. Pathol.* **6**, 289–299.

# PROTOCOLS

## 1. METAPHASE ANALYSIS FROM WHOLE BLOOD CULTURE (47)

### 1.1 MATERIAL AND EQUIPMENT FOR CULTURE

1. Sodium heparin (green-top) tube for peripheral blood collection.
2. Incubator 5% $CO_2$ 37°C.
3. Low-speed centrifuge.
4. Tissue culture flasks T25.
5. Pasteur pipets.
6. Plastic conical tubes 15 mL.
7. RPMI-1640 with HEPES.
8. Fetal bovine serum.
9. L-glutamine (200 m$M$ or 29.2 mg/mL).
10. Penicillin/streptomycin solution (10,000 U/mL and 10 mg/mL, respectively).
11. Phytohemagglutinin (PHA), lyophilized.
12. Colcemid, lyophilized, or colcemid solution 10 μg/mL.
13. Potassium chloride (KCl).
14. Absolute methanol.
15. Glacial acetic acid.
16. Vortex mixer.

### 1.2 PREPARATION OF WORKING AGENTS

*All manipulations must be done in a sterile environment using sterilized equipment.*

1. Culture medium.
   a. Supplement RPMI-1640 with fetal bovine serum and L-glutamine:
      For every 100 mL RPMI-1640, add
         10 mL fetal bovine serum.
         1 mL L-glutamine.
         1 mL penicillin/streptomycin solution.
   b. Store at 4°C (fridge) for up to 1 month after the addition of L-glutamine in tightly closed glass bottle.
   c. *Instead of RPMI-1640, other medium can be used such as minimum essential medium (MEM) with nonessential amino acids, McCoy 5A, TC199.*
2. Hypotonic solution
   a. For 100 mL of deionized water, add 0.56 g KCl (final concentration 0.075 $M$ KCl).
   b. Sterilize with 0.45-μm pore size Nalgene Filter Flask.
   c. Store at room temperature in small quantities for up to 1 month.
3. Fixative
   a. 3 parts absolute methanol, 75 mL.
      1 part glacial acetic acid, 25 mL.
   b. Prepare fresh for every experiment.
4. Colcemid solution (if colcemid is in lyophilized form).
   a. Dilute with the suggested quantity of sterile deionized water for a final concentration of 10 μg/mL.
   b. Store at 2–5°C for several months.

**Caution:** *Colcemid is a toxic chemical and should be handled with care. Avoid skin or eye contact and inspiration.*

5. PHA solution
    a. Phytohemagglutinin is supplied lyophilized; it should be dissolved in the appropriate amount of sterile distilled water according to the supplier.
    b. The solution can be kept in frozen aliquots for several weeks. Defrost an aliquot just before use.
    c. Do not leave it at room temperature for long.

## 1.3. SPECIMEN: WHOLE PERIPHERAL BLOOD

1. Obtain 10 mL of peripheral blood. Put directly in a heparin green-top tube and mix gently to avoid clotting.
2. Hemolyzed or clotted blood is unsuitable.

## 1.4. PROCEDURE FOR CULTURE SETUP

1. Mix:
    a. 113 mL supplemented medium.
    b. 10 mL whole blood.
    c. 1 mL PHA.
2. Aliquot 10 mL per T25 flask.
3. Leave flasks in vertical position for 72 h in incubator (5% $CO_2$ 37°C) with lids left loose.
4. Agitate by hand once or twice a day.

## 1.5. CELL HARVEST

1. Add 100 µL colcemid to each flask. Leave in the incubator for 20 min.
2. Transfer the contents of each flask into a 15-mL conical tube and centrifuge for 10 min at 400$g$.
3. Discard the supernatant with a Pasteur pipet.
4. Resuspend the pellet in the few drops of remaining fluid by flicking the bottom of the tube with a finger.
5. Add 10 mL of hypotonic solution 0.075 $M$ KCl (room temperature), the first 1.5 mL drop by drop. Mix by flicking the tube to ensure that the pellet is well suspended. Leave it at room temperature for 5–10 min. Observe a drop of the cell suspension under a phase contrast microscope for swelling (lymphocytes getting larger and rounder, nucleoli more prominent).
6. Centrifuge cells at 400$g$ for 10 min. Discard the supernatant except for a few drops.
7. Resuspend the pellet in the remaining fluid by tapping the tube with a finger. Add fresh fixative dropwise in the beginning with constant gentle agitation of the tube. Vortex for a while at the lowest speed. Add more fixative to a total volume of 10 mL and let the tubes sit at room temperature for at least 10 min.*
8. Centrifuge at 400$g$ for 10 min.
9. Discard the supernatant and resuspend the pellet in 10 mL of fixative
10. Perform **steps 8** and **9** another three times.
11. Store cell pellets in fixative at −20°C if not used immediately.

*It is important to get the cell pellet completely broken up. However, any harsh treatment of the cells at this stage may rupture them, resulting in incomplete metaphases on the slides. It is better to avoid passing cell suspension through narrow spaces such as pipets.*

## 1.6. MATERIAL AND EQUIPMENT FOR CHROMOSOME SPREADS

1. Acetone.
2. Concentrated HCl.
3. 95% ethanol.
4. Absolute methanol.
5. Glacial acetic acid.
6. Glass staining dishes.
7. Pasteur pipets.
8. Glass Pasteur pipet.
9. Glass slides, precleaned, Superfrost.
10. Distilled water.
11. Phase microscope.

## 1.7. PREPARATION OF REAGENTS

1. Slide cleaning:
   a. Acetone: use as it is.
   b. Acid/alcohol: equal parts of concentrated HCl and ethanol
   c. Tap water.
2. Fixative: mixture of 3:1 methanol/glacial acetic acid.

## 1.8. SLIDE CLEANING PROCEDURE

1. Clean slides by immersing for 5 min in the following series of solutions: acetone, HCl/ethanol, running tap water.*
2. The slides should be stored in a staining dish with distilled water in the refrigerator until use.

*Alternatively, slide cleaning can be done by passing through a series of 100, 95, 70, and 50% ethanol and then in running tap water for 1 h. Air-dry in the hood before use.*

## 1.9. CHROMOSOME SPREAD PROCEDURE

1. If cell pellet is stored at −20°C, remove and allow to reach room temperature.
2. Remove the supernatant and add 5 mL of fresh fixative. Break up the pellet by bubbling air in the solution with a Pasteur pipet, taking care not to scrape the sides of the conical tube with the pipet.
3. Centrifuge at 400g, discard the supernatant, and add drops of fresh fixative with a Pasteur pipet, agitating gently each time until the solution becomes slightly hazy.*
4. Remove the staining dish with the stored slides from the refrigerator and keep slides in cold water before dropping cells on them.
5. With a glass Pasteur pipet (5¾ inch), draw some drops of chromosome solution. (Alternatively, a P-1000 pipetman can be used arranged at 80 µL.)
6. Remove a slide from the staining dish and let excess water drain by touching the edge of it on a paper towel. Leave only a very thin coat of water on it.
7. Hold the slide frosted end up. Touch the tip of the pipet to the left edge of the slide besides the frosted edge. Let the fluid go onto the slide by drawing the pipet tip along the slide, and slowly changing the angle of the slide toward a vertical position, to distribute the spread evenly.**
8. Let the slide dry at a vertical position (45° angle). Note the drying time. (See critical points.)
9. Examine the slide under a phase microscope (×10 objective) to judge number and quality

of mitoses present. If cell density is too high, add a few drops of the fixative. If cell density is too low, centrifuge and resuspend in less fixative.

10. If pleased with the result, make more slides in the same way, and let them air-dry.
11. Dehydrate slides in a series of 70, 95, and 100% ethanol, 5 min in each, at room temperature. Allow them to air-dry.
12. Store slides in a slide box for 2 weeks at room temperature before hybridization.\*\*\*

*The final volume of fixative will depend on the size of the cell pellet and may need to be adjusted if metaphases are too sparse or many. It is advisable to try a volume of no more than 1 mL.*

\*\**Alternatively, using the same glass Pasteur pipet, drop 1–3 drops of the chromosome preparation on different points across the slide from above. Good results are obtained when breathing on the slide until condensation forms and then dropping the cell suspension from a distance of at least 50 cm.*

\*\*\**If slides are to be used for hybridization at a time shorter than 2 weeks they have to be pretreated in 2× SSC before dehydration in ethanols. This pretreatment artificially ages the chromosomes, making them less sensitive to over-denaturation. Prewarm 40 mL of 2× SSC, pH 7.0, in a Coplin jar to 37°C in a water bath. Place prepared metaphase slides in the jar and incubate for 30 min.*

## 1.10. IMPORTANT CLUES IN METAPHASE PREPARATIONS FOR FISH

Good mitotic index and mitotic quality is very important for successful hybridization. Some parameters during the whole procedure affect dramatically the final metaphase yield.

Colcemid, a mitotic spindle inhibitor, is the first parameter. Longer exposure to colcemid will result in a larger number but more condensed chromosomes. This is not desirable in all cases, in gene mapping, for example.

The second critical point is the hypotonic treatment using a KCl solution. Longer exposure will result in nucleus eruption with metaphase breakdown to individual chromosomes; shorter time will leave cytoplasmic remnants around metaphases, which will prevent chromosomes from spreading and will interfere with hybridization.

For all these reasons it is important to microscope the first prepared chromosome spreads in each experiment to adjust factors to the optimal.

Good preparations should have at least three complete well-spread metaphases per field at 100× magnification. Chromosomes should appear dark gray in color. If they are light gray or very black and refractile, they will not hybridize well. Also, no cytoplasmic debris should be visible around chromosomes. Chromosome quality also depends greatly on spreading technique. Temperature and humidity affect the preparation significantly, since they influence drying time of the fixative. If drying time is too long, metaphases will not spread completely, and chromosomes look refractile and may not retain good morphology after denaturation. If drying is too quick, chromosomes are difficult to see under the microscope due to low contrast and tend to come off the slide during denaturation.

Since making good chromosome spreads is somewhat of an art, it is advisable to practice both methods provided in the previous protocol and select from them. If using the first method, adjust the temperature of water in the staining dish with the slides, to warmer if drying time is too long or colder if drying time is very short.

## 2. METAPHASE ANALYSIS FROM SOLID TUMORS *(48, 49)*

### 2.1. MATERIALS AND EQUIPMENT

1. RPMI-1640.
2. Fetal bovine serum.
3. L-glutamine (200 m*M*).
4. Penicillin/streptomycin solution (10,000 U/mL and 10 mg/mL, respectively).
5. Colcemid lyophilized or colcemid solution 10 µg/mL.
6. T25 type culture flasks.
7. Collagenase II solution (1%).
8. Trypsin.
9. EDTA.
10. Insulin solution (5 µg/mL; optional).
11. Glutathione solution (10 µg/mL; optional).
12. Potassium chloride (KCl).
13. Absolute methanol.
14. Glacial acetic acid.
15. Sterile plastic centrifuge tubes, 15 mL with screw cup.
16. Sterile Pasteur pipets.
17. Inverted microscope.
18. Neubauer plaque for cell counting.

### 2.2. PREPARATION OF WORKING AGENTS

1. Culture medium.
   a. Supplement RPMI-1640 with fetal bovine serum and L-glutamine:
      For every 100 ML RPMI-1640, add
      20 mL fetal bovine serum.
      1.3 mL L-glutamine.
      1.3 mL penicillin/streptomycin solution.
2. Trypsin-EDTA solution:
   a. Trypsin (1:250) 0.25 g, EDTA 0.038 g, PBS-CMF 100 mL*
   b. Sterilize with 0.45 µm pore Nalgene filter flask. Store frozen.
      *PBS-CMF, phosphate buffered saline, calcium and magnesium free.*
3. Hypotonic solution (KCl) prepared as in **Protocol 1.2.**
4. Fixative (methanol/acetic acid) prepared as in **Protocol 1.2.**
5. Collagenase solution (1%)
   a. Collagenase II, 10 mg; culture medium, 10 mL.
   b. Sterilize by filtering through a syringe, with a disposable 0.2-µm pore size Nalgene Filter, and store frozen in small aliquots for up to 3 months.
6. Insulin solution*: 5 µg/mL insulin in culture medium or phosphate-buffered saline [PBS].
   a. Sterilize by filtering through a syringe, with a disposable 0.2-µm pore size Nalgene Filter, and store at 5°C. (To keep for a long time, prepare the solution in PBS and store frozen after sterilization.)
   b. Add 1 mL of the prepared insulin solution (5 µg/mL) per 100 mL of medium.

7. Glutathione solution: culture medium or PBS.
    a. Sterilize by filtering through a syringe, with a disposable 0.2-μm pore size Nalgene Filter, and store at 5°C. (To keep for a long time, prepare the solution in PBS and store frozen after sterilization.)
    b. Add 1 mL of the prepared glutathione solution (10 μg/mL) per 100 mL of medium.

*Insulin and glutathione use is optional. They are believed to propagate cell growth.*


## 2.3. SPECIMEN

Piece of tumor tissue is obtained and placed aseptically in 5–10 mL of culture medium, or sterile saline, or PBS in a well-sealed sterile container. The specimen should be brought to the lab as soon as possible and should be kept on ice if transportation takes long.


## 2.4. CULTURE SETUP

1. Wash the tissue sample two to three times using sterile PBS in a sterile container.
2. Place tissue sample in a sterile Petri dish without any liquid. Dissociate and discard with a sterile blade and forceps any blood clots or necrotic or normal tissue. Cut a small piece (of no more than 0.5 g) of healthy tumor tissue and mince it in tiny pieces (1–2 mm). Avoid tissue drying up by using drops of PBS.
3. Place the minced tissue in a centrifuge tube with at least 5 mL of collagenase solution (1%). Triturate the specimen by pipetting several times (at least 30) in sterile conditions using a sterile Pasteur pipet.
4. Add another 5 mL of fresh collagenase solution (1%) to the medium in the tube and incubate with the tube cup loosened, for 2–3 h. (Hard tissues may need longer incubation, up to 16 h).
5. Triturate the specimen again.
6. Centrifuge at $400g$ (800 rpm) for 10 min. Discard the supernatant. Wash cell pellet with medium and centrifuge again.
7. Discard the supernatant. Add 1 mL of medium enriched with insulin and glutathione and resuspend again well.
8. Expose a small sample of the cell solution to Trypan blue stain, to count cell number and viability using the Neubauer plaque (approximate cell density $2.5–3 \times 10^6$ cell/mL).
9. Make up the cell suspension to 10 mL and distribute equally into 2–3 T25 flasks
10. Incubate horizontally, with cups loosened for 24 h.
11. Discard medium with unattached cells the next day and put fresh medium (no more than 5 mL per flask.)
12. Observe for cell growth (inverted microscope) after the fourth day and change culture medium three times a week.
13. Cultures are usually ready for harvest (large proliferating colonies) after 7–10 days.


## 2.5. CELL HARVEST

1. Expose the selected flask to colcemid (0.01 μg/mL) for 15 h (overnight). Discard the medium, leaving the cell layer undisturbed.
2. Follow **steps 2–7** in **Protocol 1.5**.

# 3. INTERPHASE ANALYSIS ON CELL IMPRINTS— CYTOLOGY SPECIMENS

## 3.1. SPECIMEN PREPARATION—FIXATION

### 3.1.1. Disaggregation of Nuclei from Fresh and Frozen Tissues
*For discussion, see ref. 9.*

1. Cut small slices (2 × 5 × 5 mm) of fresh or frozen tissue using a scalpel blade and mince finely in a Petri dish.
2. Add 1 mL of 0.5% pepsin in 0.9% NaCl (pH 1.5) to the dish, transfer the mix to a 15 mL centrifuge tube, and incubate in a 37°C water bath for 15–30 min, with vortexing every 5 min.
3. Subsequently add 14 mL PBS and collect the nuclei by centrifugation.
4. Discard the supernatant and resuspend the nuclear pellet in the residual supernatant.
5. Apply a small drop (approx 25 µL) of the cell suspension to a glass slide and leave to air-dry.
6. Fix the specimens in methanol/acetic acid (3:1) for 20 min, dry, and store at −20°C until the hybridization procedure.

### 3.1.2. Touch Preparations of Fresh and Frozen Tissues

1. Touch a small slice of fresh or frozen tissue several times against a siliconized (Superfrost +) glass slide.
2. Fix in methanol/acetic acid (3:1) for 20 min or chilled acetone for 15 min and leave to air-dry.
3. Store slides at −20°C until the hybridization procedure.

### 3.1.3. Fine Needle Aspirates

1. Prepare smears from fine needle aspirates on siliconized or charged glass slides and leave them to air-dry.
2. Fix specimens in methanol/acetic acid (3:1) for 20 min or chilled acetone for 15 min and leave them to air-dry.
3. Store slides at −20°C until processing.

### 3.1.4. Cytologic Specimens (Pleural Fluid, Peritoneal Fluid, and others)

1. Collect cytologic specimens from pleural, peritoneal, or other biologic fluids in tubes and centrifuge.
2. Resuspend the cell pellet in PBS and centrifuge again.
3. Apply a drop of the cell suspension to a glass slide and leave to air-dry.
4. Fix the specimens in methanol/acetic acid (3:1) for 20 min, air-dry, and store at −20°C until hybridization.

## 3.2. FISH TECHNIQUE

### 3.2.1. Specimen Pretreatment

1. Prewarm 40 mL of 2× SSC/0.5% NP-40 in a Coplin jar in a 37°C water bath.
2. Place the slides in a Coplin jar and incubate for 30 min. Instead of NP-40, igepal may be used.
3. Dehydrate slides in 70, 80, and 95% ethanol at room temperature for 2 min each. Air-dry slides.

### 3.2.2. Specimen Denaturation

1. Prewarm 40 mL of denaturation solution (2× SSC/70% formamide), pH 7.0, in a glass Coplin jar at a 72°–74°C water bath.
2. Immerse slides in prewarmed denaturation solution for 2 min. Do not agitate. *Temperature must be verified by placing a clear thermometer into the Coplin jar. Overdenaturated slides may not hybridize or counterstain well.*
3. Dehydrate slides in cold (−20°C) 70, 80, and 95% ethanol for 2 min each. *The slides must be used the same day they are denatured.*

### 3.2.3. Probe Preparation

1. Prewarm DNA probe at 37°C for 5 min. Vortex gently and centrifuge briefly to collect contents in the bottom of the tube.
2. Dilute probe in the hybridization mixture. For satellite probes, the suggested hybridization mixture is composed of 2× SSC/65% formamide and 10% dextran sulfate, whereas for unique-sequence probes, it is composed of 2× SSC/50% formamide and 10% dextran sulfate. The probe is diluted at a final concentration 1–3 ng/μL.
3. Denature DNA probe-hybridization solution at 72°–74°C for 5 min in a water bath.
4. Centrifuge the probe briefly and keep in an ice bath until hybridization.

### 3.2.4. Hybridization

1. Apply probe-hybridization mixture on each specimen (20 μL is adequate for an area under a 22 × 22-mm cover slip).
2. Cover with a glass cover slip. Seal the edges with cover slip sealant if necessary.
3. Place slides within a prewarmed humidified chamber and incubate specimens at 37°C. Thirty minutes of hybridization time is adequate for satellite probes, but overnight hybridization is preferred for all other probes.
   *Hybridization temperature should be kept strictly at 37°C to avoid cross-hybridization to other satellite sequences (<37°C) or decline of signal intensity (>37°C).*

### 3.2.5. Posthybridization Washing

1. Prewarm 40 mL of 0.5× SSC, pH 7.0, in a glass Coplin jar in a 72°–74°C water bath.
2. Carefully remove the cover slip with forceps and immerse slides in the prewarmed 0.5× SSC for 5 min.
   *Wash solution temperature and time are crucial. Low temperature (<72°C) and/or shorter wash time will result in nonspecific hybridization signal. Higher temperature (>74°C) and/ or longer washing time will result in reduction of signal intensity.*
3. Transfer specimens to PBST (PBS/0.1% Tween) and wash for 2× 3 min.

### 3.2.6. Detection of the Hybridization Reaction

Detection of the hybridization reaction depends on the type of probe labeling. Fluoro-chrome-labeled DNA probes do not require further processing except for nuclear coun-terstaining (proceed to **Protocol 3.2.9.**) Fluorescence detection of biotin- or digoxigenin-labeled DNA probes is performed as follows:

### 3.2.7. Biotin-Labeled Probes

1. Incubate slides with streptavidin-FITC or streptavidin-TR 5 μg/mL in PBST at 37°C for 20 min in a humidified incubation chamber.
2. Wash slides in PBST 3× 3 min.

**Table 2**
**FISH Nuclear Counterstain According to Probe Labeling (Oncor)**

| Probe label | Counterstain (μg/mL) |
|---|---|
| Fluorescein | Propidium iodide (0.3) |
| Texas red | DAPI (0.1) |
| Fluorescein and Texas red | DAPI (0.1) |

### 3.2.8. Digoxigenin-Labeled Probes

1. Incubate slides with FITC-labeled anti-digoxigenin or TR-labeled anti-digoxigenin at 37°C for 20 min in a humidified incubation chamber (Oncor, ready to use).
2. Wash slides in PBST 3× 3 min.

### 3.2.9. Counterstain

Apply to the specimen 10–15 μL of the appropriate counterstain (**Table 2**), and cover with a glass cover slip.

### 3.2.10. Evaluation of FISH Signals

1. Score only single nonoverlapping nuclei. Avoid large clusters of cells, as it is difficult to assign the signals.
   *DNA denaturation and probe penetration is hampered in the cell clusters.*
2. Count only signals completely separated from each other (numerical abnormalities)

## 4. INTERPHASE ANALYSIS OF FORMALIN-FIXED PARAFFIN-EMBEDDED TISSUES

### 4.1. SAMPLE PREPARATION-DEPARAFFINIZATION

1. Cut 4–5-μm sections from paraffin blocks and place them on a siliconized (or superfrost +) glass slide.
2. Place the slides in an oven at 37°C for 2–3 h to dry.
3. Bake slides at 65°–70°C for 1–2 h.
4. Deparaffinize slides in xylene for 2× 15 min.
5. Rehydrate slides in graded ethanols (100, 95, 70%) 5 min each and allow them to air-dry.

### 4.2. TISSUE PRETREATMENT

1. Incubate slides in tissue pretreatment solution (provided by Oncor and prepared according to manufacturer's suggestions) preheated to 45°C for 15–20 min.
2. Wash slides twice in 2× SSC, 5 min each.
3. Incubate slides in proteinase K (250 μg/mL, preheated to 45°C) at 45°C for 15–20 min, depending on tissue type.
4. Wash slides three times in 2× SSC, for 5 min each.
5. Dehydrate slides in 70, 80, and 95% ethanol, 1 min each. Allow slides to air-dry.

*After this step, it is possible to evaluate the adequacy of pretreatment, by staining nuclei with propidium iodide/antifade, cover-slipping, and examining slides by a fluorescence microscope. In adequately digested tissue, the nuclei show bright orange/ red fluorescence with clearly defined nuclear borders and little or no background fluorescence. Underdigested tissue shows persistent green autofluorescence and poor propidium iodide staining; the proteinase K step should be repeated. Tissue slides showing overdigestion, that is, cells displaying loss of nuclear borders, should be discarded.*

## 4.3. FISH TECHNIQUE

### 4.3.1. Probe Preparation

See Protocols subheading 3.2.3.

### 4.3.2. Hybridization

Place 10–20 µL of probe on each slide (depending on tissue size) and cover with a glass cover slip.

1. Denature probe and tissue DNA simultaneously by placing the slides on a hot plate or oven at 67–69°C for 5 min.
2. Incubate slides in a humidified chamber at 37°C for 2–16 h (satellite probes) or 16 h (unique-sequence probes).

### 4.3.3. Posthybridization Wash

See Protocols subheading 3.2.5.

### 4.3.4. Detection of the Hybridization Reaction and Counterstaining

See Protocols subheadings 3.2.6. and 3.2.9.

# 7

# Tyramide Amplification Methods for *In Situ* Hybridization

## John L. Frater, MD and Raymond R. Tubbs, DO

## INTRODUCTION

The relative insensitivity of *in situ* hybridization (ISH) has been cited as an important limiting factor in the potential applications of this technique *(1)*. Therefore, a major initiative on the part of researchers who use molecular biologic techniques is the improvement of sensitivity of ISH. There are two means of achieving this: the amplification of target sequences and the amplification of signal *(2)*. The former encompasses such techniques as *in situ* polymerase chain reaction (PCR) and *in situ* reverse transcriptase (RT)-PCR, and is not discussed in this review. The latter includes catalyzed reporter deposition, a major component of which is the use of labeled tyramide as a means of signal amplification. This technique, which has subsequently been adapted for use in ISH from its initial immunohistochemical applications, is a powerful tool for increasing sensitivity in molecular biologic testing. Through the use of tyramide signal amplification™ (TSA™), the theoretical limits of ISH can be approached.

As is the case with other highly sensitive assay techniques, the increased sensitivity of TSA has inherent problems and limitations. With increased signal comes increased noise, which creates inherent difficulties. Furthermore, the technique is in a continuing state of evolution, so any review of the topic finds itself almost immediately behind the times as new protocols are continuously developed. Because of the rapid and ongoing development in this exciting area, the dissemination of new protocols via electronic means, e.g., the Internet, plays an important role. It is highly recommended that the interested reader consult the website maintained by Dr. Gerhard Hacker, *http://www.sbg.ac.at/kgg/protocols*, for information on new protocols as they arise.

The purpose of this chapter is to discuss TSA as applied to ISH. After a brief discussion of the historical development of TSA, a summary of the theoretical basis of TSA and a partial description of protocols are given. The limitations of TSA, as well as a discussion of practical troubleshooting techniques, follows.

## HISTORICAL BACKGROUND

TSA (NEN Life Science, Boston, MA) is a proprietary term protected by U.S. patents 5,731,158, 5,583,001, and 5,196,306 and foreign equivalents for catalyzed reporter

From: *Morphology Methods: Cell and Molecular Biology Techniques*
Edited by: R. V. Lloyd © Humana Press, Totowa, NJ

deposition (CARD) used to amplify an ISH signal. CARD was originally described in 1989 by Bobrow et al. *(3)* as a novel amplification method for solid phase immunoassays. Amplification is achieved by the deposition of biotin-labeled phenols at the site of interest in a reaction catalyzed by the analyte-dependent reporter enzyme horseradish peroxidase. Bobrow et al. *(4–6)* subsequently modified this technique for use in other bioassays. Although the term CARD strictly refers to the technique using any source of biotin-labeled phenols, this particular CARD technique uses tyramide as the source of the phenol moiety. Therefore, for practical purposes, all CARD techniques currently use TSA.

The subsequent modification of TSA for use in fluorescence *in situ* hybridization (FISH) was reported by Raap et al. in 1995 *(7)*. TSA was described by Raap as a two-layered antibody technique, the first layer being a peroxidase-conjugated anti-hapten immunoglobulin/streptavidin. The peroxidase substrate was tyramide. The second layer consisted of (strept)avidin conjugates and was used to detect the biotin molecules deposited in the first layer. The authors used several different tyramide conjugates, which reacted with the peroxidase, resulting in the creation of a highly reactive intermediate substance that aided and enhanced the localization of fluorochromes. Modifications of this original protocol provide the basis for all CARD methods using TSA. Other authors have described ELISA-based *(4–6)* and immunohistochemical methods *(8)* using TSA.

## THEORY

All protocols currently use the CARD method initially proposed by Bobrow and the TSA™ modification described by Raap et al. in 1995 *(3,7)*. The exact mechanism of TSA is not entirely known, but it appears to involve the following general principles, described in significant detail in a number of cogent reviews *(2,9,10)*. The object of TSA is to increase the sensitivity of ISH by an increase in detectable signal.

As mentioned in the previous section, TSA is a modification of Bobrow's CARD technique. Briefly, the TSA process consists of the addition of a fluorochrome- or hapten-conjugated tyramide to a region of RNA or DNA altered by a horseradish peroxidase-catalyzed chemical reaction. The tyramide intermediate compound is highly reactive and binds to certain amino acids lying in close proximity to the horseradish peroxidase. A basic understanding of the chemistry of the amino acid tyramine and its derivatives is helpful to understand this pivotal step in the TSA process. From the standpoint of TSA, the most important part of the tyramine ([4-(2-aminoethyl) phenol]) molecule is its phenol group (**Fig. 1**). It is thought that horseradish peroxidase acts on this site to create an unstable, highly reactive intermediate that binds the amino acid binding sites. The tyramides deposited in this fashion are then detected using fluorescence or a chromogenic method if a hapten-conjugated tyramide has been used (**Fig. 2**).

The end result of this process is an improvement in signal detection and hence increased sensitivity using TSA. Macechko et al. *(11)* demonstrated this, comparing immunologic amplification and TSA using a 40-kb cosmid probe and a 15-kb phage probe. A single fluorochrome-conjugated antibody layer was only weakly detected, whereas subsequent use of a second unlabeled secondary antibody and a third layer consisting of an antibody conjugated to the same fluorochrome resulted in a significantly enhanced signal. However, after a 1:10 dilution of the probes, the immunologically

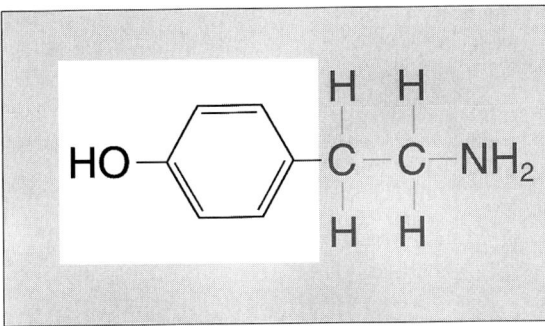

**Fig. 1.** The tyramine ([4-(2-aminoethyl) phenol]) molecule. It is postulated that the highly reactive phenol moiety (shaded region) of tyramide intermediates generated by the tyramide signal amplification process binds to amino acids in close proximity to the horseradish peroxidase molecules (see text).

amplified signal was lost, whereas it was still detectable using TSA. Thus it is possible to detect smaller targets and use less probe reagent using TSA compared with immunologic amplification methods *(11)*.

In addition to comparing the sensitivities of ISH with and without TSA, several authors have shown that different tyramide conjugates confer different sensitivities using ISH. The original TSA report of Raap et al. *(7)* described the use of four different tyramine conjugates: biotin, fluorescein, tetramethylrhodamine, and aminomethylcoumarin acetic acid. In 1997 van Gijlswijk et al. *(12)* compared the performance of several different tyramide conjugates using a middle repetitive specific probe (28S rRNA) and a single-copy gene-specific cosmid probe with a 30–40 kb target size. Nearly all conjugates examined using these probes were easily visualized. The best performing conjugates used BODIPY FL and rhodamine, although nearly all conjugates could be visualized. Only two conjugates had suboptimal results (defined as decreased signal compared with nonconjugated ISH), X-rhodamine and Texas Red. They noted that these two compounds differed from the other conjugates in having an additional sulfonic acid moiety, which conveyed an additional negative charge to these conjugates, possibly impacting performance. Additionally, the authors demonstrated that TSA could be adapted for multicolor FISH. The best results using multiple rounds of hybridization alternating with tyramide detection were obtained using 7-hydroxycoumarin-, fluorescein-, and rhodamine-tyramide conjugates sequentially. Using this approach, intermittent hydrogen peroxidase inactivation was unnecessary *(12)*.

Current estimates of the limits of detection sensitivity with ISH using tissue sections are approximately 10–20 mRNA/viral DNA copies per cell *(13)*. TSA markedly increases sensitivity, allowing detection of single RNA or DNA copies. However, the technical limitations, including the need for relatively large probes and background tissues, may impose practical constraints on the sensitivity of the method. However, TSA does appear to improve ISH sensitivity substantially, and when performed by individuals familiar with molecular techniques, it can approximate the theoretical claims regarding improved sensitivity, including single-copy gene detection.

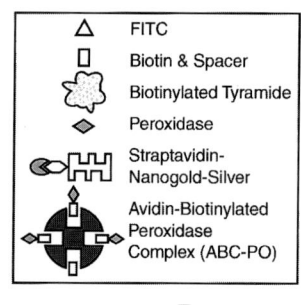

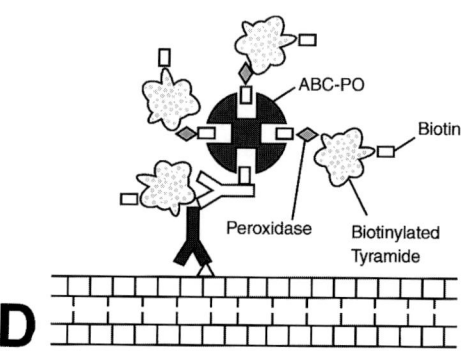

**Biotinylated Tyramide**

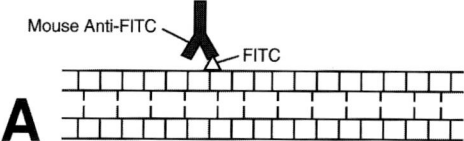

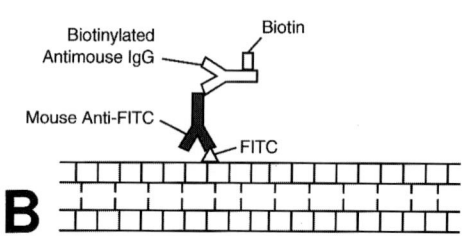

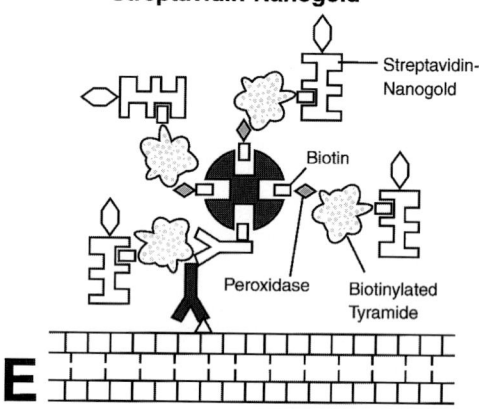

**Streptavidin-Nanogold**

**Avidin-Biotinylated Peroxidase Complex (ABC-PO)**

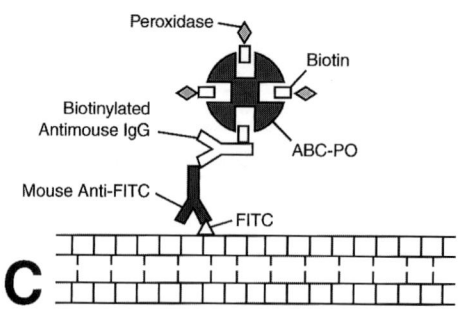

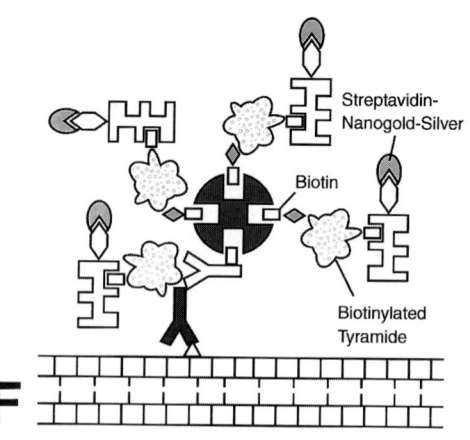

**Autometallography**

## LIMITATIONS AND CAVEATS

The practical limitations of TSA are due to a number of factors, including issues regarding visual resolution and practical considerations common to many molecular biologic techniques. A number of authors have commented on the decreased resolution of this method compared with conventional ISH. This appears to be a result of the diffusion of highly reactive intermediate substances during the course of the reaction. To minimize the loss of resolution, they recommend shortening the tyramide reaction time or tapering probe and antibody concentrations *(7,9,14)*. Others recommend adding normal horse serum, unlabeled tyramide, or polymeric substances to improve resolution *(11,15,16)*.

Because of its high sensitivity, TSA has the potential to amplify nonspecific background signal, which may result in an unfavorable signal/noise ratio. To minimize this problem, one author has recommended that routine optimization be performed for each experiment using kits that are commercially available from a number of sources, including NEN Life Science Products and Dako (Glostrup, Denmark) *(17)*. Alternatively, Speel et al. *(18,19)* have proposed a method of signal optimization with CARD signal amplification using labeled tyramides in an appropriate buffer.

The issue of endogenous peroxidase and the elimination of its interference with the TSA reaction is analogous to the problem of endogenous peroxidases in immunohistochemistry. Endogenous peroxidases in human tissue are potent enough to catalyze the TSA reaction. To avoid this unwanted reaction, the endogenous peroxidases must be blocked or quenched. Because TSA is performed after the ISH procedure, one author has stated that TSA itself needs no treatment for endogenous peroxidases in addition to that used for ISH *(17)*. This point is not unanimously agreed on, and two other authors recommend pretreatment of the sample with a solution of 3% $H_2O_2$ in methanol prior to performing the TSA *(9,20)*.

Other issues involving the use of TSA apply to all molecular biologic techniques. For instance, the use of appropriate positive and negative controls is obligatory *(9)*.

When attempting to detect multiple signals simultaneously, a related problem must be overcome. In the detection of multiple targets using TSA, a sequential application of two or three rounds of labeling and TSA are performed. It is important to inactivate

---

◀ **Fig. 2.** Diagrammatic representation of TSA and related techniques. (**A**) Hybridization of an FITC-labeled probe to the DNA or RNA analyte strand is followed (**B**) by the addition of biotin-labeled anti-FITC antibody. (**C**) Then, the avidin-biotinylated peroxidase complex is added. (**D**) The peroxidase associated with this complex is the catalyst for the addition of tyramide, resulting in the production of a biotinylated tyramide that is closely associated with the original probe. Despite the lack of a distinct antigen-antibody reaction, this can result in a fairly localized signal if performed properly. This biotinylated reporter can act as the catalyzed reporter; at this point, one may proceed to the various detection techniques or, alternatively, perform additional manipulation of the site of interest using autometallographic techniques. (**E**) As a refinement of CARD, autometallography is performed in the following fashion. After the addition of the biotinylated tyramide conjugate, the biotin is bound to streptavidin molecules that have covalently bound nanogold moieties. (**F**) Following this step, the nanogold acts as a nidus for the deposition of metallic silver, the detection of which may result in improved sensitivity.

the remaining peroxidase completely between steps. To achieve this, mild acid treatment with 0.01 *N* HCl may be used *(18,21,22)*. If this step is not performed, repeated deposition of the tyramide conjugate at the prior target site could result.

Despite the limitations of TSA, it appears to be superior to techniques such as *in situ* PCR, which has considerably greater practical limitations compared with TSA. *In situ* PCR is an example of a target amplification procedure in which pretreated cells are permeabilized to allow entrance of primers, nucleotides, and polymerase enzymes. The amplified products are detected using ISH techniques. This technique has been plagued with difficulties, including low amplification efficiency and poor reproducibility. Because of these serious limitations, this technique has been of only limited use in diagnostic pathology *(9,17)*. Additionally, the TSA product can be detected using techniques such as FISH and brightfield ISH (BRISH) *(17)*.

## POTENTIAL USES

A recent review *(17)* has summarized the potential uses of TSA. Because of the markedly increased sensitivity that can be achieved using this technique, there are numerous possibilities for TSA combined with ISH, including those listed below.

### Detection of Repetitive and Single-Copy (Limit 1–5 kb) DNA Sequences in Cell Preparations

Kerstens et al. *(14)* first demonstrated this use for TSA in preparations of normal human peripheral blood smears with labeling of chromosomes 3 (p$\alpha$3.5) and 12 (p$\alpha$ 12H8) with centromeric DNA probes. Hu$\alpha$C on chromosome 14q32 and Bcl-2 on chromosome 18q21 were labeled with biotin-14-dATP. The size range of the probes varied from 2.8 (Bcl-2) to 16 kb (Hu$\alpha$c). After TSH, detection of the resultant signals with both FISH and BRISH was successfully performed. Notably, the increased signal obviated the need for a CCD camera or signal enhancement. TSA was also successfully performed on formalin-fixed, paraffin-embedded bladder carcinoma cases using the same probes and detection techniques *(14)*.

### Detection of up to Three Different DNA Sequences Simultaneously in Cell Preparations

Initially, the use of TSA in the simultaneous evaluation of multiple signals was limited by the relatively small number of labeled tyramides described in the literature. This situation was alleviated by Hopman et al. *(10)*, who described synthesis techniques for the synthesis of numerous tyramide conjugates, including conjugates of biotin, digoxigenin, trinitropenyl, and fluorochrome. These authors compared the performance of these conjugates with a preparation of human transitional cell carcinoma cells and methanol/acetic acid-fixed lymphocyte metaphase spreads. DNA probes for chromosomes 1q12, 1q36, 1q42-43, and 4p16 labeled with biotin-11-dUTP, digoxigenin-11-dUTP, or fluorescein-12-dUTP were used. After TSA with amino coumarine acid-tyramide, rhodamine-tyramide, and fluorescein-tyramide, FISH could successfully discern three discrete signals for the three chromosome 1 probes using the TCC cell line *(10,17)*.

### Detection of Low- or Single-Copy DNA sequences of Intracellular Organisms or Viruses such as Human Papillomavirus in Tissue and Cell Preparations

Using TSA with a streptavidin-Nanogold and silver autometallography detection system. Zehbe et al. *(20)* were able to detect single human papillomavirus (HPV)-16 DNA copies in the SiHa cell line. Cheung et al. *(23)* reported the results of a comparison of three different methods for the detection of HPV in 61 formalin-fixed, paraffin-embedded cervical carcinoma specimens. Using a biotinylated probe for HPV serotypes 16/18, 65.6% of cases were positive by TSA, compared with 39.3% by conventional streptavidin-biotin-peroxidase ISH and 44.3% by streptavidin-Nanogold-silver ISH. Moreover, in examining the staining pattern of the TSA method, it was apparent that TSA was capable of detecting single copies of HPV-16/18 DNA *(23–25)*. This was first reported for FISH by Adler et al. *(26)*, compared TSA methodologies for the detection of HPV-16 in two different cell lines. He noted high sensitivity after optimization of TSA. Probe size appeared to be a factor: the number of positive cells was approximately proportional to the size of the probe used. A 293-bp probe appeared suboptimal, whereas probes $\geq$619 bp gave better results. The sensitivity of the test was excellent: detection of two copies of HPV-16 DNA was possible in one cell line (SiHa). Use of the cyanine 3 tyramide was recommended for maximization of sensitivity and best signal resolution and was particularly useful in detection of HPV DNA in the SiHa cell line, which had a low copy number *(26)* **(Fig. 3)**.

### Detection of rRNA and mRNA Within Cells

TSA has been of proven use in detection of rRNA and mRNA in systems. Reed et al. *(27)* developed a probe directed against a gene encoding a cyclin D-1 analog in human herpes virus 8 (HHV8). Because the HHV8 levels in cases of Kaposi's sarcoma are presumably quite low, highly sensitive molecular methods such as PCR-ISH and PCR were necessary to detect the virus in biopsy specimens. As discussed previously, PCR-ISH is a problematic system, and the infected cell type cannot be localized using PCR. Using TSA with a biotin-conjugated tyramide and the aforementioned probe, Reed et al. *(27)* were able to localize HHV8 successfully to a specific KS cell type. Since the detected signal diminished markedly after treatment with RNAse, the probe appears to be directed toward the gene mRNA.

### Other Uses

Although this chapter is limited to a discussion of ISH methods using TSA, it should be noted that there are several other uses for TSA in diagnostic and experimental pathology. Adapting biotinylated tyramides for use in immunohistochemistry, Adams *(8)* reported a 1000-fold increase in sensitivity in tissue compared with specimens prepared by the conventional avidin-biotinylated complex technique. Others have reported much more modest increases in sensitivity using TSA and immunohistochemistry, on the order of 5–50% compared with conventionally treated specimens *(26,28–30)*. The interested reader is directed to a recent review of these subjects *(17)*.

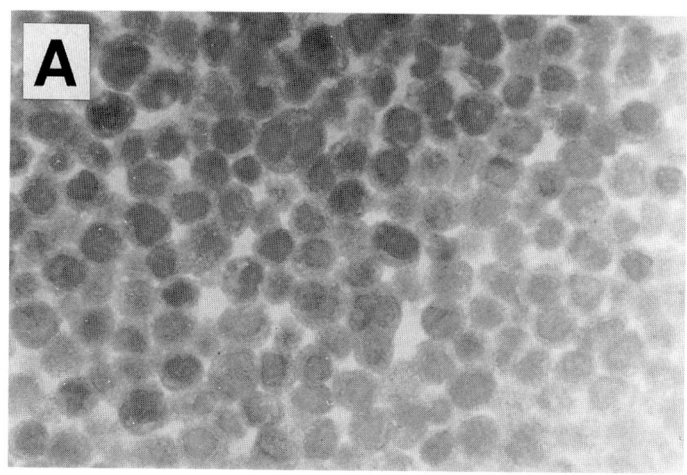

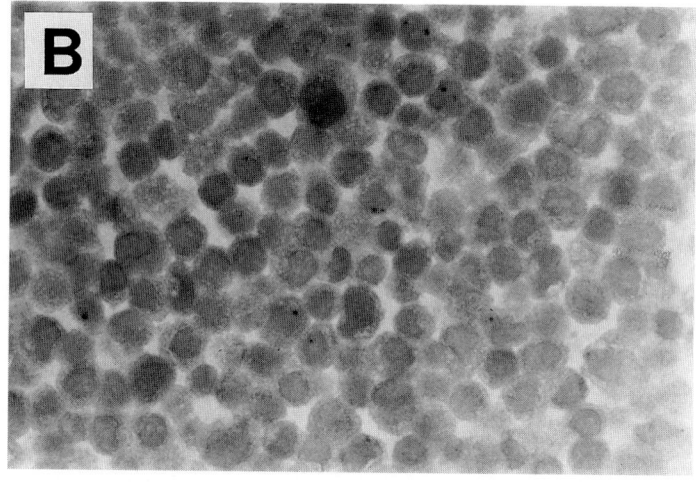

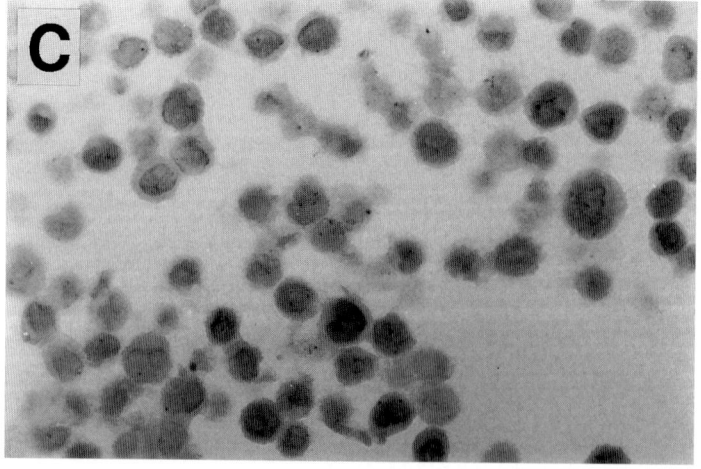

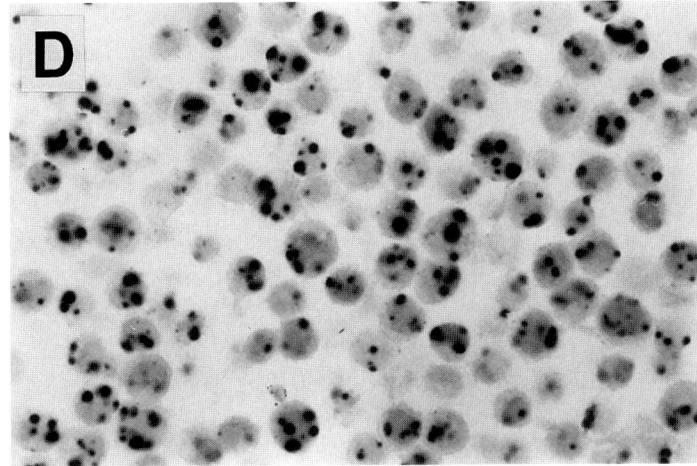

**Fig. 3.** Signal amplification achieved with CARD. (**A**) SiHa cells, known to contain 1–2 copies of HPV-16 per cell, conventional chromogenic ISH. HPV is undetected (DAPI/$H_2O_2$ detection). (**B**) SiHa cells, CARD ISH procedure; 1–2 copies of HPV identified (DAPI/$H_2O_2$ detection). (**C**) CaSki cells, conventional chromogenic ISH (DAPI/$H_2O_2$ detection). (**D**) CaSki cells, CARD ISH. Numerous and in part confluent signals are identified (DAPI/$H_2O_2$ detection).

An interesting new development that may further increase TSA sensitivity involves its use in conjunction with branched DNA (bDNA) ISH. bDNA, (also called oligodeoxy-ribonucleotide) is another method whereby signal amplification may be improved. bDNA has been tested most extensively in quantitative measurement of viral load in chronic hepatitis in the solution phase. After nucleic acid denaturation and hybridization of probes to target and solid phase performed in the usual fashion, the bDNA multimer is then hybridized to the probe. After the incorporation of alkaline phosphatase molecules into the bDNA, the system acts as a signal amplifier. This method has been successful in detection of viral nuclei acid with a detection limit of 1–3 plaque-forming units (pfu). This method has been reported to be highly reproducible and easy for experienced lab personnel to perform *(31–33)*. In combination with TSA, further increases in sensitivity are theoretically possible. Because of the reported reproducibility and ease of handling of bDNA, this may be feasible with no appreciable loss of specificity.

Flow cytometry is another modality that has been reported to benefit from TSA technology. Kaplan and Smith *(34)* compared fluorescence intensity of conventionally and TSA-treated samples from the Jurkat malignant T-cell line. They reported a 10–100-fold increase in amplification using TSA for a number of antibodies (MHC class I, CD5, CD3, CD4, CD6, CD7, CD34, CD45, MHC class II, Fas ligand, and phosphatidylserine).

TSA has been adapted for a number of sophisticated clinical and research uses, including high-resolution mapping of DNA sequences, multiple color techniques for simultaneous detection of multiple probes, and comparative genomic hybridization *(34a)*. Combined FISH/immunohistochemistry with TSA has been described, which allows simultaneous detection of glial cell line-derived neurotropic factor 2 mRNA and localization of its co-receptor Ret *(35)*. Comparative genomic hybridization may be particularly useful in the assessment of tumors. Another study, looking at the assessment of numerical chromosomal aberrations in pituitary adenomas, reveals another advantage of TSA. Results of FISH and TSA of a group of these tumors were not significantly different. However, unlike FISH, cases treated by TSA could be stored for long periods, a particularly useful property when assessing clinical specimens *(35a)*.

TSA was adapted for ultrastructural studies by Schöfer et al. in 1997 *(36)*. Using a gold-conjugated antibody similar to that described by Hainfeld *(37)* to detect the tyramide-conjugated complex, the technique was used for electron microscopic detection of spermatid DNA, skeletal muscle actin, and rDNA after ISH.

## REFERENCES

1. Tubbs, R. R., Pettay, J., Grogan, T., et al. (2000) Super sensitive *in situ* hybridization by tyramide signal amplification (TSA) and nanogold®-silver staining: the contribution of autometallography and catalyzed reporter deposition (CARD) to the rejuvenation of *in situ* hybridization, in *Gold and Silver in Molecular Morphology*, Eaton, Natick, M. A., in press.
2. Komminoth, P. and Werner, M. (1997) Target and signal amplification: approaches to increase the sensitivity of *in situ* hybridization. *Histochem. Cell Biol.* **108**, 325–333.
3. Bobrow, M. N., Harris, T. D., Shaughnessy, K. J., and Litt, G. J. (1989) Catalyzed reporter deposition, a novel method of signal amplification. Application to immunoassays. *J. Immunol. Methods* **125**, 279–285.
4. Bobrow, M. N., Shaughnessy, K. J., and Litt, G. J. (1991) Catalyzed reporter deposition, a novel method of signal amplification. II. Application to membrane immunoassays. *J. Immunol. Methods* **VOL?**, 103–112.
5. Bobrow, M. N., Litt, G. J., Shaughnessy, K. J., Mayer, P. C., and Conlon, J. (1992) The use of catalyzed reporter deposition as a means of signal amplification in a variety of formats. *J. Immunol. Methods* **150**, 145–149.
6. Bobrow, M. N., Litt, G. J. (1993) Method for the detection or quantitation of an analyte using an analyte dependent enzyme activation system. United States Patent Number 5196306.
7. Raap, A. K., van de Corput, M. P. C., Vervenne, R. A. W., van Gijlswijk, R. P. M., Tanke, H. J., and Wiegant, J. (1995) Ultra-sensitive FISH using peroxidase-mediated deposition of biotin- or fluorochrome tyramides. *Hum. Mol. Genet.* **4**, 529–534.
8. Adams, J. C. (1992) Biotin amplification of biotin and horseradish peroxidase signals in histochemical stains. *J. Histochem. Cytochem.* **40**, 1457–1463.
9. Totos, G., Tbakhi, A., Hauser-Kronberger, C., Tubbs, R. R. (1997) Catalyzed reporter deposition: a new era in molecular and immunomorphology—nanogold-silver staining and colorometric detection and protocols. *Cell Vision* **4**, 433–442.
10. Hopman, A. H. N., Ramaekers, F. C. S., Speel, E. J. M. (1998) Rapid synthesis of biotin-, digoxigenin-, trinitrophenyl-, and fluorochrome-labeled tyramides and their application for *in situ* hybridization using CARD amplification. *J. Histochem. Cytochem.* **46**, 771–777.
11. Macechko, P. T., Krueger, L., Hirsch, B., Erlandsen, S. R. (1997) Comparison of immunologic amplification vs. enzymatic deposition of fluorochrome-conjugated tyramide as detection systems for FISH. *J. Histochem. Cytochem.* **45**, 359–363.
12. Van Gijlswijk, R. P. M., Zijlmans, H. J. M. A. A., et al. (1997) Fluorescence-labeled tyramides: use in immunocytochemistry and fluorescence *in situ* hybridization. *J. Histochem. Cytochem.* **45**, 375–382.
13. Speel, E. J. M., Seremaslani, P., Roth, J., Hopman, A. H. N., Komminoth, P. (1998) Improved mRNA *in situ* hybridization on formaldehyde-fixed and paraffin-embedded tissue using signal amplification with different haptenized tyramides. *Histochem. Cell Biol.* **110**, 571–577.
14. Kerstens, H. M. J., Poddighe, P. J., Hanselaar, G. J. M. (1995) A novel *in situ* hybridization

signal amplification method based on the deposition of biotinylated tyramine. *J. Histochem. Cytochem.* **43**, 347–352.

15. Van Gijlswijk, R. P., Wiegant, J., Vervenne, R., Lasan R., Tanke, H. J., and Raap, A. K. (1996a) Horseradish peroxidase-labeled oligonucleotides and fluorescent tyramides for rapid detection of chromosome-specific repeat sequences. *Cytogenet. Cell Genet.* **75**, 258–262.

16. Van Gijlswijk, Wiegant, J., Raap, A. K., and Tanke, H. J. (1996b) Improved localization of fluorescent tyramides for fluorescence *in situ* hybridization using dextran sulfate and polyvinyl alcohol. *J. Histochem. Cytochem.* **44**, 389–392.

17. Speel, E. J. M., Hopman, A. H. N., and Komminoth, P. (1999) Amplification methods to increase the sensitivity of *in situ* hybridization: play CARD(S). *J. Histochem. Cytochem.* **47**, 281–288.

18. Speel, E. J. M., Ramaekers, F. C. S., and Hopman, A. H. N. (1997) Sensitive multicolor fluorescence *in situ* hybridization using catalyzed reporter deposition (CARD) amplification. *J. Histochem. Cytochem.* **45**, 1439–1446.

19. Speel, E. J. M., Hopman, A. H. N., and Komminoth, P. (2000) Signal amplification for DNA and mRNA *in situ* hybridization, in: In situ *hybridization protocols* (Darby, J., ed.), *Methods in Molecular Biology*, Humana, Totowa, NJ, pp. 195–216.

20. Zehbe, I., Hacker, G. W., Su, H., Hauser-Kronberger, C., Hainfeld, J. F., and Tubbs, R. (1997) Sensitive *in situ* hybridization with catalyzed reporter deposition, streptavidin-nanogold, and silver acetate autometallography: detection of single-copy human papillomavirus. *Am. J. Pathol.* **150**, 1553–1561.

21. Speel, E. J. M. (1999) Detection and amplification systems for sensitive, multiple-target DNA and RNA *in situ* hybridization: looking inside cells with a spectrum of colors. *Histochem. Cell Biol.* **112**, 89–113.

22. Teramato, N., Szekaly, L., Pokrovskaja, K., et al. (1998) Simultaneous detection of two independent antigens by double staining with two mouse monoclonal antibodies. *J. Virol. Methods* **73**, 89–97.

23. Cheung, A. L. M., Graf, A-H., Hauser-Kronberger, C. H., Dietze, O., Tubbs, R. R., and Hacker, G. W. (1999) Detection of human papillomavirus in cervical carcinoma: comparison of peroxidase, nanogold, and catylized reporter deposition (CARD)-nanogold *in situ* hybridization. *Mod. Pathol.* **12**, 689–696.

24. Sano, T., Hikino, T., Niwa, M. T. Y., et al. (1998) *In situ* hybridization with biotinylated tyramide amplification: detection of human papillomavirus DNA in cervical neoplastic lesions. *Mod. Pathol.* **11**, 19–23.

25. Tubbs, R. R., Hauser-Kronberger, C., and Hacker, G. W. (1998) Correspondence re: Sano, T., Hikino, T., Niwa, Y., et al. *In situ* hybridization with biotinylated tyramide amplification: detection of human papillomavirus DNA in cervical neoplastic lesions [Letter to the Editor]. *Mod. Pathol.* **11**, 19–23.

26. Adler, K., Erickson, T., and Bobrow, M. High sensitivity detection of HPV-16 in SiHa and CaSki cells utilizing FISH enhanced by TSA. *Histochem. Cell Biol.* **108**, 321–324.

27. Reed, J. A., Nador, R. G., Spaulding, D., Tani, Y., Cesarman, E., and Knowles, D. M. (1998) Demonstration of Kaposi's sarcoma-associated herpes virus cyclin D homolog in cutaneous Kaposi's sarcoma by colorometric *in situ* hybridization using a catalyzed signal amplification system. *Blood* **91**, 3825–3832.

28. De Haas, R. R., Verwoerd, N. P., Van Der Corput, M. P., Van Gijlswijk, R. P., Siitari, H., and Tanke, H. J. (1996) The use of peroxidase-mediated deposition of biotin-tyramide in combination with time-resolved fluorescence imaging of europium chelate label in immuno-histochemistry and *in situ* hybridization. *J. Histochem. Cytochem.* **44**, 1091–1099.

29. Kressel, M. (1997) Tyramide amplification allows anterograde tracing by horseradish peroxidase-conjugated lectins in conjunction with simultaneous immunohistochemistry. *J. Histochem. Cytochem.* **46**, 527–533.

30. Toda, Y., Kono, K., Abiru, H., et al. (1999) Applications of tyramide signal amplification system to immunohistochemistry: a potent method to localize antigens that are not detectable by ordinary method. *Pathol. Int.* **49**, 479–483.

31. Hendricks, D. A., Stowe, B. J., Hoo, B. S., et al. (1995) Quantitation of HBV DNA in human serum using a branched DNA (bDNA) signal amplification assay. *Am. J. Clin. Pathol.* **104**, 537–546.

32. Horn, T., Chang, C-A., and Urdea, M. S. (1997) Chemical synthesis and characterization of branched oligodeoxyribonucleases (bDNA) for use as signal amplifiers in nucleic acid quantification assays. *Nucleic Acids Res.* **25**, 4842–4849.

33. Urdea, M. S., Wuestehube, L. J., Laurenson, P. M., and Wilber, J. C. (1997) Hepatitis C— diagnosis and monitoring. *Clin. Chem.* **43**, 1507–1511.

34. Kaplan, D. and Smith, D. (2000) Enzymatic amplification staining for flow cytometric analysis of cell surface molecules. *Cytometry* **40**, 81–85.

34a. Joos, S., Fink, T. M., Ratsch, A., Lichter, P. (1994) Mapping and chromosome analysis; the potential of fluorescence in situ hybridization. *J. Biotech.* **35**, 135–153.

35. Zaidi, A. U., Enomoto, H., Milbrandt, J., and Roth, K. A. (2000) Dual fluorescent *in situ* hybridization and immunohistochemical detection with tyramide signal amplification. *J. Histochem. Cytochem.* **48**, 1369–1375.

35a. Buonamici, L., Serra, M., Losi, L., Eusebi, V. (2000) Application of CARD-ISH for assessment of numerical chromosome aberrations in interphase nuclei of human tumor cells. *Int. J. Surg. Pathol.* **8**, 201–206.

36. Schöfer, C., Weipoltshammer, K., Almeder, M., and Wachtler, F. (1997) Signal amplification at the ultrastructural level using biotinylated tyramides and immunogold detection. *Histochem. Cell Biol.* **108**, 313–319.

37. Hainfeld, J. F. (1987) A small gold-conjugated antibody label: improved resolution for electron microscopy. *Science* **230**, 450–453.

38. Van de Corput, M. P. C., Dirks, R. W., van Gijlswijk, R. P. M., and van de Rijke, F. M. (1998) Fluorescence *in situ* hybridization using horseradish peroxidase-labeled oligodeoxynucleotides and tyramide signal amplification for sensitive DNA and mRNA detection. *Histochem. Cell Biol.* **110**, 431–437.

39. Van de Corput, M. P. C., Dirks, R. W., van Gijlswijk, R. P. M., et al. (1998) Sensitive mRNA detection by fluorescence *in situ* hybridization using horseradish peroxidase-labeled oligodeoxynucleotides and tyramide signal amplification. *J. Histochem. Cytochem.* **46**, 1249–1259.

40. King, G., Chambers, G., Murray, G. I. (1999) Detection of immunoglobulin light chain mRNA by *in situ* hybridisation using biotinylated tyramine signal amplification. *J. Clin. Pathol. Mol. Pathol.* **52**, 47–51.

41. Yang, H., Wanner, I. B., Roper, S. D., and Chaudhari, N. (1999) An optimized method for *in situ* hybridization with signal amplification that allows the detection of rare mRNAs. *J. Histochem. Cytochem.* **47**, 431–445.

42. Schmidt, B. F., Chao, J., Zhu, Z., DeBiasio, R. L., and Fisher, G. (1997) Signal amplification in the detection of single-copy DNA and RNA by enzyme-catalyzed deposition (CARD) of the novel fluorescent reporter substrate Cy3.29-tyramide. *J. Histochem. Cytochem.* **45**, 365–373.

**PROTOCOLS**

# TSA by Rapp et al., *1995*

## INTRODUCTION

This is the first TSA protocol reported in the literature, described by Raap et al. in 1995 *(7)*. It was subsequently refined and reported in greater detail by other authors but is provided here to illustrate further the basic principles common to all TSA protocols.

## PREPARATION OF THE BIOTIN-TYRAMIDE SOLUTION

The authors performed the following procedure: a 1 mg/mL biotin-tyramide stock solution (NEN-duPont) was diluted 1:1000 in 0.2 $M$ Tris-HCl, 10 m$M$ imidazole at pH 8.8. $H_2O_2$ was then added to achieve a final concentration of 0.01%.

## SYNTHESIS OF FLUOROCHROME-LABELED TYRAMIDES

Tyramides were synthesized using 25 µL of 150 m$M$ tyramine (Aldrich) in dimethyl sulfoxide (DMSO) with 50 µL of 160 m$M$ N-hydroxysuccinimide esters of the fluorochromes (Boehringer Mannheim). Tyramine stock solutions (1 mg/mL) were prepared. The peroxidase reaction with the fluorochrome-labeled tyramides was 10 min at room temperature with a 1:1000 dilution of the stock solution in 0.2 $M$ Tris-HCl, 10 m$M$ imidazole, 0.1% $H_2O_2$ at pH 8.8.

## ISH

ISH was performed by the usual method. Experiments were performed using DNA with biotin- or digoxigenin-dUTP DNA ISH, using human chromosome spreads, and RNA ISH with HeLa rat-9G cells and *Plasmodium berghei*.

## INCUBATION

Slides were incubated with either streptavidin-peroxidase or anti-digoxin-peroxidase. The former was provided by Vector in the original report and the latter by Boehringer Mannheim. After the washing stage, the biotin-tyramide solution was incubated for 15 min.

## COMPLETION

The slides were then washed, incubated in streptavidin-FITC, washed, and mounted in Vectashield (Vector), a commercially available antifading agent.

# CARD/TSA with Nanogold

## INTRODUCTION

The following protocol is adapted from a recent review of the subject *(1)*; it uses CARD/TSA in combination with Nanogold, a technique successfully employed to detect such minute targets as single human papillomavirus (HPV) DNA copies. Nanogold was first reported by Hainfeld for use in electron microscopy and subsequently adapted for this current use *(9,23,37)*. An advantage of this technique is that it can be successfully employed in formalin-fixed, paraffin-embedded tissue sections as well as formalin-fixed cytologic specimens. As is true with the other protocols, the extreme sensitivity of this method can cause problems with excessive background staining when issues with fixation and probes arise.

## SPECIMEN PREPARATION

Specimens are deparaffinized with fresh xylene washes (2 × 15 min each) followed by rinsing in absolute EtOH (2 × 5 min). The specimens are then treated with a solution of 3% hydrogen peroxide in methanol at room temperature for 30 min, followed by rinsing in double-distilled (ultrapure) water for 10 s and phosphate-buffered saline (PBS) for 3 min. After this, the specimens are incubated with 0.1 mg/mL proteinase K (Boehringer Mannheim) in PBS at 37°C for approx 8 min. This last time may be variable, and optimization is necessary. Microwave treatment at this point may be helpful. The specimen is then washed twice with PBS, 3 min each, and then with ultrapure water for 10 s. The specimens are then hydrated with graded EtOH (50, 70, 100%) for 5 min each, followed by air-drying.

## HYBRIDIZATION

Specimens are first prehybridized with a 1:1 solution of deionized formamide and 20% dextran sulfate in 2× standard saline citrate (SSC) at 50°C for 5 min. After careful shaking of the prehybridization block, 1 drop of biotinylated DNA probe is added on the section, which is then covered with a cover slip. Entrapment of air bubbles under the cover slip is to be carefully avoided. The DNA is then denatured by heating the sections on heating blocks at 92–94°C for 8–10 min. The sections should then be incubated in a moist chamber for at least 2 h. The sections are then washed with two changes of 2× SSC for 5 min each, followed by washes with 0.5× SSC, 0.2× SSC, and then distilled water, which should remove the cover slips and prepare the sections for the next step.

## PREPARATION FOR TSA

Place the slides in Lugol's iodine solution (Merck) for 5 min. Then wash them with tap water, followed by double-distilled water. Wash with 2.5% sodium thiosulfate until the sections are colorless (this should take a few seconds). Wash the sections in distilled water for 2 min. Drain the sections, wipe the parts of the slide around the section, and apply PAP-pen (Dako) to the area of the slide adjacent to the sections. Then incubate the sections with blocking solution (4× SSC containing 5% casein sodium salt or 0.5% blocking powder [NEN Life Science]). Wash the sections in 4× SSC containing Tween-20 (Boehringer Mannheim) for 2 min.

## TSA

Incubate sections with streptavidin-biotin-peroxidase complex (Dako) at room temperature for 30 min. Wash with three changer of 4× SSC containing 0.05% Tween-20 for 2 min each, followed by two changes of PBS for 2 min each. Then incubate the sections with biotinylated tyramide at room temperature for 10 min (NEN kit) or 15 min (Dako). Follow manufacturers' recommendations for appropriate use of these products.

## NANOGOLD

Immerse the sections in PBS containing 0.1% fish-gelatin for 5 min. Then incubate the sections with streptavidin-Nanogold (Nanoprobes) diluted 1:750 in PBS containing 1% bovine serum albumin at room temperature for 60 min. Then wash in three changes of PBS-gelatin for 5 min each. After this, wash the specimens in ultrapure (EM grade) water.

## AUTOMETALLOGRAPHY

Perform silver acetate autometallography or GoldEnhance development (Nanoprobes) according to the manufacturer's instructions.

## COMPLETION

After performing autometallography, counterstain sections with hematoxylin and eosin or nuclear fast red, dehydrate, and mount using Permount or DPX. Avoid the use of Eukitt for this purpose.

# TYRAMIDE-ENHANCED FISH WITH PEROXIDASE-LABELED OLIGONUCLEOTIDES

## INTRODUCTION

Detection of cellular mRNA can be difficult for a number of reasons. First, compared with DNA, RNA degrades much more rapidly, requiring prompt specimen treatment for detection by FISH. Also, mRNA is present in relatively small quantities within the cell compared with DNA. For this reason, low sensitivity has plagued many attempts at RNA detection. Because of its greater sensitivity, FISH with radioactive probes has been successful. However, the current attempts to minimize the use of radioisotopes in the clinical laboratory setting has made this method problematic. An additional problem encountered when using small probes is an unacceptable signal/noise ratio, resulting in decreased sensitivity. The noise appears to result from nonspecific probe binding and nonspecific binding of immunologic detection layers *(38,39).*

To alleviate these problems, a method of mRNA detection using horseradish peroxidase-labeled oligodeoxynucleotides in conjunction with TSA has been developed. To reduce noise, the nonspecific deposition of immunoreactive horseradish peroxidase moieties is avoided. Instead, oligodeoxynucleotides are directly conjugated to the horseradish peroxidase molecules, resulting in greater signal localization.

A number of protocols for mRNA detection by TSA have been described *(14,38–42).* The following protocol for mRNA detection by horseradish peroxidase-labeled oligodeoxynucleotides and TSA has been described by Van de Corput *(38,39).*

## SPECIMEN PREPARATION

Two experimental protocols were described. In the first, cells from human urothelial carcinoma and osteosarcoma cell lines were cultured on uncovered glass slides.

## RNA-FISH

The target RNA was denatured prior to hybridization with 70% deionized formamide/ 2× SCC. Following this, 10 μL of hybridization mixture was added, and a cover slip was placed on each slide. Hybridization was performed in a moist chamber and lasted 30 min. Next, the cover slips were removed, and the cells were washed with 2× SSC and 40% formamide (5 min) and 2× SSC. Then the cells were again washed three times at 2 min each with 2× SSC and rinsed with TBS (100 m$M$ Tris-HCl, 150 m$M$ NaCl) with 0.05% Tween-20. The hybridization product was detected using TSA with reagent (NEN Life Science Products) with a 30-min incubation.

## CYTOPLASMIC AND NUCLEOLAR STAINING

A 40-mer 28S ribosomal RNA-specific ODN probe labeled with Texas Red was used for staining. Hybridization for 20 min was performed. The slides were then washed with 2× SSC, dehydrated, air-dried, and embedded in Vectashield (Vector, Burlingame, CA). 4,6-diamidino-2-phenylindol · 2HCl (DAPI) was used as the nuclear counterstain.

## SIGNAL DETECTION

The slides were examined using a standard microscope with a single-bandpass filter for fluorescein and Texas Red.

# 8

# Ultrastructural *In Situ* Hybridization

## Akira Matsuno, MD, PHD, Tadashi Nagashima, MD, PHD, R. Yoshiyuki Osamura, MD, PHD, and Keiichi Watanabe, MD, PHD

## INTRODUCTION

In 1969, the *in situ* hybridization (ISH) technique was introduced for the detection of ribosomal RNA gene products *(1,2)*. Many investigators have since improved this technique to identify specific genes or gene products in cells. The development of synthetic oligonucleotide probes, which could be easily designed and produced, contributed greatly to the refinement of ISH *(3)*. Nonradioactive synthesized oligonucleotide probes labeled with biotin or digoxigenin were introduced for the detection of ISH signals *(4,10)*. ISH under a brightfield light microscope (LM-ISH) has since become a widely used method for examining the tissue distribution and expression of mRNA. The LM-ISH method is lacking in the resolution required for studies on the spatial relationship between mRNA and the protein product. This type of information can only be provided by ultrastructural studies.

Ultrastructural ISH with an electron microscope (EM-ISH) is a recently developed technique and is essential for the intracellular identification of mRNA and to study the role of mRNA in protein synthesis. Ultrastructural ISH for RNA was first described by Jacob et al. *(11)*. This method has been further developed by several investigators *(12–26)*. Three different approaches have been applied by investigators to EM-ISH studies: the preembedding method *(13,17,21–26)*, the nonembedding method using ultrathin frozen sections *(14,15)*, and the postembedding method *(18–20,24,26)*. Recently, we developed a nonradioisotopic EM-ISH method using biotinylated synthesized oligonucleotide probes and applied it to the ultrastructural visualization of growth hormone (GH) and prolactin (PRL) mRNAs and pathophysiological studies in rat pituitary cells *(24–26)*. In addition, we developed a combined EM-ISH and immunohistochemistry method for the purpose of simultaneous identification of pituitary hormone and its message in the same cell *(27–33)*. In this chapter, the previously reported EM-ISH techniques including our nonradioisotopic EM-ISH method are reviewed, along with our combined immunohistochemistry and nonradioisotopic preembedding ISH methods. We also present an overview on the methods reported in the literature.

From: *Morphology Methods: Cell and Molecular Biology Techniques*
Edited by: R. V. Lloyd © Humana Press, Totowa, NJ

## ELECTRON MICROSCOPIC IN SITU HYBRIDIZATION

### Preparation of Biotinylated Synthesized Oligonucleotide Probes

We have utilized biotinylated synthesized oligonucleotide probes for the detection of pituitary hormone, GH, and PRL mRNA in pituitary cells. The sequences of the oligonucleotide probes for rat GH and PRL mRNAs are 5'-dATC GCT GCG CAT GTT GGC GTC and 5'-dGGC TTG CTC CTT GTC TTC AGG, respectively *(34)*. Both the antisense and sense oligonucleotide probes were synthesized with a DNA synthesizer (model 392, Applied Biosystems, Foster City, CA) and biotinylated by a 3'-end labeling method using ENZO's terminal labeling kit (ENZO, Farmingdale, NY), according to the manufacturer's protocol. The specificities of the biotinylated probes for both hormone mRNAs were confirmed with Northern blot hybridization, using total RNA extracted from normal male Wistar-Imamichi rat pituitary glands *(24,25)*.

As stated above, three different approaches have been applied for this EM-ISH study: the preembedding method, the nonembedding method using ultrathin frozen sections, and the postembedding method. The major concerns of EM-ISH are to maintain tissue morphology and to retain the messages. As Le Guellec et al. *(21)* have noted, ultrastructural preservation in EM-ISH using ultrathin frozen sections is poor, and specimens embedded in Lowicryl K4M exhibit poorer ultrastructural preservation than those embedded in Epon resin. To obtain satisfactory morphologic preservation, we routinely utilize 6-μm-thick frozen sections fixed in 4% paraformaldehyde for the preembedding method and tissues embedded in LR White resin for the postembedding method.

### Preembedding EM-ISH Method

1. Tissues are fixed at 4°C overnight in 4% paraformaldehyde dissolved in 0.01 *M* phosphate-buffered saline (PBS), pH 7.4.
2. After immersion in graded concentrations of sucrose dissolved in PBS at 4°C (10% for 1 h, 15% for 2 h, 20% for 4 h), tissues are embedded in Optimal Cutting Temperature (OCT) compound (Tissue-Tek, Miles Laboratories, Elkhart, IN).
3. Tissue specimens (6 μm thick) are mounted on 3-aminopropylmethoxysilane-coated slides.
4. After air-drying for 1 h, tissue sections are washed in PBS for 15 min.
5. Tissue sections are treated with 0.1 μg/mL proteinase K at 37°C for 30 min, followed by treatment for 10 min with 0.25% acetic anhydride in 0.1 *M* triethanolamine.
6. The slides are washed in 2× sodium chloride, sodium citrate (SSC) at room temperature for 3 min and then prehybridized at 37°C for 30 min. The prehybridization solution consists of 10% dextran sulfate, 3× SSC, 1× Denhardt's solution (0.02% Ficoll/0.02% bovine serum albumin [BSA]/0.02% polyvinylpyrrolidone), 100 μg/mL salmon sperm DNA, 125 μg/mL yeast tRNA, 10 μg/mL polyadenylic-cytidylic acid, 1 mg/mL sodium pyrophosphate, pH 7.4, and 50% formamide.
7. The biotinylated probe with the concentration of 0.1 ng/μL is diluted with this prehybridization solution and hybridization is carried out at 37°C overnight.
8. The slides are washed at room temperature with 2× SSC, 1× SSC, and then 0.5× SSC for 15 min, respectively.
9. After washing in PBS for 5 min, the slides are dipped in diaminobenzidine (DAB) solution (incomplete Graham-Karnovsky's solution) for 1 h.
10. The hybridization signals are detected with streptavidin-biotin-horseradish peroxidase

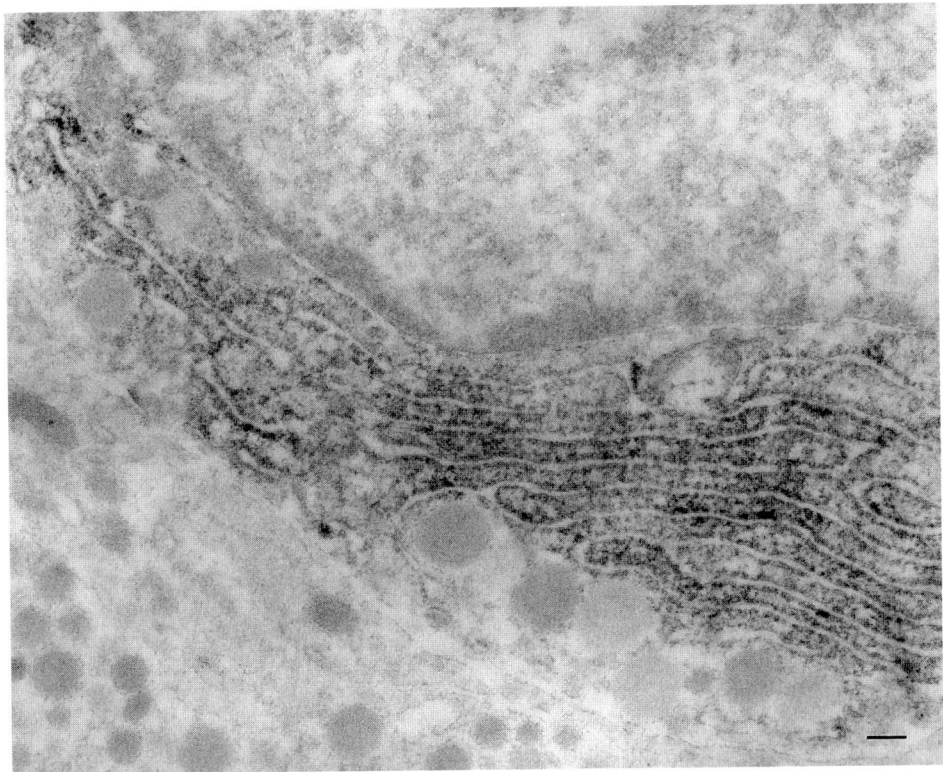

**Fig. 1.** Photograph of preembedding EM-ISH for rat GH mRNA. Rat GH mRNA is localized diffusely on the polysomes of the entire rough endoplasmic reticula (RER). Bar = 200 nm. (Reproduced with permission from ref. *30*.)

(ABC-HRP), using Vectastain's ABC kit (Vector, Burlingame, CA), and thereafter hybridization signals are developed with 0.017% $H_2O_2$ in DAB solution (complete Graham-Karnovsky's solution) for 7 min. At this stage, hybridization signals are confirmed light microscopically.

11. 2% osmium tetroxide is applied to the sections for 1 h. After dehydration with graded ethanol (50, 70, 80, 90, 95, 100%), tissue sections are embedded in Epon resin by the inverted beam capsule method.

12. Polymerization of Epon resin at 60°C for 2 days. Ultrathin sections are inspected under electron microscopy.

13. The negative control experiments should include hybridization with probes of sense or scrambled sequences and without probes, as well as pretreatment with ribonuclease A (100 µg/mL) at 37°C for 45 min before hybridization.

*Results*

This preembedding EM-ISH study with an antisense biotinylated oligonucleotide probe localized rat GH mRNA diffusely on the polysomes of the entire rough endoplasmic reticulum (RER) **(Fig. 1)**. The preembedding EM-ISH study with a sense probe for rat GH mRNA showed negative signals **(Fig. 2)**.

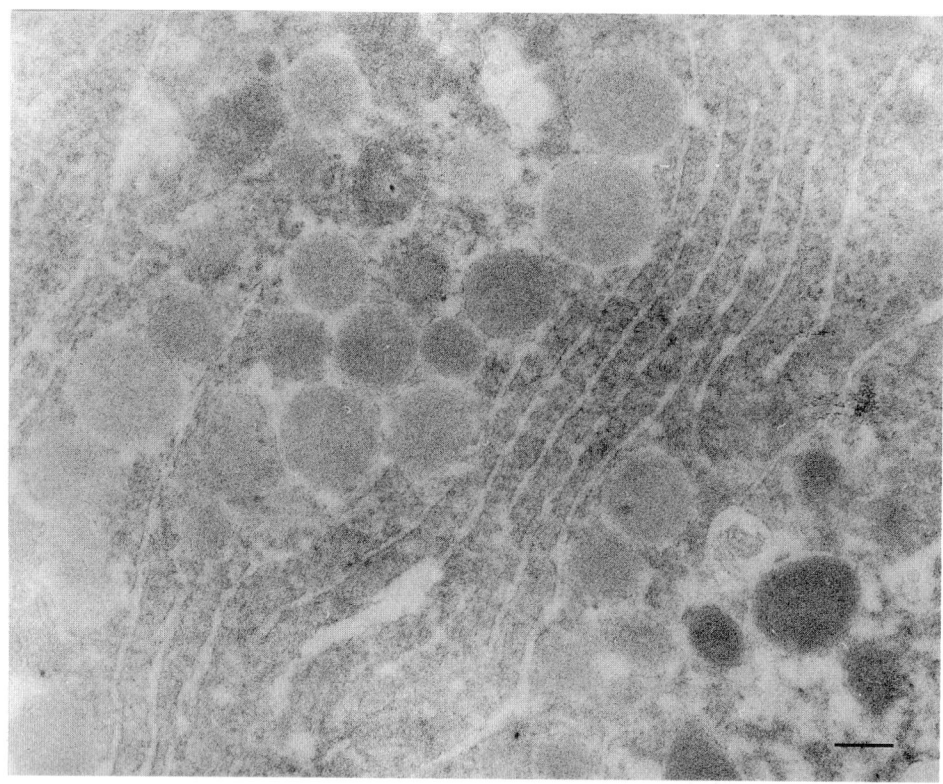

**Fig. 2.** Control study with a sense probe for rat GH mRNA. Rat GH mRNA is not localized on the RER. Bar = 200 nm. (Reproduced with permission from ref. *30.*)

*Comments*

In the preembedding method, both ultrastructure and mRNA are sufficiently preserved. The preembedding method is valuable in that hybridization signals can be confirmed light microscopically. The only drawback of this preembedding method is failure of quantification of hybridization signals that are detected enzymatically.

Pitfalls:

1. Prolonged fixation of the tissues will decrease the messages.
2. Excessive treatment with proteinase K will destroy the morphology.

### Postembedding EM-ISH Method

1. Tissues are fixed at 4°C overnight in 4% paraformaldehyde dissolved in PBS and embedded in LR White resin (Polyscience, Warrington, PA). Tissues are carefully placed at the bottom of 00 gelatin capsules (Lilly Pharmaceuticals, Indianapolis, IN), which are filled with LR White resin and sealed.
2. After polymerization at 50°C for 24 h in a vacuum oven, ultrathin sections are retrieved on nickel grids.
3. After prehybridization at 37°C for 30 min, hybridization is carried out on the grids at 37°C overnight. The hybridization solution is the same as that used for preembedding method. The concentration of the oligonucleotide probe is 1 ng/mL.

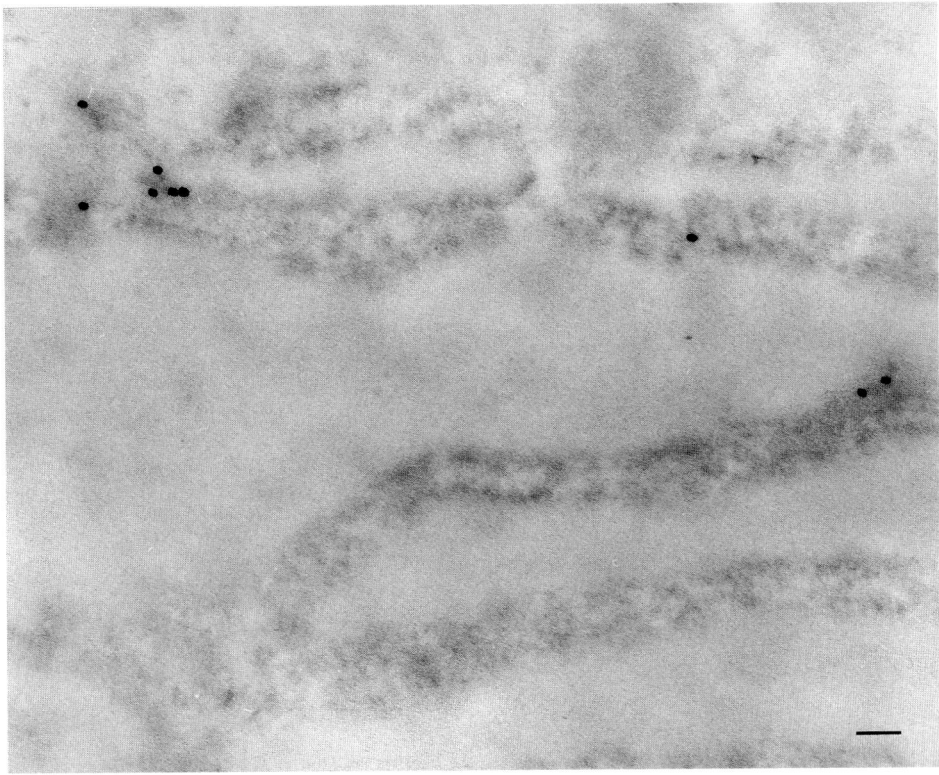

**Fig. 3.** Hybridization signals for rat GH mRNA are localized on the polysomes of the RER using 20 nm streptavidin gold. Bar = 100 nm. The hybridization signal intensity is lower than that for the preembedding method. (Reproduced with permission from ref. *30.*)

4. After hybridization, the grids are dipped in 2× SSC, 1× SSC, and then 0.5× SSC for 5 min, respectively.
5. The hybridization signals are developed for 30 min with 20 nm streptavidin gold (British Biocell, Cardiff, UK) diluted 1:50 in 1% BSA-PBS. After being dipped in PBS and distilled water and dried at room temperature, the grids are inspected under electron microscopy.
6. The control experiments are hybridizations with probes of sense or scrambled sequences and without probes.

*Results*

The postembedding EM-ISH study localized hybridization signals for rat GH mRNA on the polysomes of the RER using 20 nm streptavidin gold (**Fig. 3**). The EM-ISH study with a sense probe for rat GH mRNA showed no hybridization signals (**Fig. 4**). The hybridization signal intensity is lower than that for the preembedding method.

*Comments*

In the postembedding method using tissues embedded in LR White resin, the ultrastructure is also sufficiently preserved. Compared with the preembedding method, the postembedding method has several drawbacks: 1) message preservation is difficult during polymerization of LR White resin at relatively high temperatures for extended periods, which leads to mRNA degradation; and 2) nonspecific signals may be seen

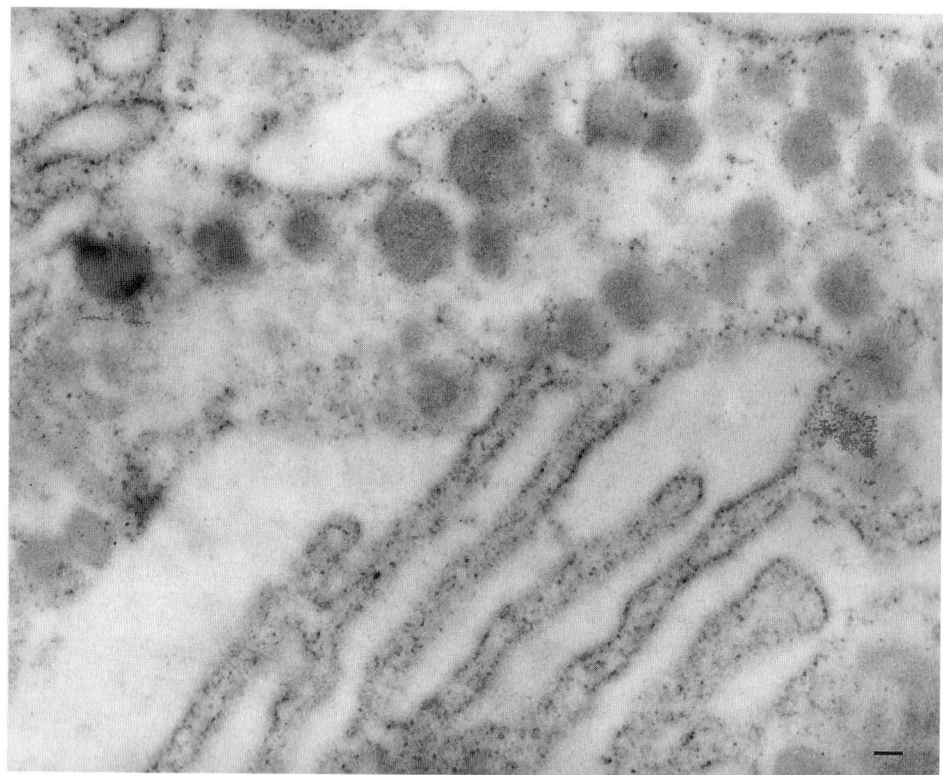

**Fig. 4.** EM-ISH with a sense probe for rat GH mRNA shows no hybridization signals. Bar = 100 nm. (Reproduced with permission from ref. *30.*)

due to the nonspecific affinity of gold particles. Quantification of hybridization signals can be obtained through counting the number of gold particles. Nevertheless, we should note that the signals may be decreased in the postembedding method.

Pitfall:

1. Prolonged fixation of the tissues will decrease the messages.

### Nonembedding EM-ISH Method Using Ultrathin Frozen Sections

Morel et al. *(14)* applied this nonembedding method for the detection of atrial natriuretic peptide synthesis in pituitary gonadotroph cells. They utilized ultrathin frozen sections of rat pituitary glands, which were cut at −120°C and mounted on collodion-coated nickel grids. Ultrathin frozen sections were incubated for 3 h at 40°C with a 2.5 μmol antisense biotinylated oligonucleotide probe. Grids were then washed twice in 2× SSC at room temperature. Hybridization signals were detected with rabbit anti-biotin serum and anti-rabbit immunoglobulin G-colloidal gold.

In general, as Le Guellec et al. *(21)* stated, the nonembedding EM-ISH method using ultrathin frozen sections may be highly sensitive; however, this method has the drawback of poor morphologic preservation.

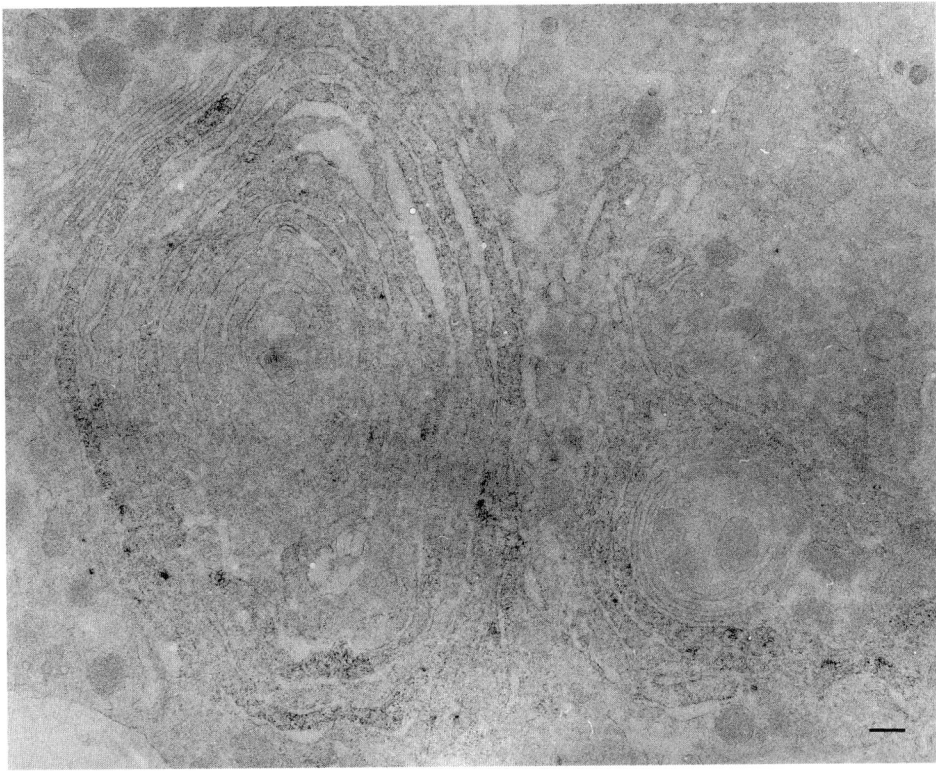

**Fig. 5.** Photograph of pituitary cells from female rats given estrogen for 7 weeks subjected to preembedding electron microscopic ISH. Rat PRL mRNA is localized frequently on the polysomes of the whirling RER. Bar = 200 nm. (Reproduced with permission from ref. *31.*)

## *Pathophysiologic Studies on Changes in Ultrastructural Expression of Rat PRL mRNA Induced by Estrogen and Bromocriptine: Preembedding Method*

Female Wistar-Imamichi rats treated intramuscularly with 5 mg estradiol dipropionate (E2 depot: Ovahormon Depot; Teikoku Zoki, Tokyo, Japan) every 4 weeks are sacrificed 3, 5, and 7 weeks after injection, with or without subcutaneous injection of 1 mg bromocriptine (Novartis, Basel, Switzerland) for 4 days. The anterior lobes removed from their pituitary glands are immediately fixed at 4°C overnight in 4% paraformaldehyde dissolved in PBS and serve for ultrastructural ISH studies. The protocol has been described above.

### *Results*

Preembedding EM-ISH studies revealed the whirling changes of the RER in the specimens of female rats given estrogen for 7 weeks and also frequent but focally localized hybridization signals of rat PRL mRNA on the polysomes of the whirling RER **(Fig. 5)**. After bromocriptine administration, rat PRL mRNA expression at the light microscopic level decreased markedly, and electron microscopic examination revealed diffuse localization of rat PRL mRNA hybridization signals on the distorted,

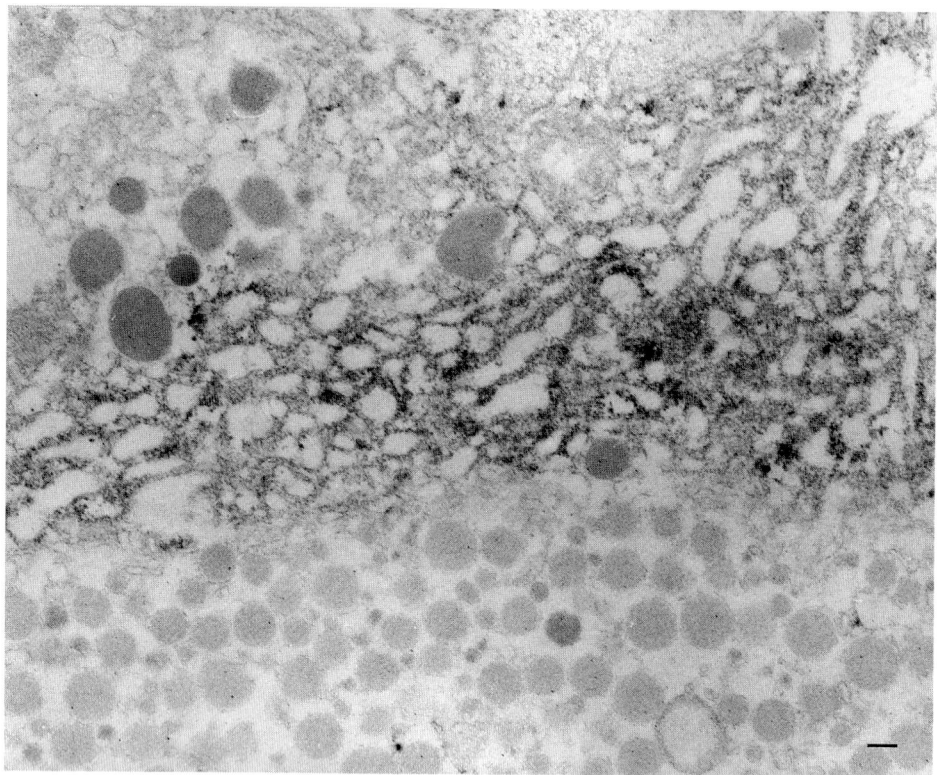

**Fig. 6.** Electron microscopic observation of pituitary cells after bromocriptine administration reveals rat PRL mRNA hybridization signals localized on the distorted, vesiculated, and partly dilated RER, and also increased number of accumulated secretory granules. Bar = 200 nm. (Reproduced with permission from ref. *30*.)

vesiculated, and partly dilated RER **(Fig. 6)**. There were also increased numbers of secretory granules, which resulted in increased PRL immunoreactivity induced by bromocriptine.

*Comments*

For quantitative analyses of PRL mRNA expression, other experiments including Northern blot hybridization and LM-ISH studies are required. Using the [32]P-labeled oligonucleotide probe for rat PRL mRNA, a 1.0-kb transcript was detected on the nitrocellulose membrane, and the PRL mRNA and β-actin mRNA hybridization signal density ratios for the pituitary glands of untreated, control female rats, those treated with estrogen alone for 3 and 7 weeks, and those treated with estrogen plus bromocriptine were evaluated densitometrically *(26)*. As shown in our previous report, the hybridization signal density of PRL mRNA was enhanced as the duration of estrogen treatment increased, and it decreased markedly after bromocriptine adminsitration *(26)*. Our previous LM-ISH studies revealed that the hybridization signals of PRL mRNA from estrogen-treated rats were more intense than those from normal rats without estrogen administration; the hybridization signal frequency increased as the duration of estrogen

**Table 1**
**Effect of Fixatives on Nucleic Acid Preservation in Paraffin Blocks**

| Good fixatives | Poor fixatives |
|---|---|
| 10% buffered formalin | B-5 |
| Ethanol | Bouin's |
| Acetone | Zenker's |
| Omnifix | Glutaraldehyde |
| | Carnoy's |

DURATION OF STORAGE OF PARAFFIN BLOCKS

As a general rule, the older the specimen age, the less amenable it is to PCR amplification. Again, it mostly affects long DNA sequences, whereas DNA segments on the order of 100 bp can be amplified from paraffin blocks embedded 40 years ago *(10)* and even from much older specimens.

It is important to stress that even after a short time in formalin, much of the DNA is partially degraded, and mostly low molecular weight DNA can be isolated. Although DNA sequences of more than 1000 bp can potentially be amplified after 24-h fixation in formalin *(6)*, for reproducible results the PCR target should be <400 bp, with the optimal size varying between 100 and 200 bp.

The procedure of DNA extraction from paraffin-embedded tissue starts with dewaxing to remove paraffin from the tissue samples. This is followed by steps to liberate the DNA from the tissue. This can be done by proteinase K digestion or by using a variety of other techniques such as boiling for 10–15 min *(11)*, sonication *(12)*, or treatment with a polyvalent chelating agent, Chelex 100 *(13)*. In many situations, complete DNA isolation is unnecessary and even undesirable because additional steps may increase the risk of contamination but do not improve the yield of DNA *(14)*.

METHOD

1. Using a fresh blade cut five paraffin sections, 10 μm thick, using a microtome and avoiding cross-contamination between samples. Push the cut sections into a 1.5-mL microcentrifuge tube using a sterile disposable tip. (To avoid cross-contamination between samples, use disposable microtome blades or a clean cutting surface after each sample; discard the first slice from a paraffin block.)
2. To deparaffinize sections, add 1 mL of xylene, vortex for 30 s, and spin down for 5 min at 13,000$g$. Pipet off the xylene.
3. Repeat **step 2** once.
4. Add 1 mL of 100% ethanol. Vortex for 30 s, spin at 13,000$g$ for 5 min, and pipet off the ethanol.
5. Add 1 mL of 80% ethanol. Vortex for 30 s, spin at 13,000$g$ for 5 min, and pipet off the ethanol.
6. Add 1 mL of 50% ethanol. Vortex for 30 s, spin at 13,000$g$ for 5 min, pipet off the ethanol, and air-dry briefly.
7. Add 100 μL of digestion buffer (50 m$M$ Tric-HCl, pH 8.0, 1 m$M$ EDTA) and proteinase K to 100 μg/mL, and incubate at 52°C for 4 h or overnight. There should be no obvious tissue debris remaining after incubation. Otherwise add an additional 50 μg/mL of proteinase K, invert the tube 10 times, and incubate for 2–3 h more.
8. Spin the tube briefly and heat at 94°C for 20 min to inactivate the proteinase K. Use 10 μL for PCR.

If a stock of DNA is required for multiple PCR reactions, the extraction can be started with 10 sections, 40 µm thick; after deparaffinization **(steps 1–6)**, the digestion and phenol-chloroform extraction can be performed as described above for DNA isolation from fresh or frozen tissue **(steps 3–12)**.

## Primer Design

The choice of primer sequences is a complex process and depends partly on the specific application. A major factor is the constraints imposed by the nature of the target sequence (the segment to be amplified) and the purpose of the assay. If the goal is to detect a polymorphism that consists of a short (≤50 nucleotides) insertion or deletion, the segment to be amplified should be relatively short, 10 times the difference in length, at most. If point mutations are to be detected by sequence analysis or restriction enzyme analysis, the primers need to be positioned to optimize these assays. If mRNA is being assayed, amplifying across an intron/exon junction ensures that the assay will not generate a false-positive from contaminating genomic DNA.

The standard primer length is 20 nucleotides, although much longer ones are used successfully. Twenty nucleotides provide enough sequence complexity to prevent random priming but not so much as to permit formation of stable hybrids with partial homology. When possible, the cytosine + guanine (GC) content in each primer should be similar and between 45% and 65%, so that melting temperatures ($T_m$, see formula below) are moderate and similar for the pair of primers. Since the annealing temperature for a primer is approx 5°C below its $T_m$, if the GC contents of a primer pair differ markedly, the annealing temperature needed for the more AT-rich primer may permit significant mispairing of the GC-rich primer, resulting in nonspecific amplification. Conversely, the temperature needed for the GC-rich primer may preclude annealing of the AT-rich primer. For example:

$T_m$ (in °C) = 81.5 + 16.6 × log[cation] + 0.41 × %(GC) − 500/length
$T_m$ for 20-mer, *50% GC*, 0.1 *M* cation = 60°C
$T_m$ for 20-mer, *65% GC*, 0.1 *M* cation = 67°C

Primer sequences should be free of intra- and intermolecular homology that could result in stable stem-loop structures or interstrand hybrids forming between two strands of the same primer or between one strand from each primer of the pair. The presence of such structures decreases the efficiency of the reaction in each cycle since the effective primer concentration is decreased. Such hybrids can be especially troublesome if they result in a recessed 3′ end since it will be extended by the polymerase, corrupting the primer(s). This corruption is amplified with each cycle, resulting in failure of the PCR assay. Other motifs to be avoided are simple sequences such as long mononucleotides (e.g., GGGGG) or dinucleotide repeats (e.g., ATATATAT) that are not flanked by a more complex sequence. It is desirable for the 3′ terminal nucleotide to be a G or C. The extra hydrogen bond helps stabilize the end of the hybrid, resulting in more efficient initiation of DNA synthesis by the polymerase. This is especially true if the primer is AT rich (≥50%).

These factors constitute a useful set of guidelines for primer design. Currently, a number of computer programs are available to assist in primer design. However, each primer and each primer pair must be tested empirically. It is not uncommon for primers to appear satisfactory on paper but to fail in amplification reactions. Below are examples of successful primer pairs. (All $T_m$s are calculated based on a 0.1 $M$ cation concentration.)

**Template**

5'-GCCGTGTTCCGGCTGTCAGCGCAGGGGCGCCCGGTTCTTTTTGTCAAGACCGACCTGTCCGGTGC/

5'-GTGTTCCGGCTGTCAGCGCA-3'→ **Upstream primer: 65% GC T$_m$ = 67°C**

3'-CGGCACAAGGCCGACAGTCGCGTCCCCGCGGGCCAAGAAAAACAGTTCTGGCTGGACAGGCCACG/

/GGAAAATGGCCGCTTTTGGATTCATCGACTGTGGCCGGCTGGGTGTGGCGGACCGCTATCAGGACATA-3'
**Downstream primer: 65% GC T$_m$ = 67°C** ←3'-ACCGCCTGGCGATAGTCCTG-5'

/CCTTTTACCGGCGAAAAGGTAAGTAGCTGACACCGGCCGACCCACACCGCCTGGCGATAGTCCTGTAT-5'

**Other Primer Pairs**

5'-TCACTATTCCCGGTGTTACAGTC-3' 23 nt, 48% GC, $T_m = 63$°C

5'-CCCTTTCAGTGTCCCACAACCT-3' 22 nt, 55% GC, $T_m = 65$°C

Template constraints largely defined this pair. To match lengths and melting temperatures better, the upper primer was extended to include TC to increase its %GC, and the lower primer was extended to include the terminal T to reduce its %GC.

5'-ACAGGGACACAGGTACACCG-3' 20 nt, 60% GC, $T_m = 65$°C

5'-TGGTCCGAAGTGACCTCAGC-3' 20 nt, 60% GC, $T_m = 65$°C

5'-GACTCTTCAGCTTTCTGACCCATCG-3' 25 nt, 52% GC, $T_m = 66$°C

5'-GATGGATCAGTAGAGCAGGGCTAC-3' 24 nt, 56% GC, $T_m = 67$°C

These primers were constrained by the need to amplify, differentially, one of three related genes. The extra nucleotide of the upper primer largely compensates for its lower GC content.

Some laboratories have access to a central oligonucleotide synthesis facility, whereas others will rely on the vast number of companies that synthesize primers. Prices are generally less than $3 per base for purified oligonucleotides, making 20-base primers very affordable. Dissolve oligonucleotides in freshly autoclaved water or TE to a concentration of 100 µ$M$, aliquot, and store at −20°C.

*Time, Temperature, and Number of Cycles*

The *denaturing* step is almost always 30 s at 94°C. The effective denaturing time is actually longer than 30 s since denaturing begins at or below 90°C. Only sequences very rich in GC (70%) will remain double-stranded after 30 s at 94°C. In these rare cases, the denaturing step can be increased to 1 min. In exceptional cases, it may be necessary to denature at ≥95°C. The *Taq* polymerases remain active even after 30–40 cycles of 30 s at 95°C, but enzyme activity will wane more quickly when the denaturing time or temperature is increased. Test experiments should be performed to determine whether more enzyme should be added before all the cycles have been completed.

*Annealing* is typically for 30 s to 1 min at a temperature 5–10°C below the $T_m$ of the primers. At the ideal annealing temperature, stable primer-template hybrids will form, with minimal formation of mismatched, partial hybrids. The polymerase is active at annealing temperatures (50–60°C), so stable hybrids are quickly elongated by several nucleotides, further stabilizing them. The temperature can then be increased without denaturing the nascent strands. Optimizing the time and temperature can be laborious, but is important. If the temperature is too high or the time too short, fewer templates are copied per cycle, reducing sensitivity. Conversely, annealing for too long or below the ideal temperature leads to mispriming, which results in amplification of spurious bands, confounding interpretation of the results. The annealing process begins as the reactions reach the primer $T_m$, so, as for the denaturing step, the effective annealing time is greater than the nominal time given in protocols.

DNA *synthesis* (often called the *elongation* step) is typically carried out for 45 s to 2 min at 72°C, which is close to the optimum temperature for most commercially available thermostable polymerases. As with the other steps, the appropriate time needs to be empirically determined. An extension time of 1 min is generally enough for products up to 2 kb in length. Longer target sequences require more time. Minimizing the length of this step reduces the time to complete all the cycles, but care should be taken to avoid the accumulation of incomplete products, which will decrease sensitivity. Longer synthesis times than necessary are less problematic, although there may be some loss of enzyme activity in later cycles.

An initial step, called the *hot start,* is usually included in PCR protocols. In hot-start PCR, one key reagent (e.g., *Taq* polymerase, dNTPs, or template) is withheld either physically or functionally from the reaction mixture until the temperature in the tube is ≥80°C. The temperature is then raised to 94°C for 2–4 min, after which the program proceeds to the first denaturation step. The hot start serves two important functions. First, it melts any nonspecific hybrids that may form between primers and single-strand template during preparation of the reactions; second, it ensures that all template DNA is completely single-stranded as the first cycle begins. The result is increased specificity and sensitivity of the PCR. This is because during preparation of the reactions, the primers will anneal nonspecifically to single-strand DNA fragments. Because *Taq* polymerase has a small amount of activity at room temperature, these hybrids are extended by *Taq* polymerase, thus forming nonspecific products that not only show up as extraneous bands upon electrophoresis, but also reduce the yield of the intended product through competition for reagents. The production of these nonspecific products will be prevented if an essential component for DNA synthesis becomes available only at the elevated temperature.

A very effective hot start can be achieved by using a polymerase/inhibitor complex that denatures to release active enzyme only at high temperatures *(15)* (Life Technologies, Bethesda, MD; Applied Biosystems, Foster City, CA). Simply preparing the reactions on ice and placing them quickly in the preheated (≥80°C) thermocycler block is not as effective but will significantly reduce background from mispriming.

The *number of cycles* needed for amplification varies with the copy number of the target DNA and nature of the starting material. Thirty cycles amplifies the starting template approx $1 \times 10^9$ fold. Thus, $10^5$ or $10^4$ initial copies of high-quality template will require 25 or 30 cycles, respectively, to provide enough material for visualization on an ethidium bromide-stained agarose gel. More cycles may be needed (35 cycles for $10^5$ initial copies) to achieve a visible band when the template is from partially degraded DNA isolated from fixed tissues. More cycles may also be needed when trying to detect rare events or ones of unknown frequency, such as mutations in a subset of cells from a tissue. However, as a general rule, the number of cycles in one PCR should not exceed 35 to avoid problems with background and nonspecific amplification.

### Reaction Components

Reactions should be assembled on ice. Several recipes exist for PCR reactions, but all consist of basic components (see below) that are readily available and easily prepared in any laboratory. A volume of reaction mix sufficient for all samples, and consisting of all components except template DNA, is prepared and then aliquoted into the reaction tubes. The DNA is added just before the tubes are placed in the thermocycler. The final volume of reactions varies from 20 to 100 µL and should not be changed once successful conditions are established.

1. Buffer and salt consist of 10 m$M$ Tris-HCl, pH 8.3, and 50 m$M$ KCl. Tris is used as the pH buffer, and KCl is important for optimal enzyme activity; however, it is inhibitory at concentrations above 50 m$M$. Many companies supply a 10× or 5× stock of reaction buffer containing these components, which is further optimized for their enzyme preparation or which contains a proprietary PCR-enhancing component. Typically, they offer either magnesium-free buffer or one that yields 1.5 m$M$ final of MgCl$_2$ concentration.
2. Magnesium (MgCl$_2$) is added at a final concentration of 1.5–6 m$M$. The optimum [Mg$^{+2}$] must be determined for each primer set. Control reactions are performed with [Mg$^{+2}$] varying in 1- or 2-m$M$ increments followed by 0.5-m$M$ steps above and below the concentration(s) giving a positive result.
3. Primer concentrations are usually 1–5 µ$M$ each. This concentration ensures that primers will always be at least 10-fold in excess relative to template, even at the end of the PCR. The best concentration must be determined empirically, as for MgCl$_2$. Since divalent cations complex with DNA readily, the primer (and template) concentration can change the effective Mg$^{+2}$ concentration. Thus, it is often necessary to retitrate Mg$^{+2}$ requirements if the primer concentrations are changed dramatically.
4. Nucleotide concentrations (dATP, dCTP, dGTP, dTTP) vary from 50 to 200 µ$M$ each. As for primers, the goal is that nucleotides not become a limiting reagent. Too great an excess will effect the Mg$^{+2}$ requirement as above, however.
5. *Taq* DNA polymerase is available from many suppliers of biotechnology reagents. It is usually supplied at 5 U/µL and is used at 1–2.5 U per reaction. One unit is defined as the incorporation of 10 ηmol of deoxynucleotide into acid-precipitable DNA in 30 min at 74°C.
6. Template DNA is added last. The amount varies according to the quality of the template and purpose of the assay. A typical amount is 20–50 ng of high-quality genomic DNA or 100 ng of DNA isolated from paraffin-embedded tissue.
7. Mineral oil (only enough to cover the reaction mix) is added on top to prevent evaporation in some thermocyclers. However, most newer models have a heated lid, which prevents condensation and obviates the need for oil.
8. Optional components include glycerol or dimethylsulfoxide (DMSO) used at 5–10% to reduce background and lower the $T_m$ of GC-rich primers. These agents change the dielectric properties of aqueous solutions and so affect base pairing of DNA. However, DMSO partially inhibits the polymerase and should be used at the minimum concentration needed. The amount of enzyme or the elongation time is increased when DMSO is used.

Below is an example of a typical reaction mix for 20 samples (18 samples plus a positive and negative control) based on a 30 µL reaction volume. Note that you should prepare enough mix for 21 tubes to accommodate pipetting losses.

|  | Final concentration | Add to reaction mix |
|---|---|---|
| 10× buffer | 1× | 63 µL |
| (100 m$M$ Tris-HCl, pH 8.3, |  |  |
| 500 m$M$ KCl) |  |  |
| 25 m$M$ MgCl$_2$ | 2 m$M$ | 50.4 µL |
| (assuming 2 m$M$ is optimum) |  |  |
| 100 µ$M$ upstream primer | 1.5 µ$M$ | 9.45 µL |
| 100 µ$M$ downstream primer | 1.5 µ$M$ | 9.45 µL |
| dNTPs mixed, each 10 m$M$ | 150 µ$M$ each | 9.45 µL |
| *Taq* polymerase (5U/µl) | 1.0 U/tube | 1.8 µL |
| dd autoclaved water |  | 444.45 µL |

Mix and put 28 µL into each reaction tube. Add 50 ng of template DNA to each tube in a total volume of 2 µL and proceed with the assay.

### Tubes and Trays for Reaction Vessels

Reaction vessels have to be chosen to be compatible with the thermocycler. Most older machines accept individual 0.5-mL microcentrifuge tubes. Although standard 0.5- or 0.6-mL microcentrifuge tubes may work adequately, *thin-walled tubes,* available from most suppliers, are preferred because they allow more efficient heat transfer so that the temperature/time profile of the reactions more closely matches that programmed.

Increasingly popular is the format patterned after 96-well microtiter plates. Several options are available for these thermocyclers. Thin-walled 0.25-mL tubes are available that come individually or as strips of eight linked tubes for easier handling. Caps are similarly supplied. The 96-well trays are easily cut to the number of wells needed, to reduce waste. A variety of lid arrangements for the trays is also available. For high through-put laboratories, thermocyclers that are configured to accept 96 samples are preferred.

## DETECTION AND ANALYSIS OF PCR PRODUCTS

### Fractionation by Gel Electrophoresis

Standard 1.5% *agarose gels* are adequate for analysis of PCR products from 150 to 1000 bp (**Fig. 2**). For fragments <500 bp, gels of 1.5–2% are best. Specially purified fractions of agarose may improve resolution, but standard products are adequate. For fragments >500 bp, the agarose concentration can be reduced to 1.0–1.2%; those larger than 1000 bp resolve well in 0.8–1.0% agarose. For fragments <1000 bp, the best resolution is obtained when TBE (45 m$M$ Tris-borate, 1 m$M$ EDTA, pH 8.3) is the electrophoresis buffer as opposed to TAE (40 mM Tris-acetate, 1 m$M$ EDTA, pH 7.8). Fragments <250 bp diffuse noticeably in agarose gels. This problem is minimized if the rate of electrophoresis is increased. Thus, it is important to run these gels at high voltage (approx 10 V/cm of gel length) and to minimize staining time. At least one

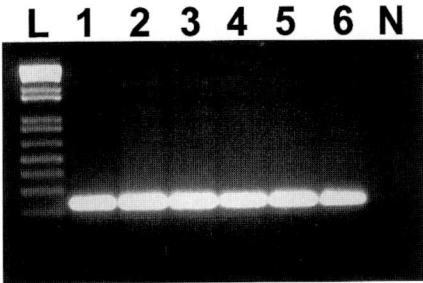

**Fig. 2.** Analysis of a 236-bp PCR product by electrophoresis in a standard 1.5% TBE agarose gel. Ten microliters of each product (30 cycles of amplification starting from 50 ng high-quality genomic DNA) were loaded into the wells. The gel was run at 10 V/cm for 60 min and washed in 0.5 µg EthBr/mL for 20 min. L, 1 Kb Plus DNA Ladder (Gibco BRL); lanes 1–6, PCR products; N, "no template" negative control.

lane per gel must be reserved for size standards, typically a 100-bp ladder, which are available from most biotech supply companies.

*Polyacrylamide gels* may be used to resolve and detect small fragments (≤200 bp). These gels are typically 10% acrylamide (plus 0.3% bis-acrylamide) and are run in TBE. In this system, diffusion of small fragments is minimal, so resolution is greater and smaller amounts of DNA can be detected. Using an acrylamide gel, differences of as little as 10 bp are easily distinguished. Thus, they must be used for detecting restriction site polymorphisms and microsatellite length polymorphisms in amplified sequences where size differences are often small.

### Stains for Detecting Amplified Products

1. *Ethidium bromide* (EthBr) intercalcates between the stacked bases of double-stranded DNA. When it does, its fluorescence under ultraviolet illumination is greatly increased. This dye has been the standard for staining DNA gels for decades. Gels are stained in a solution of 0.5 µg EthBr/ml solution for about 20 min, rinsed, and photographed. The limits of detection are about 10 ng per DNA fragment in a 3–5-mm-thick gel (sensitivity is less with thicker gels).
2. *SYBR® Green dyes* (Molecular Probes, Eugene, OR) provide greater sensitivity (about 1 ng detection limit) and less background than EthBr. They are also much less toxic. These dyes can be detected with phosphorimager lasers and thus are preferred for quantitation of PCR products. They are diluted 1/10,000 from the stock supplied by the manufacturer using water or TE. Staining procedures are essentially the same as for EthBr. The main drawback to these dyes is their comparatively high cost.
3. With *silver staining,* as little as 10 pg of DNA can be detected *(16).* This method also allows the stained gel to be used as a permanent record. The staining procedure is more complicated and generates waste requiring special disposal, however. Although more commonly used with acrylamide gels, it is also applicable to agarose gels *(17).*
4. Although fraught with problems related to radioactive contamination and containment, much greater sensitivity can be achieved by *radiolabeling* amplified products. Isotopically labeled nucleotides are added to the reactions for the last one or two cycles of amplification. The reactions are then fractionated on acrylamide gels and exposed to X-ray film or to a phosphorimager screen (usually after drying the gels) for detection of amplified products. This technique requires the preparation of labeled marker fragments so the size of amplified fragments can be correctly determined. The labeled nucleotides need to be added only during

the last few cycles, since 50% of the strands can be labeled per cycle (50% labeled after one cycle, 75% labeled after two cycles, and so forth).

## Positive Identification of Amplification Products and Detection of Mutations

Several methods exist to verify that the amplified DNA contains the desired sequence and to screen the product for mutations or polymorphism.

### Sequence Analysis

Sequence analysis of the amplified DNA provides direct confirmation of the identity of the product and allows detection of any possible mutations, but it is offset by a higher cost and time consumption. A portion of the PCR product can be precipitated with ethanol or purified from the agarose or acrylamide gel and sequenced directly using radiolabeled primers internal to the amplification primers and standard didexoy-nucleotide chain termination method. Alternatively, it can be cloned into a vector and sequenced using vector-specific primers. To facilitate cloning of PCR products, restriction sites can be incorporated into the primer sequences. Digestion of the PCR product with the appropriate restriction enzymes will facilitate "sticky end" ligation into similarly restricted vector DNA.

### Restriction Endonuclease Mapping

Another technique applicable to many but not all situations is restriction endonuclease mapping of the amplified DNA. If the recognition sequence for a particular restriction enzyme should be present in the amplified fragment, DNA from the PCR reaction can be precipitated with ethanol and resuspended in about 20 μL of TE. The appropriate buffers and enzyme are added to half the sample, and digestion is allowed to proceed. Electrophoretic fractionation of the digested and undigested samples should reveal two fragments of predicted size for the digested sample and one fragment that is the sum of the smaller two fragments in the undigested sample.

Restriction mapping also allows rapid detection of mutations that are associated with the creation or elimination of restriction sites in the amplified DNA. As an example, a G→A mutation at nucleotide 1691 of the coagulation factor V mRNA changes an arginine to a glutamine and also eliminates an *Mnl*I restriction site. This mutation is the most common cause of inherited predisposition to thrombosis, and a clinical diagnostic test has been developed that utilizes PCR and this restriction site polymorphism *(18)*. The PCR product from normal individuals yields three fragments of 163, 67, and 37 bp after digestion with the *Mnl*I enzyme, whereas the DNA from a patient carrying the mutation yields only two products, an abnormal fragment of 200 bp and a normal fragment of 67 bp.

### Southern Blot Analysis

Positive identification of the amplified DNA can be performed using Southern blot analysis if an oligonucleotide probe for the sequence between the primers is available. The PCR products are electrophoresed through 1.2% agarose gel, transferred to a nylon membrane, and hybridized with the labeled, sequence-specific oligonucleotide probe. It is important that the probe be internal and have no overlap with either of the amplification primers. The technique increases specificity and resolution of the PCR.

*Allele-Specific Oligonucleotide Hybridization*

Point mutations in the amplified DNA can be detected using allele-specific oligonucleotide (ASO) hybridization. This method is based on the fact that an oligonucleotide with one mismatch binds more weakly to its complement than does a perfectly matched oligonucleotide. Short oligonucleotides corresponding to the wild-type sequence and to one of the expected point mutations are synthesized, labeled, allowed to hybridize to dot blots of the DNA products, and washed under conditions that allow the discrimination of a single nucleotide mismatch between the probe and the target PCR product. Hybridization of the wild-type oligonucleotide detects the presence of the wild-type allele, whereas hybridization of a mutation-specific oligonucleotide shows the presence of the corresponding point mutation. The hybridization can also be performed "in reverse," with oligonucleotides immobilized on the filter and labeled PCR product hybridized to the filter. This approach allows the simultaneous testing for several anticipated point mutations with a single hybridization reaction. ASO hybridization of the PCR products can be used for detection of genetic diseases caused by a single mutation or by a limited number of specific point mutations in a target gene, such as a single point mutation of the β-globin gene in sickle cell disease *(19)* or several specific mutations in the cystic fibrosis gene *(20)*.

*Single-Strand Conformation Polymorphism*

The PCR product can be screened for unknown or random mutations using single-strand conformation polymorphism (SSCP) *(21)*. The method is based on the fact that single-stranded DNA in nondenaturing condition has a folded conformation that determines its electrophoretic mobility. The conformation is determined by the base sequence, so that a mutation or polymorphism in the sequence will lead to a different mobility in the gel. First, the target sequence is amplified and simultaneously labeled using radioactively labeled PCR primers or dNTPs. Then, the PCR product is denatured and analyzed by electrophoresis through a non-denaturing polyacrylamide gel. Any differences in the base sequence will be detected as a mobility shift and will produce a different band pattern relative to the strands of the normal, wild-type PCR product (see below). A nonradioactive SSCP has been introduced with detection of PCR products by silver staining *(22)*. SSCP is a convenient and relatively rapid method to screen for samples that are likely to possess a mutation. The selected samples can then be sequenced to determine the exact nature of each sequence variation. This can be facilitated by cutting out the aberrant band from the gel, followed by DNA purification and amplification by a second PCR.

## CONTROLS, LIMITATIONS, AND OTHER CONSIDERATIONS

### Quality Control of Reagents

The need for sensitivity and reproducibility requires that reagents, especially primers and nucleotides, be of the highest quality and that they be handled and stored to minimize degradation. Primers should be purified by high-performance liquid chromatography when possible to eliminate incomplete oligonucleotides as well as residue from synthesis reagents. Primers are lyophilized for delivery and should be reconstituted in freshly autoclaved water or TE (10 m$M$ Tris-HCl, 1 m$M$ EDTA, pH 7.5). Stocks should be made to 100 μ*M,* divided into small aliquots of 20–50 μL each, and stored at −20°C. Multiple freeze/thaw cycles degrade primers, and it is best to discard any remainder after five uses of an aliquot. A similar approach of freezing aliquots is needed for deoxynucleotides, which are available in very pure form as 100-m*M* stocks. Although simple chemicals like KCl and Tris need not be aliquotted and frozen, their purity and sterility must be ensured. Water used to make reagents and reaction mixes should also be the highest quality available and unopened since being autoclaved. It is a good practice to make a 5× or 10× buffer mix that is then divided and frozen as above. These preparations, although tedious at first, allow one to use fresh reagents that are identical in composition.

### Positive and Negative Controls

Optimizing reaction conditions, as described above, is usually done with a source of high-purity target/template DNA. Once conditions have been established, they need to be tested using DNA prepared by the same method as will be used in actual analyses. To test for the presence of inhibitory material in a DNA prep, a known number of copies of previously tested template should be amplified alone and as a "spike" of the test mix. The sensitivity of the test needs to be established as well, that is, how many copies of target sequence must be present before the PCR will amplify them to a detectable level. This determination is critical for knowing how much test sample will be needed.

Every amplification assay—that is, every time a mix is made and a set of tubes processed—must include a positive and negative control. The positive control must amplify known positive DNA that has been prepared like the test samples. The negative control must derive from the same reaction mix as the other samples but with only template omitted ("no template" control). It is most valid if this tube is the last one prepared so that any contamination that may occur will be present when the mix is sampled for the negative control. A positive result for the assay can be concluded only if the negative control is clean. Conversely, negative results can be interpreted as absence of target sequence only if amplification is detected in the positive control at the expected level. All other results must be interpreted as questionable and should be repeated for confirmation. If the negative control yields a fragment, contamination has entered the system, and all positive results are considered unreliable (even if not all samples are positive). The test must be repeated using fresh reagents after the source and extent of the contamination have been determined and eliminated (see below).

### Contamination Control

Preventing contamination of reagents and equipment is necessarily an obsession in laboratories engaged in extensive use of PCR. The capacity to amplify a single copy of DNA to a routinely detectable level is a serious matter. The best prevention protocol begins by identifying for the technologist the most likely sources and modes of transfer of contaminating DNA so that their diligence is appropriately focused. Make sure they understand how small an amount of DNA is needed to generate serious contamination. Provide careful instruction for pipettor, tip, and tube handling to prevent generation of aerosols and transfer of DNA residue.

1. Have one complete set of pipettors, tube racks, and solutions, as well as microcentrifuge, that are used only for preparing the PCR reactions and another set that is used only for manipulating samples after amplification and/or for manipulating cloned DNA of the target sequence. Most contamination comes from using a pipettor to prepare PCR reactions after it has been used to handle amplified or cloned DNA.

2. Have one lab bench or area designated as the PCR bench. This bench should be isolated from areas where amplified products are manipulated. Many labs designate a fume hood or laminar flow hood as the PCR bench or have a separate small room for the purpose. This certainly isolates and defines the area but is probably overkill. If a hood is used, it must be configured so all manipulations can be performed comfortably. A tired worker makes mistakes.

3. Regularly clean pipettors. There are a variety of procedures used for this purpose, some probably more extreme than needed. The simplest cleaning method, and one that is fully effective, is to disassemble the instrument and wash the barrel, plunger, Teflon parts, and O-rings in mild detergent. Soap and water is the best cleaning agent. There are also commercially available solutions marketed specifically for this purpose that are effective. Ethanol (70%) is often used to sterilize surfaces but is not the best cleaning agent for this purpose. Ethanol will precipitate DNA but not necessarily flush it away and can actually be counterproductive in this case. Dilute bleach can be used, but it is harsh and will shorten the life of some parts of the instrument. After cleaning, rinse the parts well in distilled water and then rinse with 70% ethanol to enhance drying. Avoid 100% ethanol. It removes plasticizers and reduces the life of the instrument.

4. Aerosol-resistant tips are widely used and greatly reduce the risk of contaminating pipettor barrels. Although they add significantly to the cost of each assay, they may be advisable for the laboratory using PCR extensively as a research or diagnostic tool.

5. PCR workstations can be purchased or constructed as an alternative to a separate bench or room. Essentially, an area is enclosed in Plexiglas with doors that open toward the user. Racks, pipettors, and other equipment designated for PCR are kept inside, and all reactions are assembled inside the enclosure. When not in use, ultraviolet lights are turned on to decontaminate exposed surfaces. Pipetters still require attention, however, since only one side can be exposed and the insides of the barrels are not exposed at all.

6. When preparing and aliquotting reagent stocks such as buffers, primers, and nucleotides, use freshly autoclaved solutions and tubes and freshly cleaned pipettors.

7. To minimize losses due to contamination and to minimize its spread, it is important to divide all reagents into small aliquots. By doing so, if a contamination occurs, it is a simple and inexpensive matter to discard all current reagents and open fresh aliquots.

## Limitations

The size of the fragment that can be amplified by PCR is limited in two ways. Although it is theoretically possible to amplify several thousand base pairs, factors intrinsic to the DNA sequence may reduce this length by an order of magnitude. Such factors could be secondary structure—such as direct or inverted repeats—high GC content, or regions of high AT content. These limitations may be overcome by using modified buffers and reaction conditions.

A limitation more likely to be encountered by pathologists is that due to sample integrity. The amount of tissue degradation that occurs before a sample is fixed as well as chemical degradation that occurs during fixation and tissue processing can reduce the average length of the fragments into which the genome has been broken. As discussed above, shorter target sequences (≤200 bp) are more likely to be successfully amplified than longer ones. Longer amplifications may be successful, but the amount of template DNA will have to be increased to increase the likelihood of having enough genome copies with the desired region intact.

The error rate of *Taq* polymerase is about $2/10^5$ nucleotides *(23)*. Although very low, this level of infidelity is enough to introduce several mutations during the course of an amplification. Such mutations do not adversely effect most applications since only a small proportion of the copies will be affected. Obviously, mutations introduced in early cycles will be represented at a greater frequency than those introduced in later cycles. If amplified fragments are to be cloned and sequenced, however, it is important to analyze several isolates to be sure that the reported sequence is accurate.

The fidelity of amplification by PCR can be increased by maximizing the annealing temperature, minimizing the $MgCl_2$ and dNTP concentrations, minimizing the annealing and extension times, and using DNA polymerase with proofreading activity *(24)*.

## Other Considerations

### Importance of Initial Cycles

The fidelity and efficiency of the first 5–10 cycles critically affect the outcome of the PCR. Any mispriming or mispairing that occurs in these cycles will be amplified throughout the reaction. Errors occurring in later cycles affect a comparatively small number of copies. Thus, if the annealing temperature is too permissive or if mispairing with some extension has occurred during sample preparation, aberrant bands or a general background smear of DNA may be produced.

### Time Extensions for Later Cycles

During later cycles of the PCR, the concentrations of primers and nucleotides decrease, and there may be some loss of enzyme integrity. The combined result is that copying may not be completed in the designated period of time, resulting in decreased yield and a lower signal-to-noise ratio in the final product. This problem can be overcome by lengthening the synthesis step of later cycles. Most thermocyclers have options for incremental extension of individual steps (e.g., 2 a cycle) to overcome this problem. Often, a final synthesis step of 5 min is added to extend any incomplete strands to full length.

### Special Reaction Buffers and Amplification Kits

Kits are available from various biotech supply companies. Often they tout a "one mix fits all" buffer system to which you need only add your primers and template DNA. Many of these kits work well. Their main drawback is that if they quit working or fail to work for a particular application, the user has difficulty trouble-shooting the PCR because a description of the components of the kit and their concentrations is rarely provided by the manufacturer. In the end, the investigator may have to develop the assay from the basics anyway.

## REVERSE TRANSCRIPTASE-POLYMERASE CHAIN REACTION

The addition of a reverse transcription (RT) step allows PCR to amplify RNA fragments. RT-PCR is essential for detection of viral RNA and chimeric transcripts of rearranged genes in hematopoetic and soft tissue tumors and for the broad area of gene expression studies. The major steps of the RT-PCR assay are as follows: 1) isolation of RNA, 2) reverse transcription of RNA into cDNA, and 3) amplification of cDNA by PCR.

## RNA Extraction

It is important to note that RNA is less stable than DNA and is susceptible to rapid degradation by RNAases that are present in cells, tissue, and the environment. Therefore, all steps of RNA handling should be performed using appropriate precautions. Clean gloves should be worn at all times to avoid introduction of RNAases from hands. Sterile, disposable nuclease-free plasticware should be used whenever possible. A set of solutions should be reserved for RNA work only. Water and most solutions should be treated with diethylpyrocarbonate (DEPC) to inhibit RNAase activity. (Add 0.1 mL DEPC/100 mL, leave overnight, and autoclave.) Autoclaving inactivates the DEPC, so avoid subsequent contamination by RNAases. Tris solution should not be treated with DEPC; instead it should be prepared using DEPC water and powder that is certified as nuclease free.

### RNA Extraction from Fresh or Frozen Tissue

Fresh tissue represents the best source of good-quality RNA. If tissue samples or cells from which RNA is to be extracted cannot be processed immediately, they should be snap frozen in liquid nitrogen and stored in liquid nitrogen or at −70°C.

RNA can be isolated by many methods. Many commonly used methods use the guanidine thiocyanate/acid phenol method as described by Chomczynski and Sacchi *(25)*. In this isolation procedure, tissue is lysed in the presence of guanidine thiocyanate, a strong protein denaturant, and a mild detergent. Under acidic conditions (pH 4), in the presence of 4 *M* guanidine thiocyanate and phenol, RNA remains soluble, whereas most proteins and small fragments of DNA will be found in the organic (phenol) phase and larger DNA fragments and some proteins remain in the interface. The method provides nearly total recovery of RNA from small quantities of cells or tissue. Several commercial kits offer a ready-to-use phenol-guanidine thiocyanate solution for the single-step method. Some of them are TRIzol Reagent (Life Technologies), TRI reagent (Molecular Research Center), RNAzol B (Cinna Scientific), or Tri-Pure Isolation Reagent (Boehringer Mannheim).

METHOD
1. Homogenize tissue sample in 1 mL of TRIzol reagent per 100 mg of tissue in a glass Teflon homogenizer or in a Falcon 2063 tube using a power homogenizer.
2. Incubate for 5 min at room temperature. Remove insoluble materials by centrifugation at 12,000*g* for 10 min at 4°C. Transfer the supernatant into a new tube.
3. Add 0.2 mL of chloroform per 1 mL of TRIzol reagent, shake the tube vigorously by hand for 15 s and incubate at room temperature for 3 min.
4. Centrifuge the sample at no more than 12,000*g* for 10 min at 4°C. The mixture should be separated into a lower phenol-chloroform phase (red), an interphase, and an upper aqueous phase (colorless). RNA remains exclusively in the aqueous phase.
5. Transfer the upper aqueous phase to a 1.5-mL microcentrifuge tube and precipitate with an equal volume of isopropanol for 10 min at room temperature.
6. Centrifuge at no more than 12,000*g* for 10 min at 4°C. Discard the supernate.
7. Wash the RNA pellet once with cold 70% ethanol. Mix pellets with ethanol by vortexing, and spin at no more than 7500*g* for 5 min at 4°C.

8. Pipet off the ethanol, and briefly dry the RNA pellet (air-dry or vacuum-dry) for 5–10 min. Avoid overdrying the pellet.

9. Dissolve RNA in 50 mL of DEPC water by passing the solution a few times through a pipet tip and incubating for 10 min at 55°C.

10. Quantitate the RNA concentration by spectrophotometry (in a 1-cm-long light path, [RNA] = $A_{260} \times 40$ µg/mL × dilution factor) and store at −70°C. Use 1–5 µg for the RT reaction.

## RNA Extraction from Paraffin-Embedded Tissue

Although RNA purified from fixed and paraffin-embedded tissue is substantially degraded, in many cases it can provide a substrate for amplification of small-size fragments by the PCR. The success of extraction and amplification of RNA from paraffin-embedded tissues depends on several factors, including the type of fixative and duration of fixation, as well as the duration of paraffin block storage. The effects of fixative type and duration of fixation on RNA yield are similar to those on DNA *(8,26)* and are discussed earlier in the chapter. However, it is important to note that RNA is more susceptible to degradation so that even after fixation in buffered formalin for less than 24 h, usable RNA can be extracted only from approximately 50% of paraffin blocks *(27)*. Duration of storage of paraffin blocks is also important. In our experience, RNA extracted from paraffin blocks stored for more than 20 years allows amplification of 120–150-bp fragments in <10% of cases.

METHOD

1. Cut 5–10 sections, 20 µm thick, using a microtome with a fresh blade and avoiding cross-contamination between samples. Push the cut sections into a microcentrifuge tube using a sterile disposable tip.

2. Deparaffinize sections by adding 1 mL of xylene, vortexing for 1 min, and spinning down at 13,000*g* for 5 min in a microcentrifuge. Pipet off the xylene.

3. Repeat **step 2** once.

4. Add 1 mL of 100% ethanol. Vortex for 30 s, spin at 13,000*g* for 5 min and pipet off the ethanol.

5. Add 1 mL of 80% ethanol. Vortex for 30 s, spin at 13,000*g* for 5 min, and pipet off the ethanol.

6. Add 1 mL of 50% ethanol. Vortex for 30 s, spin at 13,000*g* for 5 min, and pipet off the ethanol. Air-dry or vacuum-dry for 3 min.

7. Add 450 µL of digestion buffer (20 m*M* Tris-HCl, 20 m*M* EDTA, 1% SDS, pH 8.0) and proteinase K to 100 µg/mL and incubate at 55° overnight. If not completely digested, add an additional 50 µg/mL of proteinase K and incubate for 3–4 h more. (After this step, the alternative protocol can be used.*)

8. Spin at 15,000*g* for 5 min at room temperature. Retain supernatant and discard pellet. Acidify supernatant with 0.1 vol of 3 *M* sodium acetate (pH 4.8).

9. Add 500 µL water-saturated phenol, invert 20 times, and centrifuge at 15,000*g* for 5 min. Transfer the supernatant aqueous phase without disturbing the interphase to a new microcentrifuge tube.

10. Add 500 µL of phenol/chloroform/isoamyl alcohol (25:24:1), invert 20 times, and centrifuge at 15,000*g* for 5 min. Transfer the supernatant to a new tube.

11. Add 500 µL of chloroform/isoamyl alcohol (24:1), invert 20 times, and centrifuge at 15,000*g* for 5 min. Transfer the supernatant to a new tube.

12. Add 2 µL of glycogen (20 µg/µL) in each tube. Precipitate RNA with 0.8 vol of isopropanol at −20°C overnight or at −70°C for 2 h.

13. Spin at 12,000*g* for 15 min at 4°C, wash pellet with 70% ethanol, and air-dry or vacuum-dry pellet (avoid overdrying).

14. Dissolve pellets in 15 µL of DEPC water. Store at −70°C. Use 5 µL for each RT reaction.

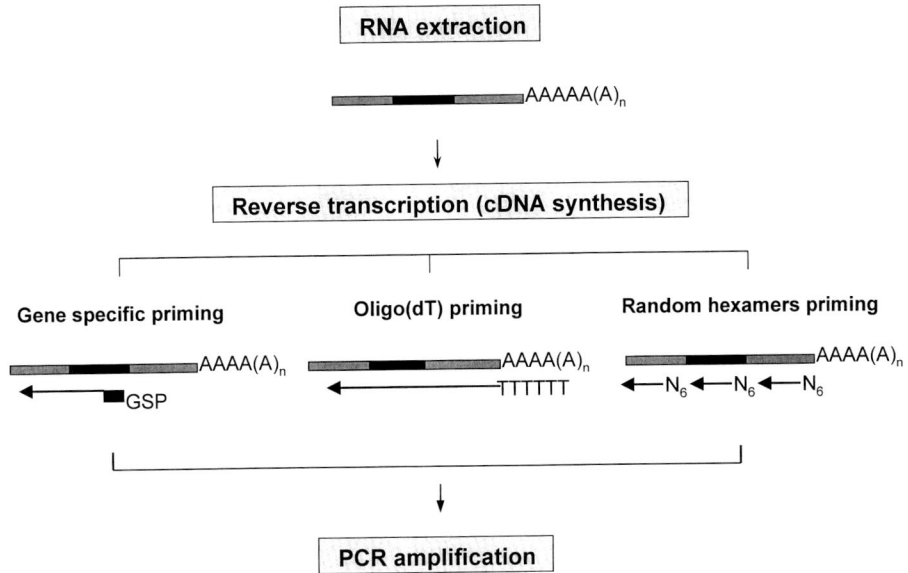

**Fig. 3.** Schematic representation of the reverse transcriptase-polymerase chain reaction (RT-PCR) using different priming methods.

*Alternatively, after digesetion with proteinase K **(step 7)**, isolation of RNA can be performed in a single step using TRIzol reagent or equivalent as follows:

8. Divide the contents of each tube into two separate 1.5-mL microcentrifuge tubes. Add to each tube 1 mL of TRIzol reagent and 200 µL of chloroform; shake tubes vigorously by hand for 15 s and incubate at room temperature for 3 min.

9. Spin at no more than 12,000g for 10 min at 4°C. Transfer the upper aqueous phase to a new tube.

10. Add 2 µL of glycogen to each tube. Precipitate RNA with 0.1 vol of 3 *M* sodium acetate and 0.8 vol of isopropanol at −20°C overnight or at −70°C for 2 h.

11. Spin at 12,000g for 15 min at 4°C, wash pellet with 70% ethanol, and air-dry or vacuum-dry pellet (avoid overdrying).

12. Dissolve pellets in 15 µL of DEPC water. Store at −70°C. Use 5 µL for each RT reaction.

### Reverse Transcription of RNA into cDNA (First-Strand cDNA Synthesis)

Reverse transcription of the RNA molecule into the first cDNA strand is usually performed as a separate reaction from PCR, or it can be coupled with PCR in the same tube by using the T*th* DNA polymerase that acts both as a thermostable RT and DNA polymerase. The former approach gives the advantage of producing a stock of cDNA that can be used for multiple PCR assays.

The basic components of the RT reaction are RNA template, RT enzyme, and primers to initiate the transcription to a complementary DNA strand. Three types of primers can be used: random hexanucleotides (hexamers), an oligo(dT) primer that binds to the poly(A) tail of mRNA, and gene-specific primers **(Fig. 3)**.

1. Random hexamers bind randomly at complementary sites in the RNA molecules to prime cDNA synthesis. This is the most nonspecific priming, and it ensures that all RNA fragments

in the extracts are templates for first-strand cDNA synthesis. This is particularly important when the integrity of RNA is compromised or unknown. This is the only method to prime RNA isolated from fixed tissues.

2. Oligo(dT) primers (12–18 nucleotides) bind to 3′ polyA tails of mRNA to prime synthesis of full-length cDNAs. This is a more specific priming method to prime selectively mRNA that constitutes only 1–2% of total RNA. The amount of cDNA is considerably less, but the complexity is greater (ribosomal RNA is not copied) than when random hexamers are used. This method may give an RT-PCR product more consistently.

3. Gene-specific primer (GSP) contains a sequence of the target cDNA and therefore primes selective transcription of the RNA of interest. This is the most specific priming method; however, some GCPs fail to prime first-strand cDNA synthesis even though they function in PCR of a DNA template.

The protocol for reverse transcription of RNA with random hexamer primers using the Gibco-BRL kit is as follows:

1. Prepare RNA/primer mixtures in a sterile, nuclease-free 0.5-mL microcentrifuge tube as follows:

    1–5 μg total RNA (or 5 μL of RNA isolated from paraffin-embedded tissue)
    random hexamers (50 ng/μL)                                    4 μL
    10 m$M$ dNTP mix                                              1 μL
    DEPC water                                               to 10 μL

2. Incubate at 65°C for 5 min and place on ice for at least 1 min.
3. Prepare the reaction mixture for each sample as follows:

    10× RT buffer                                                2 μL
    25 m$M$ MgCl$_2$                                              4 μL
    0.1 $M$ DTT                                                  2 μL
    RNAase Inhibitor                                             1 μL

4. Add 9 μL of reaction mixture to each RNA/primer mixture, mix, centrifuge briefly, and incubate at 25°C for 10 min.
5. Add 1 μL (50 U) of SuperScript II reverse transcriptase (Gibco-BRL), mix, and incubate at 25°C for 10 min.
6. Transfer to 42°C and incubate for 50 min.
7. Terminate the reaction at 70°C for 15 min. Chill on ice. Store cDNA at –20°C. Use 2 μL for PCR.

## PCR Step

After the reverse transcription reaction is complete, the amplification of cDNA by PCR is not different from amplification of double-strand DNA templates. All the general and technical considerations for PCR apply to RT-PCR as well. However, it is important to note that RT-PCR is a method that detects the *expression* of a gene, not its presence in the genome. Therefore, in some situations it may be important to ensure that PCR is amplifying cDNA and not genomic DNA molecules that can be present in RNA preparations. This can be achieved by designing primers for cDNA amplification that are complementary to different exons; the amplification will be cDNA specific because many introns are too long to be amplified by standard PCR, or it will result in a much larger size of PCR product. The risk of amplification of contaminant DNA can be eliminated by treating RNA extracts with DNAase I prior to the RT reaction.

To increase chances for successful amplification of RNA targets extracted from paraffin-embedded tissue, the smallest possible target size should be selected—typically in a range of 80–200 bp. Sensitivity of the PCR can be enhanced by Southern blotting of the product and hybridization with a radiolabeled internal oligonucleotide probe *(28)*. However, the use of a second round of PCR amplification or nested RT-PCR should be discouraged because of a high risk of cross-contamination or nonspecific amplification that would lead to false-positive results.

The important controls for each RT-PCR reaction are as follows: a negative control at the reverse transcription step in which the RT enzyme is omitted (a control for the cDNA specificity of the reaction); a standard PCR "no template" negative control; and a positive control in which RNA extracted from tissue with known expression of the target gene is added to the RT reaction. In addition, for each sample studied by RT-PCR for expression of a particular mRNA target, an amplification control is necessary to ensure that the template RNA is of sufficient integrity to allow the amplification of the target product. Usually, a portion of an ubiquitously expressed gene of a similar or larger size should be amplified by PCR using the same cDNA template to ensure that amplification is possible. Examples of such control reactions are amplification of phosphoglycerate kinase cDNA *(29)* or N-RAS cDNA (see below).

A positive result can be concluded only if the negative controls are clean. A negative result can be interpreted as absence of a target mRNA fragment in a specimen only when both conditions are met: 1) the expected level of amplification is detected in the positive control, and 2) the template mRNA showed successful amplification of another similar-size RNA fragment (positive amplification control). All other results must be interpreted as inconclusive and should be repeated.

## QUANTITATIVE PCR

One of the most exciting applications of PCR for the molecular biologist has been its use to quantitate the amount of starting DNA and, especially, RNA in a reaction. The important areas of clinical application of quantitative PCR are the assessment of residual disease after treatment of leukemia or lymphoma, the detection of viral load, and the assay for gene amplification or for loss of heterozygosity.

The principle behind quantitative PCR is simple. Reactions are "spiked" with a known amount of standard DNA of some definition. The standard is amplified at the same time as the test sequence, the PCR products are fractionated by the usual methods, the amount of amplified test DNA is compared with the amount of amplified standard DNA, and the number of initial test copies is calculated based on the relative signal intensities. Typically, a standard curve is generated using different initial numbers of standard DNA molecules to correct for nonlinear standard curves due to decreasing efficiency with increasing cycle number. Initial experiments are required to establish the number of cycles through which amplification remains logarithmic as well as the lower limit of sensitivity.

Choice of standards is a complicated process. It is best to use the same primer set so the PCR conditions are optimal for both test and standard. This requires that the standard curve be generated in separate reaction tubes *(30)*. If, however, there are factors in the test samples that inhibit PCR or if there is tube-to-tube variation, the standard curve will not adequately represent the test reactions. Standards can be amplified in the same tubes as the samples by constructing a sequence that is slightly different in size (50–100 bp) than the test sequence when amplified. Engineering a restriction site polymorphism *(31)*, an insertion *(32)*, or a deletion into a clone of the test sequence is used in a variation termed competitive PCR *(31,32)*. A clever variation of this principle whereby the same standard DNA can be used for a set of genes (primer pairs) of interest has also been described *(33)*.

The most accurate method—and the most expensive—is real-time PCR. With specially designed thermocyclers, attached optical systems, and fluorescently labeled primers or intercalating dyes, the amplification process is monitored in each tube *as it occurs*. A complete description of this method is presented in another chapter of this book.

## APPLICATIONS OF PCR

The versatility of PCR has resulted in multiple PCR-based applications, the number of which continues to increase rapidly. Many clinical diagnostic and research fields take advantage of the high sensitivity and specificity of PCR to generate millions of identical copies of specific portions of individual genes from only a few cells or from archival tissue.

The PCR amplification followed by direct sequencing or another analytic method is a common method to diagnose point mutations in genes associated with inherited diseases or in oncogenes and tumor suppressor genes associated with different types of human neoplasms. The RT-PCR, on the other hand, has become a standard method for detection of specific chromosomal rearrangements that are common in soft tissue sarcomas, hematopoetic malignancies, and some epithelial cancers. Another important area of PCR application is in the detection of infections by amplification of a defined region of DNA or reverse-transcribed RNA using primers unique for the pathogen.

Some of the practical applications of PCR were briefly discussed throughout this chapter. Here, we describe in detail two protocols of the PCR-based screening and detection of point mutations in a gene of interest and of detection of chromosomal rearrangement using RT-PCR.

### *Detection of* p53 *Mutations in DNA Extracted from Paraffin-Embedded Tissue*

When a large number of paraffin-embedded tumor samples needs to be screened for mutations, this can be done using PCR amplification followed by SSCP and sequencing of the selected specimens. The following protocol describes the detection of mutations in exon 5 of the *p53* tumor suppressor gene.

1. Review hematoxylin and eosin-stained slides and select from each case a paraffin block containing tumor tissue only or tumor with a minimal amount of adjacent normal tissue. From each block, cut 5 of 10-μm-thick sections. Alternatively, microdissect tumor away from surrounding normal tissue.
2. Perform partial DNA extraction using the protocol described above for paraffin-embedded tissue: deparaffinize with xylene, hydrate through graded ethanol solutions, digest with proteinase K overnight, and heat-inactivate proteinase K.
3. Obtain the following oligonucleotide primers for *p53* exon 5:
   upstream primer 5′-ATCTGTTCACTTGTGCCCTGACTTTC-3′
   downstream primer 5′-ACCCTGGGCAACCAGCCCTGTC-3′

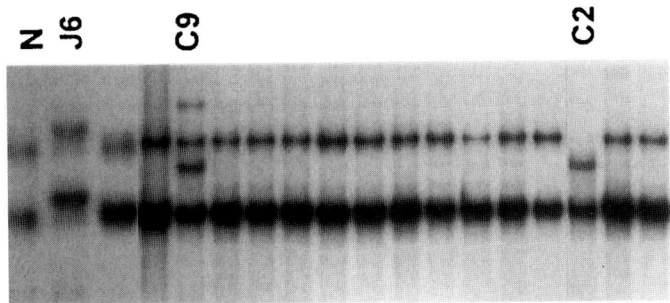

Fig. 4. SSCP analysis of PCR-amplified tumor DNA for mutations in exon 5 of the *p53* gene. N, normal DNA (negative control); J6, tumor sample with known point mutation in *p53* exon 5 (positive control); lanes 3–18, tumor samples. Samples C9 and C2 show abnormal pattern of migration in the polyacrylamide gel, suggesting the presence of mutations.

4. Assemble the following radioactive PCR in a final volume of 30 μL to amplify 18 tumor samples, one normal DNA (e.g., placental DNA) control, and one positive control (DNA from a tumor or cell line with known mutation of *p53* exon 5). (Reaction mix should be prepared for 22 samples)

|  | Final concentration | Add to reaction mix |
|---|---|---|
| 10× PCR buffer | 1× | 66 μL |
| 25 m$M$ MgCl$_2$ | 1.5 m$M$ | 39.6 μL |
| 100 μ$M$ upstream primer | 5 m$M$ | 33 μL |
| 100 μ$M$ downstream primer | 5 m$M$ | 33 μL |
| dNTPs mixed, each 10 m$M$ | 200 μ$M$ each | 13.2 μL |
| *Taq* polymerase (5 U/μL) | 1.5 U/tube | 6.6 μL |
| 1 μCi of [α–$^{32}$P]dCTP (3000 Ci/mmol) |  | 0.2 μL |
| autoclaved distilled water |  | 424.4 μL |

Mix and put 28 μL into each reaction tube. Heat the thermocycler to 80°C, insert tubes, add 100 μg of template DNA to each tube in a total volume of 2 μL (hot start), and begin the program with the following cycling profile:

| initial denaturation | | 4 min at 94°C |
|---|---|---|
| | cycle × 35 | |
| denaturation | | 30 s at 94°C |
| annealing | | 30 s at 59°C |
| elongation | | 1 min at 72°C |
| final elongation | | 5 min at 72°C |

5. When the PCR is finished, perform the SSCP analysis. Dilute 2 μL of each PCR product 1:10 in a loading dye (96% formamide, 20 m$M$ EDTA, 0.05% bromphenol blue, and 0.05% xylene cyanol). Denature samples by heating at 90°C for 4 min and immediately place on ice; load 4 μL onto freshly prepared 0.5% MDE vertical gel (J. T. Baker, Phillipsburg, NJ) in 1× TBE. Run the gel at 600 V for 16 h at room temperature. Remove the gel from glass plates, place onto Whatman paper, and dry on a gel dryer. Put dried gel into an X-ray cassette, load Kodak X-OMAT film, and expose at −70°C for 5 h. Develop and analyze for the shifted bands **(Fig. 4)**. All bands that demonstrate a pattern of migration different from normal DNA suggest the presence of mutations.

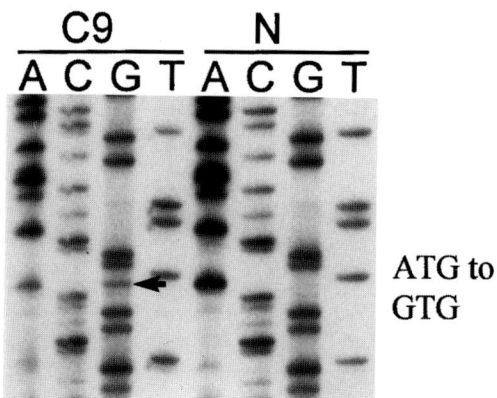

**Fig. 5.** Sequence analysis of tumor sample C9, suspected to harbor a mutation by SSCP analysis, reveals a heterozygous somatic mutation A→G in codon 160 of *p53*. N, normal DNA sequence.

6. Detect the nature of the shifted band by sequencing. Precipitate with ethanol the PCR product from samples that displayed a band shift by SSCP analysis and sequence using the *fmol* sequencing kit (Promega, Madison, WI) by following the manufacturer's instruction. The sequence analysis of two positive samples from the example SSCP gel shown above revealed a heterozygous somatic mutation of A→G in codon 160 of *p53* in sample C9 (**Fig. 5**) and C→T mutation in the third position of codon 182 in sample C2.

## Detection of Chromosomal Rearrangements Using RT-PCR

The presence of a chromosomal rearrangement can be detected by PCR with primers designed to amplify a sequence flanking the fusion point between two genes participating in a rearrangement. In most cases, RNA needs to be isolated from tumor cells, and the detection is done by RT-PCR, because in most rearrangements the breakpoints in two genes occur within introns, some of them very long, with breakpoint locations along the intron varying in each tumor. However, after intronic sequences are spliced out to make mRNA, the same chimeric mRNA is generated from the fusion gene transcript, allowing for standardization of the diagnostic procedure.

Here, we describe a protocol for detection of the RET/ELE1 rearrangement in thyroid papillary carcinomas by RT-PCR.

1. Use 100 µg of fresh or snap frozen tumor tissue to extract RNA according to the protocol supplied earlier in the chapter.
2. Use 3 µg of total RNA for the RT reaction using random hexamer priming as described earlier.
3. Synthesize two oligonucleotide primers that correspond to the RET and ELE1 genes and flank the ELE1/RET fusion point:

    ELE1 primer 5′-AAGCAAACCTGCCAGTGG-3′
    RET primer 5′-CTTTCAGCATCTTCACGG-3′

    The expected size of the amplicon is 242 bp.
4. Assemble the following PCR reaction in a final volume of 30 µL to amplify seven tumor samples, one RT negative control, one PCR negative control, and one positive control (cDNA from a sample known to have ELE1/RET rearrangement). Reaction mix should be prepared for 11 samples:

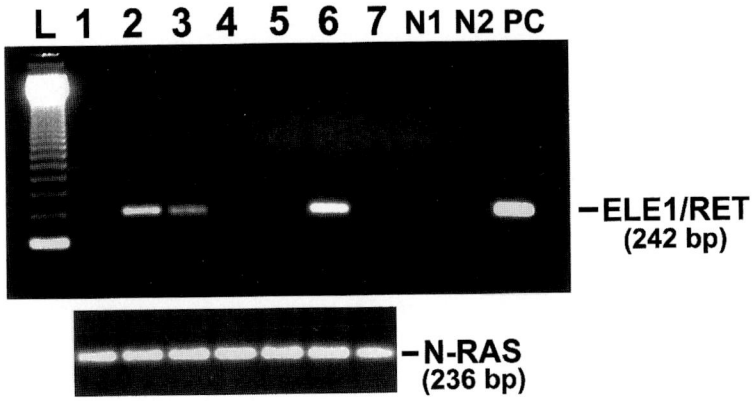

**Fig. 6.** Detection of ELE1/RET rearrangement by RT-PCR. Samples 2, 3, and 6 show the amplification of a 242-bp fragment of the chimeric mRNA and are considered positive for the rearrangement. Samples 1, 4, 5, and 7 are considered negative since they show no ELE1/RET amplification but contain RNA of sufficient quality as detected by amplification of a 236-bp fragment of N-RAS cDNA. The samples were electrophoresed in 1.5% agarose gel. L, 123-bp DNA ladder; N1, RT reaction negative control (no RT enzyme); N2, PCR "no template" negative control; PC, positive control.

|  | Final concentration | Add to reaction mix |
|---|---|---|
| 10× PCR buffer | 1× | 33 μL |
| 25 m$M$ MgCl$_2$ | 1.5 m$M$ | 19.8 μL |
| 100 μ$M$ upstream primer | 5 m$M$ | 16.5 μL |
| 100 μ$M$ downstream primer | 5 m$M$ | 16.5 μL |
| dNTPs mixed, each 10 m$M$ | 200 μ$M$ each | 6.6 μL |
| autoclaved distilled water |  | 212.3 μL |

Mix and put 27.7 μL into each reaction tube. Add to each tube 2 μL of cDNA, insert tubes in thermocycler, and start the program. Approximately 2 min after beginning of the initial denaturation step (4 min at 94°C), add 0.3 μL of *Taq* polymerase (5 U/μL) to each tube (hot start) and run 30 cycles using the following cycling profile:

| denaturation | 30 s at 94°C |
|---|---|
| annealing | 0 s at 59°C |
| elongation | 1 min at 72°C (with final elongation for 5 min at 72°C) |

5. Assemble another PCR to test the integrity of the RNA template. For this purpose, use the same cDNA samples to amplify a 236-bp sequence of the c-N-RAS using a pair of primers spanning intron 1 of the gene:
    upstream primer 5′-ATGACTGAGTACAAACTGGT-3′
    downstream primer 5′-AGGAAGCCTTCGCCTGTCCT-3′
The conditions of the assay, including optimal MgCl$_2$ concentration and annealing temperature, are identical to those used for the ELE1/H4 amplification.

6. Load 10 μL of each PCR product onto a 1.5% agarose gel. Samples that show a 242-bp band are considered to have the rearrangement **(Fig. 6)**. Samples that show no amplification for ELE1/RET, but successfully amplify N-RAS cDNA, are considered negative for the rearrangement. Positive cases can be selectively sequenced to confirm the presence of a chimeric transcript.

## REFERENCES

1. Saiki, R. K., Scharf, S. J., Faloona, F. A., et al. (1985) Enzymatic amplification of β-globin genomic sequences and restriction site analysis for diagnosis of sickle cell anemia. *Science* **230**, 1350–1354.
2. Mullis, K. B. and Faloona, F. A. (1987) Specific synthesis of DNA in vitro via a polymerase-catalyzed chain reaction. *Methods Enzymol.* **155**, 335–350.
3. Engelke, D. R., Hoener, P. A., and Collins, F. S. (1988) Direct sequencing of enzymatically amplified human genomic DNA. *Proc. Natl. Acad. Sci. USA* **85**, 544–548.
4. Wong, C., Dowling, C. E., Saiki, R. K., Higuchi, R. G., Erlich, H. A., and Kazazian, H. H. (1987) Characterization of β-thalassemia mutations using direct genomic sequencing of amplified single copy DNA. *Nature* **330**, 384–386.
5. Saiki, R. K., Gelfand, D. H., Stoffel, S., et al. (1988) Primer-directed enzymatic amplification of DNA with a thermostable DNA polymerase. *Science* **239**, 487–491.
6. Greer, C. E., Peterson, S. L., Kiviat, N. B., and Manos, M. M. (1991) PCR amplification from paraffin-embedded tissues: effects of fixative and fixation time. *Am. J. Clin. Pathol.* **95**, 117–124.
7. Shibata, D. (1994) Extraction of DNA from paraffin-embedded tissue for analysis by polymerase chain reaction: new tricks from an old friend. *Hum. Pathol.* **25**, 561–563.
8. Ben-Ezra, J., Johnson, D. A., Rossi, J., Cook, N., and Wu, A. (1991) Effect of fixation on the amplification of nucleic acids from paraffin-embedded material by the polymerase chain reaction. *J. Histochem. Cytochem.* **39**, 351–354.
9. Greer, C. E., Lund, J. K., and Manos, M. M. (1991) PCR amplification from paraffin-embedded tissues: recommendations on fixatives for long-term storage and prospective studies. *PCR Methods Appl.* **1**, 46–50.
10. Shibata, D., Martin, W. J., and Arnheim, N. (1988) Analysis of DNA sequences in forty-year old paraffin-embedded thin-tissue sections: a bridge between molecular biology and classic histology. *Cancer Res.* **48**, 4564–4566.
11. Kallio, P., Syrjanen, S., Tervahauta, A., and Syrjanen, K. (1991) A simple method for isolation of DNA from formalin-fixed paraffin-embedded samples for PCR. *J. Virol. Meth.* **35**, 39–47.
12. Heller, M. J., Robinson, R. A., Burgart, L. J., TenEyck, C. J., and Wilke, W. W. (1992) DNA extraction by sonication: a comparison of fresh, frozen, and paraffin-embedded tissue extracted for use in polymerase chain reaction assays. *Mod. Pathol.* **5**, 203–206.
13. Sepp, R., Szabo, I., Uda, H., and Sakamoto, H. (1994) Rapid techniques for DNA extraction from routinely processed archival tissue for use in PCR. *J. Clin. Pathol.* **47**, 318–323.
14. Frank, T. S., Svoboda-Newman, S. M., and Hsi, E. D. (1996) Comparison of methods for extracting DNA from formalin-fixed paraffin sections for nonisotopic PCR. *Diagn. Mol. Pathol.* **5**, 220–224.
15. Birch, D. E. (1996) Simplified hot start PCR. *Nature* **381**, 445–446.
16. Bassam, B. J., Caetano-Anolles, G., Gresshoff, P. M. (1991) Fast and sensitive silver staining of DNA in polyacrylamide gels. *Anal. Biochem.* **196**, 80–83.
17. Peats, S. (1984) Quantitation of protein and DNA in silver-stained agarose gels. *Anal. Biochem.* **140**, 178–182.
18. Liu, X. Y., Nelson, D., Grant, C., Morthland, V., Goodnight, S. H., Press, R. D. (1995) Molecular detection of a common mutation in coagulation factor V causing thrombosis via hereditary resistance to activated protein C. *Diagn. Mol. Pathol.* **4**, 191–197.
19. Sutcharitchan, P., Saiki, R., Huisman, T. H., et al. (1995) Reverse dot-blot detection of the African-American beta-thalassemia mutations. *Blood* **86**, 1580–1585.
20. Kant, J. A., Mifflin, T. E., McGlennen, R., Pice, E., Naylor, E., and Cooper, D. L. (1995) Molecular diagnosis of cystis fibrosis. *Clin. Lab. Med.* **15**, 877–898.

21. Orita, M., Suzuki, Y., Sekiya, T., and Hayashi, K. (1989) Rapid and sensitive detection of point mutations and DNA polymorphisms using the polymerase chain reaction. *Genomics* **5**, 874–879.
22. Ainsworth, P. J., Surh, L. C., and Coulter-Mackie, M. B. (1991) Diagnostic single strand conformation polymorphism (SSCP): a simplified non-radioisotopic method as applied to a Tay-Sachs B1 variant. *Nucleic Acids Res.*, **19**, 405–406.
23. Lundberg, K. S., Shoemaker, D. D., Adams, M. W., Short, J. M., Sorge, J. A., Marthur, E. J. (1991) High-fidelity amplification using a thermostable DNA polymerase isolated from *Pyrococcus furiosus*. *Gene* **108**, 1–6.
24. Cariello, N. F., Swenberg, J. A., Skopek, T. R. (1991) Fidelity of *Thermococcus litoralis* DNA polymerase (Vent) in PCR determined by denaturing gradient gel electrophoresis. *Nucleic Acids Res.* **19**, 4193–4198.
25. Chomczynski, P. and Sacchi, N. (1987) Single-step method of RNA isolation by acid guanidine thiocyanate-phenol-chloroform extraction. *Anal. Biochem.* **162**, 156–159.
26. Foss, R. D., Guha-Thakurta, N., Conran, R. M., Gutman, P. (1994) Effect of fixative time on the extraction and polymerase chain reaction amplification of RNA from paraffin-embedded tissue. *Diagn. Mol. Pathol.* **3**, 148–155.
27. Ladanyi, M. and Bridge, J. A. (2000) Contribution of molecular genetic data to the classification of sarcomas. *Hum. Pathol.* **31**, 532–538.
28. Nikiforov, Y. E., Rowland, J. M., Bove, K. E., Monforte-Munoz, H., and Fagin, J. A. (1997) Distinct pattern of ret oncogene rearrangements in morphologic variants of radiation-induced and sporadic thyroid papillary carcinomas in children. *Cancer Res.* **57**, 1690–1694.
29. Argani, P., Zakowski, M. F., Klimstra, D. S., Rosai, J., and Ladanyi, M. (1998) Detection of the SYT-SSX chimeric RNA of synovial sarcoma in paraffin-embedded tissue and its application in problematic cases. *Mod. Pathol.* **11**, 65–71.
30. Muthuchamy, M., Pajak, L., Howles, P., Doetschman, T., and Wieczorek, D. (1993) Developmental analysis of tropomyosin expression in embryonic stem cells and mouse embryos. *Mol. Cell. Biol.* **13**, 3311–3323.
31. Becker-Andre, M. and Hahlbrock, K. (1989) Absolute mRNA quantification using the polymerase chain reaction (PCR). A novel approach by a PCR aided transcript titration assay (PATTY). *Nucleic Acids Res.* **17**, 9438–9446.
32. Gilliland, G., Perrin, S., Blanchard, K., and Bunn, F. (1990) Analysis of cytokine mRNA and DNA: detection and quantitation by competitive polymerase chain reaction. *Proc. Natl. Acad. Sci. USA* **87**, 2725–2729.
33. Wang, A. M., Doyle, M. V., and Mark, D. F. (1989) Quantitation of mRNA by the polymerase chain reaction. *Proc. Natl. Acad. Sci. USA* **86**, 9717–9721.

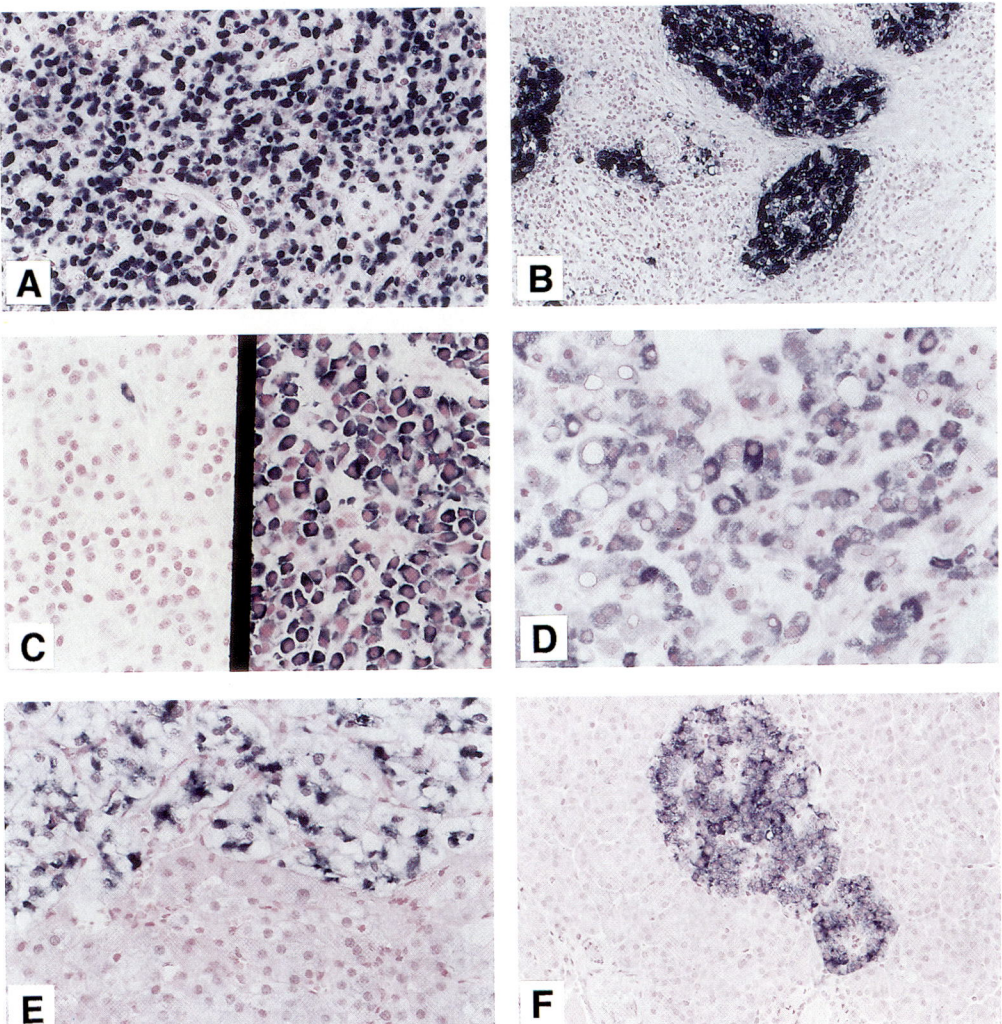

**Fig. 1. (A)** *In situ* hybridization detects EBV in a posttransplant lymphoproliferative disorder. There is diffuse staining of all lymphoid cells for EBV probe. *In situ* hybridization with
digoxigenin-labeled probes and detection by alkaline phosphatase/NBT-BCIP for all figures.
**(B)** *In situ* hybridization to detect EBV in a nasopharyngeal carcinoma metastatic to a cervical
lymph node. The presence of EBV in this clinical setting is diagnostic of a nasopharyngeal
carcinoma. **(C)** Detection of κ (left) and λ (right) light chains in a plasmacytic lymphoma. The
tumor shows λ light chain restriction with all of the tumor cells expressing this light chain. A
few residual normal plasma cells stained positively for κ light chain. **(D)** *In situ* hybridization
for albumin to diagnose a hepatocellular carcinoma metastatic to the scapula. Albumin expression is relatively specific for normal and neoplastic liver cells. **(E)** Chromogranin A and B
expression in the adrenal cortex. The adrenal cortex in the lower portion of the figure is negative for chromogranin A and B cocktail probe. **(F)** Pro-insulin mRNA expression in the normal
pancreatic islets. Most of the islet cells (60–80%) express the pro-insulin mRNA (original magnification ×250 for A, B, D, and E; ×200 for F).

*Color Plate 1 - Fig. 1, Chapter 3, see page 37.*

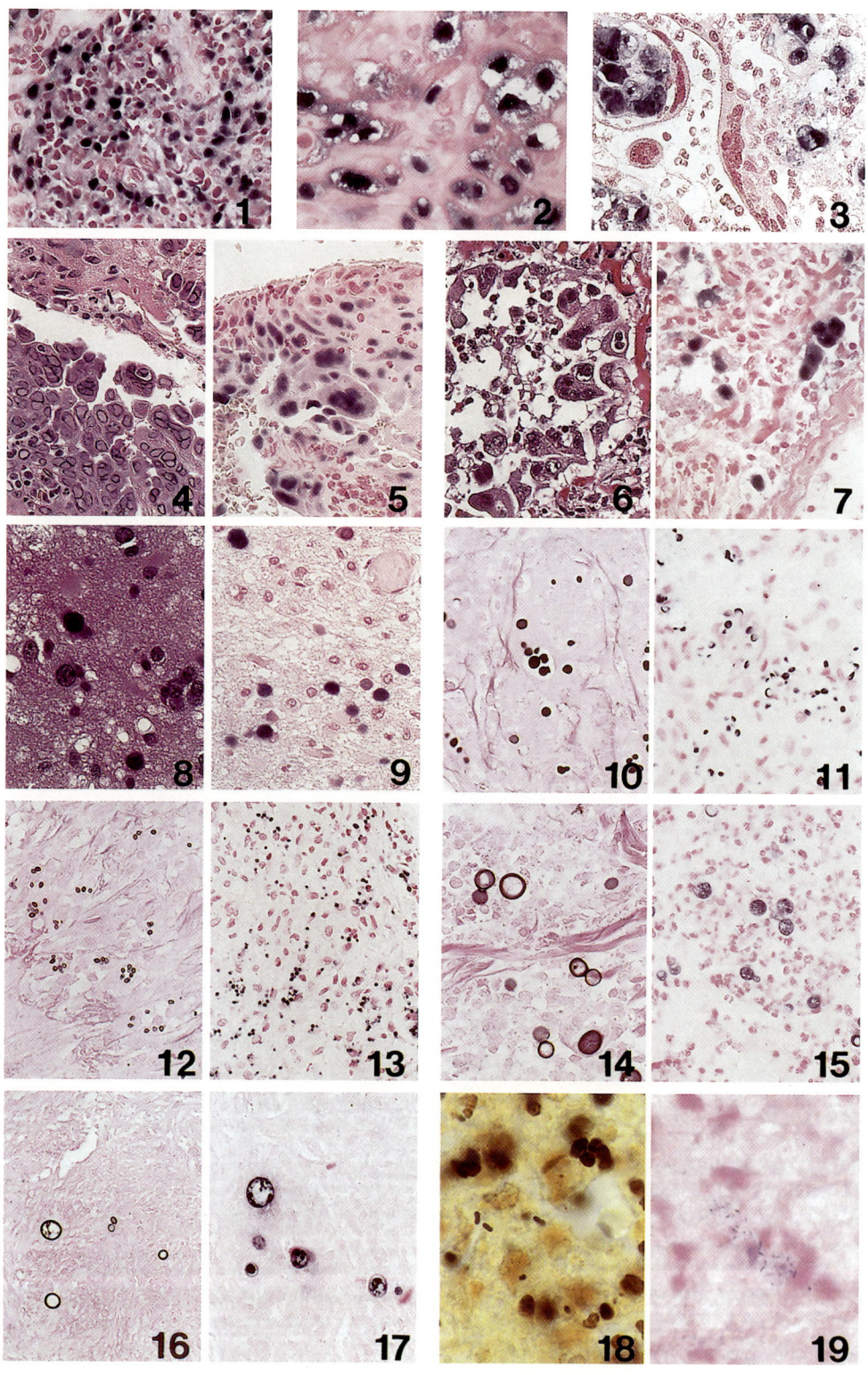

*Color Plate 2* - Figs. 1–19, Chapter 4, see pages 48 and 49.

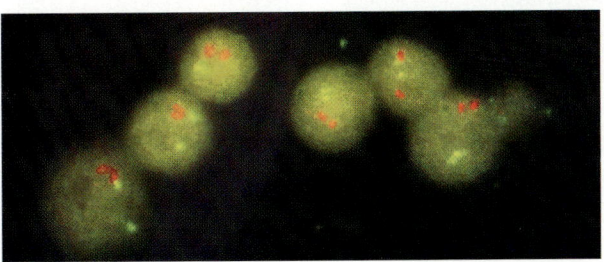

**Fig. 5.** A two-color microdeletion probe. The green signal is a marker for chromosome 15, in this case the PML gene, while the red signal shows the Prader-Willi region on proximal 15q.

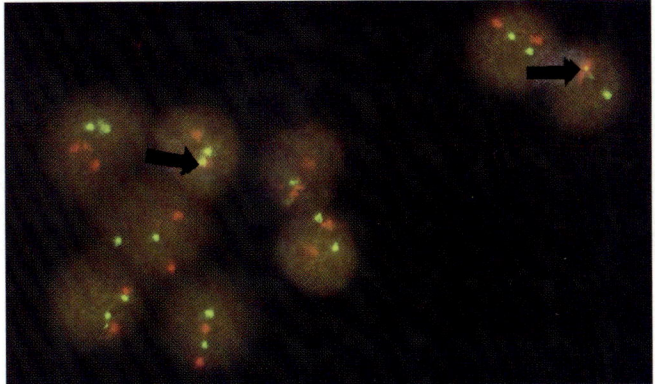

**Fig. 6.** A two-color probe for *BCR* and *ABL* genes on a touch preparation from a spleen. Note occasional fusion signals (arrows) which are suggestive of a translocation.

**Fig. 7.** Painting probe for chromosome 9. Note that the centromeres, which contain repeated sequences, do not paint. Often a faint banding pattern can be seen with paints because the distribution of unique sequences is not even.

***Color Plate 3*** *- Figs. 5–7, Chapter 5, see pages 74–79.*

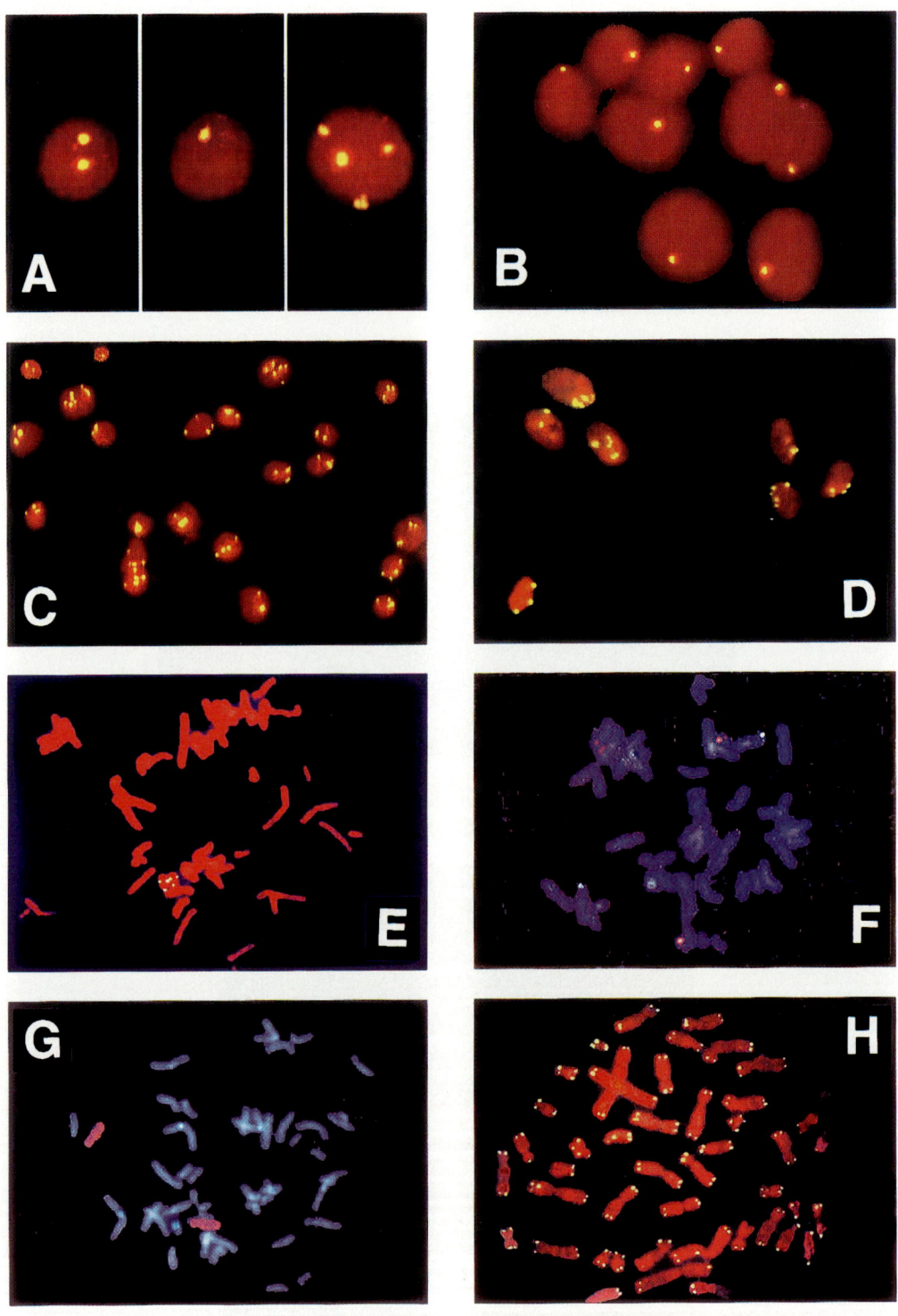

***Color Plate 4*** - *Fig. 3, Chapter 6, see pages 94, 95.*

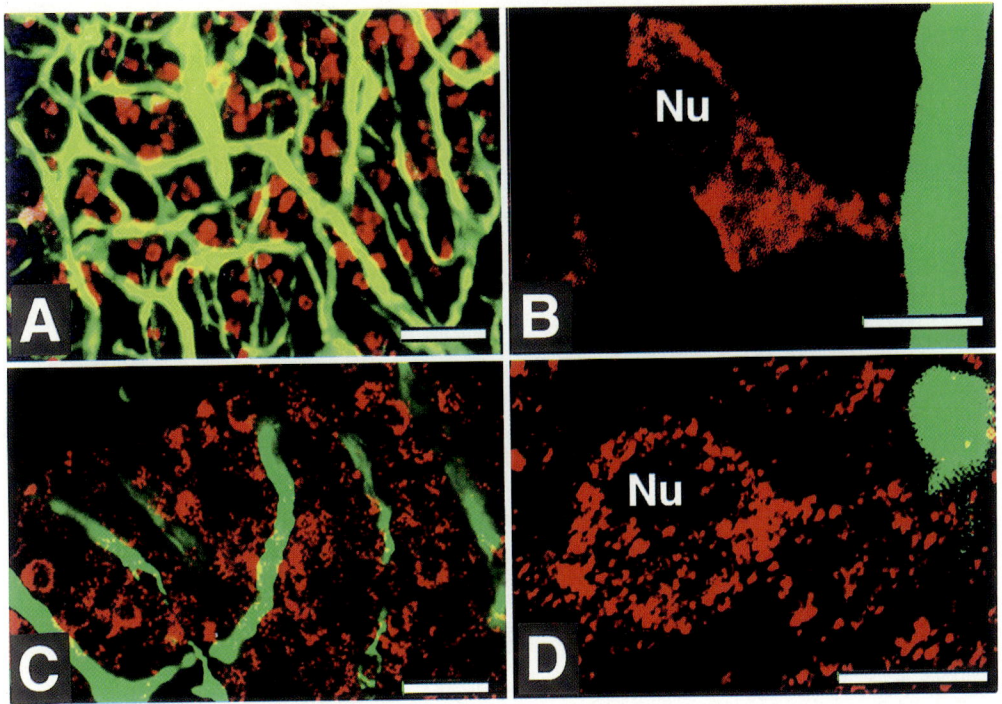

**Fig. 9, A–D.**

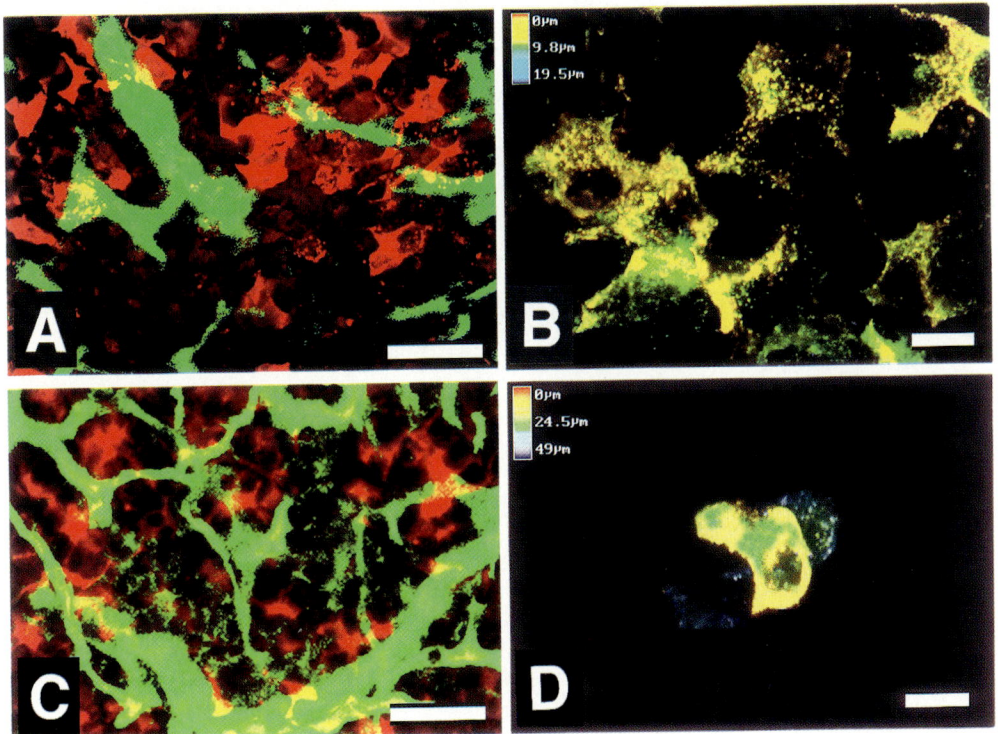

**Fig. 10, A–D.**

*Color Plate 5* - Figs. 9 and 10, Chapter 10, see pages 174–177.

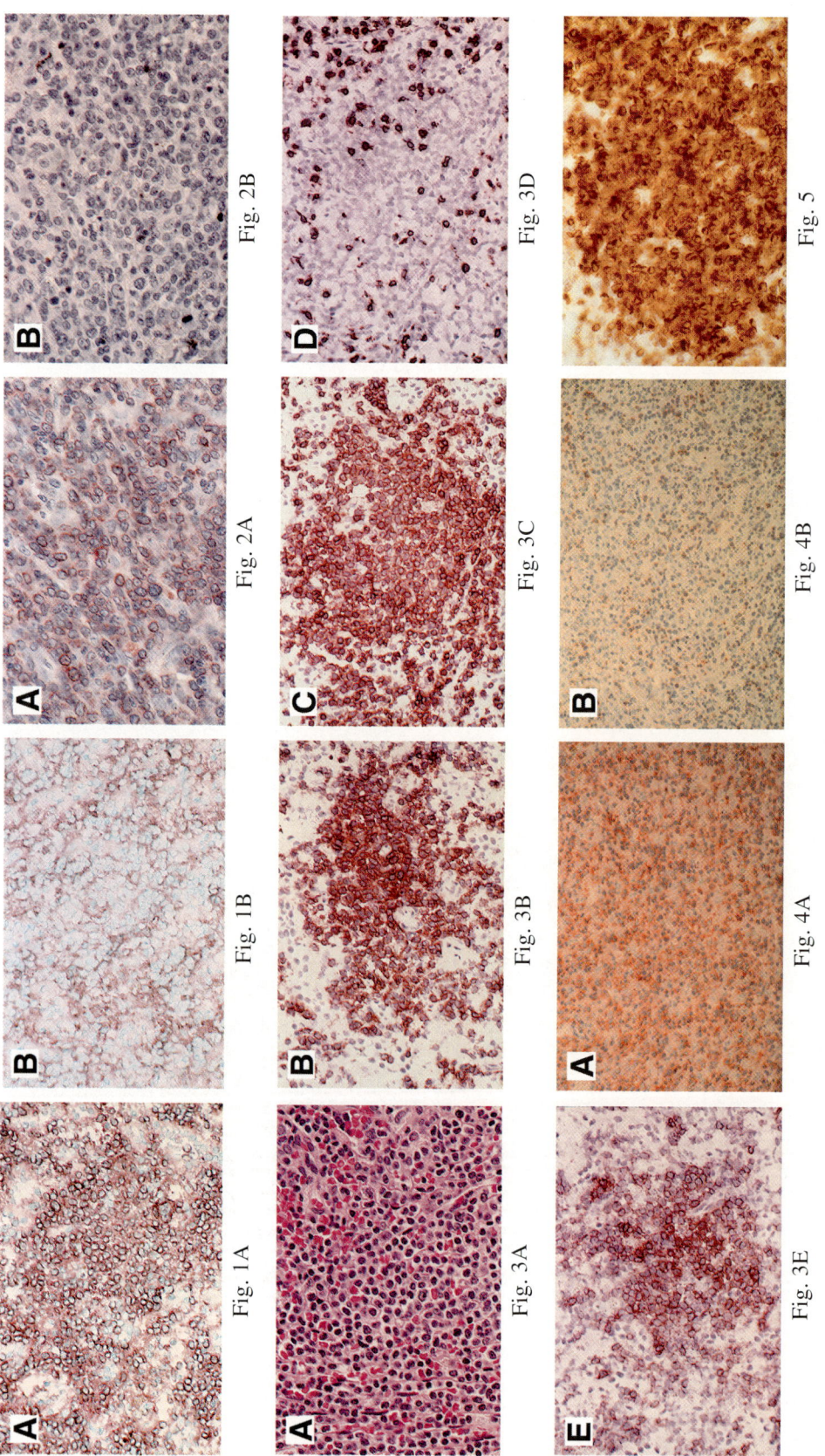

**Color Plate 6** - *Figs. 1–5, Chapter 16, see pages 279–289.*

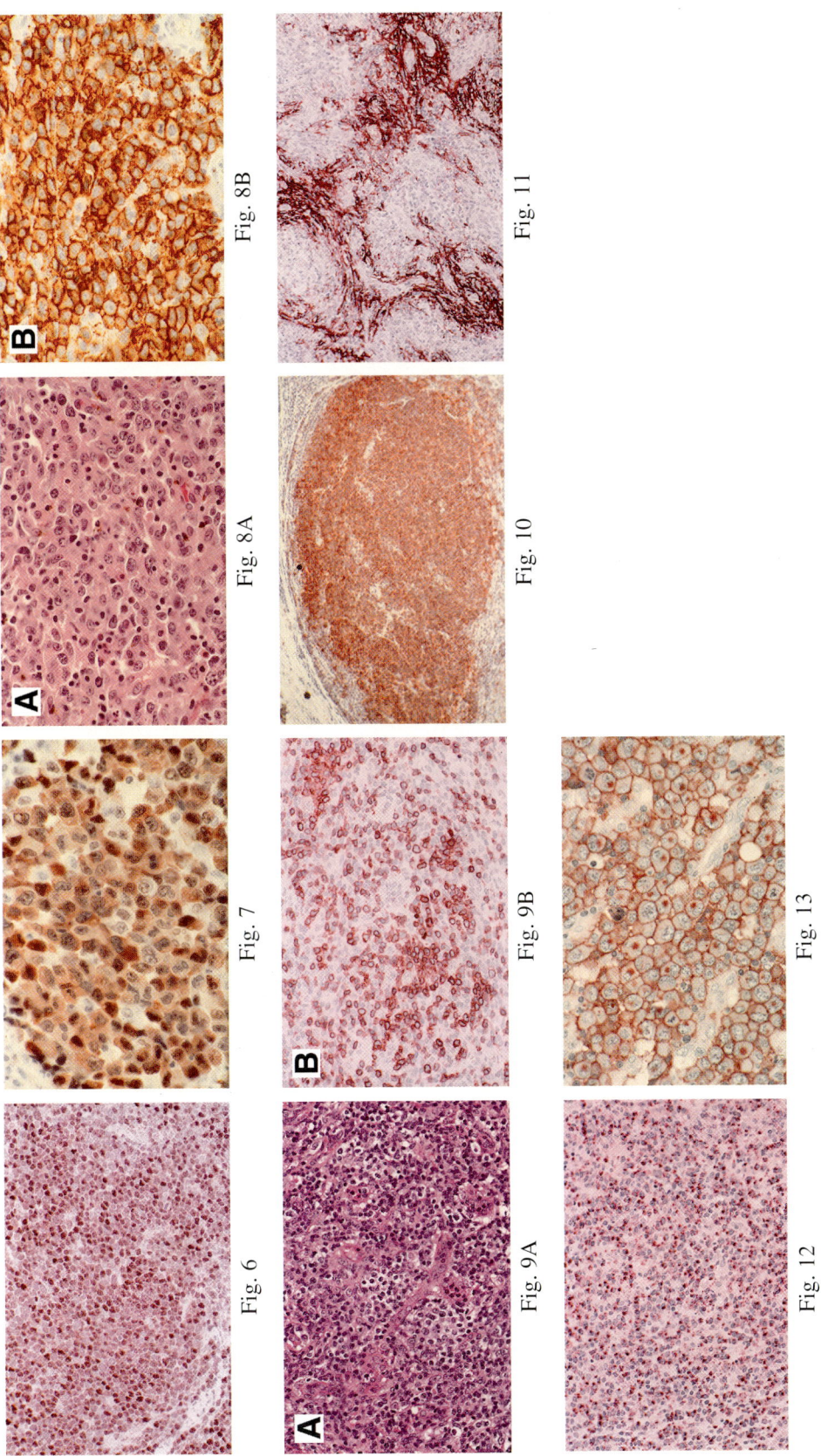

***Color Plate 7*** - *Figs. 6 – 13 from Chapter 16, see pages 270–303.*

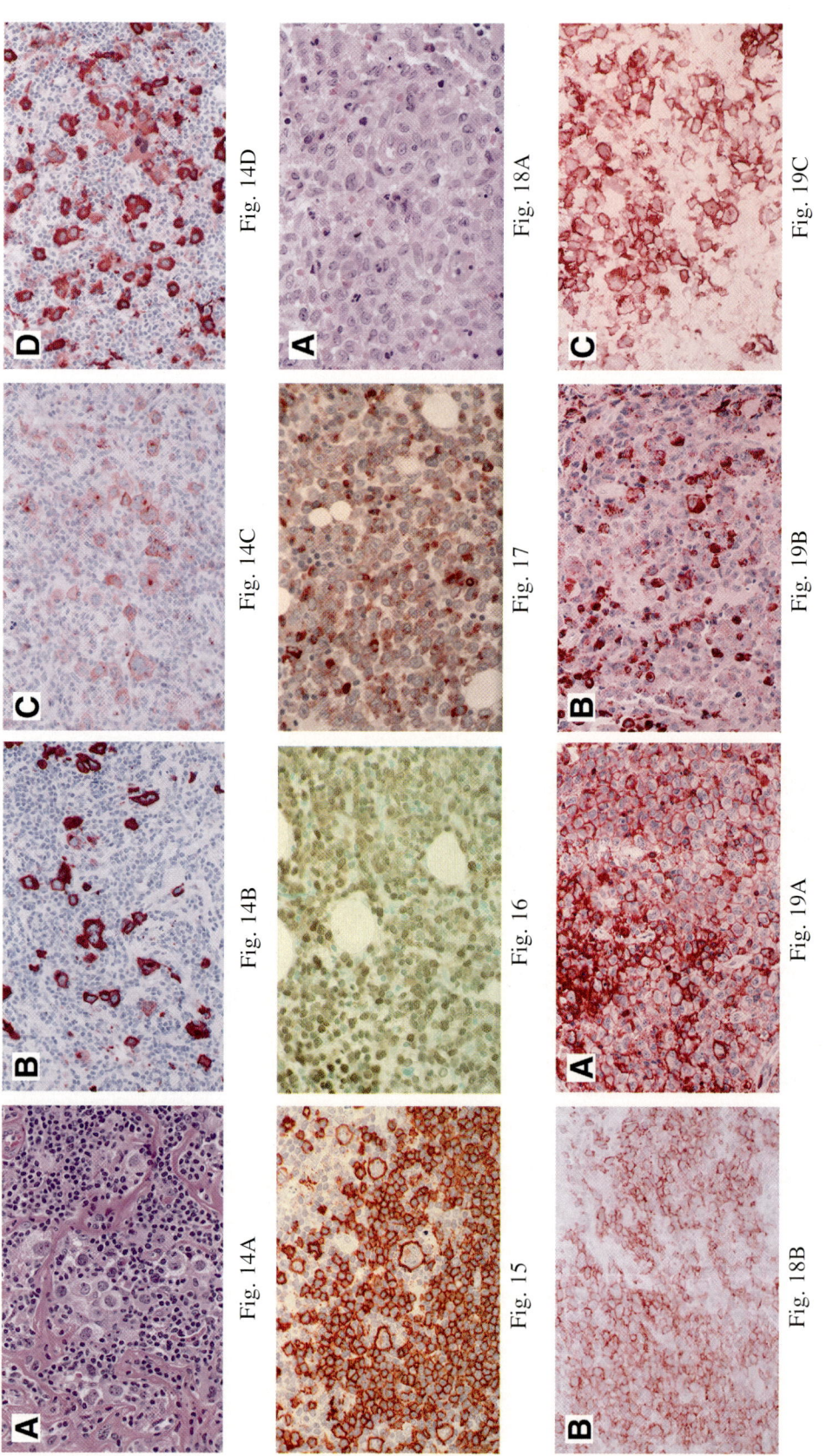

**Color Plate 8** - *Figs. 14–19, Chapter 16, see pages 304–310.*

# 12

# Detection of Nucleic Acids in Cells and Tissues by *In Situ* Polymerase Chain Reaction

## Omar Bagasra, MD, PHD Lisa E. Bobroski, MS, and Muhammad Amjad, PHD

## INTRODUCTION

The solution-based polymerase chain reaction (PCR) method for amplification of defined gene sequences has proved a valuable tool not only for basic researchers but also for clinical scientists. Using even a minute amount of DNA or RNA and choosing a thermostable enzyme from a large variety of sources, one can enlarge the amount of the gene of interest, which can be analyzed and/or sequenced. Thus genes or segments of gene sequences present only in a small sample of cells or small fraction of mixed cellular populations can be examined. However, one of the major drawbacks of the solution-based PCR technique is that the procedure does not allow the association of amplified signals of a specific gene segment with the histologic cell type(s) *(1–24)*. For example, it would be advantageous to determine what types of cells in the peripheral blood circulation or in a pathologic specimens carry the human immunodeficiency virus type 1 (HIV-1) gene or a vector used for gene therapy or an aberrant gene in a leukemia patient and what percent of leukemia cells is present after various forms of antitumor therapies.

The ability to identify under the microscope individual cells expressing or carrying specific genes of interest in a latent form in a tissue section provides a great advantage in determining various aspects of normal, as opposed to pathologic, conditions. For example, this technique could be used in determination of viral burden *(11)*, degree of gene expression *(13,21,22)*, or tumor burden *(25,26)*, before and after therapy *(24,27,32)*, in lymphomas or leukemias, in which specific aberrant gene translocations are associated with certain types of malignancy *(28)*. In the case of HIV-1 or other viral infections, one can determine the effects of therapy or putative antiviral vaccination by evaluating the number of cells still infected with viral agent post chemotherapy or vaccination *(2,5,6,20)*. Similarly, one can potentially determine the preneoplastic lesions by examining tumor suppresser genes (i.e., *p53* mutations associated with certain tumors or oncogenes or other aberrant gene sequences known to be associated with certain types of tumors) *(26,28,29)*. In the area of diagnostic pathology, determination of origin of metastatic tumors is a perplexing problem. By utilizing the proper primers for genes

From: *Morphology Methods: Cell and Molecular Biology Techniques*
Edited by: R. V. Lloyd © Humana Press, Totowa, NJ

expressed by certain histologic cell types, one can potentially determine the origin of metastatic tumors by performing reverse transcriptase (RI)-initiated *in situ* PCR *(26)*. One can also determine the subcellular localization of a gene or virus inside the cytoplasmic or nuclear compartments *(2–7,12,14,18,27,30–32)*. Both RNA and DNA signal can be detected simultaneously by utilizing primers that anneal to the splicing junctions of exons of mRNA *(8,10,13,16,17,21,22,25,33,34)*.

Our laboratories have been utilizing *in situ* PCR techniques for several years, and we have developed simple, sensitive protocols for both RNA and DNA *in situ* PCR that have proved reproducible in multiple double-blinded studies *(2–7,10,17,18,22)*. One can use this method for amplification of both DNA or RNA gene sequences. By use of multiple labeled probes, one can detect various signals in a single cell *(14,22)*. In addition, under special circumstances, one can perform immunohistochemistry and RNA and DNA amplification at a single cell level (so-called triple labeling) *(10,23)*.

To date, we have successfully amplified and detected HIV, simian immunodeficiency virus (SIV), human papillomavirus (HPV), hepatitis B virus (HBV), cytomegalovirus (CMV), Epstein-Barr virus (EBV), human herpes virus (HHV)-6, HHV-8, herpes simplex virus (HSV), and *p53* and its mutations, as well as mRNA for surfactant protein A, estrogen receptors, and inducible nitrous oxide synthesis (iNOS) gene sequences associated with multiple sclerosis, by DNA and/or RNA (RT *in-situ* PCR). We have used various tissues, including peripheral blood mononuclear cells (PBMCs), lymph node, spleen, skin, breast, lungs, placenta, sperm, cytologic specimens, Kaposi's sarcoma (KS), cultured cells, numerous others formalin-fixed, paraffin-embedded tissues and various cellular populations in the frozen sections of AIDS brains, infected with HIV-1 *(1–14,17,18,22,23,25,30)*. In the following pages, we present a detailed protocol currently being utilized in our laboratory. For more details, the readers are referred to ref. *33*.

## PREPARATION OF GLASS SLIDES

Before one can perform *in situ* reactions by this protocol, the proper sort of glass slide with a Teflon appliqué must be obtained. Then the glass surface must be treated with the proper sort of silicon compound. Both of these factors are very important, and the following discussion explains why.

### Glass Slides with Teflon-Bordered Wells

First, one should always use glass slides that are partially covered by a Teflon coating. Not only does the glass withstand well the stress of repeated heat denaturations, but it also presents the right chemical surface—silicon oxide—needed for proper silanization.

Furthermore, slides with special Teflon coatings that form individual "wells" are useful because vapor-tight reaction chambers can be formed on the surface of the slides when cover slips are attached with coatings of nail polish around the periphery. These reaction chambers are necessary because within them, proper tonicity and ion concentrations can be maintained in aqueous solutions during thermal cycling—conditions that are vital for proper DNA amplification. The Teflon coating serves a dual purpose in this regard. First, the Teflon helps keep the two glass surfaces slightly separated, allowing for reaction chambers about 20 μm in height to form between. Second, the Teflon border helps keep the nail polish from entering the reaction chambers when the polish is being applied. This is important, because any leakage of nail polish into a reaction chamber can compromise the results in that chamber. Even if one is using an advanced thermal cycler with humidification, use of the Teflon-coated slides is still recommended, as the hydrophobicity of the Teflon combined with the pressure applied by a cover slip helps spread small volumes of reaction cocktail over the entire sample region, without forcing much fluid out the periphery.

To prepare glass slides properly, follow this procedure:

1. Prepare the following 2% AES solution just prior to use: 3-aminopropyltriethoxysilane (AES; Sigma A-3648) 5 mL and acetone 250 mL.
2. Put solution into a Coplin jar or glass staining dish and dip glass slides in 2% AES for 60 s. Allow to dry for 1–2 h at room temperature or overnight, if not in a hurry.
3. Dip slides five times into a different vessel filled with 1000 mL of distilled water.
4. Repeat **step 3** three to four times, changing the water each time.
5. Air-dry in a laminar flow hood from a few hours to overnight, and then store slides in a sealed container at room temperature. Try to use slides within 15 days of silanization; 250 mL of AES solution is sufficient to treat 200 glass slides.

## PREPARATION OF TISSUE

### Cell Suspensions

To use peripheral blood leukocytes, first isolate cells on a Ficoll-Hypaque density gradient. Tissue culture cells or other single-cell suspensions can also be used. Prepare all cell suspensions with the following procedure:

1. Wash cells with 1× phosphate-buffered saline (PBS) twice.
2. Resuspend cells in PBS at $2 \times 10^6$ cells/mL.
3. Add at least 10 μL of cell suspension to each well of the slide using a P20 micropipet, and spread across surface of slides.
4. Air-dry slide in a laminar flow hood.

### Adherent Cultured Cells

Several types of slides are designed to support *in situ* PCR after they are attached on the glass slides. We favor the FALCON CultureSlide setup. The cells are grown on this type of slides. If certain primary cell cultures require attachment factors or growth media, these could be used in conjunction with this cell culture system. After appropriate confluency (usually >60%), cells are washed with 1× PBS, gently, then heat-fixed, and fixed with 2% paraformaldehyde (PFA), overnight.

### Paraffin-Fixed Tissue

Routinely fixed paraffin tissue sections can be amplified quite successfully. This permits the evaluation of individual cells in the tissue for the presence of a specific RNA or DNA sequence. For this purpose, tissue sections are placed on specially designed slides that have single wells (described further in the **Materials and Methods** section, page 224). In our laboratory, we routinely use placental tissues, central nervous system (CNS) tissues, cardiac tissues, and so on, which are sliced to a 5–6 μm thickness.

Other laboratories prefer to use sections up to 10-µm thickness, but in our experience, amplification is often less successful with the thicker sections, and multiple cell layers can often lead to difficult interpretation due to superposition of cells. However, if one is using tissues that contain particularly large cells—such as ovarian follicles—then thicker sections may be appropriate.

1. Place tissue section on the glass surface of the slide.
2. Incubate the slides in an oven at 60–80°C (depending on type of paraffin used to embed the tissue) for 1 h, to melt the paraffin.
3. Dip the slides in EM grade xylene solution for 5 min, and then in EM grade 100% ethanol for 5 min. (EM grade reagents are benzene-free.) Repeat these washes two or three times, to rid the tissue of paraffin completely.
4. Dry the slides in an oven at 80°C for 1 h.

### Frozen Sections

It is possible to use frozen sections for *in situ* amplification; however, the morphology of the tissue following the amplification process is generally not as good as with paraffin sections. It seems that the cryogenic freezing of the tissue, combined with the lack of paraffin substrate during slicing, compromises the integrity of the tissue. Usually, thicker slices must be made, and the tissue "chatters" in the microtome. As any clinical pathologist will relate, definitive diagnoses are made from paraffin sections, and this rule of thumb seems to extend to the amplification procedure as well.

It is very important to use tissues that were frozen in liquid nitrogen or placed on dry ice immediately after they were harvested before autolysis began to take place. If tissues were frozen slowly by placing them at −70°C, then eventually ice crystals will form inside the tissues, creating a gap that will distort the morphology.

To use frozen sections, use as thin a slice as possible (down to 4–6 µm), apply to the slide, dehydrate for 10 min in 100% methanol (the exception to methanol is when surface antigens are lipoprotein, which will denature in methanol; then use 2% PFA or other reagent), and air-dry in a laminar flow hood. Then proceed with the heat treatment described below.

## IN SITU AMPLIFICATION: *LOW ABUNDANCE OF DNA AND RNA TARGETS*

### Basic Preparation, All Protocols

For all sample types, the following steps comprise the basic preparatory work that must be done before any amplification-hybridization procedure. The flowchart is depicted in **Fig. 1**.

### Heat Treatment

Place the slides with attached tissue or cells on a heat block at 105°C for 0.5–2 min, to stabilize the cells or tissue on the glass surface of the slide. This step is absolutely critical, and one may need to experiment with different periods to optimize the heat treatment for specific tissues. Our laboratory routinely uses 90 s for DNA target sequences and 30–45 s for RNA sequences.

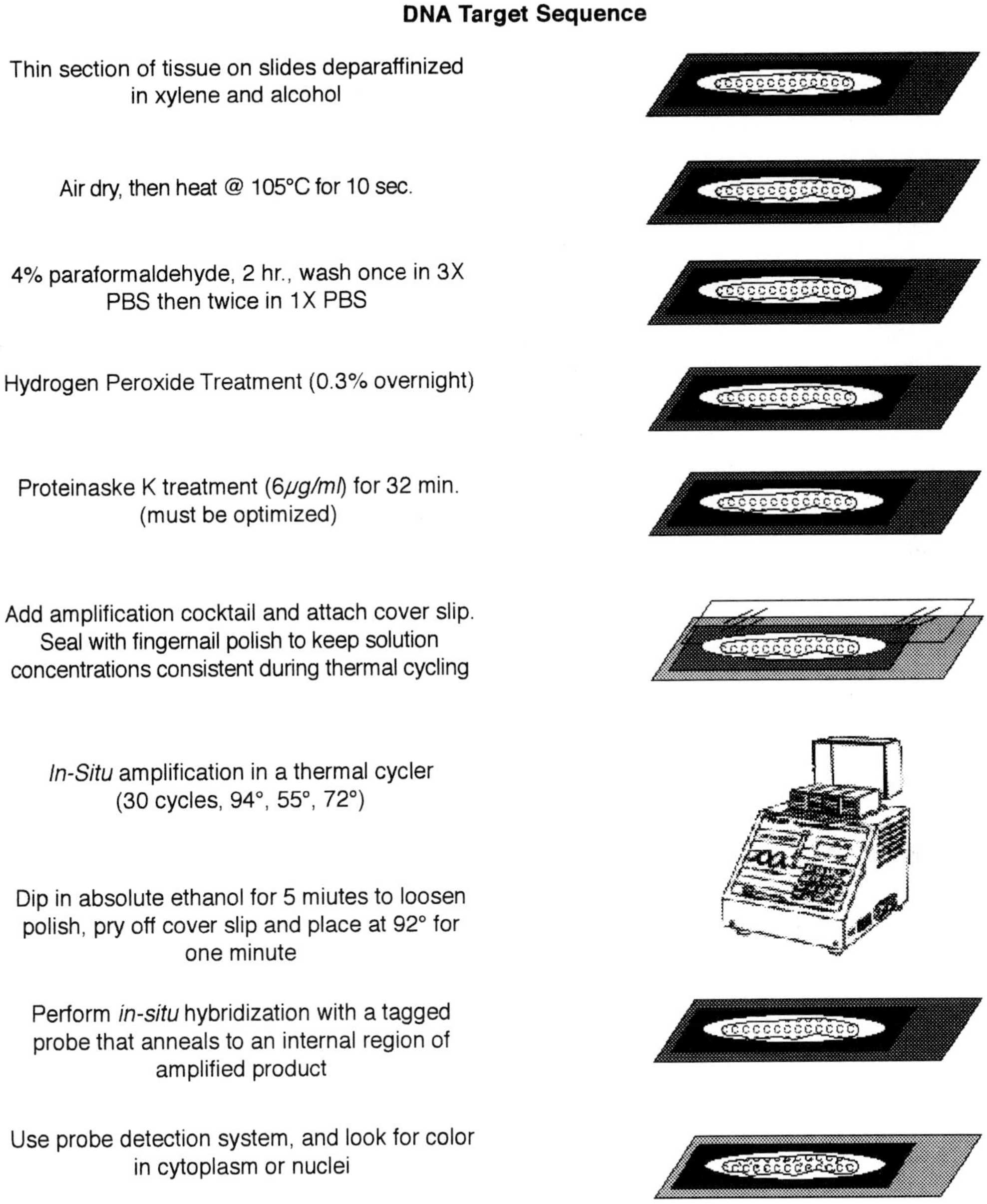

**DNA Target Sequence**

Thin section of tissue on slides deparaffinized
in xylene and alcohol

Air dry, then heat @ 105°C for 10 sec.

4% paraformaldehyde, 2 hr., wash once in 3X
PBS then twice in 1X PBS

Hydrogen Peroxide Treatment (0.3% overnight)

Proteinaske K treatment (6$\mu$g/m*l*) for 32 min.
(must be optimized)

Add amplification cocktail and attach cover slip.
Seal with fingernail polish to keep solution
concentrations consistent during thermal cycling

*In-Situ* amplification in a thermal cycler
(30 cycles, 94°, 55°, 72°)

Dip in absolute ethanol for 5 miutes to loosen
polish, pry off cover slip and place at 92° for
one minute

Perform *in-situ* hybridization with a tagged
probe that anneals to an internal region of
amplified product

Use probe detection system, and look for color
in cytoplasm or nuclei

**Fig. 1.** Overview of *in situ* PCR protocol: amplification and hybridization of DNA target
sequence. See text for details and modifications.

*Fixation and Washes*

1. Place the slides in a solution of 2% PFA in PBS (pH 7.4) for 2 h at room temperature. Use
   of the recommended Coplin jars or staining dishes facilitates these steps.
2. Wash the slides once with 3× PBS for 10 min, agitating periodically with an up and
   down motion.
3. Wash the slides with 1× PBS for 10 min, agitating periodically with an up and down motion.
   Repeat once with fresh 1× PBS.

4. At this point, slides with attached tissue can be stored at −80°C until use. Before storage, dehydrate with 100% ethanol.

If biotinylated probes or peroxidase-based color development are to be used, the samples should further be treated with a 0.3% solution of hydrogen peroxide in PBS, to inactivate any endogenous peroxidase activity. Once again, incubate the slides overnight—either at 37°C or at room temperature. Then, wash the slides once with PBS.

If other probes are to be used, proceed directly to the following proteinase K digestion, which is perhaps the most critical step in the protocol.

### Proteinase K Treatment (The Most Rate-Limiting Step)

1. Treat samples with 6 μg/mL proteinase K in PBS for 5–60 min at room temperature or at 55°C. To make a proper solution, dilute 1.0 mL of proteinase K (1 mg/mL) in 150 mL of 1× PBS.
2. After 5 min, look at the cells under the microscope at 400×. If the majority of the cells of interest exhibit uniform-appearing, small, round "salt and pepper" dots on the cytoplasmic membrane, then stop the treatment immediately with **step 3**. Otherwise, continue treatment for another 5 min, and reexamine.
3. After proper digestion, heat slides on a block at 95°C for 2 min to inactive the proteinase K.
4. Rinse slides in 1× PBS for 10 s.
5. Rinse slides in distilled water for 10 s.
6. Air-dry.

### Optimizing Digestion

The time and temperature of incubation should be optimized carefully for each cell line or tissue section type. Too little digestion, and the cytoplasmic and nuclear membranes will not be sufficiently permeable to primers and enzyme, and amplification will be inconsistent. Too much digestion, and the membranes will either deteriorate during repeated denaturation or worse, signals will leak out. In the first case cells will not contain the signal, and high background will result. In the last case, many cells will show pericytoplasmic staining or "rim staining" representing the leaked signals going into the cells not containing the signals. Attention to detail here can often mean the difference between success and failure, and this procedure should be practiced rigorously with extra sections before continuing on to the amplification steps.

In our laboratory, proper digestion parameters vary considerably with tissue type. Typically, lymphocytes will require 5–10 min at 25°C or room temperature, CNS tissue will require 12–18 min at room temperature, and paraffin-fixed tissue will require 15–30 min at room temperature. However, these periods can vary widely, and the appearance of the "salt and pepper dots" is the important factor. Unfortunately, the appearance of the "salt and pepper dots" is less prominent in paraffin sections.

The critical importance of these dots should not be underestimated, since a extra 2–3 min of treatment after the appearance of dots will result in leakage of signals.

An alternative to the observation of "dots" method is to select a constant time and treat slides in varying amounts of proteinase K. For example, treat slides for 15 min in 1–6 μg/mL of proteinase K.

### RT Variation: In Situ RNA Amplification

One has two choices in order to detect an RNA signal. The first and more elegant method is simply to use primer pairs that flank spliced sequences of mRNA, as these particular sequences will be found only in RNA will be split into sections in the DNA (see Fig. 3). Thus, by using these RNA-specific primers, one can skip the following DNase step and proceed directly to reverse transcription.

The second, more brutal, yet often necessary approach is to treat the cells or tissue with a DNase solution subsequent to the proteinase K digestion. This step destroys all of the endogenous DNA in the cells so that only RNA survives to provide signals for amplification.

Note: All reagents for RT *in situ* amplification should be prepared with RNase-free water (i.e., diethylpyrocarbonate [DEPC]-treated water). In addition, the silanized glass slides and all glassware should be RNase-free, which we ensure by baking the glassware overnight in an oven before use in the RT amplification procedure.

*DNase Treatment*

Prepare an RNase-free DNase solution:

40 m$M$ Tris HCl, pH 7.4.
6 m$M$ MgCl$_2$.
2 m$M$ CaCl$_2$.
100 U/µL final vol. of DNase (use RNase-free DNase, such as 1000 U/µL RQ1 DNase cat. no. 776785 from Boehringer).

1. Add 10–15 µL of solution to each well.
2. Incubate the slides at 37°C in a humidified chamber for 1 h. If one is using liver tissue, this incubation should be extended for an additional hour.
3. After incubation, rinse the slides with a similar solution that was prepared without the DNase enzyme.
4. Wash the slides twice with DEPC-treated water.

Note: Some cells are particularly rich in ribonuclease; in this circumstance, add the following ribonuclease inhibitor to the DNase solution: 1000 U/mL placental ribonuclease inhibitor (e.g., RNAsin) plus 1 m$M$ dithiotheitol (DTT). Also, some investigators prefer to use a long incubation period with a lower concentration of DNase (1 U/mL for 18 h).

*Reverse Transcriptase Reaction*

Next, one wishes to make DNA copies of the targeted RNA sequence so that the signal can be amplified. The following are typical cocktails for the RT reaction:

1. If using AMVRT or MMLVRT enzyme:

| | |
|---|---|
| 10× reaction buffer* | 4.0 µL |
| 10 m$M$ dATP | 2.0 µL |
| 10 m$M$ dCTP | 2.0 µL |
| 10 m$M$ dGTP | 2.0 µL |
| 10 m$M$ dTTP | 2.0 µL |
| RNasin at 40 U/µL (Promega)** | 0.5 µL |
| 20 µ$M$ downstream primer | 1.0 µL |
| AMVRT 20 U/µL | 0.5 µL |
| DEPC water | 6.0 µL |
| Total volume | 20.0 µL |

*5× reaction buffer: 250 m$M$ Tris-HCl pH 8.3, 375 m$M$ KCl, 15 m$M$ MgCl$_2$.

2.  If using Superscript II enzyme from BRL:

| | |
|---|---|
| 5× reaction buffer (as supplied with enzyme) | 4.0 μL |
| 10 m*M* dATP | 2.0 mL |
| 10 m*M* dCTP | 2.0 mL |
| 10 m*M* dGTP | 2.0 mL |
| 10 m*M* dTTP | 2.0 μL |
| RNasin at 40 U/μL (Promega)** | 0.5 μL |
| 20 μ*M* downstream primer | 1.0 μL |
| Superscript II, 200 U/μL | 0.5 μL |
| 0.1 *M* dTT | 2.0 μL |
| DEPC water | 4.0 μL |
|     Total volume | 20.0 mL |

   a.  Add 10–15 μL of either cocktail to each well. Carefully place the cover slip on top of the slide.
   b.  Incubate at 42°C or 37°C for 1 h in a humidified atmosphere.
   c.  Incubate slides at 92°C for 2 min.
   d.  Remove cover slip and wash twice with distilled water. Proceed with the amplification procedure, which is the same for both DNA- and RNA-based protocols.
**RNAsin inhibits ribosomal RNases—use for optimal yields.

## RT Enzymes

Avian myoblastosis virus reverse transcriptase (AMVRT) and Moloney murine leukemia virus reverse transcriptase (MMLVRT) give comparable results in our laboratory. Other RT enzymes will probably work also. However, it is important to read the manufacturer's descriptions of the RT enzyme and to make certain that the proper buffer is used.

An alternative RT enzyme is available that lacks RNase activity. Called Superscript II, it is available from BRL Lifesciences, and it is suitable for RT of long mRNAs. It is also suitable for routine RT amplification, and in our laboratory it has proved to be more efficient than the two enzymes described above.

## Primers and Target Sequences

In our laboratory, we simply use antisense downstream primers for our gene of interest, as we already know the sequence of most genes we study. However, one can alternatively use oligo (dT) primers first to convert all mRNA populations into cDNA and then to perform the *in situ* amplification for a specific cDNA. This technique may be useful when one is performing amplification of several different gene transcripts at the same time in a single cell. For example, if one is attempting to detect various cytokine expressions, one can use an oligo (dT) primer to reverse transcribe all the mRNA copies in a cell or tissue section. Then one can amplify more than one type of cytokine and detect the various types with different probes that develop into different colors (see **Multiple Signals and Multiple Labels in Individual Cells**, page 220.)

In all RT reactions, it is advantageous to reverse-transcribe only relatively small fragments of mRNA (<1500 bp). Larger fragments may not completely reverse-transcribe due to the presence of secondary structures. Furthermore, the RT enzymes—AMVRT and MMLVRT, at least—are not very efficient in transcribing large mRNA fragments. However, this size restriction does not apply to amplification reactions that are exclusively DNA, for the polymerase enzyme copies nucleotides better. In *in situ* DNA reactions, we routinely amplify genes up to 300 bp.

The following are several additional points one should keep in mind:

1. The length for both sense and antisense primers should be 14–22 bp.
2. At the 3' ends, primers should contain GC-type base pairs (e.g., GG, CC, GC, or CG) to facilitate complementary strand formation.
3. The preferred GC content of the primers is 45–55%.
4. Try to design primers so they do not form intra- or interstrand base pairs. Furthermore, the 3'-ends should not be complementary to each other, or they will form primer dimers.
5. One can design an RT primer so that it does not contain secondary structures.

*Annealing Temperatures*

Annealing temperatures for RT and for DNA amplification can be chosen according to the following formula:

$$T_m \text{ of the primers} = 81.5°C + 16.6 (\log M) + 0.41 (G + C\%) - 500/n$$

where $n$ = length of primers and M = molarity of the salt in the buffer, usually 0.047 $M$ for DNA reactions and 0.070 $M$ for RT reactions (see below).

If using AMVRT, the value will be lower, according to the following formula:

$$T_m \text{ of the primers} = 62.3°C + 0.41 (G + C\%) - 500/n$$

A simpler version could be used for primers of 18 bp or less:

$$T_m = 4°C \text{ (no. of GC pairs)} + 2°C \text{ (no. of AT pairs)}$$

Usually, primer annealing is optimal at 2°C above its $T_m$. However, this formula provides only an approximate temperature for annealing, since base stacking, near-neighbor effect, and buffering capacity may play a significant role for a particular primer.

Optimization of the annealing temperature should be carried out first with solution-based reactions. It is important to know the optimal temperature before attempting to conduct *in situ* amplification, as *in situ* reactions are simply not as robust as solution-based ones. We hypothesize that this is because primers do not have easy access to DNA templates inside cells and tissues, as numerous membranes, folds, and other small structures can keep primers from binding homologous sites as readily as they do in solution-based reactions.

There are two addition ways to determine the real annealing temperatures:

1. To utilize a recently developed thermocycler designed for determination of actual annealing temperature called the Robocycler (Stratagene, La Jolla, CA).
2. To utilize so-called touchdown PCR *(22)*.

The logic for determining the correct annealing temperature for *in situ* PCR is as follows:

Recently, M.J. Research (Watertown, MA) have devised a thermocycler that has the capacity to perform *in situ* gene amplification both in slides and in solution (tubes) simultaneously, in the same block. Also, their DNA-Engine Twin-Tower MS 16×2 can perform multiple functions, and *in situ* hybridization could be carried out if one inserts a wet Kimwipe paper in one of the slide slots. In addition, Perkin-Elmer also has recently introduced an excellent thermocycler. These kinds of thermocyclers can be very useful in the *in situ* amplification of a gene of interest.

## Amplification Protocol, All Types

Prepare an amplification cocktail containing the following: 1.25 $\mu M$ of each primer, 200 $\mu M$ (each) dNTP, 10 m$M$ Tris-HCl (pH 8.3), 50 m$M$ KCl, 1.5 m$M$ MgCl$_2$, 0.001% gelatin, and 0.05 U/$\mu$L *Taq* polymerase. The following is a convenient recipe that we use in our laboratory:

| | |
|---|---|
| 25 $\mu M$ forward primer (i.e., SK 38 for HIV-1) | 5.0 $\mu$L |
| 25 $\mu M$ reverse primer (i.e., SK 39 for HIV-1) | 5.0 $\mu$L |
| 10 m$M$ each dNTP | 2.5 $\mu$L |
| Taq pol (Ampli-Taq 5 U/$\mu$L)* | 1.0$\mu$L |
| 10× PCR buffer | 10.0 $\mu$L |
| H$_2$O | 76.5 $\mu$L |
| Total volume | 100.0 $\mu$L |

*Note: Other thermostable polymerase enzymes have also been used quite successfully.

1. Layer 10–15 $\mu$L of amplification solution onto each well with a P20 micropipet so that the whole surface of the well is covered with the solution. Be careful—do not touch the surface of the slide with the tip of the pipet.
2. Add a glass cover slip (22 × 60 mm) and carefully seal the edge of the coverslip to the slide with two coats of clear nail polish. If using tissue sections, use a second slide instead of a cover slip (see discussions below on attaching cover slips and hot start).
3. Allow nail polish to dry completely until not sticky to the touch. (An alternative to nail polish is a self-seal, which, when mixed directly into the amplification cocktail, seals the slides at its edges, limiting evaporation of the cocktail. We have used this self-seal in our laboratory extensively, and it has given excellent results. It is available from M.J. Research.)
4. Place slides in a thermocycling instrument.
5. Run 30 cycles of the following amplification protocol:

Optimal annealing temperature
   (see below)

| | |
|---|---|
| 94°C | 30 s |
| 45°C | 1 min |
| 72°C | 1 min |

These times/temperatures will probably require optimization for the specific thermocycler being used. Furthermore, the annealing temperature should be optimized, as described below. These particular incubation parameters work well with SK38 and SK39 primers for the HIV-1 gag sequence, when amplified in an M.J. Research PTC-100-60 or PTC-100-16MS thermal cycler.

6. After the thermal cycling is complete, dip slides in 100% ethyl alcohol (EtOH) for at least 5 min, to dissolve the nail polish. Pry off the cover slip using a razor or other fine blade—the cover slip generally pops off quite easily. Scratch off any remaining nail polish on the outer edges of the slide so that fresh cover slips will lie evenly in the subsequent hybridization/detection steps.
7. Place slides on a heat block at 92°C for 1 min—this treatment helps immobilize the intracellular signals.
8. Wash slides with 2× standard saline citrate (SSC) (see **Materials and Methods**, page 224) at room temperature for 5 min.

   The amplification protocol is now complete and one can proceed to the labeling/hybridization procedures.

Notes on attaching cover slip/top slide: Be certain to paint the polish carefully around the entire periphery of the cover slip or the edges of the dual slide, as the polish must treatment increased. After bromocriptine treatment, PRL mRNA expression decreased considerably *(26)*.

In this study we found a difference in localization of GH and PRL synthesis on the polysomes of the RER. Hybridization signals of GH mRNA are distributed diffusely on the entire RER, whereas those of PRL mRNA are scattered and distributed focally on the RER even after stimulation by estrogen. After bromocriptine treatment, hybridization signals of PRL mRNA are shown to be distributed diffusely on the RER. These differences and alterations of mRNA distribution on the RER may be difficult to explain; however, these phenomena are possibly evoked by the dynamics of signal recognition particles and their receptors on the RER. EM-ISH is an important technique for clarifying the intracellular localization of mRNA and the exact site of specific hormone synthesis on the surface of the RER. As shown in the studies on changes in ultrastructural expression of PRL mRNA induced by estrogen and bromocriptine, EM-ISH can serve for morphologic and pathophysiologic investigation on mRNA expression induced by some stimulatory factors.

## COMBINED IMMUNOHISTOCHEMISTRY AND PREEMBEDDING EM-ISH

As for mRNA preservation, the preembedding EM-ISH method using frozen sections fixed in 4% paraformaldehyde has more advantages over the postembedding EM-ISH method using tissues embedded in LR White resin. Frozen sections fixed in 4% paraformaldehyde have better morphologic preservation than immediately frozen sections. Based on this assessment *à propos* of maintaining tissue morphology and retaining the messages, we utilize the preembedding EM-ISH method using 6-μm-thick frozen sections fixed in 4% paraformaldehyde for the simultaneous detection of mRNA and encoded protein.

### Method

**Steps 1–11** are the same as described in the section on the preembedding EM-ISH method.

12. Polymerization of Epon resin at 60°C for 2 days.
13. The ultrathin sections are attached to nickel grids.
14. To retrieve the immunoreactivity of the targeted protein, the ultrathin sections embedded in Epon resin are etched with 10% $H_2O_2$ for 30 min or with 4% sodium periodate for 10 min, followed by washing with distilled water.
15. Immunohistochemical staining is carried out at room temperature for 1 h using the appropriate antibody.
16. The grids are washed with 0.1 $M$ phosphate buffer (PB), and the immunoreaction is visualized at room temperature for 1 h with 20 nm protein A colloidal gold (E. Y. Laboratories, San Mateo, CA) diluted 1:40 in 0.1 $M$ PB.
17. After being washed with 0.1 $M$ PB and distilled water, the grids are dried at room temperature and observed under an electron microscopy.
18. The immunohistochemical negative control experiment is substitution of normal serum for the antibody.

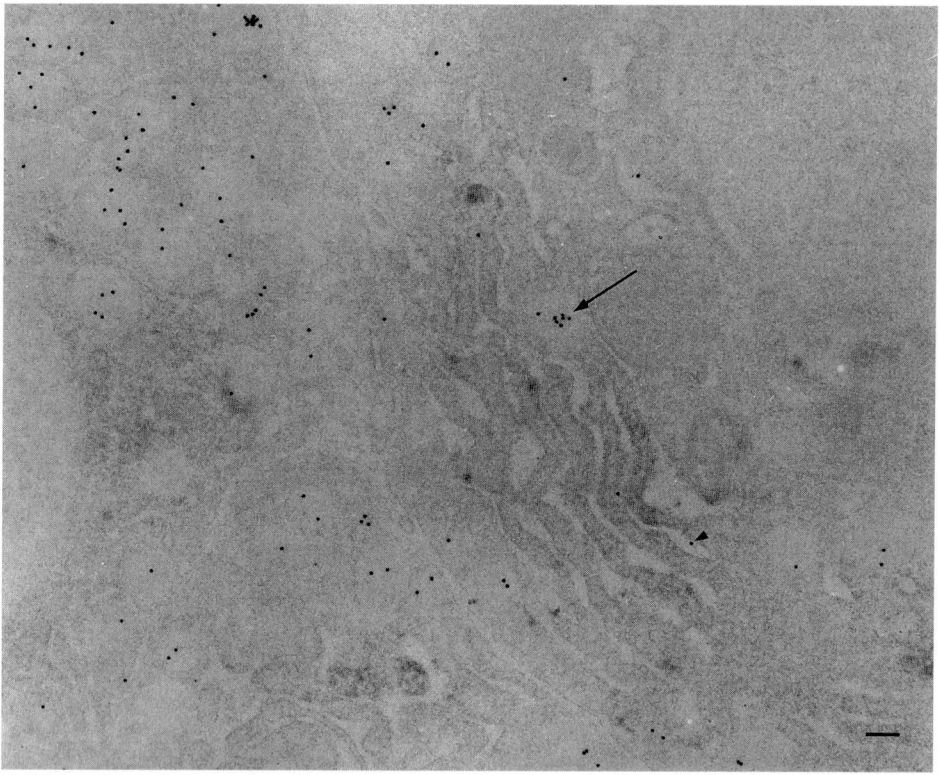

**Fig. 7.** Combined immunohistochemistry and preembedding EM-ISH at the ultrastructural level. Positive signals for rat GH mRNA are localized as osmium black on the polysomes of the RER using biotinylated antisense oligonucleotide probe. Rat GH protein is identified mainly on the secretory granules with 20 nm protein A colloidal gold particle (arrow) is also identified in the cisternae of the RER (arrowhead). Bar = 200 nm. (Reproduced with permission from ref. *30.*)

*Results*

As shown in the previous section, EM-ISH with an antisense probe for rat GH mRNA revealed diffuse localization on the polysomes of the entire RER. Subsequent immunohistochemical staining using anti-rat GH antibody (monkey, diluted 1:100 with PBS), supplied by the National Institute of Diabetes and Digestive and Kidney Diseases (NIDDK, Bethesda, MD), and 20 nm protein A colloidal gold identified rat GH on the secretory granules **(Fig. 7)**. Colloidal gold signals for GH were distributed mainly on the secretory granules and were also identified in the cisternae of the RER **(Fig. 7)**. The immunoreactivity retrieved after the etching process with 10% $H_2O_2$ was similar and comparable to that with 4% sodium periodate. Immunohistochemical control experiments substituting normal monkey serum for anti-GH antibody combined with the preembedding EM-ISH method revealed no reactions of protein A colloidal gold particles on the secretory granules **(Fig. 8)**.

*Comments*

The only flaw with this combined method is the deosmification and degradation of the signals of mRNA caused by the etching process using $H_2O_2$ or sodium

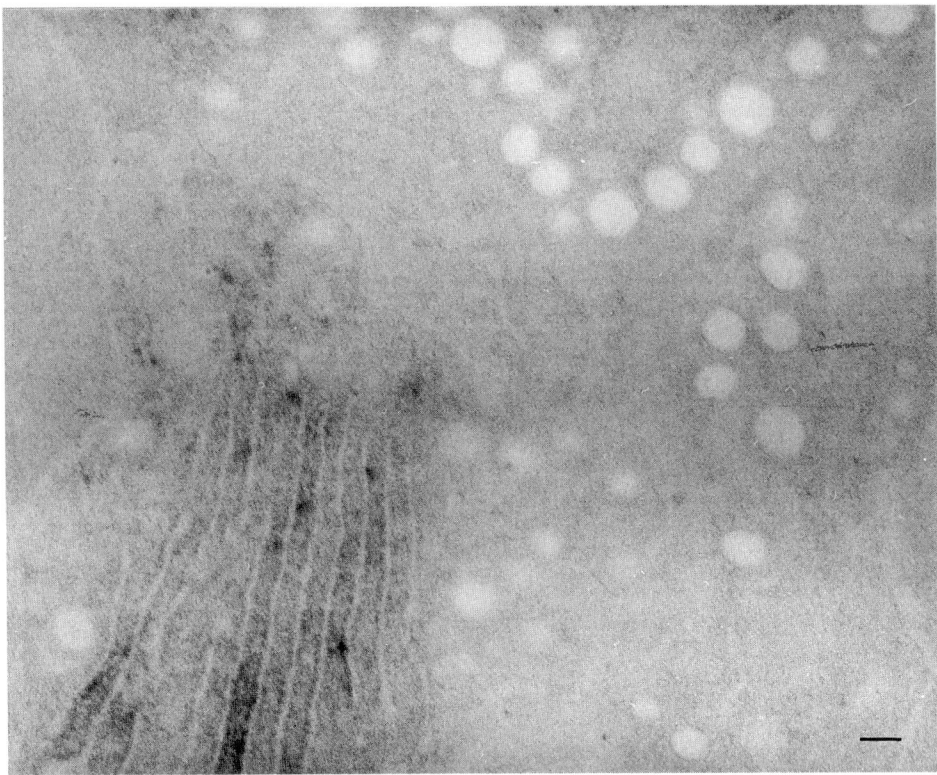

**Fig. 8.** Immunohistochemical control experiments at the ultrastructural level with the substitution of normal monkey serum for anti-rat GH antibody. No positive reactions of protein A colloidal gold particles are observed on the secretory granules, whereas positive signals for rat GH mRNA are identified as osmium black on the polysomes of the RER. Bar = 200 nm. (Reproduced with permission from ref. *30.*)

periodate. To solve this problem, we have recently used LR White resin for tissue embedment *(29).*

Pitfalls:

1. Prolonged fixation of the tissues will decrease the messages.
2. Excessive treatment with proteinase K will destroy the morphology.
3. Excessive etching will shade off the reaction products (osmium black) of mRNA.

### Modified Method Using LR White resin

**Steps 1–10** are the same as described in the section on the preembedding EM-ISH method.

11. 2% osmium tetroxide is applied to the sections. After dehydration with graded ethanol (50, 70, 80, 90, 95, 100%), tissue sections are embedded in LR White resin using the inverted beam capsule method.
12. Polymerization at 50°C for 2 days in a vacuum oven.
13. The ultrathin sections are attached to nickel grids.
14. Immunohistochemical staining is carried out at room temperature for 1 h using the appropriate antibody.

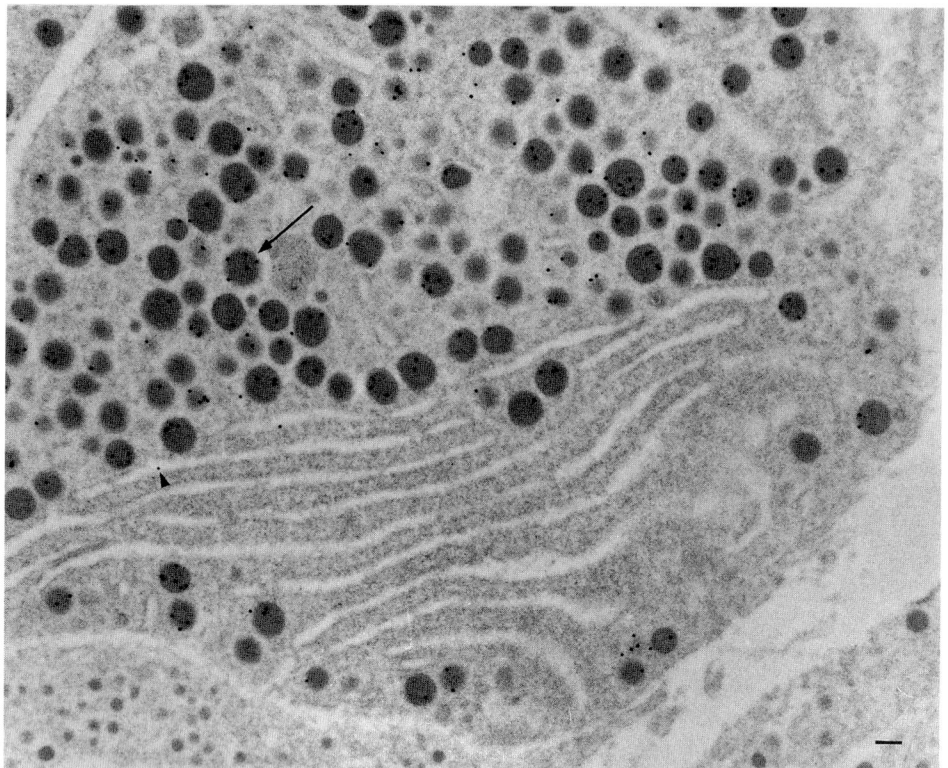

**Fig. 9.** Using LR White resin for tissue embedment, rat GH mRNA is also localized on the polysomes of the entire RER. Subsequent immunohistochemical staining identifies rat GH protein both on the secretory granules (arrow) and in the cisternae of the RER (arrowhead). Bar = 200 nm. (Reproduced with permission from ref. *31*.)

15. The grids are washed with 0.1 *M* PB, and the immunoreaction is visualized at room temperature for 1 h with 20 nm protein A colloidal gold diluted 1:40 in 0.1 *M* PB.
16. After being washed with 0.1 *M* PB and distilled water, the grids are dried at room temperature and observed under an electron microscopy.
17. The immunohistochemical negative control experiment is substitution of normal serum for the antibody.

*Results*

EM-ISH using LR White resin for tissue embedding also localized rat GH mRNA on the polysomes of the entire RER **(Fig. 9)**. Subsequent immunohistochemical staining identified rat GH protein both on the secretory granules and in the cisternae of the RER **(Fig. 9)**.

*Comments*

As shown in our previous reports *(24,25,27,28)*, GH mRNA is distributed diffusely on the RER. In these reports using tissues embedded in Epon resin, somewhat heterogeneous electron density was observed in GH mRNA expression. In LR White resin-embedded tissues, GH mRNA was also distributed diffusely on the RER, its electron density being homogeneous. The cause of this difference in electron density of heterogeneity and homogeneity cannot be determined, but it may be attributed to the characteristic difference of both resins. As the manufacturer stated, the hydrophilic and low lipid solvent character of LR White resin is known to promote excellent visualization of membrane and cytosol structures and thus may be helpful for preservation of the reaction products in LR White resin-embedded tissues.

Pitfalls:

1. Prolonged fixation of the tissues will decrease the messages.
2. Excessive treatment with proteinase K will destroy the morphology.
3. Embedding in LR White resin using the inverted beam capsule method may be difficult technically.

*Significance*

There are two major problems to be resolved in the ultrastructural double-staining method to visualize mRNA and encoded protein simultaneously in the same cell. One is to retain the messages and the other is to maintain the immunoreactivity of the encoded protein in the same cell. Based on the aforementioned assessment *à propos* of maintaining tissue morphology and retaining the messages, we employ the preembedding EM-ISH method using 6-μm-thick frozen sections fixed in 4% paraformaldehyde for the simultaneous detection of mRNA and encoded protein. The immunoreactivity can be retrieved by the etching process with $H_2O_2$ or sodium periodate even after modification such as osmification and embedding in Epon resin, and tissues embedded in Epon resin can serve for simultaneous ultrastructural detection of messages and encoded proteins. The only problem is the deosmification and degradation of the mRNA signals, which are caused by etching with $H_2O_2$ or sodium periodate. To solve this problem, we have recently used LR White resin for tissue embedding *(29)*. In LR White resin-embedded tissues, retrieval of immunoreactivity with $H_2O_2$ or sodium periodate is not required, and therefore degradation of mRNA signals can be avoided.

To our knowledge, only five reports, except for our previous ones *(27–33)*, describing the simultaneous ultrastructural detection of mRNA and encoded protein have been published, in each of which the postembedding EM-ISH method using colloidal gold particle was utilized *(16,35–38)*. However, in these postembedding EM-ISH studies, the relatively frequent nonspecific reactions of colloidal gold particles used for the detection of mRNA were observed in the cisternae of the RER. The EM-ISH method for Lowicryl K4M-embedded tissues is generally supposed to be faulty in morphologic preservation. Decreased message preservation was observed in postembedding EM-ISH studies using LR White-embedded tissues *(24,38)*. From the viewpoint of specific hybridization signal detection and morphologic preservation, the combined immunohistochemistry and preembedding EM-ISH method is considered superior to the previously reported postembedding EM-ISH method and is successful and satisfactory for simultaneous ultrastructural identification of mRNA and encoded protein.

Translation of mRNA for secreted proteins is initiated on free ribosomes and then translocated to the polysomes on the RER with the aid of signal recognition particles once the signal peptide is produced. Synthesized proteins are secreted into the luminal space of the RER and subsequently transported to secretory granules via the Golgi apparatus. This ultrastructural double-staining method for mRNA and encoded protein can provide an important clue for elucidating the intracellular correlation of mRNA translation and secretion of translated protein.

## REFERENCES

1. Gall, J. G. and Pardue, M. L. (1969) Formation and detection of RNA-DNA hybrids in cytological preparations. *Proc. Natl. Acad. Sci. USA* **63**, 378–383.
2. John, H. A., Birnstiel, M. L., and Jones, K. W. (1969) RNA-DNA hybrids at the cytological level. *Nature* **223**, 582–587.
3. Lewis, M. E., Sherman, T. G., and Watson, S. J. (1985) In situ hybridization histochemistry with synthetic oligonucleotides: strategies and methods. *Peptides* **6**, 75–87.
4. Guitteny, A. F., Fouque, B., Mougin, C., Teoule, R., and Bloch, B. (1988) Histological detection of messenger RNAs with biotinylated synthetic oligonucleotide probes. *J. Histochem. Cytochem.* **36**, 563–571.
5. Hankin, R. C. and Lloyd, R. V. (1989) Detection of messenger RNA in routinely processed tissue sections with biotinylated oligonucleotide probes. *Am. J. Clin. Pathol.* **92**, 166–171.
6. Larsson, L. I. (1989) In situ hybridization using biotin-labeled oligonucleotides: probe labeling and procedures for mRNA detection. *Arch. Histol. Cytol.* **52**, 55–62.
7. Farquharson, M., Harvie, R., and McNicol, A. M. (1990) Detection of messenger RNA using a digoxigenin end labeled oligodeoxynucleotide probe. *J. Clin. Pathol.* **43**, 424–428.
8. Pringle, J. H., Ruprai, A. K., Primrose, L., et al. (1990) In situ hybridization of immunoglobulin light chain mRNA in paraffin sections using biotinylated or hapten-labeled oligonucleotide probes. *J. Pathol.* **162**, 197–207.
9. Schmitz, G. G., Walter, T., Seibl, R., and Kessler, C. (1991) Nonradioactive labeling of oligonucleotides in vitro with the hapten digoxigenin by tailing with terminal transferase. *Anal. Biochem.* **192**, 222–231.
10. Shorrock, K., Roberts, P., Pringle, J. H., and Lauder, I. (1991) Demonstration of insulin and glucagon mRNA in routinely fixed and processed pancreatic tissue by in-situ hybridization. *J. Pathol.* **165**, 105–110.
11. Jacob, J., Todd, K., Birnstiel, M. L., and Bird, A. (1971) Molecular hybridization of ³H-labelled ribosomal RNA with DNA in ultrathin sections prepared for electron microscopy. *Biochim. Biophys. Acta* **228**, 761–766.
12. Webster, H. F., Lamperth, L., Favilla, J. T., Lemke, G., Tesin, D., and Manuelidis, L. (1987) Use of a biotinylated probe and in situ hybridization for light and electron microscopic localization of Po mRNA in myelin-forming Schwann cells. *Histochemistry* **86**, 441–444.
13. Guitteny, A. F. and Bloch, B. (1989) Ultrastructural detection of the vasopressin messenger RNA in the normal and Brattleboro Rat. *Histochemistry* **92**, 277–281.
14. Morel, G., Chabot, J. G., Gossard, F., and Heisler, S. (1989) Is atrial natriuretic peptide synthesized and internalized by gonadotrophs? *Endocrinology* **124**, 1703–1710.
15. Morel, G., Dihl, F., and Gossard, F. (1989) Ultrastructural distribution of growth hormone (GH) mRNA and GH intron 1 sequences in rat pituitary gland: effects of GH releasing factor and somatostatin. *Mol. Cell. Endocrinol.* **65**, 81–90.
16. Singer, R. H., Langevin, G. L., and Lawrence, J. B. (1989) Ultrastructural visualization of cytoskeletal mRNAs and their associated proteins using double-label in situ hybridization. *J. Cell Biol.* **108**, 2343–2353.
17. Wolber, R. A., Beals, T. F., and Maassab, H. F. (1989) Ultrastructural localization of herpes simplex virus RNA by in situ hybridization. *J. Histochem. Cytochem.* **37**, 97–104.
18. Jirikowski, G. F., Sanna, P. P., and Bloom, F. E. (1990) mRNA coding for oxytocin is present in axons of the hypothalamo-neurohypophyseal tract. *Proc. Natl. Acad. Sci. USA* **87**, 7400–7404.

19. Le Guellec, D., Frappart, L., and Willems, R. (1990) Ultrastructural localization of fibronectin mRNA in chick embryo by in situ hybridization using [35]S or biotin labeled cDNA probes. *Biol. Cell* **70**, 159–165.

20. Le Guellec, D., Frappart, L., and Desprez, P. Y. (1991) Ultrastructural localization of mRNA encoding for the EGF receptor in human breast cell cancer line BT20 by in situ hybridization. *J. Histochem. Cytochem.* **39**, 1–6.

21. Le Guellec, D., Trembleau, A., Pechoux, C., Gossard, F., and Morel, G. (1992) Ultrastructural nonradioactive in situ hybridization of GH mRNA in rat pituitary gland: preembedding vs ultrathin frozen sections vs postembedding. *J. Histochem. Cytochem.* **40**, 979–986.

22. Trembleau, A., Calas, A., and Fevre-Montange, M. (1990) Ultrastructural localization of oxytocin mRNA in the rat hypothalamus by in situ hybridization using a synthetic oligonucleotide. *Brain Res. Mol. Brain Res.* **8**, 37–45.

23. Pomeroy, M. E., Lawrence, J. B., Singer, R. H., and Billings-Gagliardi, S. (1991) Distribution of myosin heavy chain mRNA in embryonic muscle tissue visualized by ultrastructural in situ hybridization. *Dev. Biol.* **143**, 58–67.

24. Matsuno, A., Ohsugi, Y., Utsunomiya, H., et al. (1994) Ultrastructural distribution of growth hormone, prolactin mRNA in normal rat pituitary cells: a comparison between preembedding and postembedding methods. *Histochemistry* **102**, 265–270.

25. Matsuno, A., Teramoto, A., Takekoshi, S., et al. (1994) Application of biotinylated oligonucleotide probes to the detection of pituitary hormone mRNA using Northern blot analysis, in situ hybridization at light and electron microscopic levels. *Histochem. J.* **26**, 771–777.

26. Matsuno, A., Ohsugi, Y., Utsunomiya, H., et al. (1995) Changes in the ultrastructural distribution of prolactin and growth hormone mRNAs in pituitary cells of female rats after estrogen and bromocriptine treatment, studied using in situ hybridization with biotinylated oligonucleotide probes. *Histochem. Cell Biol.* **104**, 37–45.

27. Matsuno, A., Utsunomiya, H., Ohsugi, Y., et al. (1996) Simultaneous ultrastructural identification of growth hormone and its messenger ribonucleic acid using combined immunohistochemistry and non-radioisotopic in situ hybridization: a technical note. *Histochem. J.* **28**, 703–707.

28. Matsuno, A., Nagashima, T., Takekoshi, S., et al. (1998) Ultrastructural simultaneous identification of growth hormone and its messenger ribonucleic acid. *Endocr. J.* **45 (suppl.)**, S101–S104.

29. Matsuno, A., Ohsugi, Y., Utsunomiya, H., et al. (1998) An improved ultrastructural double-staining method of rat growth hormone and its mRNA using LR White resin: a technical note. *Histochem. J.* **30**, 105–109.

30. Matsuno, A., Nagashima, T., Osamura, R. Y., and Watanabe, K. (1998) Application of ultrastructural in situ hybridization combined with immunohistochemistry to pathophysiological studies of pituitary cell: technical review. *Acta Histochem. Cytochem.* **31**, 259–265.

31. Matsuno, A., Itoh, J., Osamura, R. Y., Watanabe, K., and Nagashima, T. (1999) Electron microscopic and confocal laser scanning microscopic observation of subcellular organelles and pituitary hormone mRNA: application of ultrastructural in situ hybridization and immunohistochemistry to the pathophysiological studies of pituitary cells. *Endocr. Pathol.* **10**, 199–211.

32. Matsuno, A., Nagashima, T., Ohsugi, Y., et al. (2000) Electron microscopic observation of intracellular expression of mRNA and its protein product: technical review on ultrastructural in situ hybridization and its combination with immunohistochemistry. *Histol. Histopathol.* **15**, 261–268.

33. Matsuno, A., Nagashima, T., Osamura, R. Y., and Watanabe, K. (2000) Electron microscopic in situ hybridization and its combination with immunohistochemistry, in *Molecular Histochemical Techniques (Springer Lab Manual)* (Koji, T., ed.), Springer, New York, pp. 204–221.

34. Lloyd, R. V., Jin, L., and Chandler, W. F. (1991) In situ hybridization in the study of pituitary tissues. *Pathol. Res. Pract.* **187**, 552–555.

35. Escaig-Haye, F., Grigogiev, V., Sharova, I., Rudneva, V., Buckrinskaya, A., and Fournier, J. G. (1992) Ultrastructural localization of HIV-1 RNA and core proteins. Simultaneous visualization using double immunogold labelling after in situ hybridization and immunocyto-chemistry. *J. Submicrosc. Cytol. Pathol.* **24**, 437–443.

36. Egger, D., Troxler, M., and Bienz, K. (1994) Light and electron microscopic in situ hybridization: nonradioactive labeling and detection, double hybridization, and combined hybridization-immunocytochemistry. *J. Histochem. Cytochem.* **42**, 815–822.

37. Gingras, D. and Bendayan, M. (1995) Colloidal gold electron microscopic in situ hybridization: combination with immunocytochemistry for the study of insulin and amylase secretion. *Cell Vision* **2**, 218–225.

38. Morey, A. L., Ferguson, D. J. P., and Fleming, K. A. (1995) Combined immunocytochemistry and nonisotopic in situ hybridization for the ultrastructural investigation of human parvovirus B19 infection. *Histochem. J.* **27**, 46–53.

# 9

# Quantitation of *In Situ* Hybridization Analysis

## Lars-Inge Larsson, DSC

## INTRODUCTION

*In situ* hybridization is a powerful method for detecting specific DNA and RNA sequences in cells and tissue sections. Major areas of applications include cytogenetics, microbiology, and studies of gene expression and regulation. The methods are based on the fact that the purine and pyrimidine bases that make up DNA (adenine, A; guanine, G; thymine, T; and cytosine, C) or RNA (A, G, C, and uracil, U) can form specific base pairs (A-T, A-U, C-G) that link two complementary strands of DNA or RNA to each other. Thus, provided the sequence of interest (target) is available or known, it is possible to synthesize and label the complementary sequence and use it as a probe for *in situ* hybridization detection of the target. Since the reaction will contain only four (DNA-DNA, RNA-RNA) or, at most, five (DNA-RNA) bases as reactants, it is possible to define conditions mathematically under which only totally complementary sequences will hybridize. The strength of the hybridization is temperature dependent, and the melting point $T_m$ defines the temperature at which half the hybrids formed will dissociate or "melt." The $T_m$ depends on the degree of complementarity (lack of mismatches), the number of bases (length of the probe), the frequency of GC base pairs (which bind more strongly than A-T or A-U), and the ionic strength. Addition of formamide is often used to lower the $T_m$.

Formulas incorporating all these factors make it possible to define the experimental conditions needed for a highly specific (stringent) detection of the target with little or no risk of cross-hybridization to related sequences *(1)*. However, in practice, factors other than the mathematically definable base-pairing will contribute to nonspecific staining, and one should be aware that this method is also not without its caveats and pitfalls. These sources of error can, however, be reasonably well identified and circumvented by the use of appropriate controls and supportive biochemical procedures. Thus, *in situ* techniques have a great appeal and impact on many basic scientific as well as diagnostic disciplines.

Every examination of an object involves a comparison with other objects and, hence, a quantitation. Descriptions of cytochemical reactions usually involve expressions like strong, medium, and weak reactivity, implying a visual quantitation that is unsubstantiated by real measurements. This can be problematic because humans are ill-equipped

From: *Morphology Methods: Cell and Molecular Biology Techniques*
Edited by: R. V. Lloyd © Humana Press, Totowa, NJ

**Fig. 1.** A demonstration of the imperfection in our perception of intensities. The four larger squares are of different intensities; the centrally placed smaller squares are of exactly the same intensity. However, depending on whether the small squares are placed against a darker or lighter background, we perceive them as being either lighter or darker, respectively.

to handle visual evaluation of intensities or gray values. As illustrated in **Fig. 1**, our eyes are much better at judging differences in contrast than differences in absolute intensity. This may have an evolutionary background. Thus, if we encounter a bear in the forest it is more valuable to observe its movements (= changes in contrast) than its absolute gray value. The latter is best left to computer-based image analysis, as is the quantitative evaluation of *in situ* hybridization results. In addition, such analyses may contribute additional valuable information including the number, size, form factor, perimeter, and so forth of bears and cells.

Quantitative *in situ* hybridization is often the only way to study changes in defined cell populations. Thus, it is not always possible to microdissect tissue regions or cells for chemical analysis. Accordingly, the literature abounds with examples of quantitation by *in situ* hybridization techniques. The vast majority of such studies report on relative quantitations only. As we discuss further on, many factors cause the efficiency of hybridization to be relatively small, and this may prevent calculations of the true numbers of target sequences that were present in the living cell. Thus, quantitations are usually relative and refer either to relative intensities or to numbers of hybridizable copies. Such relative comparisons have proved to yield highly useful information, which, if needed, can be supplemented with chemical quantitations. In the following, I discuss different aspects of quantitation as illustrated in the chart depicted in **Fig. 2**. These considerations primarily apply to detection of RNA expression, since this is the area in which almost all quantitative work has been done.

## TISSUE PRETREATMENT

*In situ* hybridizations require that cells and tissues be fixed as soon as possible. Thus, RNA in particular is very vulnerable to degradation. For quantitative studies, standardized tissue handling and rapid fixation is of utmost importance.

The purpose of the fixation for *in situ* hybridization is threefold: 1) to inhibit nucleases that may degrade the target, 2) to immobilize the target, and 3) to preserve the morphology. To obtain a rapid and uniform fixation of the tissue, we prefer to use perfusion fixation. If this is not possible, thin slices (1 mm thick) of tissue are cut and immersed in fixative. Handling of clinical material is problematic as it is often difficult to standardize. To correct for variations in tissue quality, reference hybridizations to ubiquitous RNAs like ribosomal RNAs have been suggested *(2)*. A problem inherent

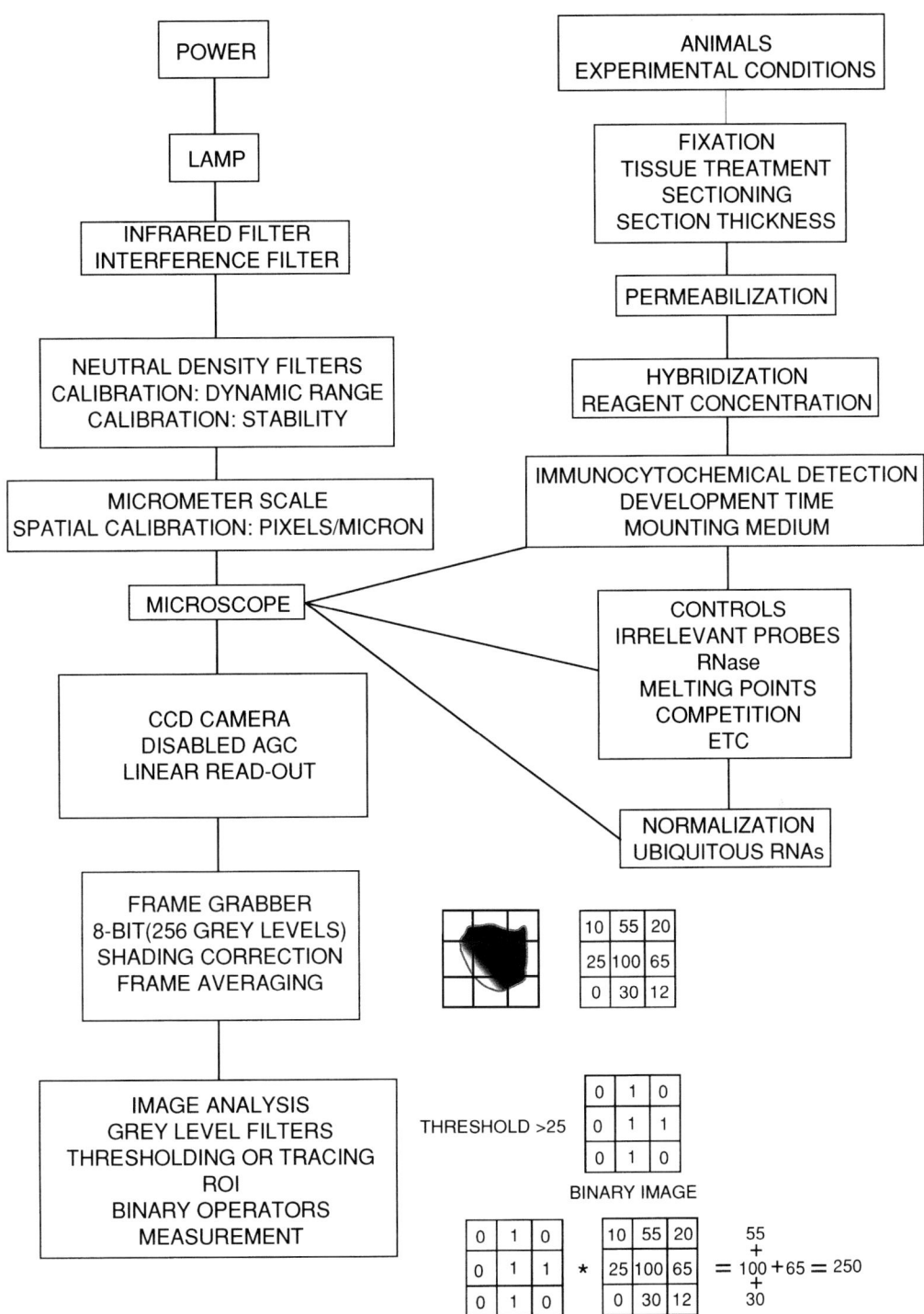

**Fig. 2.** Flow chart summarizing essential steps in quantitation of *in situ* hybridization as discussed in the text. At the lower right corner is a simplified cartoon of a cell that is digitized, thresholded into a binary image, and measured as described in the **Image Analysis** section.

in the use of reference RNAs is that housekeeping and ribosomal RNAs may also change under experimental and pathologic conditions *(3)*.

Although in the past many different fixatives have been used for *in situ* hybridization, there is now general consensus that formaldehyde-based fixatives are preferable. At the conditions of temperature, concentration, and pH commonly used, formaldehyde does not react chemically with nucleotides but acts by trapping target sequences in a crosslinked protein network.

The fixed tissue may be embedded either in paraffin or in plastic, cut on a Vibratome, or be frozen for cryostat sectioning. Formaldehyde-fixed cultured or isolated cells may be stored dry or in 70% ethanol. In addition, methods for freezing unfixed tissues directly for cryostat sectioning or for freeze-drying have been used *(4–10)*. Whatever method is chosen, the ensuing sectioning must meticulously avoid contaminating the tissue with RNases. Use of contaminated microtome knives is one obvious risk. A second risk is connected with the use of proteins like albumin, gelatin, or egg white for attaching sections to slides. Many of these adhesives are contaminated with RNases and should be avoided. The use of silane-coated slides, which are expensive to buy but may be inexpensively manufactured in the laboratory *(9)*, circumvents this problem. Needless to say, close attention should be paid to the potential contamination of laboratory equipment and water with RNases. In our experience, water quality may represent a considerable problem and many workers prefer to add diethylpyrocarbonate (DEPC) to their distilled water, to inhibit RNases. One should be aware that handling of this chemical poses a certain health hazard and that appropriate routines for destroying it by autoclaving are necessary *(11)*. If tissue sections are used for hybridization, due attention should be paid to the reproducibility of the section thickness, as this parameter influences both background and signal *(12)*.

The next step of the procedure usually involves permeabilization. Thus, depending on the tissue and the conditions of fixation and embedding, target sequences may become more or less inaccessible to the probes. Fixation in conventional 10% buffered formalin combined with paraffin embedding may totally mask target sequences, whereas the same fixation combined with cryostat sectioning causes much less masking *(7)*. Moreover, in certain cell types like keratinocytes, masking will be more severe than in other cells, due to the content of keratins and other cell-specific proteins.

Accordingly, it is not possible to establish one single permeabilization protocol that will work for all tissues and pretreatment procedures. Popular methods for permeabilization include the use of proteolytic enzymes like proteinase K or pepsin. Digestion with these enzymes must be carefully optimized since underdigestion produces a suboptimal signal, and overdigestion may lead to extraction of target. Thus, use of these methods often results in a compromise between unmasking and loss of target sequences. This is one of the reasons why accurate enumeration of endogenous target copies is difficult or impossible. A further problem rests with the use of proteinase K. Being a natural product, this enzyme is contaminated with RNases, and it is therefore recommended to let it autodigest (and thus proteolytically to destroy nucleases) before use. This is not necessary with pepsin, which is used in 0.2 *M* hydrochloric acid, where nucleases are inactive. However, pepsin performs best in certain tissues, and proteinase K is best in others *(11)*. This may reflect differences between cellular proteins and restricts our options for permeabilization. Despite their shortcomings, proteases are needed for

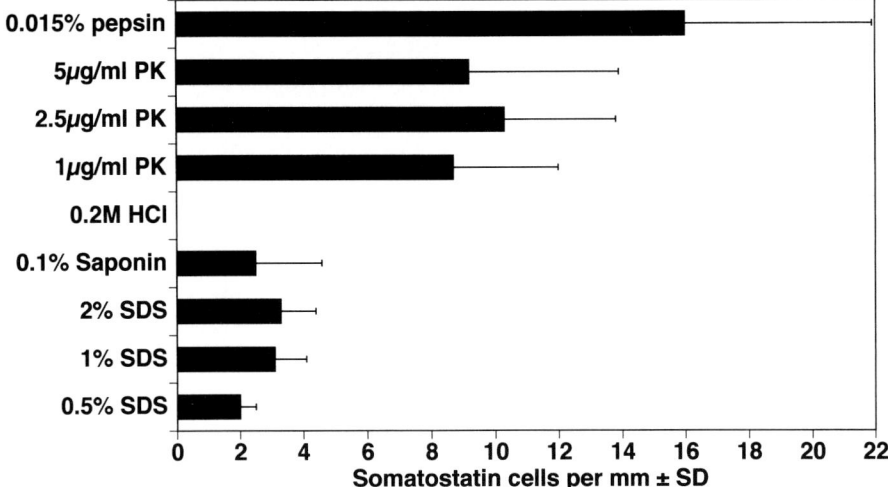

**Fig. 3.** Effectiveness of different permeabilization methods (pepsin, proteinase K [PK], HCl, saponin, and sodium dodecyl sulphate [SDS]) on detectability of somatostatin mRNA-positive cells in formalin-fixed and paraffin-embedded rat gastric mucosa. Note that in this type of tissue only the proteases yield effective permeabilization.

obtaining optimal results in most tissues (**Fig. 3**). After protease digestion, many workers postfix sections briefly in formaldehyde to prevent losses of target molecules. In cases of weakly fixed cells and tissues, no or weaker permeabilizing agents like hydrochloric acid or sodium dodecyl sulfate may be used *(8)*.

Different methods are used after permeabilization to minimize nonspecific binding of the probe. These include acetylation of positively charged tissue amino groups, which otherwise may bind the probe ionically. In addition, prehybridization with inert DNA and/or RNA are often used. In many cases, these procedures are needed, but it may be worthwhile to check whether they really are helpful. Sometimes prehybridization may in fact increase background.

## PROBE SELECTION AND LABELING

After the pretreatments discussed above, sections are hybridized with labeled probes. The choice of probe and labeling method is dictated by availability, expertise, and purpose. Certain rules apply, however. Thus, with double-stranded targets (genomic DNA), double-stranded probes are preferred, whereas single-stranded probes should be used for single-stranded targets. In the case of double-stranded probes and targets, it is necessary first to denature them (melt the complementary strands apart) and then to prevent both target and probe from reannealing before hybridization occurs. Hybridization to both strands of the target will produce a stronger signal. Conversely, in the case of single-stranded targets like mRNA, only the antisense strand of a double-stranded probe will be able to hybridize, whereas the other (sense) strand will merely contribute to the background. Moreover, in sections and cells, the probe molecules have to diffuse for a considerable distance before hybridizing to their target. This carries the risk that double-stranded probes may reanneal before hybridizing. For all these reasons, single-stranded probes are preferred for mRNA hybridizations. Three types of probes are used for this purpose: 1) synthetic oligodeoxynucleotide (oligo) probes, 2) RNA probes, and 3) single-stranded polymerase chain reaction (PCR)-generated DNA probes.

Oligo-probes are synthesized on DNA synthesis machines and may be labeled either during or after synthesis. Radioactive labeling is performed after synthesis either by enzyme-catalyzed incorporation of radioactive phosphor in the 5'-end or of radioactive nucleotides in the 3'-end *(13)*. The latter reaction is catalyzed by terminal deoxynucleotidyl transferase (terminal transferase) and is the most common approach for labeling oligoprobes. The nucleotides may be labeled with several different isotopes, including $^3$H, $^{35}$S, $^{32}$P, or $^{33}$P. Of these, $^{35}$S and $^{33}$P are preferred. $^3$H is seldom used since the energy of its radiation is so low that much of it is absorbed within the tissue before reaching the autoradiographic emulsion. Consequently, $^3$H produces a very precise, but insensitive, localization. The opposite applies to $^{32}$P, which produces radiation of such high energy that the precision of localization suffers. Thus, $^{32}$P may be used for low-resolution work for which cellular precision is not needed but high sensitivity is required. A further concern relates to the shelf life of these probes. Thus, whereas $^3$H has a very long shelf life, $^{32}$P-labeled probes can at most be stored for a week. $^{35}$S and $^{33}$P have better precision and longer shelf life than $^{32}$P and possess a higher sensitivity than $^3$H. Accordingly, these two isotopes are the ones usually employed for radioactive *in situ* hybridization and quantitation *(14)*.

These isotopes have been extensively used for quantitative *in situ* hybridizations in the central nervous system. In the gastrointestinal tract, $^{35}$S-labeled probes have caused problems by binding nonspecifically to a population of connective tissue cells *(15,16)*. Although this source of error can be detected by appropriate controls, it is nevertheless disturbing. Interestingly, the nonspecific binding is $^{35}$S-specific and is not seen with nonradioactively labeled probes *(15)*. It may reflect reactions with disulphide-rich proteins.

Nonradioactive, chemical labeling of oligo-probes may be carried out either during or after synthesis. Our studies have demonstrated that such nonenzymatic labeling can be highly advantageous *(7,17)*. It results in sizeable quantities of probes that contain only the specified number of reporter molecules and, accordingly, produce highly reproducible results. In contrast, incorporation of nonradioactively labeled nucleotides in the 3'-end by terminal transferase produces a highly heterogeneous mixture of probes that may vary severalfold in tail length *(7)*. This reduces the sensitivity of detection *(7)*. Moreover, the quality of different batches of terminal transferase may vary, leading to problems with reproducibility that are not encountered with chemically labeled probes.

In our routine, oligo-probes averaging 30 nucleotides in length (30-mers) are labeled with three or four biotin or digoxigenin molecules in the 3'-end *(5)*. Addition of more markers per oligoprobe does not seem to improve sensitivity. Although biotin is often an excellent nonradioactive marker for *in situ* hybridization, it produces problems in tissues that are rich in biotin, like the pancreas. Problems may also occur with cells cultured in biotin-rich media. Although methods and commercial kits for blocking endogenous biotin exists, they have not been entirely satisfactory. Therefore, use of digoxigenin as a marker is preferred when biotin background is a problem. Alternatively, other markers like fluorescein isothiocyanate (FITC) may be used alone or in combinations with biotin or digoxigenin for multiple hybridizations. Digoxigenin has stood the test of time as a reporter yielding very low background even after prolonged developments.

People buying oligo-probes from commercial vendors should take care to specify that the oligos are intended for use as probes and not as PCR primers. The reason for this is that PCR primers can be fabricated cheaply with dilute chemistry in the DNA synthesis machines, whereas oligo-probes need to be of better quality. Methods for selecting oligonucleotide probes have been discussed elsewhere *(5,18,19).*

RNA probes are usually produced by *in vitro* transcription. They can be generated from sequences cloned into suitable vectors or may be generated by PCR using primers that incorporate short synthetic RNA polymerase promoter sequences *(20,21).* Needless to say, they are more vulnerable to degradation than DNA probes. However, this is not a major issue since RNase-free conditions are needed anyway for mRNA hybridization. During *in vitro* transcription, the probes may be labeled with either radioactively (usually $^{35}$S) or nonradioactively (usually digoxigenin) tagged nucleotides *(22,23).* Subsequently, they are usually hydrolyzed to fragments of 100–200 bp, as this may improve penetration into tissues *(23).* The hybrids that form in the tissue are resistant to RNase (cf. RNase protection assays), and nonhybridized probes that are trapped nonspecifically in the tissue can be removed by RNase A digestion after hybridization *(21,22).* This may dramatically reduce background.

PCR-generated single-stranded DNA probes may be produced by assymetrical PCR but are more efficiently produced by the λ exonuclease method *(24).* They may be labeled with radioactively or nonradioactively tagged nucleotides during PCR. Apart from better stability, they have no specific advantages over RNA probes and are less commonly used.

The selection between labels and probes is dictated by personal experience and by specific requirements. For instance, a molecular biologist who wants to have a quick look at the expression pattern of a newly cloned gene would probably reach that goal most easily by producing a $^{35}$S-labeled RNA probe. At the other extreme, clinical analyses of chromosome abberations will need the better resolution and faster results provided by nonradioactive hybridization. For quantitative work, the resolution needed may also vary. Thus, studies of the central nervous system frequently employ low-resolution analyses of variations of gene expression in macroscopically defined nuclei. These studies do not require the high resolution offered by the nonradioactive approach. In contrast, studies at the cellular level generally benefit from nonradioactive methods that also permit the use of determinations of cell size and number.

Quantitative *in situ* hybridization results may be misleading if analyses of cell numbers and sizes are not undertaken *(25).* In the past, many authors have claimed that radioactive methods possess a higher sensitivity than nonradioactive techniques. With the recent demonstration by the Singer laboratory *(26)* of single-copy detection by nonradioactive hybridization, this objection no longer seems to hold. Although, admittedly, single-copy detection is not routinely attainable by either radioactive or nonradioactive techniques, advances in nonradioactive methods including the use of catalyzed reporter deposition (CARD) techniques *(27)* in conjunction with sensitive digitized image acquisition promises that ultrasensitive detection of hybridization signals may become routine *(28–30)*. For detection of low copy numbers, it is, however, imperative that not only the sensitivity of the detection system, but also the preservation and presentation of target be highly optimized.

Short oligonucleotide probes carry less label than the longer RNA probes and are therefore less sensitive. However, they are better at detecting small mismatches in sequences than longer probes. Thus, the presence of mismatches (particularly in the middle of the sequence) has a much greater impact on the $T_m$ of an oligonucleotide than of a longer probe *(31)*. Moreover, oligos can be conveniently fabricated and labeled in quantity. The relative lack of sensitivity of oligo-probes can to some extent be countered by producing multiple oligos complementary to different regions of the target to be investigated *(5,26,32,33)*. There are, however, certain constraints imposed on this approach. Thus, particularly for quantitative purposes, the multiple oligos need to be carefully matched with respect to $T_m$. With short target sequences this may be difficult to achieve. An alternative to conventional oligos is the use of synthetic peptide nucleic acids (PNA), which bind very avidly to target sequences. So far, PNA probes have been used mostly for DNA hybridization; with respect to quantitation, they have been used for estimating telomere lengths in chromosomes *(34–37)*.

## HYBRIDIZATION AND WASHING CONDITIONS

For quantitation, probes must usually be added at saturating concentrations **(Fig. 4)**. This is determined by titrating the amount of probe added until the hybridization signal reaches a plateau, whereafter further addition of probe produces no further increase in signal *(12)*. If hybridizations are carried out at low to medium stringency followed by washing procedures at high stringency, a paradoxic decrease in the hybridization signal is sometimes observed on addition of surplus amounts of probe *(38)*. This may be because under low stringency conditions, two or more probe molecules simultaneously hybridize to the same target and subsequently are washed off at high stringency. To avoid this pitfall, we both hybridize and wash at high stringency. As discussed in the following section, saturating probe concentrations may not always be possible. In such cases, kinetic measurements have to be performed *(39,40)*.

## DETECTION SYSTEMS

*In situ* hybridization signals revealed by either autoradiographic or nonautoradiographic techniques can be quantitated. In the neurosciences, a low-resolution approach for autoradiographic detection is often employed. This method makes use of film sheets that are placed in contact with the hybridized sections. Following a suitable exposure time (days to weeks), the sheets are processed for development under standardized conditions and used for densitometry. The density obtained will be proportional to the degree of blackening of the film. Measurements should be referred to as density since the autoradiographic silver grains do not absorb light but block its transmission. In this type of approach, cellular resolution is not obtained. However, different groups of nerve cells can be identified and average changes in them recorded. In addition, automatic imagers can be used for this approach *(41,42)*. For obtaining a higher resolution, conventional autoradiography using liquid emulsion is employed.

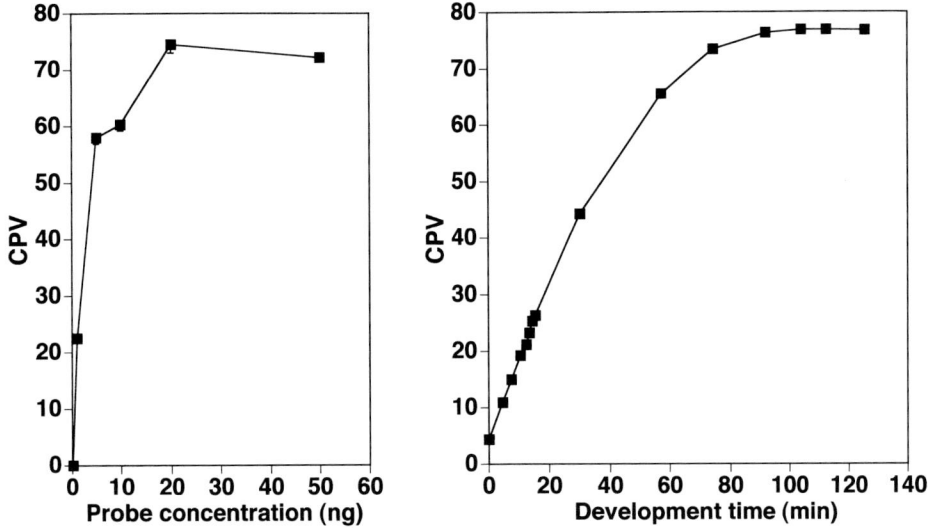

**Fig. 4. Left:** Effects of increasing probe concentrations (biotin-labeled 24-mer oligodeoxy-nucleotide probe complementary to rat proopiomelanocortin mRNA) on the signal (CPV, cor-rected pixel values; see ref. *12*) in rat pituitary melanotrophic cells. Note that probe saturation occurs at concentrations above 20 ng. Detection was achieved with the ASAP method and development in BCIP-NBT. **Right:** Timed development of rat melanotrophic cells hybridized under the same conditions. Note the clear linear kinetics that prevail initially during development and that are followed later by a plateau phase.

Both approaches have been carefully evaluated *(14,41,43–45)*. After conventional autoradiography, the signal can be quantitated by automated or manual grain counting. Alternatively, density measurements can be employed. Although grain counting proce-dures can be carried out in either brightfield or darkfield, it has been suggested that density measurements are best carried out in brightfield *(43)*. Importantly, the autoradio-graphic signal can be expressed in terms of radioactivity by using commercial or home-made radioactive standards that are exposed and developed in parallel with the test sections. Analyses of the standards often reveal nonlinear relationships between the amount of radioactive standards and the signal (grain number/density), which necessi-tates use of curve-fitting programs *(14)*. Provided the specific activity of the probe is known, it is possible to express the radioactivity as number of hybridized probe mole-cules. Although this is attractive, one should remember that this figure need not reflect the true number of target molecules that was originally present in the living cell, but only reflects the number of "hybridizable copies"!

There are also some limitations to the determination of probe-specific activity. Thus, probes labeled using terminal transferase will, as discussed above, exhibit great heterogeneity in tail lengths, and the specific activity will represent an average of this value. Since overlabeled probes hybridize less efficiently to tissues *(7)*, there is a risk that the overall specific activity does not reflect the specific activity of the population of probe molecules that actually hybridize. Although this concern does not detract from the validity of relative quantitations and comparisons, it represents one more obstacle to calculations of the true number of endogenous target sequences. An alternative

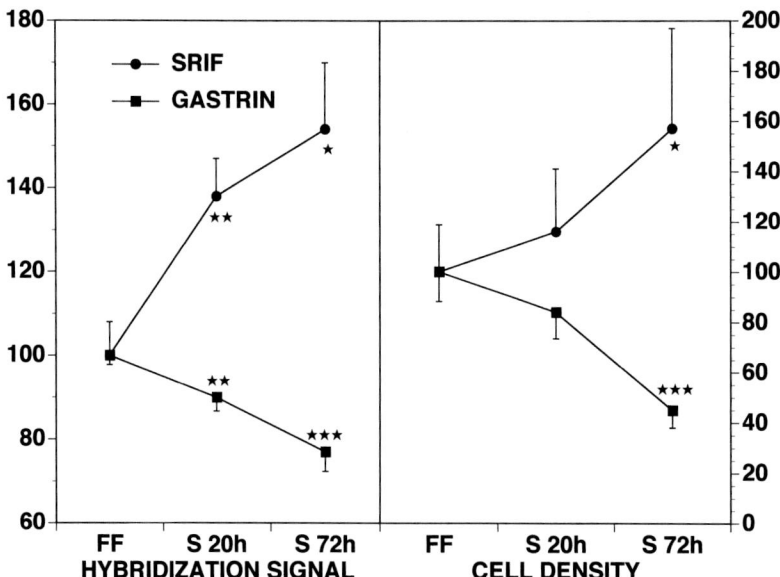

**Fig. 5.** Determination of gastrin and somatostatin (SRIF) hybridization intensities (left) and cell numbers in antrophyloric mucosa of freely fed (FF) rats and in rats that have starved (S) for 20 and 72 h. The data are expressed as percent changes ± SEM relative to the levels in freely fed animals. Note that both the hybridization intensity and numbers of gastrin mRNA-positive cells decrease with starvation, whereas the reverse applies to somatostatin mRNA-positive cells. The magnitude of these changes would be grossly underestimated were only the hybridization intensities measured.

method for calculating how many probe molecules have hybridized to the section is to hybridize with different mixtures of labeled and unlabeled probes *(46).*

Many studies have documented that it is also possible to quantitate nonradioactive hybridization using enzymes or fluorescent reporter molecules as markers *(3,4,12,34, 35,37,39,47–51).* As in conventional hybridization cytochemistry, the nonautoradio-graphic approach is more complicated. However, in addition to advantages already mentioned, it also yields superior morphologic definition that allows determinations of cell sizes and cell numbers. The importance of this aspect is often overlooked in autoradiographic applications **(Fig. 5)** *(25).*

Many studies have employed quantitations using alkaline phosphatase as a marker. The reason for this is that the pioneering work by Butcher *(52,53)* on quantitation of this enzyme made it possible to apply his principles directly to the hybridization analysis *(12).* Moreover, several popular *in situ* hybridization methods employ alkaline phosphatase as a reporter enzyme for nonradioactive hybridization detection. The developer mostly used for alkaline phosphatase detection is bromochloroindolyl phosphate-nitroblue tetrazolium (BCIP-NBT). This developer yields formazan reaction products that exist in a red-absorbing and a blue-absorbing form. The red product forms first during development and is responsible for the reddish coloration of weakly reactive sites, whereas the blue product forms later and imparts a deep blue color to strongly reactive sites. The mixture of the two produces a purple color of sites of intermediate reactivity. To quantitate both forms accurately, it is necessary to measure at the isobestic wavelength (585 nm) *(52,53)*. Developed sections should not be dehydrated but mounted in glycerin buffer, since the red product is soluble in ethanol.

Detection of nonradioactive probes may involve either direct methods encompassing, e.g., alkaline phosphatase-conjugated streptavidin or anti-digoxigenin or more sensitive multilayered immunocytochemical methods. More recently, CARD methods leading to peroxidase-catalyzed precipitation of biotin- or fluorescence-labeled tyramides have been adapted to *in situ* hybridization *(5,28–30)*.

From the quantitative viewpoint, it is advantageous to use as simple a protocol as possible since every single step of multilayer methods needs to be carefully controlled. In practice, this must be balanced against the sensitivity required, and quantitations in our laboratory are routinely made using a three-step APAAP (alkaline phosphatase-anti-alkaline phosphatase) *(54)* or a more sensitive ASAP (as-soon-as-possible or anti-biotin-streptavidin-alkaline phosphatase method; *17)*, modified from the method of Guerin-Reverchon et al. *(55)*. Hopefully, advances in fluorescence-based detection will make these multilayer methods obsolete.

The detection system needs to be carefully titrated with respect to concentration and time of application of the reagents so that reagent concentrations are not limiting and do not contribute to unnecessary background *(12)*. It is also important to control the development time carefully. Thus, development in alkaline phosphatase initially follows linear kinetics, leading to a steadily increasing signal. Eventually, a plateau is reached **(Fig. 4)**. This plateau is first reached by cells containing high target concentrations and later by cells containing lower concentrations. It is therefore important to perform pilot experiments to time the development so that all reactive cells are in the linear phase. This can be done by placing a drop of developer on top of the section, covering with a cover slip, using a spacer, and measuring intensities at specified intervals. No cover-slipping is needed if a water immersion objective is used. We have used a Nikon 20× water immersion objective for this for over 6 years without experiencing any deterioration in its performance. Following every development cycle, the front lens of the objective is rinsed with redistilled water. The BCIP-NBT developer is light sensitive, and care must be taken to expose it to light only when measurements are done. This can be accomplished by blocking the light from the lamp with, e.g., a black disk. Shutting the microscope lamp on and off should be avoided, as this may destabilize the light source.

In a more elegant approach, a mechanical stage on the microscope can be preset to defined positions, which are automatically recalled at defined intervals. This approach has been used to measure the initial kinetics of development for *in situ* hybridizations *(39)*. Such measurements may be better than end-point measurements of development, particularly if saturating probe concentrations have not been reached *(40)*. However, they do tend to make the analysis rather cumbersome since measurements have to be done during development. Consequently, this limits the number of samples that can be analyzed. Fortunately, in our experience, end-point measurements are sufficient under saturating probe conditions and can be made on stored specimens *(12)*.

When the optimal development time has been established (usually leading to much paler specimens than would be considered optimal for conventional microscopy), batches of experimental slides are developed together in Coplin jars that are protected from light by aluminum foil. Visual inspection of these slides during development should be avoided. Development is stopped in a buffer containing EDTA for chelating the divalent cations needed for alkaline phosphatase activity, and the sections are mounted in a water-miscible medium.

## IMAGE ANALYSIS

As illustrated in **Fig. 2**, the image analysis system consists of a microscope, a CCD camera, and a frame grabber (image digitizer) connected to a computer equipped with software for image acquisition and analysis. Ready-made, turnkey systems may be bought from commercial vendors. However, prices may be high. It is also important to realize that many firms supplying the hardware have little or no background for knowing the specific requirements for, e.g., *in situ* hybridization analysis. The advantages of building the system oneself are, first, that it will be cheaper, and second, that obsolete parts (like the computer) may be exchanged without the need for upgrading the entire system. This is certainly an important money saver, as requirements may change with the evolution of new detection systems. It is a cause of considerable concern that some commercial vendors make it a practice to manufacture new systems that are totally incompatible with existing versions. Finally, building the system oneself is arguably the best introduction to image analysis. In the following, I try to cover several points that are important to both home-made and commercial systems. The points discussed are pertinent to *in situ* hybridization and immunocytochemical analysis but do not take other image-processing requirements into account.

The microscope to be used should include a stabilized (or otherwise stable) light source. The stability of the light source should be checked by repeated measurements (using, e.g., sets of neutral density filters from Kodak-Eastman, Rochester, NY) over time. An infrared filter should be placed in front of the light source. Although it is not registered by the eye, infrared radiation may considerably disturb the image, as seen by the CCD camera. Additional filters required include monochrome, bandpass filters at the optimal wavelength required (e.g., 585 ± 10 nm, for work with BCIP-NBT). The neutral density filters referred to above are also used for checking the camera readout and for calibrating the system in OD units. As already emphasized, the 585-nm bandpass filter is required due to the two peaks of absorbances exhibited by the BCIP-NBT reaction product. However, to obey Lambert-Beer's law *(56)*, even reaction products having a single absorption peak must be measured in monochromatic light using appropriate bandpass filters. Thus, before measurements are commenced, at least three sets of filters need to be available.

Additional minor, but necessary, equipment includes a micrometer scale engraved on a microscope slide for spatially calibrating the system so that measurements can be expressed in micrometers rather than in pixels.

The microscope is fitted with a mount (usually a C-mount) for connecting the CCD camera. In cases of microscopes with only one camera exit, a discussion tube may be used for obtaining an extra exit, e.g., for conventional photomicrography. The CCD camera should be carefully fitted in the mount. It is not advisable to mount and unmount the camera repeatedly, as this may result in considerable contamination with dust, in the worst case on the chip itself. The CCD camera may be either monochrome or color.

However, as already pointed out, measurements should be carried out in monochromatic light, so the acquisition of a color camera would have to be motivated by requirements other than quantitation of *in situ* hybridizations. If a color CCD camera is selected, we would recommend a three-chip camera. Thus, one-chip cameras interpret the color information on only one chip, which leads to poorer image results. For measurements in brightfield, ordinary CCD cameras can be used. However, such cameras have been developed for purposes other than image analysis such as surveillance work. Therefore, they are often equipped with an automatic gain function, which comes into play during conditions of low light and increases the signal. This, of course, is disastrous to image analysis work, and the automatic gain of such a camera must be disabled. If the camera possesses manual gain, this should be set to linear. Moreover, the gamma function of the camera should also be linear. All these functions can and should be evaluated by the neutral density filters referred to above, to prove that the system provides a linear readout over the relevant densities.

A number of more expensive cameras have been developed especially for image analysis and fluorescence work. These cameras often possess the ability to cool the CCD chips, which reduces noise under low light conditions. While this is not strictly necessary for the applications discussed here, such cameras may be worth considering due to their greater versatility and potential for ultrasensitive fluorescence applications. Good discussions of the uses and pitfalls of digitized image acquisition are given in refs. *56–58.*

The camera signal is led into a frame grabber or image digitizer. The function of the latter is to break down the image into a mosaic of small fields, pixels, which are associated with a gray value and positional coordinates. In this way the original image is broken down to something reminscent of a spreadsheet with numbers, indicating gray values in positions corresponding to the original image elements (**Fig. 2**). This digital image is interpreted by the computer and displayed as a gray scale image to the user. Obviously, the more pixels available, the better the original image will be represented. This is determined by the resolution of the system and is expressed as pixels per µm. The resolution can be directly calibrated by a micrometer scale and depends on the quality of the camera. With modern CCD cameras, very high resolutions can be obtained. This is not always desirable since large images take up many megabytes of memory and are correspondingly slower to process and analyze. In practice, therefore, one often works with images of around $512 \times 512$ pixels, which are convenient to store and process rapidly.

During image acquisition, it is usually necessary to correct for unevenness of illumination. This is done by acquiring a blank field, which then is used for shade correction using one of several possible algorithms. Acquisition of low light levels, e.g., of fluorescent objects, may result in the appearance of electronic noise in the image. This is minimized by using cooled CCD cameras and the use of image averaging. Averaging is achieved by adding a number of images to each other and dividing by their number. Since the objects of interest will remain in place, while the noise will occur in haphazard positions, averaging of a sufficient number of images will considerably reduce the noise. Additional problems with blooming, glare, and so forth may occur during image acquisition *(56).*

Once the image has been digitized, a number of different programs can be used for analyzing it. Typically, these analyses will include measurements of the area (number of pixels or µm$^2$) and integrated and mean density of defined objects. Objects (often referred to as regions of interest [ROIs] can be defined manually by tracing with the mouse or automatically with a "magic wand" that selects contiguous pixels within a user-definable tolerance zone of gray values. Most common, however, is the use of thresholding operations. These usually work through sliders that will define a certain zone of the grayscale (e.g., 27–81 on a scale from 0 to 255). Pixels with gray values falling within this user-defined zone will become highlighted (e.g., in red) during the thresholding. As a consequence, it may be difficult to decide whether the highlighted pixels correspond exactly to the desired ROI. This is made much easier by simultaneously displaying the original grayscale image on the monitor. Regrettably, several image analysis programs do not offer this valuable option.

In addition to manual or interactive thresholding, it is, under certain circumstances, possible to define automatic, and thus, objective conditions for thresholding. Under most circumstances, including *in situ* hybridization measurements, automatic thresholding is, however, not possible. To avoid the observer bias that may occur during manual thresholding *(59)*, coded specimens should be analyzed by one single trained operator. The conditions for thresholding can be made easier by filtering the grayscale image to enhance its contrast and definition. Several such filters are standard components of image analysis programs.

Once thresholding is completed, the positional information (*x-y* coordinates) of the ROI pixels can be stored in a bitmap image. This is a binary image in which the pixels of interest are given the value of 1 and the rest are given the value of 0. Several image analysis programs provide a set of operators that work on binary images through dilation and erosion of the pixels. In this way, holes in the original binary image can be closed, edge-bound objects removed, and shapes defined through masked propagation (dilation of a seed [eroded] image using another image as a mask, behind the contours of which the seed image cannot be propagated).

Elegant use of these simple operators can achieve a very precise definition of the objects to be measured. In their simplest form, measurements can be returned by multiplying the pixels of the binary image with the pixels of the original (unperturbed) gray level image. Thus, pixels with the same *x-y* coordinates will be multiplied with each other. Since the ROI pixels have the value of 1 (and the rest the value of 0) in the binary image, only the corresponding gray values will remain in the gray value image. The histogram of this image can be used for obtaining the density of the ROI in pixel values **(Fig. 2)**. This represents a simplified account of image measurement routines, and most commercial image analysis software programs include far more sophisticated procedures that also return positional information and other parameters. In addition, background readings should be done and subtracted from the measurements of the hybridized cells. In many cases, background measurements can be made on nearby negative cells or section areas. This may not be possible with ubiquitous RNAs, studies of which may necessitate background readings from control sections (e.g., pretreated with RNases before hybridization or hybridized with labeled non-sense or scrambled probes). For studies of the linearity of the hybridization system, cytochemical models can be used *(12,60)* **(Fig. 6)**. If the system is calibrated, the pixel values can be recalculated to numbers of hybridizable copies of target, OD, or radioactivity units.

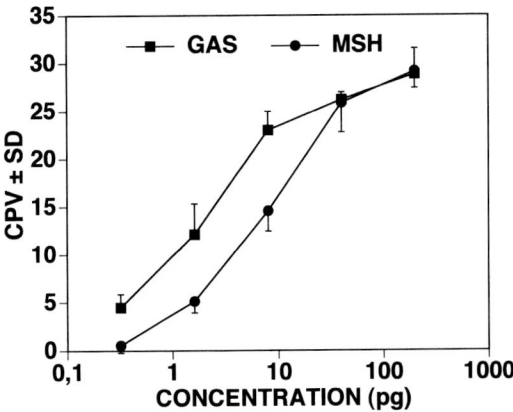

**Fig. 6.** Cytochemical models using sense sequences of MSH or gastrin immobilized on glass slides as described *(60)* and subsequently hybridized with complementary biotinylated probes followed by alkaline phosphatase-based detection. Note that the signal increases in parallel with increasing concentrations of target.

The number of ROI pixels will also be known and can be used to calculate mean density, area, perimeter, form factor, and so on.

Different image analysis packages are available for different needs. One excellent, introductory possibility is the shareware program NIH Image (available from http:// zippy.nimh.nih.gov or http://rsb.info.nih.gov/NIH-Image/download.html). Unfortunately, this program is only available for use on Macintosh computers. It can be used alone or in combination with more heavyweight programs like TCL (now SCIL) Image (TNO Institute of Applied Physics, Delft, The Netherlands). We have used these programs together on the Macintosh platform for well over 10 years. The NIH Image program has been used for controlling the frame grabber with respect to shade correction and frame averaging, whereas most measurements have been performed using the TCL Image package, which permits some very sophisticated filtering routines on the gray level as well as on binary images. The TCL/SCIL Image program *(61)* is available on many platforms including DOS/Windows.

Another program that is available on both Macintosh and PCs is IP Lab, which we have had excellent experiences with on the Macintosh platform (IPLab Spectrum, Signal Analytics, Vienna, Virginia). This program can be used for analysis of both color and grayscale images. Many of the larger image analysis programs take some time to learn. This is not the case with the free NIH Image program, which constitutes the main image analysis package in many laboratories. As an alternative, some workers use the Adobe Photoshop program for image analysis *(62,63)*. This program is available for both PCs and Macintoshes. A very worthwhile alternative is the Image-Pro Plus program (Media Cybernetics, Silver Spring, MD). Compared with other heavyweight image analysis packages, this program seems to be extremely easy to use. Needless to say, the above concerns relate only to image analysis programs that we have tested personally, and many more are available on the market. Fortunately, many demo versions are now available so one can test the program for free before buying it.

## ACKNOWLEDGMENTS

I thank the Danish Medical Research Council and the Danish Cancer Society for support.

## REFERENCES

1. Fitzpatrick-McElligott, S., Lewis, M. E., Tyler, M., Baldino, F. Jr., and Davis, L. G. (1988) In situ hybridization with radiolabeled synthetic oligodeoxynucleotide probes. *Dupont Biotechnol. Update* **3**, 2–3.
2. Yoshii, A., Koji, T., Ohsawa, N., and Nakane, P. K. (1995) In situ localization of ribosomal RNAs is a reliable reference for hybridizable RNA in tissue sections. *J. Histochem. Cytochem.* **43**, 321–327.
3. Guiot, Y. and Rahier, J. (1997) Validation of non-radioactive in situ hybridization as a quantitative approach of messenger ribonucleic acid variations. A comparison with Northern blot. *Diagn. Mol. Pathol.* **6**, 261–266.
4. Guiot, Y. and Rahier, J. (1995) The effects of varying key steps in the non-radioactive hybridization protocol: a quantitative study. *Histochem. J.* **27**, 60–68.
5. Hougaard, D. M., Hansen, H., and Larsson, L-I. (1997) Non-radioactive in situ hybridization for mRNA with emphasis on the use of oligodeoxynucleotide probes. *Histochem. Cell Biol.* **108**, 335–344.
6. Larsson, L-I., Christensen, T., and Dalbøge, H. (1988) Detection of proopiomelanocortin mRNA by in situ hybridization using a biotinylated ligodeoxynucleotide probe and avidin-alkaline phosphatase histochemistry. *Histochemistry* **89**, 109–116.
7. Larsson, L-I. and Hougaard, D. M. (1990) Optimization of non-radioactive in situ hybridization: image analysis of varying pretreatment, hybridization and probe labelling conditions. *Histochemistry* **93**, 347–354.
8. Lawrence, J. B. and Singer, R. H. (1985) Quantitative analysis of in situ hybridization methods for the detection of actin gene expression. *Nucleic Acids Res.* **13**, 1777–1799.
9. Rentrop, M., Knapp, B., Winter, H., and Schweizer, J. (1986) Aminoalkylsilane-treated glass slides as support for in situ hybridization of keratin cDNAs to frozen tissue sections under varying fixation and pretreatment conditions. *Histochem. J.* **18**, 271–276.
10. West, M. J., Ostergaard, K., Andreassen, O. A., and Finsen, B. (1996) Estimation of the number of somatostatin neurons in the striatum: an in situ hybridization study using the optical fractionator method. *J. Comp. Neurol.* **370**, 11–22.
11. Larsson, L-I. and Hougaard, D. M. (1993) Non-radioactive in situ mRNA hybridization using synthetic oligonucleotides: principles, combination with immunocytochemistry and quantitation. *Neurosci. Protocols* **20**, 1–18.
12. Larsson, L-I., Traasdahl, B., and Hougaard, D. M. (1991) Quantitative non-radioactive in situ hybridization. Model studies and studies on pituitary proopiomelanocortin cells after adrenalectomy. *Histochemistry* **95**, 209–215.
13. Collins, M. L. and Hunsaker, W. R. (1985) Improved hybridization assays employing tailed oligonucleotide probes: a direct comparison with 5′-end-labeled oligonucleotide probes and nick-translated plasmid probes. *Anal. Biochem.* **151**, 211–224.
14. Baskin, D. G. and Stahl, W. L. (1993) Fundamentals of quantitative autoradiography by computer densitometry for in situ hybridization with emphasis on $^{33}$P. *J. Histochem. Cytochem.* **41**, 1767–1776.
15. Larsson, L-I., Tingstedt, J-E., Madsen, O. D., Serup, P., and Hougaard, D. M. (1995) The LIM-homeodomain protein Isl-1 segregates with somatostatin but not with gastrin expression during differentiation of somatostatin/gastrin precursor cells. *Endocrine* **3**, 519–524.

16. Panula, P. and Wasowicz, K. (1993) Action of antiulcer drugs. *Science* **262**, 1454–1455.

17. Larsson, L-I. and Hougaard, D. M. (1993) Sensitive detection of rat gastrin mRNA by in situ hybridization with chemically biotinylated oligodeoxynucleotides: validation, quantitation and double-staining studies. *J. Histochem. Cytochem.* **41**, 157–163.

18. Lewis, M. L., Sherman, T. G., and Watson, S. J. (1985) In situ hybridization histochemistry with synthetic oligonucleotides: strategies and methods. *Peptides* **6**, 75–87.

19. Stahl, W. L., Eakin, T. J., and Baskin, D. G. (1993) Selection of oligonucleotide probes for detection of mRNA isoforms. *J. Histochem. Cytochem.* **41**, 1735–1740.

20. Brink, P. E. and Grimm, P. C. (1994) Rapid nonradioactive in situ hybridization for interleukin-2 mRNA with riboprobes generated using the polymerase chain reaction. *J. Immunol. Methods* **167**, 83–89.

21. Sitzmann, J. H. and LeMotte, P. K. (1993) Rapid and efficient generation of PCR-derived riboprobe templates for in situ hybridization histochemistry. *J. Histochem. Cytochem.* **41**, 773–776.

22. Komminoth, P., Merk, F. B., Leav, I., Wolfe, H. J., and Roth, J. (1992) Comparison of $^{35}$S- and digoxigenin-labeled RNA and oligonucleotide probes for in situ hybridization. *Histochemistry* **98**, 217–228.

23. Schaeren-Wiemers, N. and Gerifn-Moser, A. (1993) A single protocol to detect transcripts of various types and expression levels in neural tissue and cultured cells: in situ hybridization using digoxigenin-labelled cRNA probes. *Histochemistry* **100**, 431–440.

24. Hannon, K., Johnstone, E., Craft, L. S., et al. (1993) Synthesis of PCR-derived, single-stranded DNA probes suitable for in situ hybridization. *Anal. Biochem.* **212**, 421–427.

25. McCabe, J. T. and Bolender, R. P. (1993) Estimation of tissue mRNAs by in situ hybridization. *J. Histochem. Cytochem.* **41**, 1777–1783.

26. Femino, A. M., Fay, F. S., Fogarty, K., and Singer, R. H. (1998) Visualization of single RNA transcripts in situ. *Science* **280**, 585–590.

27. Adams, J. C. (1992) Biotin amplification of biotin and horseradish peroxidase signals in histochemical stains. *J. Histochem. Cytochem.* **40**, 1457–1463.

28. Kerstens, H. M. J., Poddighe, P. J., and Hanselaar, G. J. M. (1995) A novel in situ hybridization signal amplification method based on deposition of biotinylated tyramide. *J. Histochem. Cytochem.* **43**, 347–352.

29. Raap, A. K., van de Corput, M. P. C., Vernenne, R. A. W., van Gijlswijk, R. P. M., Tanke, H. J., and Wiegant, J. (1995) Ultrasensitive FISH using peroxidase-mediated deposition of biotin- or fluorochrome tyramides. *Hum. Mol. Genet.* **4**, 529–534.

30. Van Giljswijk, R. P. M., Wiegant, J., Raap, A. K., and Tanke, H. J. (1996) Improved localization of fluorescent tyramides for fluorescence in situ hybridization using dextran sulphate and polyvinyl alcohol. *J. Histochem. Cytochem.* **44**, 389–392.

31. Long, A. A., Mueller, J., Andre-Schwartz, J., Barrett, K. J., Schwartz, R., & Wolfe, H. (1992) High-specificity in-situ hybridization. *Diagn. Mol. Pathol.* **1**, 45–57.

32. Lloyd, R. V. and Jin, L. (1995) In situ hybridization analysis of chromogranin A and B mRNAs in neuroendocrine tumors with digoxigenin-labeled oligonucleotide probe cocktails. *Diagn. Mol. Pathol.* **4**, 143–151.

33. Trembleau, A. & Bloom, F. E. (1995) Enhanced sensitivity for light and electron microscopic in situ hybridization with multiple simultaneous non-radioactive oligodeoxynucleotide probes. *J. Histochem. Cytochem.* **43**, 829–841.

34. de Pauw, E. S., Verwoerd, N. P., Duinkerken, N., et al. (1998) Assessment of telomere length in hematopoietic interphase cells using in situ hybridization and digital fluorescence microscopy. *Cytometry* **32**, 163–169.

35. Hultdin, M., Gronlund, E., Norrback, K., Erikson-Lindstrom, E., Just, T., and Roos, G.

(1998) Telomere analysis by fluorescence in situ hybridization and flow cytometry. *Nucleic Acids Res.* **26**, 3651–3656.

36. Poon, S. S., Martens, U. M., Ward, R. K., and Lansdorp, P. M. (1999) Telomere length measurements using digital fluorescence microscopy. *Cytometry* **36**, 267–278.

37. Rufer, N., Dragowska, W., Thornbury, G., Roosnek, E., and Landsdorp, P. M. (1998) Telomere length dynamics in human lymphocyte subpopulations measured by flow cytometry. *Nat. Biotechnol.* **16**, 743–747.

38. Larsson, L.-I. (1997) Quantitative in situ hybridization. *Endocr. Pathol.* **8**, 3–9.

39. Leeuw, T. and Pette, D. (1994) Kinetic microphotometric evaluation of in situ hybridization for mRNA of slow myosin heavy chain in type I and C fibres of rabbit muscle. *Histochemistry* **102**, 105–112.

40. Cash, E. and Brahic, M. (1986) Quantitative in situ hybridization using initial velocity measurements. *Anal. Biochem.* **157**, 236–240.

41. Vizi, S. and Gulya, K. (2000) Calculation of maximal hybridization capacity ($H_{max}$) for quantitative in situ hybridization. A case study for multiple calmodulin mRNAs. *J. Histochem. Cytochem.* **48**, 893–904.

42. Laniece, P., Charon, Y., Cardona, A., et al. (1998) A new high resolution radioimager for the quantitative analysis of radiolabelled molecules in tissue section. *J. Neurosci. Methods* **86**, 1–5.

43. Jonker, A., de Boer, P. A. J., van den Hoff, M. J. B., Lamers, W. H., and Moorman, A. F. M. (1997) Towards quantitative in situ hybridization. *J. Histochem. Cytochem.* **45**, 413–423.

44. Palfi, A., Hatvani, L., and Gulya, K. (1998) A new quantitative film autoradiographic method of quantifying mRNA transcripts for in situ hybridization. *J. Histochem. Cytochem.* **46**, 1141–1149.

45. Singer, R. H., Lawrence, J. B., & Villnave, C. (1986) Optimization of in situ hybridization using isotopic and non-isotopic detection methods. *Biotechniques* **4**, 230–250.

46. Gerfen, C. R., McGinty, J., and Young, W. S. III. (1991) Dopamine differentially regulates dynorphin, substance P, and enkephalin expression in striatal neurons: in situ hybridization histochemical analysis. *J. Neurosci.* **11**, 1016–1031.

47. Albalwi, M., Hammond, D. W., Goepel, J. R., Hough, R. E., and Goyns, M. H. (1999) Semi-quantitative fluorescence in situ hybridization analysis indicates that the myc protein is consistently stabilized both before and after transformation of low-grade follicular center to high-grade diffuse large cell lymphoma. *Lab. Invest.* **79**, 707–715.

48. Cheng, L., Bucana, C. D., and Wei, Q. (1996) Fluorescence in situ hybridization method for measuring transfection efficiency. *Biotechniques* **21**, 486–491.

49. Higo, N., Oishi, T., Yamashita, A., Matsuda, K., and Hayashi, M. (1999) Quantitative non-radioactive in situ hybridization study of GAP-43 and SCG10 mRNAs in the cerebral cortex of adult and infant macaque monkeys. *Cereb. Cortex* **9**, 317–331.

50. Rauch, J., Wolf, D., Craig, J. M., Hausmann, M., and Cremer, C. (2000) Quantitative microscopy after fluorescence in situ hybridization—a comparison between repeat-depleted and non-depleted DNA probes. *J. Biochem. Biophys. Methods* **44**, 59–72.

51. Salchi, M., Barron, M., Merry, B. J., and Goyns, M. H. (1999) Fluorescence in situ hybridization analysis of the fos/jun ratio in ageing brain. *Mech. Ageing Dev.* **107**, 61–71.

52. Butcher, R. G. (1972) Precise cytochemical measurement of neotetrazolium formazan by scanning and integrating microdensitometry. *Histochemie* **32**, 171–190.

53. Butcher, R. G. (1978) The measurement in tissue sections of the two formazans derived from nitroblue tetrazolium in dehydrogenase reactions. *Histochem. J.* **10**, 739–744.

54. Mason, D. Y. (1985) Immunocytochemical labelling of monoclonal antibodies by the APAAP immunoalkaline phosphatase technique: In *Techniques in Immunocytochemistry*, vol. 3 (Bullock, G. B. and Petrusz, P., eds.), Academic, New York, pp. 25–42.

55. Guerin-Reverchon, I., Chardonnet, Y., Chignol, M. C., & Thivolet, J. (1989) A comparison

of methods for detection of human papillomavirus DNA by in situ hybridization with biotinylated probes on human carcinoma cell lines. *J. Immunol. Methods* **123**, 167–176.

56. Chieco, P., Jonker, A., Melchiorri, C., Vanni, G., and van Noorden, C. J. F. (1994) A users guide for avoiding errors in absorbance image cytometry: a review with original experimental observations. *Histochem. J.* **26**, 1–19.

57. Oberholzer, M., Osatreicher, M., Christen, H., and Bruhlmann, M. (1996) Methods in quantitative image analysis. *Histochem. Cell Biol.* **105**, 333–355.

58. Shotton, D. M. (1995) Electronic light microscopy: present capabilities and future prospects. *Histochem. Cell Biol.* **104**, 97–137.

59. Jagoe, R., Steel, J. H., Vucicevic, V., et al. (1991) Observer variation in quantification of immunocytochemistry by image analysis. *Histochem. J.* **23**, 541–547.

60. Larsson, L-I and Hougaard, D. M. (1994) Glass slide models for immunocytochemistry and in situ hybridization. *Histochemistry* **101**, 325–331.

61. Ten Kate, T. K., Van Balen, R., Smeulders, A. W. M., Groen, F. C. A., and De Boer, G. A. (1990) SCILIAM, a multi-level interactive image processing environment. *Pattern Recog. Lett.* **11**, 429–441.

62. Lehr, H-A, Mankoff, D. A., Corwin, D., Santeusanio, G., and Gown, A. M. (1997) Application of Photoshop-based image analysis to quantification of hormone receptor expression in breast cancer. *J. Histochem. Cytochem.* **45**, 1559–1565.

63. Matkowskyj, K. A., Schonfeld, D., and Benya, R. V. (2000) Quantitative immunohistochemistry by measuring cumulative signal strength using commercially available software Photoshop and Matlab. *J. Histochem. Cytochem.* **48**, 303–311.

# 10
# Confocal Laser Scanning Microscopy

**Akira Matsuno,** MD, PHD, **Johbu Itoh,** MD, PHD,
**Tadashi Nagashima,** MD, PHD, **R. Yoshiyuki Osamura,** MD, PHD,
**and Keiichi Watanabe,** MD, PHD

## INTRODUCTION

Immunoelectron microscopy and electron microscopic *in situ* hybridization are undoubtedly the best methods for following the dynamic changes of subcellular organelles; however, these techniques require specific tissue preparation and equipment. More recent developments include a more refined and sophisticated technique, confocal laser scanning microscopy (CLSM), which was originally described by Minsky in 1957 *(1)* and has since been applied to the field of medical biology. In early experiments, only fluorescent signals were detectable by CLSM *(2–9)*; however, recent innovations have made possible the visualization of nonfluorescent signals such as horseradish peroxidase (HRP) and diaminobenzidine (DAB) signals by CLSM *(10,11)*. Moreover, the combination of CLSM and the image analysis system (IAS) *(12)* has allowed us to visualize subcellular organelles three-dimensionally in routinely processed light microscopic specimens. We applied CLSM to specimens prepared for light microscopy *(12)* and demonstrated the intracellular identification of subcellular organelles and pituitary hormone mRNA, comparable to that of electron microscopy *(13,14)*. We also applied CLSM to the study of tumor angiogenesis *(15)* and the microvessel environment of hormone-secreting cells *(16)*. The visualization of subcellular organelles, mRNA and protein products, as well as three-dimensional images of microvessel environment of hormone-secreting cells is discussed in this chapter.

## GENERAL PRINCIPLES

The fundamentals of CLSM date back to 1957, when Minsky *(1)* described this new type of microscopy. As commonly applied, CLSM has a viable pinhole, and its modifying scan angles allow the adjustment of magnification. A schematic example of the pinhole of CLSM is given in **Fig. 1** The laser light is focused via the scanner through the tube lens and the objective, illuminating a small spot in the specimen. Emitted light emanating from the focal plane and the planes above and below is directed via the scanner to the dichroic beam splitter (dichroic mirror), where it is decoupled and

From: *Morphology Methods: Cell and Molecular Biology Techniques*
Edited by: R. V. Lloyd © Humana Press, Totowa, NJ

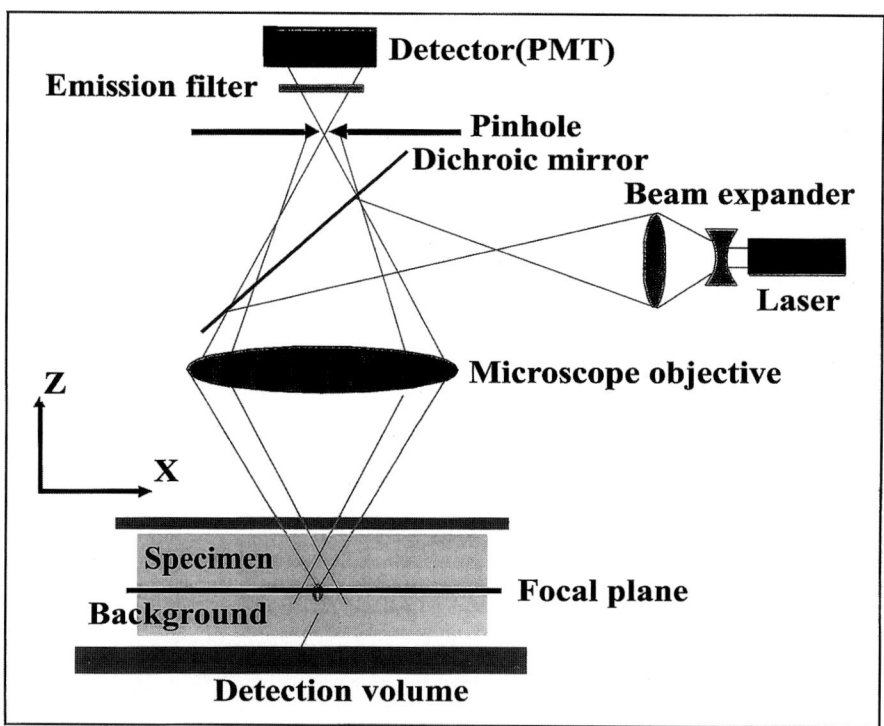

**Fig. 1.** Schematic drawing of ray path in CLSM. PMT, photomultiplier.

directed onto a detector (photomultiplier [PMT]. A pinhole in front of the photomultiplier is positioned at the crossover of the light beams emerging from the focal point.

## Image Analyzing System

The CLSM system we have utilized is composed of LSM-410 and LSM operating software, version 3.95 (Carl Zeiss, Jena, Germany). For example, DAB reaction products are illuminated by a 488-nm wavelength helium-neon laser ray. Optical reflectance signals labeled by DAB are observed using a dichroic beam splitter (NT 80/20/543, Carl Zeiss), without an emission filter. Optical tomographic images (Z-images) for three-dimensional reconstruction are acquired by plan-apochromat (×63, oil, N.A. 1.4; Carl Zeiss) and plan-neofluar (×10, N.A. 0.30; ×20, N.A. 0.50; Carl Zeiss) objective lenses. Z-directional movement for optical sectioning of entire specimens is controlled by stepping the motor unit for an axial scanning at 0.3–0.5 μm focus steps. CLSM image resolutions are 512 × 512 and/or 1024 × 1024 pixels (8 bits, 256 gray levels). Image processing is accomplished with LSM software, version 3.98, through various processes such as contrast normalization, line enhancement, edge enhancement, noise reduction, and image combination. After computer-assisted three-dimensional imaging by LSM systems, these depicted images are stored in the hard disk memory and/or magnetic optical disk EDM-230C (Sony, Tokyo, Japan). Final processed data for digitized and reconstructed images are printed with a Pictrostat digital 400, a digital printer (Fuji, Tokyo, Japan) and also recorded on color reversal 35-mm film (Ektachrome 64 Professional, Eastman Kodak, Rochester).

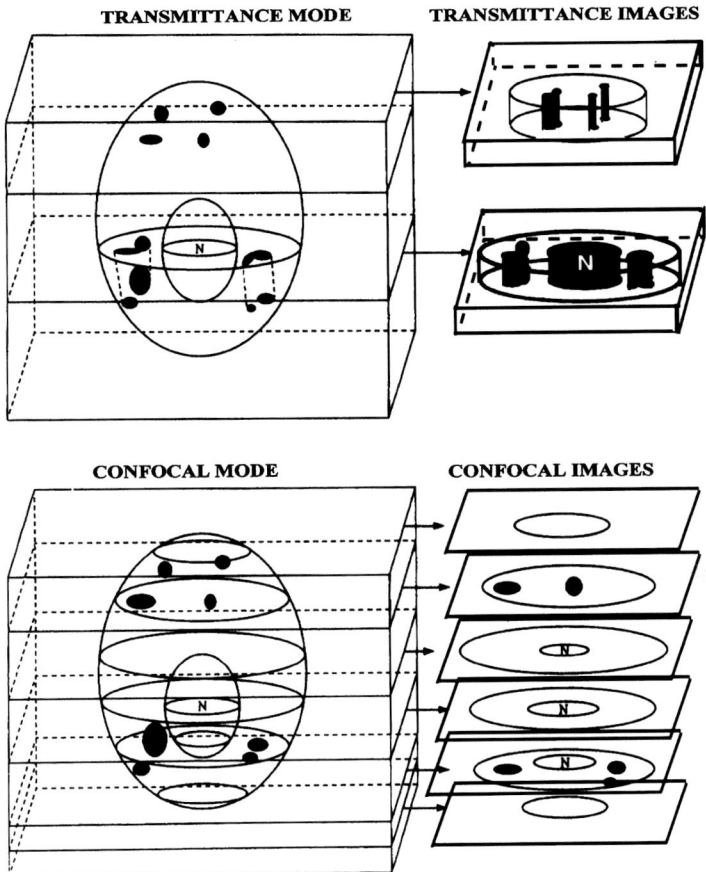

**Fig. 2.** Schematic drawing of transmittance and confocal images. The reflectance confocal image is an individual dimensional image obtained at any optional cross section, whereas the transmittance image is a projected or integrated two-dimensional image from three-dimensional images.

## Transmittance Images and Reflectance Confocal Images

With the CLSM system, one can observe transmittance images, reflectance confocal images, and combined images, as well as phase contrast and differential interference contrast (DIC) images. The reflectance confocal image is an individual dimensional image obtained at any optional cross-section, whereas the transmittance image is a projected or integrated two-dimensional image from three-dimensional images (**Fig. 2**). One of the important advantages of CLSM is the enhancement of weakly positive images, barely observed under conventional light microscopy. The combination of CLSM and IAS *(12)* allows us to visualize subcellular organelles.

## Probes

Nonfluorescent as well as fluorescent signals are detectable under CLSM *(2–9)*. The non-fluorescent signals, including HRP, DAB, osmium black, azo dye, nitroblue tetrazolium (NBT), and heavy metals, have valuable advantages over fluorescent signals since these nonfluorescent signals are permanently preserved and allow repeated or retrospective examination.

## OBSERVATION OF SUBCELLULAR ORGANELLES, MRNA AND PROTEIN PRODUCTS

Subcellular organelles and intracellular distribution of mRNA and protein products can be observed under electron microscopy, which has been described in Chapter 8. Another tool is CLSM, which we applied to the specimens prepared for light microscopy and by which we demonstrate intracellular identification of subcellular organelles and pituitary hormone mRNA *(13,14)*.

### Observation of Subcellular Organelles and Protein Products

*Materials and Protocols*

The immunohistochemically stained paraffin sections can be used for CLSM observation of subcellular organelles and protein products that are developed with HRP-DAB. The following is the protocol for the detection of pituitary hormones and subcellular organelles.

1. Pituitary tissue specimens were immunostained by the indirect peroxidase method with antibodies against human pituitary hormones, such as anti-growth hormone (GH) antibody (rabbit, polyclonal, 1:400 diluted in bovine serum albumin [BSA]/0.01 $M$ phosphate-buffered saline, pH 7.4 [PBS]; from DAKO, Carpinteria, CA).
2. A secondary antibody, anti-rabbit immunoglobulin labeled with HRP, was applied to the sections, and positive immunoreactions were visualized with HRP and DAB.
3. CLSM observation of the immunostained tissue specimens was as described above.

*Results*

CLSM observation of a paraffin section of a human somatotroph adenoma immunostained with anti-GH antibody revealed secretory granules positively immunostained for GH **(Fig. 3)**.

### Observation of Subcellular Organelles and mRNA

*Materials and Protocols*

In this section, we describe the identification of prolactin (PRL) mRNA and subcellular organelles in rat pituitary cells treated with estrogen. The following is the protocol for this experiment.

1. Female Wistar-Imamichi rats treated intramuscularly with 5 mg estradiol dipropionate (E2 depot: Ovahormon Depot; Teikoku Zoki, Tokyo, Japan) every 4 weeks were sacrificed after 7 weeks.
2. The anterior lobes removed from the pituitary glands were immediately fixed at 4°C overnight in 4% paraformaldehyde dissolved in PBS.
3. After immersion in graded concentrations of sucrose dissolved in PBS at 4°C (10% for 1 h, 15% for 2 h, 20% for 4 h), tissues were embedded in Optimal Cutting Temperature (OCT) compound (Tissue-Tek, Miles, Elkhart, IN).
4. Tissue specimens (6 μm thick) were mounted on 3-aminopropylmethoxysilane-coated slides.
5. After air drying for 1 h, tissue sections were washed in PBS for 15 min.

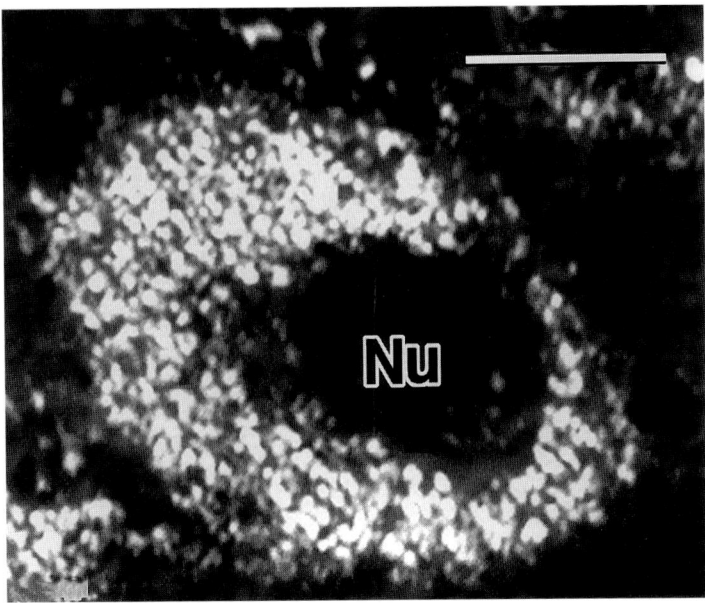

**Fig. 3.** CLSM observation of a paraffin section of a human somatotroph adenoma immuno-stained with anti-GH antibody revealed secretory granules positively immunostained for GH. Nu, nucleus. Bar = 5 μm.

6. Tissue sections were treated with 0.1 μg/mL proteinase K at 37°C for 30 min, followed by treatment for 10 min with 0.25% acetic anhydride in 0.1 $M$ triethanolamine.
7. The slides were washed in 2× sodium chloride sodium citrate (SSC) at room temperature for 3 min and then prehybridized at 37°C for 30 min. The prehybridization solution consisted of 10% dextran sulfate, 3× SSC, 1× Denhardt's solution (0.02% Ficoll/0.02% BSA/0.02% polyvinylpyrrolidone), 100 μg/mL salmon sperm DNA, 125 μg/mL yeast tRNA, 10 μg/mL polyadenylic-cytidylic acid, 1 mg/mL sodium pyrophosphate, pH 7.4, and 50% formamide.
8. The biotinylated oligonucleotide probe at a concentration of 0.1 ng/μL was diluted with this prehybridization solution, and hybridization was carried out at 37°C overnight. The sequence of antisense synthesized oligonucleotide probe for rat PRL mRNA and the details of biotinylation were described in our previous reports *(14,17–22)*.
9. The slides were washed at room temperature with 2× SSC, 1× SSC, and then 0.5× SSC for 15 min.
10. After washing in PBS for 5 min, the slides were dipped in DAB solution (incomplete Graham-Karnovsky's solution) for 1 h.
11. The hybridization signals were detected with streptavidin-biotin-horseradish peroxidase (ABC-HRP), using Vectastain's ABC kit (Vector, Burlingame, CA), and thereafter hybridization signals were developed with 0.017% $H_2O_2$ in DAB solution (complete Graham-Karnovsky's solution) for 7 min. At this stage, hybridization signals were confirmed light microscopically.
12. CLSM observation of the immunostained tissue specimens was as described above.

*Results*

With the CLSM transmittance mode, diffuse positive signals of PRL mRNA were observed, evidently in the cytoplasm, and the whirling structure of the rough endoplasmic

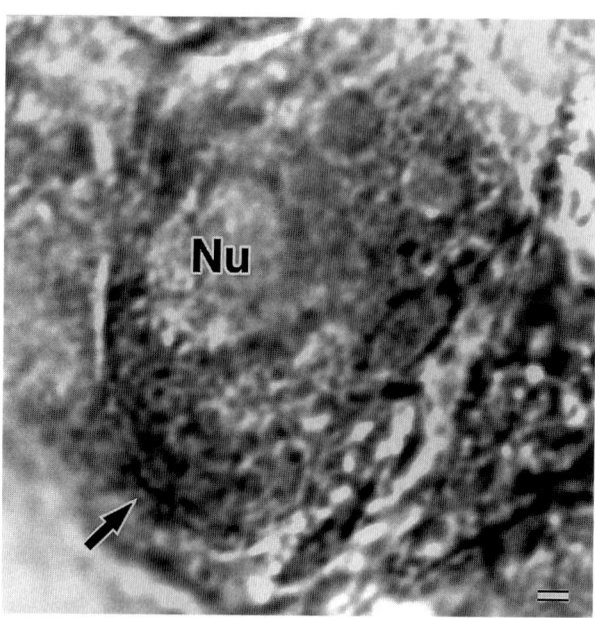

**Fig. 4.** With CLSM transmittance mode, diffuse positive signals of PRL mRNA were observed, evidently in the cytoplasm, and the whirling structure of RER (arrow) was also noted. Nu, nucleus. Bar = 1 μm. (Reproduced with permission from ref. *14.*).

reticulum (RER) was also noted **(Fig. 4)**. With the CLSM reflectance confocal mode combined with IAS, positive signals of PRL mRNA were observed **(Fig. 5)**. With combined transmittance and reflectance confocal images, positive signals of PRL mRNA and granular polysome-like structures were noted on the whirling RER in the cytoplasm **(Fig. 6)**. Under high-power magnification, these signals were interpreted to be similar and comparable to ISH signals of PRL mRNA on the whirling RER observed under EM **(Fig. 7)** *(14)*.

## Comments

The ultrastructural resolution of CLSM may be insufficient compared with that of electron microscopy. Nevertheless, a reflectance confocal image combined with a transmittance image has merit in that it can provide an individual dimensional image at any optional cross-sections of a routinely processed light microscopic specimen. Another merit of CLSM is the large number of cells that can be observed, whereas only a limited number of cells can be observed under electron microscopy. CLSM can identify, at the ultrastructural level, ISH signals of mRNA, or immunohistochemical signals of protein together with subcellular organelles; these are similar and comparable to those observed under electron microscopy *(13,14)*. CLSM is available for the study of the spatial relationship of mRNA and encoded protein and may be another useful tool for the ultrastructural ISH study of mRNA.

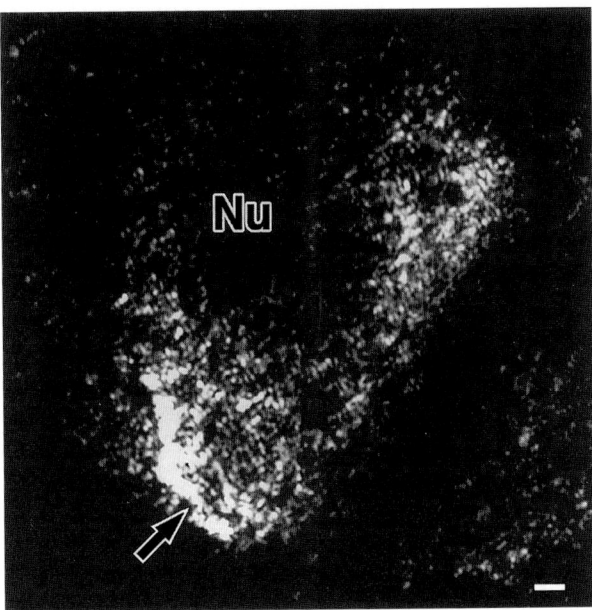

**Fig. 5.** With CLSM reflectance confocal mode combined with IAS (LSM software version 3.98), positive signals of PRL mRNA (arrow) were observed. Nu, nucleus. Bar = 1 μm (Reproduced with permission from ref. *14.*)

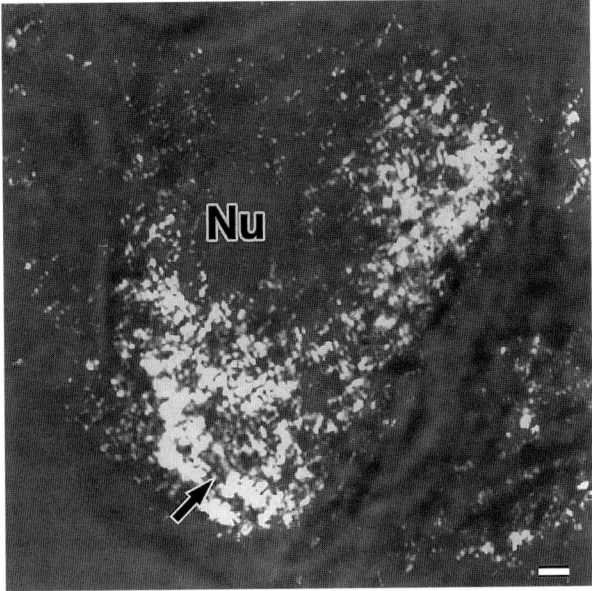

**Fig. 6.** With a combination of transmittance images and reflectance confocal images, positive signals of PRL mRNA and granular polysome-like structures were noted on the whirling RER in the cytoplasm (arrow). Nu, nucleus. Bar = 1 μm. (Reproduced with permission from ref. *14.*)

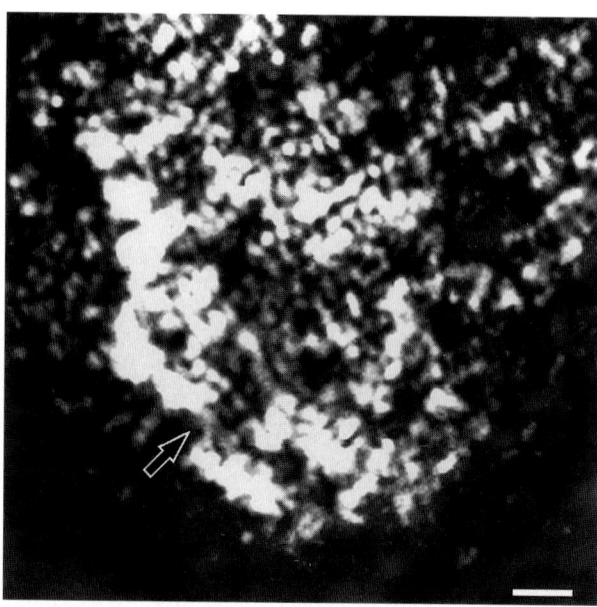

**Fig. 7.** Under high-power magnification, these signals were interpreted as being similar and comparable to ISH signals of PRL mRNA on the whirling RER observed under EM (arrow). Bar = 1 μm. (Reproduced with permission from ref. *14.*)

## OBSERVATION OF THE MICROVESSEL ENVIRONMENT

### *Three-Dimensional Analysis*

Much research has focused on the relationship between hormone-secreting endocrine cells and their microvessels *(23,24)*, as analyzed two-dimensionally in light or electron microscopic thin sections. Several techniques for identifying the vasculature have been reported *(25–42)*. However, these methods are not satisfactory for three-dimensional analysis of the microvascular environment of the endocrine cells. CLSM can reconstruct the microcirculation three-dimensionally and can serve for study of the relationship between hormone-secreting endocrine cells and the microvessel environment. We describe the techniques for visualizing this relationship in the simultaneous and three-dimensional form under CLSM using the fluorescein 5-isothiocyanate (FITC)-conjugated gelatin injection method and immunofluorescence-labeled antibody. The following is the protocol for three-dimensional analysis of hormone-secreting endocrine cells and the microvessel environment.

### *Materials and Methods*

#### *Injection of FITC-Gelatin Conjugate*

1. We combined and modified the fluorescent dye conjugated-polypeptide method *(43)* and the high molecular weight fluorescein-labeled dextran method *(44)*. FITC was obtained from Dojindo (Kumamoto, Japan) and gelatin from Sigma (St. Louis, MO).
2. FITC was conjugated to gelatin by a brief reaction at 37°C and pH 11.0. Typically, 10–50 mg of FITC was dissolved in 1 mL dimethylsulfoxide (DMSO; Sigma), pH 11.0.
3. A 20% gelatin solution made with distilled water, pH 11, and DMSO containing FITC were mixed for conjugation at 37°C overnight.

4. Then FITC-conjugated gelatin was dialyzed in 0.01 $M$ PBS, pH 7.4, at 37°C in the dark for 1 week.
5. Aliquots of the dialyzed FITC-conjugated gelatin in a total volume of 200 mL were injected into rats.

*Tissue Preparation*

1. Female adult Wistar rats were used after bilateral adrenalectomy or adrenocorticotropic hormone (ACTH) intramuscular administration (0.5 mg/day) for 2 weeks. Nontreated rats served as control.
2. The abdominal and chest walls of the animals were opened under diethyl ether inhalation anesthesia. To demonstrate microvessels, 200 mL of dialyzed FITC (30 mg/mL) conjugated gelatin solution was injected into the left ventricle of the heart, cutting off the right atrium.
3. After complete perfusion over approximately 3–5 min, the animals were fixed in cold 4% paraformaldehyde (PFA; Wako Pure Chemical, Osaka, Japan) added to 15% picric acid (Wako) solution (0°C) over approximately 25–30 min.
4. Then wet pituitary glands were resected and immediately fixed overnight in graded cold PFA (0.5–8%) containing 15% picric acid overnight at 4°C.
5. For simultaneous visualization of microvessel networks and immunohistochemical localization of ACTH and GH, the pituitary tissues were cut into 2–3-mm-thick blocks, placed on 50 × 70-mm coverslips (Matsunami Glass, Osaka, Japan), and mounted in 0.05 $M$ Tris-HCl buffer, 1,4-diazabicyclo[2,2,2]octane (DABCO; Wako).
6. The pituitary tissue specimens were used for frozen sections. These frozen sections were obtained from tissues that had been fixed overnight in 4% PFA dissolved in 0.01 $M$ PBS, pH 7.4, immersed in graded concentrations of sucrose in PBS (10% for 1 h, 15% for 2 h, and 20% for 4 h), embedded in OCT compound, and stored at −80°C until use.
7. Frozen sections (30 μm thick) were mounted on 3-aminopropylmethoxysilane-coated slides.

*Immunohistochemistry*

An immunohistochemical analysis of hormone-secreting endothelial cells was performed by the use of indirect immunofluorescent and an immunoperoxidase labeled antibody, as described by Nakane in 1975 *(45)*.

1. The first antibody employed was a rabbit polyclonal anti-ACTH antibody (1:200; DAKO, Santa Barbara, CA) or a rabbit polyclonal anti-GH antibody (1:200; DAKO).
2. The second antibody employed was tetramethylrhodamine isothiocyanate (TRITC)-conjugated anti-rabbit IgG (swine, 1:10; DAKO) or HRP-conjugated anti-mouse IgG (sheep, 1:50; Amersham, Poole, UK).
3. The peroxidase coloring reaction was performed with DAB. The nuclei were then counterstained with 5% methyl green (MG) buffered by 0.1 mol/L Veronal acetate at pH 4.0 (Chroma, Kögen, Norway).

*Confocal Laser Scanning Microscopy System*

CLSM was employed on thick sections (1–2 mm) to elucidate the relationship between hormone-secreting endothelial cells and the microvessel network by stereo pair, colored depth focus, projection, and three-dimensional cutting at reconstruction models. The system included a 488-nm argon laser (for FITC) and/or a 543-nm helium-neon laser (for TRITC and DAB). Optical fluorescence signals of microvessel network patterns labeled by FITC were observed using a dichroic beam splitter (NT 80/20/543; Carl Zeiss) and emission filter (BP510-525; Carl Zeiss). Optical fluorescence signals of ACTH-secreting cells labeled with TRITC were observed using a dichroic beam splitter (NT 80/20/543; Carl Zeiss) and emission filter (LP570; Carl Zeiss). Optical reflectance signals of GH cells labeled by HRP were observed using a dichroic beam splitter (NT 80/20/543; Carl Zeiss) and no emission filter. Double- or triple-labeled volumetric data sets of these images taken with CLSM can be digitized and made susceptible to image analysis manipulation and computer-assisted three-dimensional reconstructions using LSM software.

Optical tomographic imaging (Z-images) for three-dimensional reconstruction of the hormone-secreting endocrine cells and their microvessels was done with a plan-apochromat (×63, oil immersion, NA 1.4; Carl Zeiss) and a plan-neofluar (×10, NA 0.30; ×20, NA 0.50; Carl Zeiss) objective lens. Z-directional movement for optical sectioning of entire specimens was controlled by a stepping motor unit for axial scanning at 0.5–20-µm focus steps. The image resolution was 512 × 512 and/or 1024 × 1024 pixels (8 bits, 256 gray levels).

## Results

### Distribution of Three-Dimensional Microvessels as Determined by Fluorescent CLSM

By CLSM, FITC fluorescence-conjugated gelatin injected into the left ventricle of the heart was observed to enter the capillaries of the control rat pituitary glands (**Fig. 8**). The CLSM images in the fluorescence confocal mode identified straight vessels, arcs, branches, loops, and well-developed networks of capillaries in the anterior pituitary gland three dimensionally.

### Immunohistochemical Localization of ACTH and GH in Nontreated Control Rats by CLSM (*See* **Color Plate 5** following page 208.)

Double three-dimensional imaging with the visualization of hormone-producing cells by TRITC- and/or HRP-labeled antibodies demonstrated the shape of ACTH and GH cells and their relationship to the surrounding capillaries (**Fig. 9**). The lumina of the capillaries were labeled yellowish green by FITC. ACTH (**Fig. 9a,b**) and GH cells (**Fig. 9c,d**) fluoresced red. Many stellate cells were positive for ACTH, and microvessels were clearly visible in the pituitary glands. In ACTH cells, cells were frequently attached to the capillaries, and long cytoplasmic processes terminated at the capillaries (**Fig. 9a**). A high-power view of the image in **Fig. 9a** reveals more clearly the relationship between microvessel networks and ACTH-positive cells (**Fig. 9b**). The terminals of the ACTH processes and the capillary lumina are separated by approximately 600 nm. This suggests endothelial cells and basement membranes. The reaction products of ACTH appeared granular (**Fig. 9b**). In contrast, GH cells were generally rounded and frequently lay in free spaces among the FITC fluorescence-labeled capillaries (**Fig. 9c**). GH also appeared granular in the cytoplasm (**Fig. 9d**).

DAB (osmium black) also gave reflectance signals at a wavelength of 543 nm. DAB labeling for GH was applied to the double three-dimensional imaging with FITC fluorescence. CLSM detected fluorescence of FITC at a depth of focus of 2000 µm (**Fig. 8**) and fluorescence of TRITC and reflectance signals for DAB at 300 µm (**Fig. 9a,c**).

### Immunohistochemical Localization of ACTH in Adrenalectomy and ACTH Administration Rat Anterior Pituitary Gland by CLSM

In the adrenalectomized rats, the anterior pituitary gland showed markedly increased numbers of ACTH-immunoreactive cells (**Fig. 10a**). The positive cells revealed that

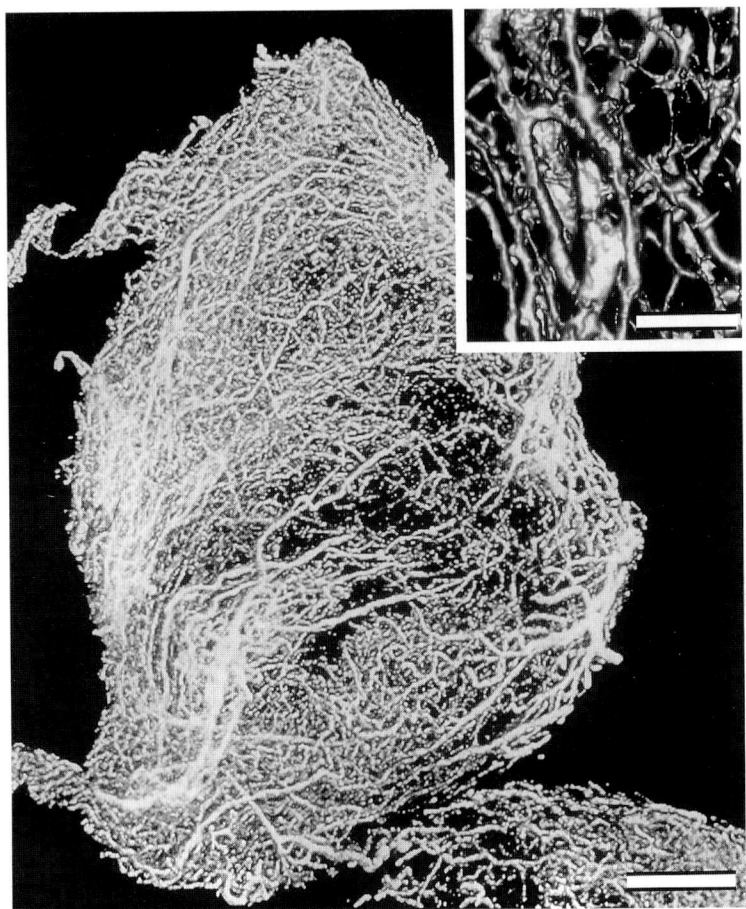

**Fig. 8.** Three-dimensional depth projection reconstructed image of microvessels in normal rat pituitary glands. The microvessel networks are clearly visible at the maximal focus depth of 2000 μm. Bar = 0.5 cm. Upper right corner: high-power magnification, bar = 100 μm.

ACTH immunoreactivity was closely associated with the microvasculature. At high power, the ACTH-positive cells appeared stellate and showed frequent slender cytoplasmic processes **(Fig. 10b)**. On the other hand, if the rats were given ACTH daily for 2 weeks, the ACTH-immunoreactive cells decreased in number and became atrophic **(Fig. 10c,d)**.

## Comments

Three-dimensional imaging by CLSM has clearly demonstrated well-developed capillary networks and a relationship with the hormone-secreting endocrine cells. In addition to normal pituitary vessels that form morphologic patterns, straight vessels that parallel each other, parallel vessels that crosslink, arcs (incompletely closed loops), arcs with branching, closed loops, and networks of closed loops **(Fig. 8)** have been identified. The presence of networks of closed loops has been found to have the strongest association with hormone-secreting endocrine cells of any parameter, including cell shape, cell

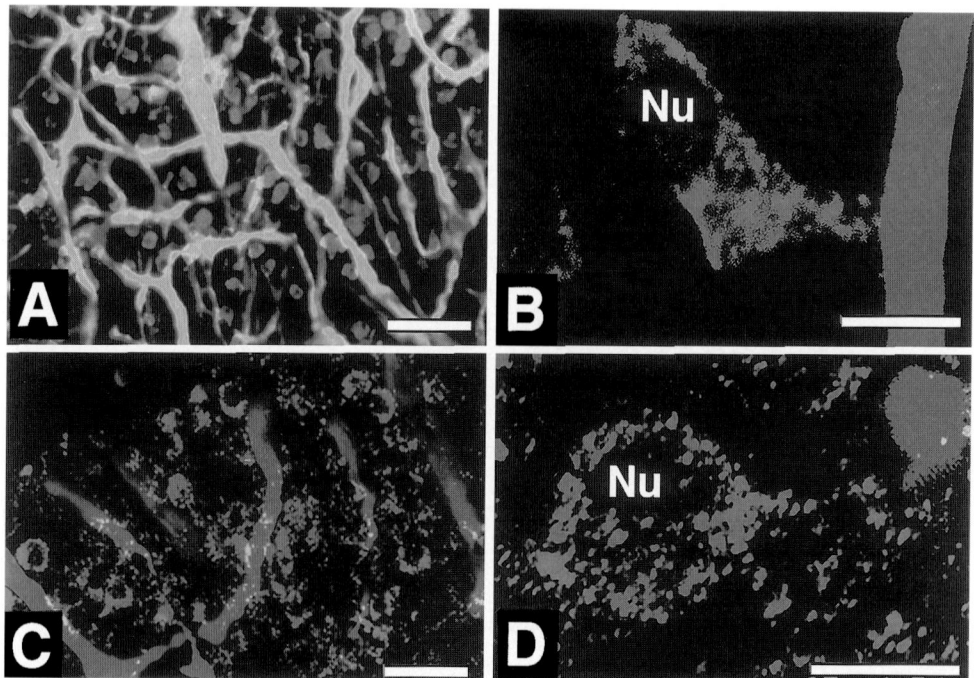

**Fig. 9.** Three-dimensional projection images of ACTH and GH cells and combined signals of fluorescent (FITC) and nonfluorescent (DAB) probe in normal rat pituitary glands. Red: ACTH cells (anti-ACTH-TRITC), fluorescence mode (A, B); GH cells (anti-GH-HRP-DAB) reflectance mode (C, D). Green: microvessels (FITC-conjugated gelatin), fluorescence mode. **(A)** ACTH cells are closely related to the microvessel networks. Low power. **(B)** ACTH signals show a granular pattern in the cytoplasm. High power. **(C)** GH cells are clearly visible three-dimensionally by CLSM reflectance mode. Low power. **(D)** GH signals appear granular in the cytoplasm three-dimensionally by CLSM reflectance mode. High power. Nu, nucleus. Bars = 25 μm in A and C; 5 μm in B and D. (*See* **Color Plate 5** following page 208.)

size, and cell type. The presence of parallel vessels with crosslinking is also strongly associated with the development or processing of hormone-secreting endocrine cells **(Fig. 9)**.

Our technique, which combined FITC-gelatin injection with immunohistochemistry using HRP- and/or fluorescence-labeled antibody as described here, was designed to visualize the microcirculatory patterns in three dimensions and to provide a simultaneous three-dimensional visualization of the microcirculation and hormone-secreting endocrine cells processing ACTH or GH by an FITC-conjugated gelatin injection method.

The critical points for the demonstration of the capillaries seemed to lie in 1) warming of the FITC fluorescence-conjugated gelatin and 2) peristaltic infusion of the solution to prevent thrombotic occlusion of the capillaries. For demonstration of ACTH and GH, effusion fixation by picric acid paraformaldehyde solution (Zamboni) preserved antigenicity as well as cell shape and cytoplasmic structures.

The technique of combined double immunolabeling and/or histochemical labeling, CLSM, and computer-assisted three-dimensionally reconstructed analysis does have advantages. First, as mentioned above, each of the labels used was detectable in deeper

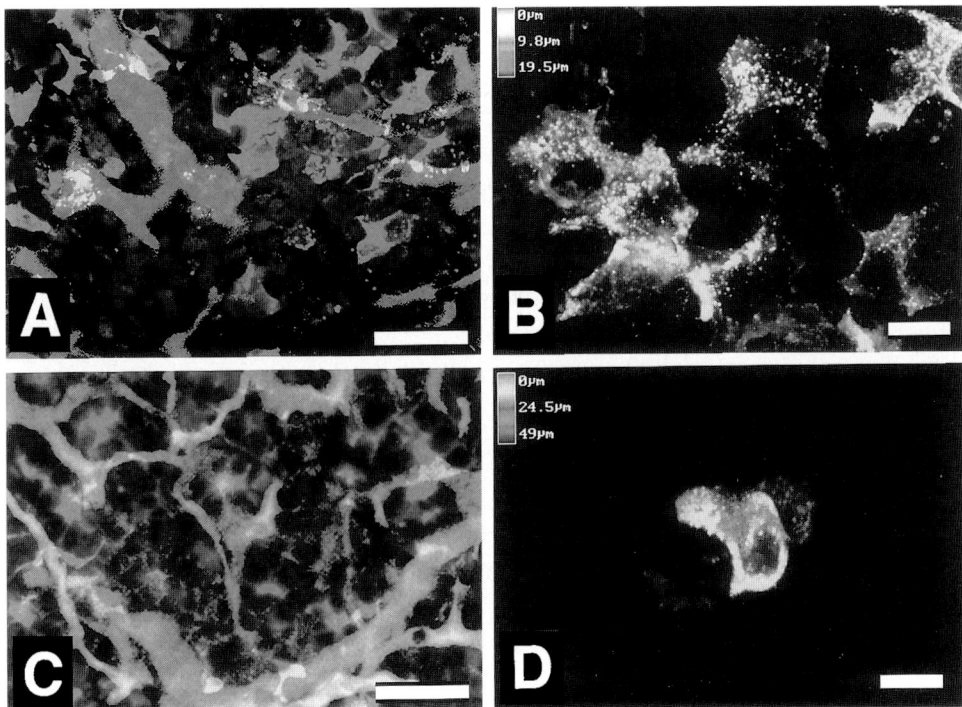

**Fig. 10.** Three-dimensional reconstructed images of bilaterally adrenalectomized and ACTH-treated rat pituitary glands. (**A**) Three-dimensional projection image of ACTH cells in the rat pituitary gland from a bilaterally adrenalectomized rat. ACTH-immunoreactive cells are markedly increased in number. The ACTH-positive cells are closely associated with microvessels. Red: ACTH cells (anti-ACTH-TRITC). Green: microvessels (FITC-conjugated gelatin). (**B**) Three-dimensional depth colored focus image of ACTH cells in bilaterally adrenalectomized rat pituitary gland. ACTH-positive cells show frequent slender cytoplasmic processes. (**C**) Three-dimensional projection image of ACTH cells in ACTH-administered rat pituitary gland. The ACTH-immunoreactive cells were decreased in number and became atrophic. Red: ACTH cells (anti-ACTH-TRITC). Green: microvessels (FITC-conjugated gelatin). (**D**) Three-dimensional depth colored focus image of ACTH cells in ACTH-administered rat pituitary gland. The ACTH-immunoreactive cells were decreased in number with a decrease in their cell volumes and contained condensed and increase numbers of cytoplasmic and enlarged secretory granules. Bars = 25 μm in a and c; 10 μm in b and d. (*See* **Color Plate 5** following page 208.)

levels and was able to reveal the relationship between vessels and tissues in the pituitary gland. Second, unlike scanning electron microscopy by cast corrosion methods, which make it impossible to observe the relationship between casts and tissues, this technique permits the three-dimensional study of samples from low- to high-power magnification. The technique described here is most suitable for the study of small lateral areas. To create a three-dimensionally reconstructed image, one must overlap data not only along the *x*- and *y*-axes but also along the *z*-axis, a procedure that is tedious and requires extensive computer resources.

By this three-dimensional method, ACTH cells were clearly depicted as polygonal cells with long cytoplasmic processes, which terminated on the capillaries. This shape of ACTH cells has frequently been described as unique and characteristic. GH cells were round and were located among the capillaries. It should be emphasized that this method not only elucidates the shape and relationship to the capillaries but also depicts a granular localization on CLSM, suggestive of secretory granules.

Our experimental study to visualize simultaneously the immunohistochemical localization of ACTH and GH and the microvessel network using CLSM with the fluorescence probes FITC, TRITC, and DAB revealed a clear characteristic intracellular localization of ACTH and GH in the rat anterior pituitary glands on thick specimens. These images showed both the immunohistochemical localization of ACTH and GH and the pattern of the microvessel network. With this technique, FITC injection combined with immuno-histochemistry simultaneously revealed differences in the localization or distribution. Permanently preserved sections allow repeated or retrospective examinations as well as observation of organs in vivo.

This method clearly showed the relationship between the microvasculature and the hormone-secreting endocrine cells on thick light microscopic specimens. It also permitted recognition of subcellular sites of the microvasculature and hormone-secreting endocrine cells, obviating most of the cumbersome procedures associated with pathologic diagnosis in tissues containing tumor microvessels.

## REFERENCES

1. Minsky, M. (1957) Microscopy apparatus. U.S. Patent No. 3013467.
2. Arndt-Jovin, D. J., Robert-Nicoud, M., and Jovin, T. M. (1990) Probing DNA structure and function with a multi-wavelength fluorescence confocal laser microscope. *J. Microsc.* **157**, 61–72.
3. Arndt-Jovin, D. J., Robert-Nicoud, M., Kaufman, S. J., and Jovin, T. M. (1985) Fluorescence digital imaging microscopy in cell biology. *Science* **230**, 247–256.
4. Bauman, J. G., Bayer, J. A., and van Dekken, H. (1990) Fluorescent in-situ hybridization to detect cellular RNA by flow cytometry and confocal microscopy. *J. Microsc.* **157**, 73–81.
5. Hozak, P., Novak, J. T., and Smetana, K. (1989) Three-dimensional reconstructions of nucleolus-organizing regions in PHA-stimulated human lymphocytes. *Biol. Cell* **66**, 225–233.
6. Michel, E. and Parsons, J. A. (1990) Histochemical and immunocytochemical localization of prolactin receptors on Nb2 lymphoma cells: applications of confocal microscopy. *J. Histochem. Cytochem.* **38**, 965–973.
7. Takamatsu, T. and Fujita, S. (1988) Microscopic tomography by laser scanning microscopy and its three-dimensional reconstruction. *J. Microsc.* **149**, 167–174.
8. Tao, W., Walter, R. J., and Berns, M. W. (1988) Laser-transected microtubules exhibit individuality of regrowth; however, most free new ends of the microtubules are stable. *J. Cell Biol.* **107**, 1025–1035.
9. White, J. G., Amos, W. B., and Fordham, M. (1987) An evaluation of confocal versus conventional imaging of biological structures by fluorescence light microscopy. *J. Cell Biol.* **105**, 41–48.
10. Itoh, J., Utsunomiya, H., Komatsu, N., Takekoshi, S., Osamura, R. Y., and Watanabe, K. (1992) A new application of confocal laser scanning microscopy (C-LSM) to observe subcellular organelles utilizing non fluorescent probe (osmium black). *Histochem. J.* **24**, 550.
11. Robinson, J. M. and Batten, B. E. (1989) Detection of diaminobenzidine reactions using scanning laser confocal reflectance microscopy. *J. Histochem. Cytochem.* **37**, 1761–1765.
12. Itoh, J., Osamura, R. Y., and Watanabe, K. (1992) Subcellular visualization of light microscopic specimens by laser scanning microscopy and computer analysis: a new application of image analysis. *J. Histochem. Cytochem.* **40**, 955–967.

13. Itoh, J., Sanno, N., Matsuno, A., Itoh, Y., Watanabe, K., and Osamura, R. Y. (1997) Application of confocal laser scanning microscopy (CLSM) to visualize prolactin (PRL) and PRL mRNA in the normal and estrogen-treated rat pituitary glands using non-fluorescent probes. *Microsc. Res. Tech.* **39**, 157–167.

14. Matsuno, A., Itoh, J., Osamura, R. Y., Watanabe, K., and Nagashima, T. (1999) Electron microscopic and confocal laser scanning microscopic observation of subcellular organelles and pituitary hormone mRNA: application of ultrastructural in situ hybridization and immunohistochemistry to the pathophysiological studies of pituitary cells. *Endocr. Pathol.* **10**, 199–211.

15. Itoh, J., Yasumura, K., Takeshita, T., et al. (2000) Three-dimensional imaging of tumor angiogenesis. *Anal. Quant. Cytol. Histol.* **22**, 85–90.

16. Itoh, J., Kawai, K., Serizawa, A., Yasumura, K., Ogawa, K., and Osamura, R. Y. (2000) A new approach to three-dimensional reconstructed imaging of hormone-secreting cells and their microvessel environments in rat pituitary glands by confocal laser scanning microscopy. *J. Histochem. Cytochem.* **48**, 569–578.

17. Matsuno, A., Ohsugi, Y., Utsunomiya, H., et al. (1994) Ultrastructural distribution of growth hormone, prolactin mRNA in normal rat pituitary cells: a comparison between preembedding and postembedding methods. *Histochemistry* **102**, 265–270.

18. Matsuno, A., Teramoto, A., Takekoshi, S., et al. (1994) Application of biotinylated oligonucleotide probes to the detection of pituitary hormone mRNA using Northern blot analysis, in situ hybridization at light and electron microscopic levels. *Histochem. J.* **26**, 771–777.

19. Matsuno, A., Ohsugi, Y., Utsunomiya, H., et al. (1995) Changes in the ultrastructural distribution of prolactin and growth hormone mRNAs in pituitary cells of female rats after estrogen and bromocriptine treatment, studied using in situ hybridization with biotinylated oligonucleotide probes. *Histochem. Cell Biol.* **104**, 37–45.

20. Matsuno, A., Nagashima, T., Osamura, R. Y., and Watanabe, K. (1998) Application of ultrastructural in situ hybridization combined with immunohistochemistry to pathophysiological studies of pituitary cell: technical review. *Acta Histochem. Cytochem.* **31**, 259–265.

21. Matsuno, A., Nagashima, T., Ohsugi, Y., et al. (2000) Electron microscopic observation of intracellular expression of mRNA and its protein product: technical review on ultrastructural in situ hybridization and its combination with immunohistochemistry. *Histol. Histopathol.* **15**, 261–268.

22. Matsuno, A., Nagashima, T., Osamura, R. Y., and Watanabe, K. (2000) Electron microscopic in situ hybridization and its combination with immunohistochemistry, in *Molecular Histochemical Techniques (Springer Lab Manual)* (Koji, T., ed.), Springer, New York, pp. 204–221.

23. Lakkakorpi, T. J., Yang, M., and Rajaniemi, H. J. (1994) Processing of the LH/CG receptor and bound hormone in rat luteal cells after hCG-induced down-regulation as studied by a double immunofluorescence technique in conjunction with confocal laser scanning microscopy. *J. Histochem. Cytochem.* **42**, 727–732.

24. Rummelt, V., Gardner, L. M., Folberg, R., et al. (1994) Three-dimensional relationships between tumor cells and microcirculation with double cyanine immunolabeling, laser scanning confocal microscopy, and computer-assisted reconstruction: an alternative to cast corrosion preparations. *J. Histochem. Cytochem.* **42**, 681–686.

25. Strong, L. H. (1964) The early embryonic pattern internal vascularization of the mammalian cerebral cortex. *J. Comp. Neurol.* **123**, 121–138.

26. Conradi, N. G., Engvall, J., and Wolff, J. R. (1980) Angioarchitectonics of rat cerebral cortex during pre- and postnatal development. *Acta Neuropathol. (Berl.)* **50**, 131–138.

27. Elias, K. A. and Weiner, R. I. (1984) Direct arterial vascularization of estrogen-induced prolactin-secreting anterior pituitary tumors. *Proc. Natl. Acad. Sci. USA* **81**, 4549–4553.

28. Kimura, K., Tojo, A., Nanba, S., Matsuoka, H., and Sugimoto, T. (1990) Morphometric

analysis of arteriolar diameters in experimental nephropathies: application of microvascular casts. *Virchows Arch.* **417**, 319–323.

29. Yoshida, Y. and Ikuta, F. (1984) Three-dimensional architecture of cerebral microvessels with a scanning electron microscope: a cerebrovascular casting method for fetal and adult rats. *J. Cereb. Blood Flow Metab.* **4**, 290–296.

30. Nakamura, K. and Masuda, T. (1981) Scanning electron microscopy of corrosion cast of rat adrenal vasculatures with emphasis on medullary artery under ACTH administration. *Tohoku J. Exp. Med.* **134**, 203–213.

31. Bonner-Weir, S. and Orci, L. (1982) New perspectives on the microvasculature of the islets of Langerhans in the rat. *Diabetes* **31**, 883–889.

32. Marin-Padilla, M. (1985) Early vascularization of the embryonic cerebral cortex: Golgi and electron microscopic studies. *J. Comp. Neurol.* **241**, 237–249.

33. Novikoff, A. B. and Goldfischer, S. (1961) Nucleosidediphosphatase activity in the Golgi apparatus and its usefulness for cytological studies. *Proc. Natl. Acad. Sci. USA* **47**, 802–810.

34. Bell, M. A. and Scarrow, W. G. (1984) Staining for microvascular alkaline phosphatase in thick celloidin sections of nervous tissue: morphometric and pathological applications. *Microvasc. Res.* **27**, 189–203.

35. Holthöfer, H., Virtanen, I., Kariniemi, A. L., Hormia, M., Linder, E., and Miettinen, A. (1982) *Ulex europaeus* I lectin as a marker for vascular endothelium in human tissues. *Lab. Invest.* **47**, 60–66.

36. Minamikawa, T., Miyake, T., Takamatsu, T., and Fujita, S. (1987) A new method of lectin histochemistry for the study of brain angiogenesis. Lectin angiography. *Histochemistry* **87**, 317–320.

37. Hoyer, L. W., de los Santos, R., and Hoyer, J. R. (1973) Antithemophilic factor antigen. Localization in endothelial cells by immunofluorescence microscopy. *J. Clin. Invest.* **52**, 2737–2744.

38. Jaffe, E. A., Hoyer, L. W., and Nachman, R. L. (1973) Synthesis of antihemophilic factor antigen by cultured human endothelial cells. *J. Clin. Invest.* **52**, 2757–2764.

39. Stemerman, M. B., Pitlick, F. A., and Dembitzer, H. M. (1976) Electron microscopic immunohistochemical identification of endothelial cells in the rabbit. *Circ. Res.* **38**, 146–156.

40. Ghandour, M. S., Langley, O. K., and Varga, V. (1980) Immunohistological localization of gamma-glutamyl transpeptidase in cerebellum in light and electron microscope levels. *Neurosci. Lett.* **20**, 125–129.

41. Auerbach, R., Alby, L., Grieves, J., et al. (1982) Monoclonal antibody against angiotensin-converting enzyme: its use as a marker for murine, bovine, and human endothelial cells. *Proc. Natl. Acad. Sci. USA* **79**, 7891–7895.

42. Barsky, S. H., Baker, A., Siegal, G. P., Togo, S., and Liotta, L. A. (1983) Use of anti-basement membrane antibodies to distinguish blood vessel capillaries from lymphatic capillaries. *Am. J. Surg. Pathol.* **7**, 667–677.

43. Strottmann, J. M., Robinson, J. B., Jr., and Stellwagen, E. (1983) Advantages of preelectrophoretic conjugation of polypeptides with fluorescent dyes. *Anal. Biochem.* **132**, 334–337.

44. D'Amato, R., Wesolowski, E., and Smith, L. E. (1993) Microscopic visualization of the retina by angiography with high-molecular-weight fluorescein-labeled dextrans in the mouse. *Microvasc. Res.* **46**, 135–142.

45. Nakane, P. K. (1975) Recent progress in the peroxidase-labeled antibody method. *Ann. NY Acad. Sci.* **254**, 203–211.

# 11
# Polymerase Chain Reaction

## Yuri E. Nikiforov, MD, PHD and Philip N. Howles, PHD

## INTRODUCTION

The polymerase chain reaction (PCR) was introduced in 1985 by Saiki et al. *(1)*, who recognized that sequence-specific oligonucleotides could be used to prime synthesis of complementary DNA strands by any of several DNA polymerases and that these primers could be chosen so as to replicate both strands of the DNA positioned between their cognate sequences in the template *(1,2)*. Initial studies with the technique were laborious and thus limited in scope. Thermostable DNA polymerases were not yet commercially available, so the experimenter had to add fresh polymerase after the DNA denaturing step of each cycle *(1,2)*. The process was not automated, so a student or technician was needed to transfer the reactions between water baths or heating blocks set for each temperature of the cycle. The original report was largely a "proof of principle" study in which a portion of the human β-globin gene was amplified and then analyzed for the presence of normal versus sickle cell coding sequences *(1,3)*. In 1988, the use of a thermostable DNA polymerase, isolated from *Thermus aquaticus*, was used in this procedure instead of the *Escherichia coli* Klenow fragment *(4,5)*. This modification constituted the major advance needed to make the technique practical and commercially viable. Various preparations of heat-stable polymerases from *T. aquaticus*, as well as other thermophilic bacteria, were soon commercially available, as were several models of programmable instruments for rapid and reliable heating and cooling of samples. The utility of PCR was immediately recognized by molecular biologists, geneticists, and scientists developing diagnostic procedures for human genetic diseases and forensic investigations. The ability to amplify directly and analyze specific portions of individual genes from only a few cells or from archival tissue has revolutionized the field of pathology as well as other clinical disciplines that involve genetic analysis. The importance of PCR was acknowledged by the scientific community at large when Kary Mullis was awarded the 1993 Nobel Prize in Chemistry for his contribution to its development.

This chapter describes the methodology of designing and executing PCR assays, including typical protocols of nucleic acid extractions from fresh and paraffin-embedded tissue, as well as various controls that are necessary and problems that are typically encountered. Examples of typical protocols of specific applications of PCR to diagnose

From: *Morphology Methods: Cell and Molecular Biology Techniques*
Edited by: R. V. Lloyd © Humana Press, Totowa, NJ

disease and to investigate pathogenic mechanisms are included. The information is by no means comprehensive, and the reader is referred to several books and reviews that are more detailed.

## GENERAL PRINCIPLES

The main principal of PCR is an exponential amplification of a target DNA sequence that results in the generation of millions of identical copies of a DNA fragment starting from just a few copies. With the addition of a reverse transcription step, PCR can also be applied to amplify RNA fragments. The assay is simple in general design and is summarized in **Fig. 1**. Short, unique segments of DNA that are positioned on either side of the sequence of interest are chosen. These pieces of DNA, the primers, are typically 20–25 nucleotides long. The length of DNA to be amplified, the template, varies from as little as 100 to a few thousand nucleotides. The primers are synthesized such that when the pair forms hybrids with their cognate sequences in the template, their 3′ ends are oriented toward each other. As the reaction proceeds, the DNA polymerase elongates the primers, with the result that an exact copy is made of the template DNA between the two primers. The DNA fragments are denatured (made single-stranded) by heating the reactions to 94°C. Upon cooling to the appropriate temperature (see below), primers anneal to the appropriate sequence, either the original template or one of the copies. The temperature is raised to 72°C (the optimum temperature for the common thermostable polymerases) to allow DNA synthesis, and the process is repeated. In an ideal reaction, a sequence of 30 cycles of polymerization results in approx $1 \times 10^9$ copies of the original template.

### *Template DNA: Source, Preparation, and Amount*

The source of DNA for amplification and methods of isolation may vary according to the specific application. In general, adequate template DNA can be prepared from almost any source, provided that degradation has not been excessive. The amount of degradation tolerated increases as the length of the segment to be amplified decreases, reflecting the decreasing probability of having a double-stranded break between the primers in every copy of template as the distance between the primers decreases. The amount of DNA used per reaction varies according to the application. When the template is high-quality genomic DNA, only 1000–10,000 copies are needed per reaction. For DNA isolated from mammalian cells ($3.3 \times 10^9$ bp/genome), 10,000 copies will require 35 ng of DNA. If the template is from partially degraded DNA, as prepared from fixed tissue, the amount of DNA may need to be increased up to 100–200 ng for each reaction.

#### *DNA Extraction from Fresh or Snap Frozen Tissue*

DNA extracted from fresh or snap frozen tissue is an ideal source of template DNA for PCR amplification. Therefore, an attempt to freeze at least a small piece of tissue should always be undertaken when subsequent PCR analysis is anticipated. Isolation of high molecular weight genomic DNA from fresh or frozen tissue involves digestion by proteases followed by phenol-chloroform extraction and precipitation of genomic

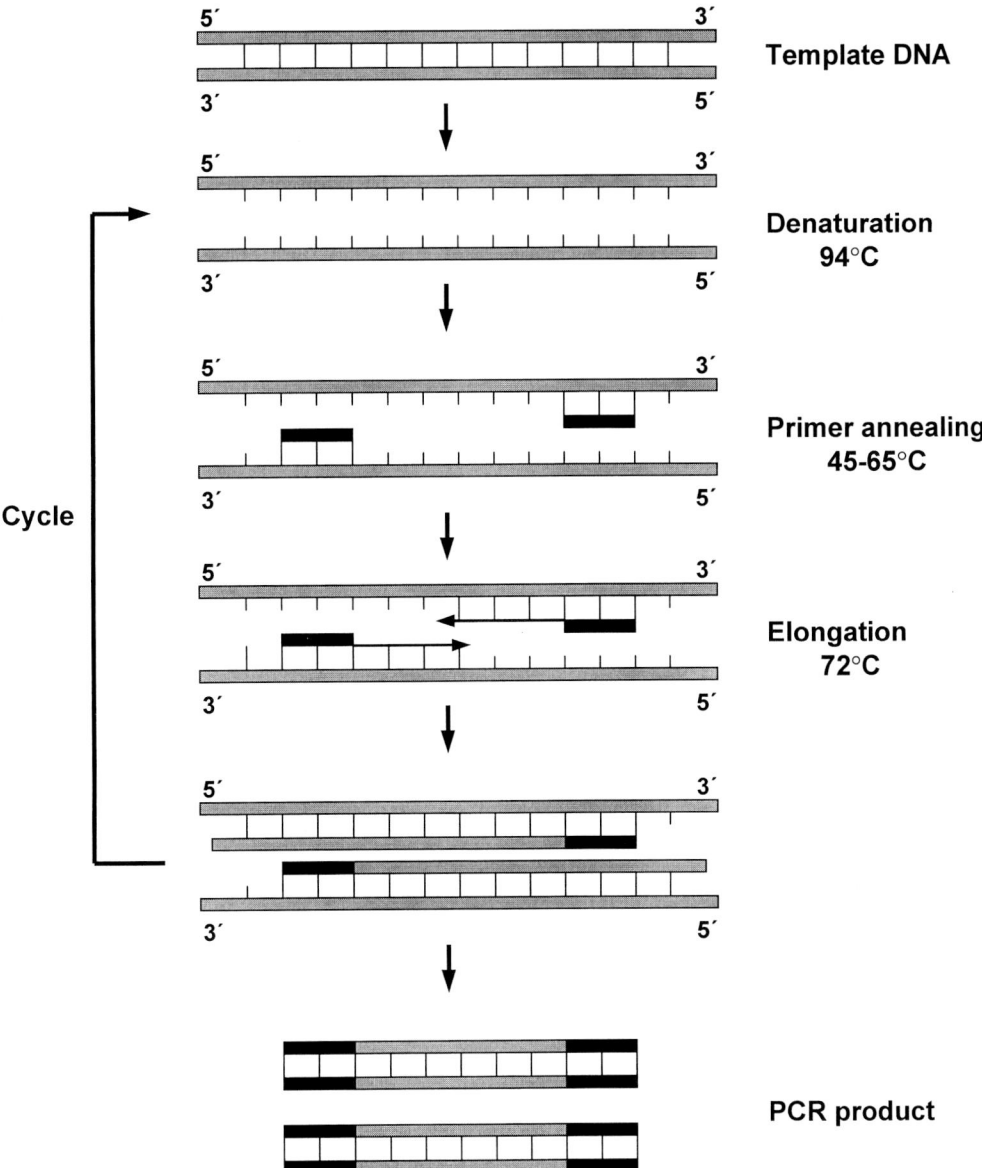

**Fig. 1.** Principle of the polymerase chain reaction (PCR).

DNA in high salt buffers with isopropanol. A number of commercial kits offer convenient alternatives, minimizing handling and exposure to chemicals, offset by a higher cost.

METHOD

1. Harvest fresh tissue using sterile scissors or disposable blade and immediately freeze in liquid nitrogen. Store at −70°C until use.
2. Transfer 100 µg of tissue to a sterile 1.5-mL microcentrifuge tube and homogenize with a

plastic micropestle in 500 μL of extraction buffer (10 m*M* Tris-HCl, pH 8.0, 100 m*M* EDTA, pH 8.0, 0.5% sodium dodecyl sulfate [SDS]). Larger amounts of tissue may require a power homogenizer (Polytron or equivalent).

3. Add proteinase K (Boehringer Mannheim, Indianapolis, IN) to 100 μg/mL, mix, and incubate at 52°C overnight. There should be no obvious tissue debris remaining after incubation. Otherwise, add an additional 50 μg/mL of proteinase K, invert tube 10 times, and incubate for 2–3 h more.

4. Add 500 μL buffered phenol, invert 20 times, and centrifuge at 13,000*g* for 5 min. Transfer the supernatant aqueous phase without interface to a new microcentrifuge tube.

5. Add 500 μL phenol/chloroform/isoamyl alcohol (25:24:1), invert 20 times, and centrifuge at 13,000*g* for 5 min. Transfer the supernatant aqueous phase without interface to a new tube.

6. Repeat **step 5** once.

7. Add 500 μL chloroform/isoamyl alcohol (24:1) and repeat extraction. Transfer the supernatant to a new tube.

8. Add 0.2 vol (final) of 10 *M* ammonium acetate. Precipitate the DNA by adding an equal volume of isopropanol and incubating at −20°C for 10 min. The DNA can be collected by either "hooking" it with a glass Pasteur pipet or centrifugation at 10,000*g* for 10 min. After centrifugation, a translucent DNA pellet should be faintly visible.

9. Discard the supernatant and wash the pellet with 500 μL cold 70% ethanol. Spin at 10,000*g* for 1 min and discard the ethanol.

10. Air-dry the pellet and resuspend the dried pellet in 100 μL TE buffer (10 m*M* Tris-HCl, pH 8.0, 1 m*M* EDTA, pH 8.0) by passing the solution a few times through a pipet tip. Solubilize the DNA at 4°C overnight or at 37°C for 2 h.

11. Quantitate the DNA concentration by spectrophotometry (in a 1-cm-long light path, [DNA] = $A_{260} \times 50$ μg/mL × dilution factor), and keep at 4°C for short-term or at −20°C for long-term storage.

## *DNA Preparation from Paraffin-Embedded Tissue*

In most cases, the DNA isolated from fixed tissues, although degraded in size, can be analyzed by PCR. The success of extraction and amplification of DNA from paraffin-embedded tissues is determined by several factors, the most important of which are the type of fixative and duration of fixation. Most tissues routinely processed in pathology laboratories are fixed in 10% buffered formalin for 12–24 h before embedding in paraffin. These provide a reliable source of DNA for PCR amplification. However, it is important to know the factors that dramatically affect the yield of PCR templates.

### Type of Fixative

Fixative type has a direct bearing on the quality of extracted DNA. The fixatives that produce good-quality of DNA or those that, on the contrary, lead to rapid degradation of DNA are shown in **Table 1** *(6,7)*.

### Duration of Fixation

Generally, extended fixation time is detrimental to PCR analysis of the materials. In the case of formalin, the optimal fixation time is less than 24 h. Longer fixation leads to a progressive reduction in the yield of DNA that can be extracted from the tissue *(8)*. This is especially the case for long amplicons, although DNA fragments 100–200 bp long can be amplified after as much as a 30-day formalin fixation *(9)*. If prolonged fixation cannot be avoided, 95% ethanol should be used to preserve the high quality of DNA *(6)*.

completely seal the cover slip-slide assembly in order to form a small reaction "chamber" that can contain the water vapor during thermal cycling. For effective sealing, do not use colored polish or any other nail polish that is especially "runny"—our laboratory prefers to use Wet & Wild Clear nail polish. Proper sealing is very important, for this keeps reaction concentrations consistent through the thermal cycling procedure, and concentrations are critical to proper amplification. However, be certain to apply the nail polish very carefully so that none of the polish gets into the actual chamber where the cells or tissues reside. If any nail polish does enter the chamber, discard that slide— the results will be questionable. Please bear in mind that the painting of nail polish is truly a learned skill; therefore, it is strongly recommended that researchers practice this procedure several times with mock slides before attempting an experiment.

In the case of thick tissue sections, it is best to use another identical blank slide for the cover instead of a cover slip. Apply the amplification cocktail to the appropriate well of the blank slide, place an inverted tissue-containing slide atop the blank slide, and seal the edges as described. Invert the slide once again so that the tissue-containing slide is on the bottom. This technique can be modified to accommodate a hot start (see below).

### Hot Start Technique

There is much debate as to whether a hot start helps to improve the specificity and sensitivity of amplification reactions. In our laboratory, we find the hot start adds no advantage in this regard, rather, it adds only technical difficulty to the practice of the *in situ* technique. However, recently a variation of the hot start has been reported *(42,75)*. In this procedure, one simply uses anti-*Taq* antibody in the PCR cocktail (containing *Taq*), which keeps the *Taq* enzyme in the cocktail "blocked" until the first cycle of 92°C when anti-*Taq* antibody is denatured and restores the full *Taq* activity. This modification essentially serves the same function as the hot start procedure but without its difficulties *(15)*.

### Thermal Cyclers

Various thermocycler technologies will work in this application; however, some instruments work much better than others. In our laboratory, we use two types: a standard, block-type thermocycler that normally holds 60 0.5-mL tubes but that can be adapted with aluminum foil, paper towels, and a weight to hold four to six slides. We also use dedicated thermocyclers that are specifically designed to hold 12 or 16 slides. We understand that other labs have used stirred-air, oven-type thermocyclers quite successfully; however, we have also heard that there are sometimes problems with the cracking of glass slides during cycling. Thermocyclers dedicated to glass slides are now available from several vendors, including Barnstead Thermolyne of Iowa, Coy Corporation of Minnesota, Hybaid of England, Perkin-Elmer of California, and M.J. Research of Massachusetts. Our laboratory has used an M.J. Research PTC-100-16MS, PTC-100-16MS, and DNA-Engine Twin-Tower 16×2 quite successfully. Recently, this company has combined the slide and tubes into a single block, allowing the simultaneous confirmation of *in situ* amplification in a tube. Furthermore, newer designs of thermal cycler incorporate humidification chambers. However, we do not yet have sufficient experience with this technology to verify whether they can eliminate the need for sealing the slides with nail polish during thermal cycling. Nonetheless, the humidified instruments are especially useful in the RT and hybridization steps, for which a humidified incubator is otherwise needed.

We suggest that you follow the manufacturer's instructions on the use of your own thermocycler, bearing in mind the following points:

1. Glass does not easily make good thermal contact with the surface on which it rests. Therefore, a weight to press down the slides and/or a thin layer of mineral oil to fill in the interstices will help thermal conduction. If using mineral oil, make certain that the oil is well smeared over the glass surface so that the slide is not merely floating on air bubbles beneath it.
2. The top surfaces of slides lose heat quite rapidly through radiation and convection; therefore, use a thermocycler that envelopes the slide in an enclosed chamber (as in some dedicated instruments), or insulate the tops of the slides in some manner. Insulation is particularly critical when using a weight on top of the slides, for the weight can serve as an unwanted heat sink if it is in direct contact with the slides.
3. Good thermal uniformity is imperative for good results—poor uniformity or irregular thermal change can result in cracked slides, uneven amplification, or completely failed reactions. If adapting a thermocycler that normally holds plastic tubes, use a layer of aluminum foil to spread out the heat.

## DIRECT INCORPORATION OF NONRADIOACTIVE LABELED NUCLEOTIDES

Several nonradioactive labeled nucleotides are available from various sources (i.e., dCTP-biotin, digoxin II-dUTP, and so on). These nucleotides can be used to label amplification products directly, and then the proper secondary agents and chromogens can be used to detect the directly labeled *in situ* amplification products (see below). However, in our opinion—as well as in the opinion of several other laboratory groups— the greatest specificity is only achieved by conducting amplification followed by subsequent *in situ* hybridization. In the direct labeling protocols, nonspecific incorporation can be significant, and even if this incorporation is minor, it still leads to false-positive signals similar to nonspecific bands in gel electrophoresis following solution-based DNA or RT amplification. Therefore, we strongly discourage the direct incorporation of labeled nucleotides as part of an *in situ* amplification protocol.

The only exception to this recommendation is when one is screening a large number of primer pairs for optimization of a specific assay—then direct incorporation may be useful. To perform such screenings, add to the amplification cocktail detailed earlier the following: 4.3 $\mu M$ labeled nucleotide—either 14-biotin dCTP or 14-biotin dATP or 11-digoxigenin dUTP—along with cold nucleotide to achieve a 0.14 m$M$ final concentration. Also, if one has worked out the perfect annealing system, using either the Robocycler or an equivalent system, then one can use direct incorporation without fear of nonspecific labeling, which we have discussed elsewhere in detail *(8)*.

## MULTIPLE SIGNALS AND MULTIPLE LABELS IN INDIVIDUAL CELLS

DNA, mRNA, and protein can all be detected simultaneously in individual cells *(14,22)*. One can label proteins by rhodamine-labeled antibodies. Then, one can perform both RNA and DNA *in situ* amplification in the cells. If one is using primers for spliced mRNA and if these primers are not going to bind any sequences in DNA, then both

DNA and RT amplification can be carried out simultaneously; of course one still needs to perform the RT step but this time without pre-DNAse treatment. Subsequently, products can be labeled with different kinds of probes, resulting in different colors of signal. For example, proteins can have a rhodamine-labeled probe, mRNA can show a fluorescein isothiocyanate (FITC) signal (FITC-conjugated probe, >20 different fluorochromes are available), and DNA can be labeled with a biotin-peroxidase probe or a fluorochrome with different color emissions. Each will show a different signal within an individual cell *(14,22)*.

## HYBRIDIZATION

The *in situ* hybridization (ISH) technique has been successfully applied in both the research and clinical settings. However, one single, easy to use universal procedure has not been developed. Therefore, specific needs of the diagnostic or research goals must be considered in choosing a suitable protocol.

In comparing ISH with other methods, one has to realize what is being detected. For example, the immunocytochemical method localizes protein within a cell or on a cell surface and therefore identifies gene expression. However, these assays cannot yield useful information on posttranslational processing of the gene product or differentiate between the uptake and storage of the protein and the site of synthesis of the protein. In addition, one needs several hundred copies of the proteins to be able to identify an expression signal. Also, most of the protein is destroyed by formalin fixation methods, making it difficult to identify the protein in most of the pathologic specimens. In the mRNA extraction methods that utilize the isolation of nucleic acids from cells (filter hybridization assays), there can be a dilution of the target found in only a few cells by many cells with little or no target. These methods provide no distribution information. They provide only an average measurement of the nucleic acid target present in the mixed cell population. Therefore ISH is a very powerful technique when the target is focally distributed within a single cell or a certain histologic cell type within a tissue. Consequently, ISH has greater sensitivity than filter assays if the gene expression is taking place in a small subpopulation of cells. The major limitation of ISH is relative insensitivity, compared with *in situ* PCR. By utilizing ISH, one can detect as few as 20 copies of mRNA. However, success to that degree of sensitivity is limited to only few highly specialized laboratories and to a few specific genes. More realistically, detection of >100 copies/cell would be an achievable goal for a laboratory not specialized in ISH. To detect a single copy of an integrated gene or to detect a very low level of gene expression (few copies of mRNA), one can amplify the gene sequences *in situ* by DNA or RNA (RT) *in situ* PCR and then utilize ISH to detect the amplicans.

Analyzing gene expression by ISH after RT can provide information on the site of mRNA synthesis, which provides information about the cellular origin of protein synthesis and demonstrates the amounts of synthesis (level of gene expression). This permits an understanding of the cell types involved in the synthesis of certain proteins in certain gene regulation and cell types infected by various viral or other infectious agents. In addition, by combining immunohistochemistry and RT, differential expression of a gene in different cell types or different stages of development can be analyzed at the microscopic level.

## Choice of Probes for ISH

Many ISH protocols employ [3]H- or [35]S-labeled nucleic acid probes, followed by autoradiographic detection. Although this method can be very sensitive and [3]H-labeled probes generate well-resolved autoradiographic signals, it is time consuming and technically difficult. Other high radiation emitting isotopes can be used, but they give nonspecific background.

Nonisotopic methods for ISH offer the advantages of probe stability, sensitivity, spatial resolution, and great time saving. The nonisotopic adaptations are generally simpler and faster than autoradiography, and the sensitivity of nonradioactive methods has increased over the years as the parameters influencing the hybridization efficiency and signal specificity have become more optimized.

Factors contributing to increased use of nonisotopic methods include faster color development time, chemically stable probes with no special disposal requirements, and different labeling and detection systems that can be used to facilitate the analysis of several probes simultaneously.

## ISH Procedure

Prepare a solution containing: 20–50 pM/μL of the appropriate probe, 50% deionized formamide, 2× SSC buffer, 10× Denhardt's solution, 0.1% sonicated salmon sperm DNA, and 0.1% sodium dodecyl sulfate (SDS). The following is a convenient recipe:

| | |
|---|---|
| probe (biotinylated, or digoxigenin) | 2 μL |
| deionized formamide | 50 μL |
| 20× SSC* | 10 μL |
| 50× Denhardt's solution | 20 μL |
| 10 mg/ml ssDNA* | 10 μL |
| 10% SDS | 1 μL |
| H$_2$O | 7 μL |
| Total volume | 100 μL |

*See **Materials and Methods**, page 224, for preparation of 20× SSC buffer; the salmon sperm should be denatured at 94°C for 10 min, before it is added to the hybridization buffer.

Note: 2% bovine serum albumin (BSA) can be added if one is observing nonspecific binding. For this purpose, one can add 10 μL of 20% BSA solution and reduce the amount of water.

1. Add 10 μL of hybridization mixture to each well and add cover slips.
2. Heat slides on a block at 95°C for 5 min to denature the double-stranded DNA.
3. Incubate slides at 48°C for 2–4 h in a humidified atmosphere.
   Note: The optimal hybridization temperature is a function of the $T_m$ (melting temperature) of the probe. This must be calculated for each probe, as described earlier. However, the hybridization temperatures used should not be too high. If that circumstance occurs, then the formula for the hybridization solution should be modified and instead of 50% formamide, 40% formamide should be substituted (described further in the *in situ* hybridization section of ref. *15*.)

### Posthybridization for Peroxidase-Based Color Development

1. Wash slides in 1× PBS twice for 5 min each time.
2. Add 10–15 µL of streptavidin-peroxidase complex (1 mg/mL stock diluted 1:30 in 1× PBS, pH 7.2). Gently apply the cover slips.
3. Incubate slides at 37°C for 1 h.
4. Remove cover slip, and wash slides with 1× PBS twice for 5 min each time. Mix chromogen: 5 mL 50 m$M$ acetate buffer, 25 µL 30% $H_2O_2$ 250 µL 3′-amino-9-ethylene carbazole (AEC).
5. Add to each well 100 µL of AEC in the presence of 0.03% hydrogen peroxide in 50 m$M$ acetate buffer (pH 5.0).
6. Incubate slides at 37°C for 10 min to develop the color—this step should be carried out in the dark. After this period, observe slides under a microscope. If color is not strong, develop for another 10 min.
7. Rinse slides with tap water and allow to dry.
8. Add 1 drop of 50% glycerol in PBS and apply the cover slips.
9. Analyze with optical microscope—positive cells will be stained a brownish red.

## VALIDATION AND CONTROLS

The validity of *in situ* amplification-hybridization should be examined in every run. Attention here is especially necessary in laboratories first using the technique, because occasional technical pitfalls lie on the path to mastery. In an experienced laboratory, it is still necessary to validate the procedure continuously and to confirm the efficiency of amplification. To do this, we routinely run two or three sets of experiments in multiwell slides simultaneously, for we must not only validate amplification, but we must also confirm the subsequent hybridization/detection steps as well.

In our lab, we frequently work with HIV. A common validation procedure we will conduct is to mix HIV-1-infected cells plus HIV-1-uninfected cells in a known proportion (i.e., 1:10, 1:100 and so on); then we confirm that the results are appropriately proportionate. To examine the efficiency of amplification, we use a cell line that carries a single copy or two copies of cloned HIV-1 *(1–14)*, and then look to see that proper amplification and hybridization has occurred.

In all amplification procedures, we use one slide as a control for nonspecific binding of the probe. Here we hybridize the amplified cells with an unrelated probe. We also use HLA-DQα and β-actin probes and primers with human PBMCs and other tissue sections as positive controls, to check various parameters of our system.

In case one is using tissue sections, a cell suspension lacking the gene of interest can be used as a control. These cells can be added on top of the tissue section and then retrieved after the amplification procedure. The cell suspension can then be analyzed with the specific probe to see whether the signal from the tissue leaked out and entered the cells floating above.

We suggest that researchers carefully design and employ appropriate positive and negative controls for their specific experiments. In the case of RT *in situ* amplification, one can use β-actin, β-globulin, HLA-DQα, and other endogenous-abundant RNAs as the positive markers. Of course, one should always have an RT-negative control for RT *in situ* amplificaiton, as well as DNAse and non-DNAse controls. Controls without *Taq* polymerase plus primers and without primers should always be included.

## MATERIALS AND METHODS

### Slides

Heavy, Teflon-coated glass slides with three 10-, 12-, or 14-mm diameter wells for cell suspensions, or single oval wells for tissue sections, are available from Cel-line Associates of New Field, New Jersey (1-800-662-0973), or Erie Scientific of Portsmouth, New Hampshire (1-800-258-0834). These specific slide designs are particularly useful, for the Teflon coating serves to form distinct wells, each of which serves as a small reaction "chamber" when the cover slip is attached. Furthermore, the Teflon coating helps to keep the nail polish from entering the reaction chamber, and multiple wells allow for both a positive and negative control on the same slide.

### Coplin Jars, Glass-Staining Dishes, and Solutions

Suitable vessels for washing, fixing, and staining 4–20 glass slides simultaneously are available from several vendors, including Fischer Scientific, Sigma, and many others.

#### 2% Paraformaldehyde

1. Take 12 g paraformaldehyde (Merck ultra pure art. no. 4005) and add to 600 mL 1× PBS.
2. Heat at 65°C for 10 min.
3. When the solution starts to clear, add 4 drops 10 $N$ NaOH and stir.
4. Adjust to neutral pH and cool to room temperature.
5. Filter on Whatman's no. 1 filter paper.

#### 10× PBS Stock Solution, pH 7.2–7.4.

Dissolve 20.5 g $NaH_2PO_4$ $H_2O$ and 179.9 g $Na_2HPO_4$ $7H_2O$ (or 95.5 g $Na_2$ $HPO_4$) in about 4 L of double-distilled water. Adjust to the required pH (7.2–7.4). Add 701.3 g NaCl, and make up to a total volume of 8 L.

#### 1× PBS

Dilute the stock 10× PBS at 1:10 ratio (i.e., 100 mL 10× PBS and 900 mL of water for 1 L). Final concentration of buffer should be 0.01 $M$ phosphate and 0.15 $M$ NaCl.

#### 0.3% Hydrogen Peroxide ($H_2O$) in PBS

Dilute stock 30% hydrogen peroxide ($H_2O_2$) at a 1:100 ratio in 1× PBS for a final concentration of 0.3% $H_2O_2$.

### Proteinase K

Dissolve powder from Sigma in water to obtain 1 mg/mL concentration. Aliquot and store at −20°C.

#### Working Solution

Dilute 1 mL of stock (1 mg/mL) into 150 mL of 1× PBS.

#### 20× SSC

Dissolve 175.3 g of NaCl and 88.2 g of sodium citrate in 800 mL of water. Adjust the pH to 7.0 with a few drops of 10$N$ solution of NaOH. Adjust the volume to 1 L with water. Sterilize by autoclaving.

*2× SSC*

Dilute 20× SSC; 100 mL of 20× SSC and 900 mL of water.

*Streptavidin Peroxidase*

Dissolve powder from Sigma in PBS to make a stock of 1 mg/mL. Just before use, dilute stock solution in sterile PBS at a 1:30 ratio.

*Color Solution*

Dissolve one AEC (Sigma) tablet in 2.5 mL of *N,N,*-dimethyl formamide. Store at 4°C in the dark.

| Working solution: | |
|---|---|
| 50 m*M* acetate buffer, pH 5.0 | 5 mL |
| AEC solution | 250 µL |
| 30% $H_2O_2$ | 25 µL |

Make fresh before each use, keeping solution in the dark.

*Preparation of 50 m*M *Acetate Buffer, pH 5.0*

Add 74 mL of 0.2 *N* acetic acid (11.55 mL glacial acid/L) and 176 mL of 0.2 *M* sodium acetate (27.2 g sodium acetate trihydrate in 1 L) to 1 L of deionized water and mix.

In Situ *Hybridization Buffer (for 5 mL)*

| | |
|---|---|
| formamide | 2500 mL |
| salmon sperm DNA (ssDNA) (10 mg/mL) | 500 µL |
| 20× SSC | 500 µL |
| 50× Denhardt's solution | 1.00 mL |
| 10% SDS | 50 µL |
| water | 450 µL |
| Total volume | 5 mL |

*Heat-denature ssDNA at 94°C for 10 min before adding to the solution.

## REFERENCES

1. Bagasra, O. (1990) Polymerase chain reaction in situ. *Amplifications* **March**, 20–21.
2. Bagasra, O., Hauptman, S. P., Lischner, H. W., Sachs, M., and Pomerantz, R. J. (1992) Detection of HIV-1 provirus in mononuclear cells by in situ PCR. *N. Engl. J. Med.* **326**, 1385–1391.
3. Bagasra, O., Seshamma, T., and Pomerantz, R. J. (1993) Polymerase chain reaction in situ: intracellular amplification and detection of HIV-1 proviral DNA and other specific genes. *J. Immunol. Methods* **158**, 131–145.
4. Bagasra, O. and R. J. Pomerantz. (1993) HIV-1 provirus is demonstrated in peripheral blood monocytes in vivo: a study utilizing an in situ PCR. *AIDS Res. Hum. Retroviruses* **9**, 69–76.
5. Bagasra, O., Seshamma, T., Oakes, J., and Pomerantz, R. J. (1993) Frequency of cells positive for HIV-1 sequences assessed by in situ polymerase chain reaction. *AIDS* **7**, 82–86.
6. Bagasra, O., Seshamma, T., Oakes, J., and Pomerantz, R. J. (1993) High percentages of CD4-positive lymphocytes harbor the HIV-1 provirus in the blood of certain infected individuals. *AIDS* **7**, 1419–1425.

7. Bagasra, O., Farzadegan, H., Seshamma, T., Oakes, J., Saah, A., and Pomerantz, R. J. (1994) Human immunodeficiency virus type 1 infection of sperm in vivo. *AIDS* **8**, 1669–1674.

8. Bagasra, O. and Pomerantz, R. J. (1994) In situ polymerase chain reaction and HIV-1, in *Clinics of North America* (Pomerantz, R.J., ed.), W. B. Saunders, publishers, Philadelphia, pp. 351–366.

9. Bagasra, O., Hui, Z., Bobroski, L., Seshamma, T., Saikumari, P., and Pomerantz, R. J. (1995) One step amplification of HIV-1 mRNA and DNA at a single cell level by in situ polymerase chain reaction. *Cell Vision* **2**, 425–1573.

10. Bagasra, O., Michaels, F., Mu, Y., et al. (1995) Activation of the inducible form of nitric oxide synthetase in the brains of patients with multiple sclerosis. *Proc. Natl. Acad. Sci. USA* **92**, 12041–12045.

11. Bagasra, O. (1996) Use of in situ PCR for measuring viral burden. *AIDS Reader* **6**, 43–47.

12. Bagasra, O. and Pomerantz, R. J. (1997) Human herpesvirus 8 DNA sequences in CD8 T-cells [Letter]. *J. Infect. Dis.* **176**, 541.

13. Bagasra, O. and Pomerantz, R. J. (1995) Detection of HIV-1 in the brain tissue of individuals who died from AIDS, in (Sarkar, G., ed.), *PCR in Neuroscience*, Academic, San Diego, pp. 339–357.

14. Bagasra, O., Seshamma, T., Pastanar, J. P., and Pomerantz, R. (1995) Detection of HIV-1 gene sequences in the brain tissues by in situ polymerase chain reaction, in *Technical Advances in AIDS Research in the Nervous System* (Majors, E., ed.), Plenum, New York, pp. 251–266.

15. Bagasra, O., Seshamma, T., Hansen, J., and Pomerantz, R. (1995) In situ polymerase chain reaction and hybridization to detect low abundance nucleic acid targets, in *Current Protocols in Molecular Biology* (Ausubel, et al.), Chapter 14, pp. 1–49.

16. Bagasra, O., Michaels, F., Zheng, Y. M., et al. (1995) Activation of the inducible form of nitric oxide synthetase in the brains of patients with multiple sclerosis. *Proc. Natl. Acad. Sci. USA* **92**, 12041–12045.

17. Bagasra, O., Lavi, U., Bobroski, L., Khalili, K., Pestaner, J. P., and Pomerantz, R. J. (1996) Cellular reservoirs of HIV-1 in the central nervous system of infected individuals: identification by the combination of in situ PCR and immunohistochemistry. *AIDS* **10**, 573–585.

18. Bagasra, O. and Amjad, M. (2000) Protection against retroviruses are owing to a different form of immunity: an RNA-based molecular immunity hypothesis. *Appl. Immunochem. Mol. Morphol.* **8**, 133–146.

19. Bobroski, L., Alexander, U., Bagasra, O., et al. (1999) Mechanism of vertical transmission of HIV-1: role of intervillous space, *Appl. Immunochem. Mol. Morphol.* **7**, 271–279.

20. Bobroski, L., Alexander, U., Bagasra, O., et al. (1999) Localization of human herpes virus type 8 (HHV-8) in the Kaposi's sarcoma tissues and the semen specimens of HIV-1 infected and uninfected individuals by utilizing in situ polymerase chain reaction. *J. Reprod. Immunol.* **41**, 149–160.

21. Embretson, J., Zupanic, M., Beneke, T., et al. (1993) Analysis of human immunodeficiency virus-infected tissues by amplification and in situ hybridization reveals latent and permissive infections at single-cell resolution. *Proc. Natl. Acad. Sci. USA* **90**, 357–361.

22. Embretson, J., Zupancic, M., Ribas, J. L., et al. (1993) Massive covert infection of helper T lymphocytes and macrophages by HIV during the incubation period of AIDS. *Nature* **62**, 359–362.

23. Hooper, D. C., Bagasra, O., Marini, J. C., et al. (1997) Prevention of experimental allergic encephalomyelitis by targeting nitric oxide and peroxynitrite: implications for the treatment of multiple sclerosis. *Proc. Natl. Acad. Sci. USA* **94**, 2528–2533.

24. Patterson, B. K., Till, M., Otto, P., et al. (1993) Detection of HIV-I DNA and messenger

RNA in individual cells by PCR-driven in situ hybridization and flow cytometry. *Science* **260**, 976–979.

25. Hsu, T-C., Scott, K., Seshamma, T., Bagasra, O., and Walsh, P. N. (1998) Molecular cloning of platelet factor XI, an alternative splicing product of the plasma factor XI. *J. Biol. Chem.* **273**, 13,787–13,793.

26. Pereira, R. F., Halford, K. W., O'Hara, M. D., et al. (1995) Cultured stromal cells from marrow serve as stem cells for bone, lung and cartilage in irradiated mice. *Proc. Natl. Acad. Sci. USA* **92**, 4857–4861.

27. Qureshi, M. N., Barr, C. E., Seshamma, T., Pomerantz, R. J., and Bagasra, O. (1994) Localization of HIV-1 proviral DNA in oral mucosal epithelial cells. *J. Infect. Dis.* **171**, 190–193.

28. Pellett, P. E., Spira, T., Bagasra, O., et al. (1999) Multi-center comparison of polymerase chain reaction detection of human herpesvirus 8 DNA in semen. *J. Clin. Microbiol.* **37**, 1293–1301.

29. Mehta, A., Maggioncalda, J., Bagasra, O., et al. (1995) In situ PCR and RNA hybridization detection of herpes simplex virus sequences in trigeminal ganglia of latently infected mice. *Virology* **206**, 633–640.

30. Lattime, E. C., Mastrangelo, M. J., O. Bagasra, and Berd D. (1995) Expression of cytokine mRNA in human melanoma tissue. *Cancer Immunol. & Immunother.* **41**, 151–156.

31. Pestaner, J. P., Bibbo, M., Bobroski, L., Seshamma, T., and Bagasra, O. (1994). Potential of in situ polymerase chain reaction in diagnostic cytology. *Acta Cytol.* **38**, 676–680.

32. Sullivan, D. E., Bobroski, L. E., Bagasra, O., Finney, M. (1997) Self-seal reagent: evaporation control for molecular histology procedure without chambers, clips or fingernail polish. *Biotechniques* **23**, 320–225.

33. Bagasra, O. and Hansen, J. (1997) *In Situ PCR Techniques.* John Wiley & Sons, New York.

34. Nuovo, G. J. (1994) *PCR In Situ Hybridization Protocols and Applications*, 2nd ed., Raven, New York.

# 13

# Immunohistochemistry: *Theory and Practice*

## Patrick C. Roche, PHD and Eric D. Hsi, MD

## INTRODUCTION

In the daily practice of surgical pathology, immunohistochemistry (IHC) is essential for the diagnosis and classification of neoplasms. The technique is no longer limited to large university-based or reference laboratories and has become widespread in community hospital laboratories. The propagation of commercially available, automated immunostainers has played a major role in increasing the capability to perform IHC and has enabled smaller laboratories to produce quality immunostains without prior technical expertise and without significant increases in personnel. The availability of a large selection of primary antibodies that can be used on formalin-fixed, paraffin-embedded tissue sections has also contributed to increased use and sensitivity of the method.

The ruling of the Food and Drug Administration (FDA) on the classification of immunohistochemical reagents and kits *(1)* is a recent and very significant event familiar to most who deal with IHC on a daily basis. Implementation of the ruling resulted in the classification of the majority of immunohistochemical reagents as Analyte Specific Reagents" (ASRs) *(2,3)* and class I medical devices, thus exempting them from premarket notification. The rationale for this decision is based on the fact that IHC staining results are incorporated into a surgical pathology report as one part of the entire diagnostic evaluation. The immunostains are not usually "stand-alone" results. Estrogen receptor and progesterone receptor, however, are considered class II devices. These immunostains do not have routine morphologic correlates but do have substantial and widely accepted scientific validation. Class III devices include those immunostains that are not part of the surgical pathology diagnostic process and may result in an independent report to a physician. These stains require premarket notification and specific FDA approval. The HercepTest™ for determination of Her2 protein overexpression as an indication for Herceptin® (trastuzumab) therapy is an example.

The FDA ruling allows IHC laboratories to continue operating essentially as they have in the past. Reagent manufacturers must now label the antibodies they sell for diagnostic use (and which do not have FDA clearance) as ASRs. The onus is on the manufacturers to follow "good manufacturing practices" and to make certain that antibody reagents have consistent high quality and the specificity that is claimed.

From: *Morphology Methods: Cell and Molecular Biology Techniques*
Edited by: R. V. Lloyd © Humana Press, Totowa, NJ

Surgical pathology reports incorporating IHC results, however, are now obligated to include a statement indicating that responsibility for assuring quality of the immunostains rests with the individual laboratory and not the manufacturer of the reagent *(4)*. The laboratory director therefore has the ultimate responsibility to generate high-quality immunostains and to document their performance.

In this chapter, current and practical procedures essential for performing quality immunostains in the modern automated IHC laboratory are discussed.

## FROZEN SECTION IMMUNOHISTOCHEMISTRY: TISSUE AND SECTION PROCESSING

Tissue must be frozen rapidly in a relatively cryoprotected fashion to prevent ice crystal formation and to preserve antigens and cellular morphology optimally. Properly snap frozen specimens can also be used for molecular genetic studies and for mRNA extraction and analysis. All fresh tissue should be considered infectious and should be handled with gloves using universal precautions. Tissue specimens can be collected and transported in 4°C balanced salt solution or tissue culture media such as minimal essential media (MEM) or RPMI-1640 prior to freezing. Specimens in liquid media should be handled and transported at 4°C to slow autolysis and should be delivered to the laboratory for freezing within 1 h of surgical excision. Tissues for freezing should be trimmed of excess adipose and connective tissue and divided with a scalpel or scissors into portions that are no larger than 1.5 cm$^2$ in surface area and 0.5 cm in thickness.

In preparation for freezing and storage, all containers and slides (i.e., imprints) need to be labeled with patient name, ID #, date, and tissue source. A 24 × 24-mm disposable base mold (Allegiance Healthcare Corporation, McGaw Park, IL) is a convenient receptacle for the actual tissue freezing, and PolyCons (4 cm diameter × 1.5 cm height; Madan Plastics, Cranford, NJ) are convenient and useful plastic containers for efficient storage of frozen specimens. Snap freezing is performed by immersion of the specimen in an isopentane (2-methylbutane) bath that has been precooled with either liquid nitrogen (−135°C to −140°C) or dry ice (−30°C). A convenient alternative to liquid nitrogen or dry ice is a refrigerated (lowest temperature −52°C) tissue freezing bath (Histobath™, Shandon, Pittsburgh, PA) containing isopentane. The Histobath is also designed for constant operation and therefore is always ready for freezing samples.

Before freezing, quickly rinse the tissue in saline or culture medium and then blot on a fresh absorbent towel (do not use gauze) to remove excessive liquid. Dispense a small amount of Tissue-Tek OCT compound (Sakura Finetek U.S.A., Torrance, CA) into the bottom of a disposable base mold, carefully position and orient the tissue in the OCT layer, and then cover completely with OCT until the base mold's lower chamber has been filled. Grip the edge of the base mold with a surgical clamp or forceps, and slowly submerge the base mold into the precooled isopentane bath until it is completely submerged. Allow the sample to freeze for 20–30 s and then remove from the isopentane bath. The frozen OCT/tissue block is then extruded from the base mold using a twisting action at the ends of the base mold, and excess freezing compound is trimmed away with a scalpel or razor blade. The frozen OCT/tissue block is then placed in a labeled PolyCon and stored in a −70°C to −80°C ultracold freezer.

In preparation for cutting frozen tissue sections, specimens stored in an ultracold freezer should be placed in the cryostat approximately 20–30 min before sectioning to equilibrate to the cryostat's temperature (approx −20°C). To retain sections on the slides throughout the immunostaining process, it is necessary to use glass slides that have been treated to increase their adhesiveness. Erie Scientific Company (Portsmouth, NH) manufactures slides that are treated with aminoalkyl silane and known as Super-Frost® Plus. The slides are generically referred to as "charged" or "silanized" slides, and they bind tissue sections through a combination of electrostatic and hydrophobic interactions *(5)*. Glass slides can also be treated with poly-L-lysine to increase their adhesive properties *(6)*, but such slides are not as commonly used. Frozen sections are cut on a cryostat at 4–5 μm thickness and immediately placed on appropriately labeled slides. Sections are dried in a 37°C oven for 15 min, and briefly fixed by immersion in 4°C high-performance liquid chromatography grade acetone (in a refrigerator or cooler) for 10 min. Slides are then removed from the acetone and dried with a fan for 10 min. Dried sections can be placed in a storage box and kept at room temperature in an electronic dessicator for up to 7 days if staining is not to be performed immediately. If longer storage of sections is required, slides should be tightly wrapped tightly in aluminum foil and stored in an ultracold freezer at −70°C to −80°C. When ready for staining, slides that have been stored in the freezer must be equilibrated to room temperature *before* they are unwrapped. Equilibration with room temperature prior to unwrapping prevents moisture from condensing or actually freezing on the section, both of which can destroy morphology and compromise immunoreactivity.

Immediately before immunostaining, sections are fixed again in a freshly prepared 1% paraformaldehyde phosphate-buffered saline (PBS) solution for 10 min. The 1% paraformaldehyde solution is made by dilution of the 10% stock with PBS. (Do not use Tris-buffered saline or any other buffers containing free amino groups.) All sections for immunohistochemistry, *except* the ones that are to be stained for κ and λ light chains, are immersed for 10 min at room temperature in the 1% paraformaldehyde solution. Slides are then rinsed in several changes of tap water and placed in a Tris-buffered saline solution for 5 min. Sections destined to be immunostained for κ and λ light chains are fixed a second time in 4°C acetone for 5 min, fan-dried for 10 min, and then hydrated for 1.5 min in PBS. The κ/λ sections are then briefly fixed in 1% paraformaldehyde for **1 min**, rinsed in tap water, and subsequently placed in Tris-buffered saline for 5 min. The shorter fixation time in 1% paraformaldehyde for slides that are to be immunostained for κ and λ light chains is essential for optimal detection of cell surface-associated immunoglobin. Manual and automated IHC can both be performed on frozen sections, but the dilution and rinsing buffers should contain only a minimal amount (0.025%) of nonionic detergent (e.g., Tween 20) to preserve nuclear and cellular detail.

## PARAFFIN SECTION IMMUNOHISTOCHEMISTRY:
## TISSUE AND SECTION PROCESSING

Tissue fixation is perhaps the single most important variable influencing the outcome of paraffin section immunostains, and its influence can not be overstated. The routine fixative of choice in most pathology laboratories is neutral-buffered formalin. It is relatively inexpensive, easy to prepare, and produces excellent morphologic preservation without shrinkage artifacts. The mechanism by which formaldehyde fixes and preserves tissues is not completely understood but does involve formation of crosslinks both within and between protein molecules via hydroxymethylene bridges *(7)*. Calcium ions have also been implicated in crosslink formation *(8)*. Crosslink formation is considered to be responsible for masking antibody-binding epitopes and to be a major contributor to the lack of sensitivity for some immunostains in paraffin sections. Alcohol fixatives are reported to be less damaging to tissue antigens and have been promoted as substitutes for formalin *(9)*. There has, however, been little movement away from neutral-buffered formalin as the routine fixative in the daily practice of pathology. Instead, research and development have increasingly focused on methods for antigen recovery in formalin-fixed tissues as means to improve immunostaining.

Despite the inherent detrimental effects on antigen preservation, strict adherence to a proper and *consistent* protocol for fixation with neutral-buffered formalin is crucial for producing high-quality immunostains. Formaldehyde can be obtained commercially as a concentrated (37–40%) solution, and 10% neutral-buffered formalin is made by a 1:10 dilution of the concentrate with PBS (50 m*M* phosphate, 0.9% NaCl, pH 7.2–7.4), resulting in a final formaldehyde concentration of 3.7–4.0%. Ideally, tissue samples that are approximately $1.0 \times 1.0 \times 0.3$ cm are placed in a cassette and immersed in freshly prepared formalin for a minimum of 12 h and a maximum of 24 h. In reality, larger pieces of tissue are usually put up, but the tissue should not be crammed into the cassette, and the volume of formalin should be 20 times the volume of tissue. The smaller the size, the better and more thorough the fixation. If the specimens cannot be processed to paraffin blocks after 24 h in formalin, transfer them to 70% ethanol for holding until the samples can be put on the processor. Prolonged contact with formalin causes excessive crosslink formation and increasing loss of antigenicity *(10–12)*. Once tissues are properly fixed and embedded in a paraffin block, they appear to retain antigenicity for decades *(13)*.

Paraffin tissue sections for IHC are routinely cut on a microtome at 4–6 μm, floated out on a 42–48°C water bath, and picked up on "charged" slides. The water bath should not contain any additives when charged slides are used, as they will compromise the adhesive properties of the slides. Storage of cut paraffin sections for later use in IHC is still a subject of debate. The issue is the reported loss of reactivity of certain antigens, including p53 and Her2, when cut paraffin sections are stored at room temperature *(14,15)*. This phenomenon may be more pronounced and exacerbated when tissue is fixed in *alcohol-supplemented* formalin instead of routine neutral-buffered formalin *(13)*. For many laboratories, positive control sections or sections for research studies must be cut in advance and stored for subsequent use. If for practical reasons this is the case, sections should not be heated or "melted down" prior to storage. Sections should be stored in a closed slide box in a cool location, or if possible at 4°C. Immediately prior to staining, the sections can be placed in a 60°C oven for 60 min to melt down the excess paraffin and promote stronger adherence to the slide. Sections are then ready for deparaffinization, hydration through graded ethanols, and blocking of endogenous peroxidase activity.

Perhaps the most significant advance in paraffin section IHC in the past decade has been the development and refinement of heat-induced epitope retrieval, also referred to as *antigen retrieval*. The impact of this innovative and revolutionary technique has been profound, and the reader is referred to Chapter 14 for an in-depth discussion of antigen retrieval methods in immunohistochemistry.

## DETECTION SYSTEMS

Numerous methods and reagents are commercially available for detection of bound primary antibodies by IHC, but enzyme-based techniques using either horseradish peroxidase or alkaline phosphatase predominate. For most automated stainers in clinical laboratories, peroxidase methods are favored. Examples of peroxidase-based immunohistochemistry include the peroxidase-antiperoxidase (PAP) method, the enzyme-labeled streptavidin-biotin (LSAB) technique, and the avidin-biotin complex (ABC) method. These are all three-layer procedures involving a primary antibody, a secondary antibody with appropriate specificity for the primary, and the enzyme-containing tertiary reagent. In the case of the PAP method, the secondary antibody is unlabeled and added in excess so that it binds both the primary antibody and the soluble enzyme-antibody (PAP) complex that comprises the third layer. Consequently, with the PAP technique the antibody species in PAP complex must match the primary antibody. For the LSAB and ABC techniques, the secondary antibody is covalently conjugated with biotin (biotinylated) and need only recognize and bind the primary antibody. The third layer of enzyme-labeled streptavidin or avidin-biotin complex then binds tightly to the biotin.

The long-established "indirect method" utilizes a secondary antibody that is directly conjugated with peroxidase or alkaline phosphatase to bind to the primary antibody. This two-step method is simple and quick but also considerably less sensitive than the three-step detection systems described above. More recently, secondary antibodies conjugated with a dextran polymer backbone containing a large number of enzyme molecules (about 100) have been introduced (EnVision™ Systems, Dako, Carpinteria, CA). With this new generation of secondary antibody conjugates, it is possible to generate two-layer immunostaining protocols that have sensitivities approaching the three-layer methods *(16)*. An added advantage is that the system is biotin free and therefore does not display the nonspecific staining associated with biotin-binding proteins or endogenous biotin.

There are also catalyzed signal amplification or catalyzed reporter deposition detection methods that generate extreme sensitivity and to which most clinical immunohistochemists are not accustomed *(17)*. Typically these are peroxidase- and tyramide-based amplification techniques; they are discussed in detail in Chapter 15.

Selection of a chromogen is also an important factor to consider in immunohistochemistry. Peroxidase systems routinely use 3,3′-diaminobenzidine (DAB) (brown) or 3-amino-9-ethylcarbazole (AEC) (red) as chromogens. For alkaline phosphatase systems, 5-bromo-4-chloro-3-indolyl phosphate/nitroblue tetrazolium (BCIP/NBT) may be used to yield a blue-purple product or fast red/naphthol AS-TR phosphate to produce a red/pink product. Chromogen choice can be influenced by personal preference (red vs blue vs brown) and degree of contrast with the counterstain (hematoxylin vs methyl green). Other considerations include incubation time and sensitivity (peroxidase/DAB reacts faster, with a shorter incubation time, than phosphatase/FAST red), safety issues (DAB is a carcinogen), or technical issues (AEC is soluble in organic solvents and cannot be used with organic solvent-based permanent mounting medium). DAB is a reliable chromogen that produces crisp, well-localized reactions that are permanent. At the Mayo Clinic, DAB is used for nuclear antigens in paraffin sections and for all antigens in frozen section immunostains related to lymphoma phenotyping. AEC (red) has traditionally been the chromogen of choice at the Mayo Clinic for membrane and cytoplasmic antigens in paraffin sections because of the excellent color contrast with a light hematoxylin counterstain. However, a drawback to the use of AEC is its solubility in organic solvents and therefore its requirement for aqueous mounting media.

## PRIMARY ANTIBODIES

Selecting a primary antibody for IHC use can be a difficult task. Frequently, different antibodies from several different sources are available against a particular antigen; often all of them claim to work for IHC. Frequently, though, there are assessments in the literature that can provide guidance in this decision. For example, different antibodies directed against p53 have been compared for their reactivity in formalin-fixed tissues and have demonstrated differences in reactivity that impact the prognostic significance of p53 antigen expression in breast cancer *(18)*. The polyclonal antibody pAb1801 and the monoclonal antibody DO7 were reported to be more effective than others tested. A comprehensive review of differences in antibody reactivity to particular antigens is beyond the scope of this discussion. However, it is critically important that the immunohistochemist be familiar with the recent literature specific to an individual antibody clone to avoid, or at least be informed of, potential problems.

Most antibodies used in a clinical immunohistochemistry laboratory are commercially available, and their specificity is well characterized. As ASRs, which most IHC reagents are, good manufacturing practices are assumed to have been followed, and most antibodies will perform "as advertised." Usually, peer-reviewed literature also exists and can be used to gather a realistic review of antibody performance. Such information will usually also provide a good reference point from which to start primary antibody dilutions and antigen retrieval methods. Many manufacturers now offer their ASR antibodies as prediluted reagents "optimized" for paraffin section IHC. Predilutes are becoming a common practice but can limit flexibility when developing a new stain in a particular laboratory. The main concern is that the antibody dilution will be such that it only works on optimally fixed tissues with the manufacturer's own detection system. For cases in which suboptimally processed tissue is evaluated or a less sensitive detection system is used, the preset dilution can be too great. Whenever possible, undiluted primary antibody is preferable.

The immunohistochemist should also pay attention to whether the primary antibody is polyclonal or monoclonal and note the species and immunoglobulin isotype. Polyclonal antibodies generally have high-affinity clones within them and can be significantly diluted (often >1:1000). On the other hand, polyclonals are also more likely to generate background staining and can require extra or prolonged blocking steps. Knowing the species and isotype of monoclonals will avoid problems in which an inappropriate secondary antibody is used. An example in which isotype makes a difference is the

LeuM1 clone (anti-CD15), an IgM mouse monoclonal antibody that has been shown to have superior performance when an anti-IgM secondary antibody is used.

## OPTIMIZING CONDITIONS

Numerous factors can be manipulated and adjusted in the process of optimizing an IHC stain. The most commonly adjusted factors are antibody dilution, duration of primary antibody incubation, choice and concentration of secondary reagent, antigen retrieval buffer, incubation temperature, choice of detection system, and addition of amplification steps. "Checkerboard" approaches for optimizing stains in which each variable is systematically altered while all others remain constant have been advocated *(19)*. However, the large number of factors just listed would require an unrealistic number of slides to be stained and evaluated. Since most automated immunostainers have limited temperature control and usually have recommended secondary reagents and detection chemistries, the immunohistochemist is left primarily with adjustments of primary antibody dilution, primary antibody incubation times, and choice of antigen retrieval method. Given these limitations, the number of test slides required to evaluate a particular antibody is more manageable and readily accomplished.

Although it is usually true that the lower the dilution the more intense the staining for a given incubation time, the use of highly concentrated antibodies may result in a prozone effect that actually decreases staining intensity. Background or nonspecific staining can also become a problem when too high a concentration of primary antibody is used. Longer incubation times (1 h to overnight) can be incorporated to permit greater dilutions of primary antibodies, but some automated immunostainers limit the adjustment (<30 min) that can be made to incubation times. If prolonged incubations are required, they are best performed "off-line."

As mentioned previously, the most significant advance in IHC in the last 10 years has been the development of heat-induced antigen retrieval methods *(20,21)*. Antigen retrieval allows antibodies that previously performed only in frozen tissues to be reactive in formalin-fixed tissues; it has greatly expanded the number of antibodies that can now be applied to paraffin-embedded tissue *(22)*. The increase in sensitivity afforded by antigen retrieval can also result in cost savings for the laboratory since higher dilutions of primary antibodies can be used, and therefore more tests per vial can be performed. Most antibodies will show an increase in staining intensity when antigen retrieval is used. Many antibodies perform better in alkaline (EDTA or Tris-based) buffer, but some antibodies prefer an acidic (citrate-based) buffer *(23)*. Conditions must be determined empirically, as there is no way to predict the behavior of a particular antibody. There are, however, several published studies that have looked at a broad range of antibodies, and these can be a valuable resource for the immunohistochemist *(22–25)*.

The increased sensitivity for detecting antigens that is afforded by antigen retrieval techniques or signal amplification methods is also associated with increased nonspecific staining. In particular, endogenous biotin or biotin-binding activity can be retrieved or detected and will result in false-positive granular cytoplasmic staining *(26)*. Biotin or biotin-binding activity can be eliminated by blocking steps that involve application of free avidin after the primary antibody step, followed by application of free biotin

*(26,27)*. Blocking kits are commercially available from several vendors, including Dako and Vector Laboratories (Burlingame, CA). Endogenous peroxidase activity in horseradish peroxidase-based systems and endogenous phosphatase in alkaline phosphatase-based systems are other potential causes of false-positive staining. Both activities can be eliminated by blocking steps that incorporate incubation with hydrogen peroxide or levamisole, respectively *(28,29)*.

## CONCLUSIONS

This chapter has presented practical methods and considerations that are essential for the production of quality immunohistochemical stains. Protocols for tissue freezing and formalin fixation have been discussed since consistent and proper execution of these basic procedures is critical for maintaining tissue morphology and antigenicity. The increasing acceptance and utilization of automated immunostainers is promoting standardization of IHC, both within an individual laboratory and among laboratories using similar reagents and equipment. National proficiency testing programs, such as those conducted by the College of American Pathologists, also encourage and support standardization of staining procedures (including issues such as specimen processing, antigen retrieval, detection systems, and antibody selection) and interpretation. Immunohistochemistry laboratories can only benefit from participation in these programs and surveys.

## REFERENCES

1. Food and Drug Administration. (1998) Medical devices; classification/reclassification of immunohistochemistry reagents and kits—FDA. Final rule. *Fed. Reg.* **63**, 30132–30142.
2. Taylor, C. R. (1999) FDA issues final rule for classification and reclassification of immunohistochemistry reagents and kits [Editorial]. *Am. J. Clin. Pathol.* **111**, 443–444.
3. Taylor, C.R. (1998) Report from the Biological Stain Commission: FDA issues final rule for classification/reclassification of immunohistochemistry (IHC) reagents and kits. *Biotech. Histochem.* **73**, 175–177.
4. Swanson, P. E. (1999) Labels, disclaimers, and rules (oh, my!). Analyte-specific reagents and practice of immunohistochemistry [editorial]. *Am. J. Clin. Pathol.* **111**, 445–448.
5. Rentrop, M., Knapp, B., Winter, H., and Schweizer. (1986) Aminoalkylsilane-treated glass slides as support for in situ hybridization of keratin cDNAs to frozen tissue sections under varying fixation and pretreatment conditions. *Histochem. J.* **18**, 271–276.
6. Jacobson, B. S. and Branton, D. (1977) Plasma membrane: rapid isolation and exposure of the cytoplasmic surface by use of positively charged beads. *Science* **195**, 302–304.
7. Werner, M., Von Wasielewski, R., and Komminoth, P. (1996) Antigen retrieval, signal amplification and intensification in immunohistochemistry. *Histochem. Cell Biol.* **105**, 253–260.
8. Morgan, J. M., Navabi, H., Schmnid, K. W., and Jasani, B. (1994) Possible role of tissue-bound calcium ions in citrate-mediated high-temperature antigen retrieval. *J. Pathol.* **174**, 301–307.
9. Bostwick, D. G., al Annouf, N., and Choi, C. (1994) Establishment of the formalin-free surgical pathology laboratory. Utility of an alcohol-based fixative. *Arch. Pathol. Lab. Med.* **118**, 298–302.
10. Battifora, H. (1991) Assessment of antigen damage in immunohistochemistry. The vimentin internal control. *Am. J. Clin. Pathol.* **96**, 669–671.

11. Battifora, H., and Kopinski, M. (1986) The influence of protease digestion and duration of fixation on the immunostaining of keratins. A comparison of formalin and ethanol fixation. *J. Histochem. Cytochem.* **34**, 1095–100.

12. Fox, C. H., Johnson, F. B., Whiting, J., and Roller, P. P. (1985) Formaldehyde fixation. *J. Histochem. Cytochem.* **33**, 845–853.

13. Roche, P. C. (1998) Antigen stability in stored paraffin sections. *CAP Today* **12**, 59–61.

14. Prioleau, J. E., and Schmitt, S. J. (1995) p53 antigen loss in stored paraffin slides. *N. Engl. J. Med.* **332**, 1521–1522.

15. Jacobs, T. W., Prioleau, J. E., Stillman, I. E., and Schmitt, S. J. (1996) Loss of tumor marker-immunostaining intensity on stored paraffin slides of breast cancer. *J. Natl. Cancer Inst.* **88**, 1054–1059.

16. Sabattini, E., Bisgaard, K., Ascani, S., et al. (1998) The EnVision++ system: a new immunohistochemical method for diagnostics and research. Critical comparison with the APAAP, ChemMate, CSA, LABC, and SABC techniques. *J. Clin. Pathol.* **51**, 506–511.

17. King, G., Payne, S., Walker, F., and Murray, G. I. (1997) A highly sensitive detection method for immunohistochemistry using biotinylated tyramine. *J. Pathol.* **183**, 237–241.

18. Horne, G. M., Anderson, J. J., Tiniakos, D. G., et al. (1996) p53 protein as a prognostic indicator in breast carcinoma: a comparison of four antibodies for immunohistochemistry. *Br. J. Cancer* **73**, 29–35.

19. Shi, S. R., Cote, R. J., Yang, C., et al. (1997) Development of an optimal protocol for antigen retrieval: a 'test battery' approach exemplified with reference to the staining of retinoblastoma protein (pRB) in formalin-fixed paraffin sections. *Histopathology* **31**, 400–407.

20. Shi, S. R., Cote, R. J., and Taylor, C. R. (1996) Antigen retrieval immunohistochemistry: past, present, and future. *Biotech. Histochem.* **71**, 190–196.

21. Shi, S. R., Key, M. E., and Kalra, K. L. (1991) Antigen retrieval in formalin-fixed, paraffin-embedded tissues: an enhancement method for immunohistochemical staining based on microwave oven heating of tissue sections. *J. Histochem. Cytochem.* **39**, 741–748.

22. Cuevas, E. C., Bateman, A. C., Wilkins, B. S., et al. (1994) Microwave antigen retrieval in immunohistochemistry: a study of 80 antibodies. *J. Clin. Pathol.* **47**, 448–452.

23. Pileri, S. A., Roncador, G., Ceccarelli, C., et al. (1997) Antigen retrieval techniques in immunohistochemistry: comparison of different methods. *J. Pathol.* **183**, 116–123.

24. Cattoretti, G., Pileri, S., Parravicini, C., et al. (1993) Antigen unmasking on formalin-fixed, paraffin-embedded tissue sections. *J. Pathol.* **171**, 83–98.

25. von Wasielewski, R., Werner, M., Nolte, M., Wilkens, L., and Georgii, A. (1994) Effects of antigen retrieval by microwave heating in formalin-fixed tissue sections on a broad panel of antibodies. *Histochemistry* **102**, 165–172.

26. Bussolati, G., Gugliotta, P., Volante, M., Pace, M., and Papotti, M. (1997) Retrieved endogenous biotin: a novel marker and a potential pitfall in diagnostic immunohistochemistry. *Histopathology* **31**, 400–407.

27. Wood, G. S. and Warnke, R. (1981) Suppression of endogenous avidin-binding activity in tissues and its relevance to biotin-avidin detection systems. *J. Histochem. Cytochem.* **29**, 1196–1204.

28. Li, C. Y., Ziesmer, S. C., and Lazcano-Villareal, O. (1987) Use of azide and hydrogen peroxide as an inhibitor for endogenous peroxidase in the immunoperoxidase method. *J. Histochem. Cytochem.* **35**, 1457–1460.

29. Ponder, B. A., and Wilkinson, M. M. (1981) Inhibition of endogenous tissue alkaline phosphatase with the use of alkaline phosphatase conjugates in immunohistochemistry. *J. Histochem. Cytochem.* **29**, 981–984.

# 14

# Antigen Retrieval Technique for Immunohistochemistry

*Principles, Protocols, and Further Development*

## Clive R. Taylor, MD, PHD and Shan-Rong Shi, MD

## INTRODUCTION

Immunohistochemistry (IHC) has a long history stemming from the work of Coons and colleagues in 1941 *(1)*, who developed an immunofluorescence technique to detect bacteria. In the early 1970s, extensive efforts were made to render immunoperoxidase methods more widely available to routine formalin-fixed, paraffin-embedded tissues through a series of technical developments, including increasingly sensitive detection systems, and various pretreatment methods prior to the immunostaining procedure *(2–9)*. The basic philosophy of applying IHC for routine formalin-fixed tissues was set forth by Taylor and Kledzik *(2)* in 1981.

However, IHC methods that included unmasking procedures such as enzymatic digestion, although widely applied, did not improve IHC staining of the majority of antigens, as pointed out by Leong et al. *(10)*. A major drawback of enzymatic digestion was that it proved difficult to control the optimal digestion conditions for individual tissue sections when they were stained with different antibodies. Difficulties in standardizing this method provided a powerful incentive to the development of a new technique, with the requirements that it should be more effective, more widely applicable, and easier to use than enzymatic digestion. In addition, to be acceptable, any new method should enhance immunohistochemical staining of routine paraffin sections in a reproducible and reliable manner.

In this context, the early experience of one of the authors *(3–5)*, led to the realization that the application of IHC to "routine formalin paraffin sections" would represent a new era for the demonstration of antigens in the types of "routinely" processed tissues that abound in university pathology laboratories. Indeed, the method opened a new vista of research and diagnostic possibilities, yielding a level of morphologic detail that allowed pathologists to correlate immunologic findings directly with traditional cytologic and histologic criteria. This advance, in turn, provided Shi with the impetus for development of an antigen retrieval (AR) technique. The first seeds in the development of a retrieval approach for routinely processed celloidin-embedded human temporal bone

From: *Morphology Methods: Cell and Molecular Biology Techniques*
Edited by: R. V. Lloyd © Humana Press, Totowa, NJ

sections sprouted in the early 1980s when Shi studied under Dr. Schuknecht's guidance at the Eastern National Temporal Bone Bank in Boston. Attempts to perform IHC study on archival formalin-fixed tissue sections were actively supported by Dr. Shuknecht. Subsequently the work involved BioGenex for a period and then extended into a broader range of applications, culminating in our studies at the University of Southern California in the early 1990s, which provide much of the basis of this chapter.

A key point that we considered in developing the AR technique was whether the crosslinkages of protein caused by formalin fixation are reversible or irreversible. The answer was, in essence, already available in the serial studies of Fraenkel-Conrat and co-workers published in 1940s *(11–13);* we had only to rediscover it! Their work indicated that hydrolysis of crosslinkages between formalin and protein may be limited by certain amino acid side chains, such as imidazol and indol. However, these crosslinkages could be disrupted by high-temperature heating (120°C) or strong alkaline treatment. These observations formed the basis for the development of AR in both heating and nonheating methods. A dramatic enhancement of immunostaining on archival formalin-fixed, paraffin-embedded tissue sections could be achieved either by boiling the tissue sections in water solution, or by employing sodium hydroxide-methanol solutions for immersing celloidin-embedded human temporal bone sections prior to immunostasining *(14,15).*

The high-temperature heating AR technique has been widely adopted as a simple, effective, and reliable pretreatment for routine IHC in surgical pathology as well as analytical morphology *(16–27).* Various modifications in terms of heating method and AR solution have been proposed since 1991. In addition to microwave and conventional heating methods *(14),* Shin et al. *(28)* described a hydrated autoclaving method to enhance the immunoreactivity of tau protein in brain tissue fixed in formalin. This work was followed by other studies *(29–32)* with good results. Subsequently, domestic pressure cookers and steamers were applied for the high-temperature heating procedure *(33–37).* Metal salt solutions and distilled water were used as the earliest AR solutions *(14).* The increase of immunostaining intensity resulting from the use of metal salt AR solutions has subsequently been demonstrated by several studies *(16,22,24,38–49).* A single article *(50)* that reported a poor result using a metal salt solution for AR-IHC appears to have been based on inappropriate comparisons, as analyzed in detail elsewhere *(51).* However, a major drawback of metal salts, particularly lead salts, is the potentially toxic effect, as emphasized by Suurmeijer and Boon *(16).* It is, therefore, preferable to avoid using lead or other toxic metal salts, if other kinds of solution may serve with equal effect in AR-IHC staining.

The use of citrate buffer solution for the AR heating method was documented by Cattoretti et al. *(52),* followed by Gown et al. *(18),* Suurmeijer and Boon *(16),* Leong and Milios (19), Taylor et al. *(22),* Swanson *(53),* Cuevas et al. *(23),* and others *(51,54–64).* Cattoretti et al. *(17)* extended their investigation to 256 antibodies using a citrate buffer of pH 6.0 as the AR solution, with a microwave heating protocol as previously documented *(14).* Meanwhile, other chemicals, such as urea, glycine-HCl buffer, Tris-HCl, phosphate buffer (PBS), periodic acid, aluminum chloride, and other metal salts, as well as detergent solutions, have subsequently been essayed as AR solutions *(17,22,42,44,65–71).* In addition, commercial AR solutions such as target unmasking fluid (TUF) have also been proposed *(72).* The chemical nature of the solution has been credited by some as a major factor that influences the effect of AR-IHC. This idea is a particularly interesting issue for commercial products, for which key ingredients are kept secret. Confronted with this form of limited "anarchy" *(73)*, a critical question must be addressed in any effort to achieve greater uniformity of the AR technique, namely, what are the major factors that influence the effectiveness of AR-IHC? The following discussion of major factors that influence the effectiveness of AR-IHC is presented to promote a better understanding of effects, in order to standardize and further develop the AR technique using a more scientific approach.

Since the AR technique was developed in 1991, the heating AR technique has been widely applied as a simple and extremely effective pretreatment for IHC on routinely formalin-fixed, paraffin-embedded tissues *(14)*. During this period, hundreds of original articles and more than a dozen comprehensive reviews have been published worldwide. An international symposium on AR-IHC was held on July 25th, 1998 at the Fifth Joint Meeting of The Japan Society of Histochemistry and Cytochemistry and The Histochemical Society in San Diego, California. At this symposium, speakers and audiences from Japan, Europe, and the United States reviewed the major issues concerning the development, clinical application, and factors that influence the effectiveness of AR-IHC.

Studies of the major factors that influence the effect of AR-IHC provide a basis for establishing a "test battery" approach, by which various factors may be rapidly evaluated in order to identify the optimal protocols for "maximal retrieval" level. This simple approach may allow some standardization of AR-IHC on routinely processed tissues, as discussed below.

## HEATING AR METHOD

Currently, the high-temperature heating AR method is the predominant technique for enhancement of IHC staining, particularly for routinely formalin-fixed, paraffin-embedded tissue sections. Extensive use of the method has, however, raised a number of interesting issues. The major factors that influence the effects of AR-IHC are summarized below.

### Heating, the Most Important Factor

As previously emphasized in the literature *(14,30,43,51,55,56,58–64,74,75)*, the heating conditions appear to be the most important factor based on the following evidence:

1. A significant enhancement of immunohistochemical staining can be achieved by using high-temperature heating of routinely processed paraffin-embedded tissue sections in pure distilled water *(14,76,77)*;

2. Higher temperatures in general yield superior results *(37,55,56,59–62)*. One of our recent studies demonstrated that an optimal result for AR-IHC is correlated with the product of the heating temperature $(T)$ and the time $(t)$ of AR heating treatment $(T \times t)$ *(78)*; a similar strong intensity of AR-IHC staining was generated by the following heating conditions $(T \times t)$: 100°C × 20 min, 90°C × 30 min, 80°C × 50 min, and 70°C × 10 h *(37,78)*. In general, the lower the temperature of heating used for AR, the longer the time required to reach an equivalent intensity of AR-IHC. That the same intensity of AR-IHC can be achieved by adjusted different heating conditions, as well as by several different heating devices, may

provide some flexibility in reaching a satisfactory result, according to the characteristics of the specimen (embedding material, thickness of the tissue section, different kind of tissue, and so on) and the equipment available.

3. An equivalent intensity of AR-IHC can be obtained using different buffers as AR solutions, if the pH values of AR solutions are monitored in a comparable manner, demonstrating that individual specific chemical constituents are not necessary factors in yielding a satisfactory result *(37,79)*.

4. Our early experience that even prolonged exposure of paraffin sections in citrate buffer solution (or indeed any buffer) without heating gave no noticeable AR effect has subsequently been confirmed by numerous studies *(37,51,60–62,74)*.

## pH of the AR Solution

The advocacy of various investigators for a variety of buffer solutions as AR solutions has nevertheless provided food for thought. One hypothesis proposed by our group in 1993 stated that the effectiveness of AR-IHC may be influenced by the pH value of the AR solution; in essence, the pH value may be more critical than the chemical composition of the AR solution. We tested the hypothesis that the pH of the AR solution may influence the quality of immunostaining by using seven different AR buffer solutions at a series of different pH values ranging from 1 to 10. We evaluated the staining of monoclonal antibodies to cytoplasmic antigens (AE1, HMB45, neuron-specific enolase [NSE]), nuclear antigens (MIB1, proliferating cell nuclear antigen [PCNA], estrogen receptor [ER]), and cell surface antigens (MT1, L26, epithelial membrane antigen [EMA] on routinely formalin-fixed, paraffin-embedded sections under different pH conditions with microwave AR heating for 10 min. The pH value of the AR buffer solution was carefully measured before, immediately after, and 15 min after the AR procedure. From this study, we drew the following conclusions:

1. There were three types of patterns reflecting the influence of pH. First, several antigens showed no significant variation utilizing AR solutions with pH values ranging from 1.0 to 10.0 (L26, PCNA, AE1, EMA, and NSE); second, other antigens (MIB1, ER) showed a dramatic decrease in the intensity of the AR-IHC at middle-range pH values (pH 3.0–6.0) but strong AR-IHC results above and below these critical zones; and third, still other antigens (MT1, HMB45) showed negative or very weak focally positive immunostaining with a low pH (1.0–2.0) but excellent results in the high-pH range.

2. Among the seven buffer solutions at any given pH value, the intensity of AR-IHC staining was very similar, with the single exception that Tris-HCl buffer tended to produce better results at higher pH, compared with other buffers.

3. Optimization of the AR system should include optimization of the pH of the AR solution.

4. A high-pH AR solution, such as Tris-HCl or sodium acetate buffer at pH 8.0–9.0, may be suitable for most antigens.

5. Low-pH AR solutions are most useful for nuclear antigens such as retinoblastoma protein (RB), ER, and androgen receptor.

6. Focal weak false-positive nuclear staining may be found when using low pH AR solution; the use of negative control slides is important to exclude this possibility *(79)*.

Evers and Uylings *(80)* also found that AR-IHC is pH and temperature dependent. They tested two antibodies, microtubule-associated protein-2 (MAP-2) and SMI-32, and indicated that the optimal pH values were pH 4.5 for MAP-2 and pH 2.5 for SMI-32. In addition, they demonstrated that the use of 4% $AlCl_3$ as the AR solution for SMI-32 could achieve a similar result to that obtained by using citrate buffer of pH

2.5, from which they concluded that it is not important what kind of solution is used as long as the pH is at an appropriate level. They subsequently demonstrated that microwave AR in Tris-HCl buffer of pH 9.0 at full-power heating yielded the best results for most antibodies tested in neuroscience research *(81)*, a finding that supported our previous conclusion *(58,60,65)*. Unfortunately, the issue is complicated by the observation that not all antigen/antibody pairings in IHC are affected optimally for the same pH value; indeed, as noted above, we have reported three separate groupings, with nuclear antigens, for example, showing greatest recovery at low pH *(58,79,82)*.

## Chemical Composition of the AR Solution

Metal salts (lead, zinc, and so on) were used in the first reported AR solution. Based on earlier studies concerning zinc formalin fixation of tissue sections *(83)*, we speculated that the metal salt solution might effectively influence the structure of protein by playing a role in refixation of the retrieved antigens. Another possible effect of metal salt solution may be molarity of the salt AR solution, as pointed out by Suurmeijer and Boon *(16)*. They investigated the effect of molarity using aluminum chloride solution with concentrations of 0.5%, 1%, 2%, and 4% for AR-IHC staining of vimentin on routinely processed paraffin tissue sections. They found that the optimal staining for vimentin in 10 different tissue components was achieved with 4% aluminum chloride. Suurmeijer *(39)* compared AR results with the use of 0.01 and 0.3 $M$ aluminum chloride and found that different antigens showed significantly different responses to changes in molarity.

From our more recent studies regarding AR-IHC under the influence of pH *(79,82)*, we found that although the pH is critical, the chemical composition of the AR solution is less important but may also play a role as a cofactor in AR-IHC in some circumstances. For example, utilizing antibodies to RB, including a polyclonal antibody (RB-WL-1) and a monoclonal antibody (PMG3-245), we found that a better staining result was obtained using acetate buffer solution of pH 1.0–2.0 with microwave heating at 100°C for 10 min, compared with citrate buffer at the same pH value. Imam et al. *(68)* reported that the use of glycine-HCl solution at pH 3.6 as the AR solution yielded stronger immunostaining with antibodies to some antigens such as androgen receptor, ER, Ki-67, MIB1, EMA, MT-1, actin, and so forth, compared with the staining obtained by using citrate buffer pH 6.0, raising the possibility that the chemical composition of glycine-HCl may play a role in AR treatment. One example of the possible influence of chemical component in the AR solution was provided by the work of Hazelbag et al. *(69)*, who demonstrated that a simple detergent solution could yield results similar to those obtained by using the commercial AR solution TUF for a variety of antibodies to keratin: however, as noted previously, the composition of TUF is a trade secret. On the other hand, Katoh and Breier *(76)* demonstrated that equally good immunostaining of p53 by AR-IHC could be obtained with normal saline, citrate buffer, or distilled water.

In conclusion, although the chemical components of the AR solution may be cofactors that influence the effectiveness of AR-IHC, so far we have not identified any particular component essential to achieve satisfactory positive staining. The potential functions of chemical components in the AR solution include 1) secondary fixation after "unfixation" by high-temperature heating, 2) stabilization of antigens during heating or strong alkaline hydrolysis, 3) maintenance of optimal molarity, and 4) unknown effects in

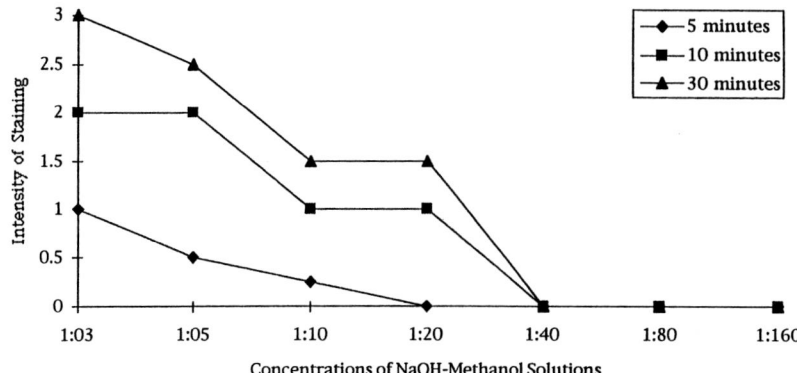

**Fig. 1.** One of the factors that influence the effect of non-heating AR-IHC used for celloidin-embedded human tissues: correlation among immunostaining reactivity, the concentration of NaOH-methanol solution, and the immersion time of celloidin sections. The stronger the concentration, the shorter the immersion time. (Reproduced with permission from ref. *62.*)

reconfiguring the "unfixed" protein, thereby recovering antigenicity. Analysis of these separate factors may allow for development of new AR solutions applicable to those cell surface antigens that do not respond well to current retrieval methods. Most recently, Pileri et al. *(71)* compared three AR solutions using a pressure cooker and found that EDTA-NaOH (pH 8.0) was the best, Tris-HCl buffer (pH 8.0) was intermediate, and citrate buffer (pH 6.0) produced the poorest results. It seems that under certain heating and pH conditions, chemical components may influence the effect of AR-IHC for some antibodies, in ways that are still not understood.

## NONHEATING AR METHOD

Penido, Tseng, and Kao (unpublished data) conducted a study of standardization and development of AR-IHC using routinely celloidin-embedded tissues of human temporal bone, kidney, liver, spleen, adenoids, lung, skin, and intestine and baboon temporal bone. A total of 603 celloidin sections were used. Factors that influenced the results of the nonheating AR method may be summarized as follows:

1. Concentration of AR solution (NaOH-methanol) and immersion time. A series of concentrations of NaOH-methanol solutions ranging from 1:3 to 1:160 diluted by 100% methanol was carefully analyzed. The results are summarized in **Fig. 1**. In general, the immunostaining reactivity was correlated with both the concentration and the immersion time in NaOH-methanol solution. Since the stronger concentration of NaOH-methanol may damage the fine structure of inner ear tissue, more diluted AR solution with elongated immersion time may be preferable. In addition, the stronger NaOH-methanol solution may result in detachment of the sections from the slides.
2. Methanol is the preferred solvent. One comparative study of AR-IHC on celloidin tissue sections employed three solutions; NaOH-methanol, NaOH-ethanol, and NaOH-water at the same ratio of 1:5. After immersing the slides in each AR solution for 30 min, the slides were rinsed for 20 min in two changes of methanol, ethanol, or distilled water, respectively. The immunostaining reactivity obtained by using the NaOH-methanol solution showed the strongest intensity, and the reactivity obtained by ethanol showed a moderate intensity. Immunostaining reactivity was not detected for the celloidin tissue sections treated by the

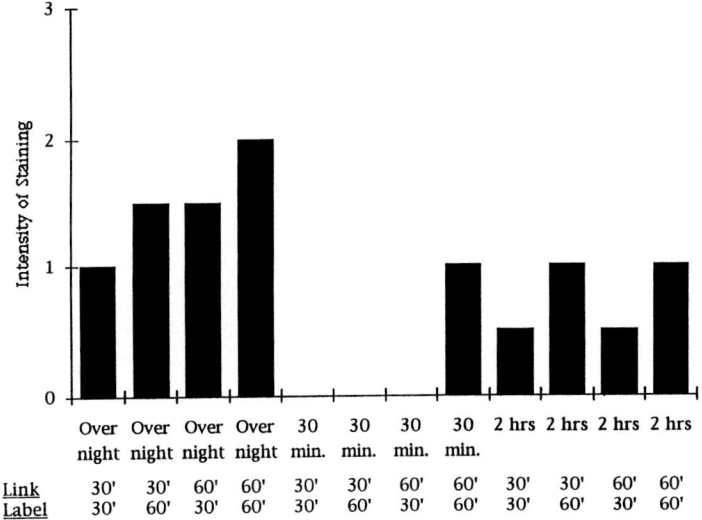

**Fig. 2.** Comparison of immunostaining intensity between different incubation times of primary antibody, link, and label moiety. An elongated incubation time is required for each step of the incubation to achieve the best result. (Reproduced with permission from ref. *62*.)

NaOH-water solution. However, a similar strong intensity was achieved by rinsing the slides in methanol with the three different AR solutions.

Methanol plays an important role in removing celloidin from the celloidin-embedded tissue, an important factor in improving access of antibody to antigen.

3. Optimal incubation time of immunostaining on celloidin tissue section after AR. A comparison of AR-IHC for celloidin tissue sections using different incubation times is summarized in **Fig. 2**. The study demonstrated that an elongated time of incubation is necessary to achieve satisfactory results, a finding similar to that incorporated in our protocol documented previously *(15,74,84,85)*, i.e., overnight incubation for primary antibody and 60 min for link and label reagents, respectively. One reason why a longer incubation time is necessary may relate to the thickness of the celloidin section (20 μm).

## Test Battery Approach

A preliminary test of the AR protocol, examining two major factors, heating condition $(T \times t)$ and pH value, is performed to establish an optimal protocol for the antibody/antigen combination.

Typically, three levels of heating conditions and 3 pH values, low, moderate, and high, may be applied to screen for a potential optimal protocol. The test battery can also be performed in two sequential steps. In the first step, we test three AR solutions at different pH value as listed above, with one standard temperature (100°C for 10 min), to find the optimal pH value of the AR solution. In the second step, we test for the optimal heating conditions based on the established pH value.

We have demonstrated that different heating methods (including microwave, pressure cooker, steam, and autoclave heating) *(37)* can be evaluated in a similar fashion and adjusted to yield equivalent intensity of staining by AR-IHC. Nevertheless, there may be some unpredictable correlations between heating condition and certain pH values. In our preliminary studies, when using middle- or high-range pH solutions, the higher the temperature, the better the staining for most nuclear and cytoplasmic antibodies that are tested. In contrast, when using a low-pH solution, overly intense heating (such as in a pressure cooker) may yield a poor result. For example, the strongest intensity of staining for MIB1 or pRB on archival paraffin sections was achieved either with regular microwave heating and low-pH solution or high-pH solution with intense heating (pressure cooker or elongated regular microwave heating method). Low pH with intense heating gave poor results *(86)*.

The test battery thus serves as a rapid screening approach to identify an optimal protocol for each antibody/antigen to be tested. The goal is to establish the "maximal retrieval" level for formalin-masked antigens with a variety of fixation times, in order to standardize immunostaining results. In one study of 14 antibodies using the test battery method, we demonstrated that the strongest intensity of AR-IHC for most antibodies tested was achieved by using either a low-pH buffer solution with regular microwave heating conditions or a high pH buffer with intense heating conditions (such as autoclave heating or microwave heating for elongated heating time) *(86)*. In addition, the use of a test battery may identify some previously "false-negative" AR-IHC staining, as demonstrated by a special AR protocol for thrombospondin-IHC staining *(60,87)*.

## TECHNICAL NOTES AND TROUBLESHOOTING

### Application of AR-IHC in Special Fields

#### Flow Cytometry of Archival Paraffin-Embedded Tissue

The microwave heating AR method has been successfully applied for archival paraffin-embedded tissues to enhance the detection efficiency of flow cytometry *(88–91)*. Two different protocols have been documented:

1. Heating treatment prior to an enzyme digestion method was recommended by Leers et al. *(89)*. They demonstrated that this method allows high-resolution DNA analysis of routinely processed paraffin-embedded tissues with twice the efficiency of recovery of single cells compared with the standard method without heating. In addition, the keratin-positive fraction and the fluorescence intensity were increased.
2. Enzyme digestion followed by heating treatment was adopted to achieve enhancement of flow cytometry on paraffin-embedded tissue sections, as exemplified by Redkar and Krishan *(91)*. They demonstrated a successful result for flow cytometric analysis of estrogen and progesterone receptors in formalin-fixed, paraffin-embedded human breast cancers. However, the heating time must be adjusted to obtain reasonable DNA histograms when using their protocol.

#### Diagnostic Cytology

Boon et al. *(92)* demonstrated that the microwave AR technique could be applied to cervical smears initially stained by the Papanicolaou method, with satisfactory results for MIB1 nuclear staining, in order to screen the positive diagnostic cells using the PAPNET computer system.

#### Plastic (Epoxy, Epon, or Methyl Methacrylate Resin)-Embedded Tissue Thin Sections

Numerous articles have been published since Suurmeijer and Boon *(16)* documented the application of AR in plastic tissue sections in 1993. AR-IHC is readily adapted to bone marrow samples, with the advantages of clear background and intense signal for reliable, high-quality IHC. This improved quality of staining is particularly desirable for the demonstration of neoplastic cells in regenerative marrow after chemotherapy, as well as the detection of residual disease after treatment. Brorson *(93)* described a comparative study to demonstrate how AR affects the yield of immunogold labeling on epoxy sections embedded with different amounts of accelerator, leading to the conclusion that the combination of an increased amount of accelerator during tissue processing for epoxy embedding, together with the AR heating method, provides enhancement of IHC staining on epoxy sections. In our experience, epon-embedded tissue sections benefit from a "de-epon" procedure using NaOH-ethanol or methanol after AR heating treatment. Under these conditions, a shorter heating time and/or lower heating temperature may be possible for some antibodies tested, resulting in superior morphologic detail *(62)*.

Recently, Hand and Church *(94)* documented a successful AR protocol using a pressure cooker for 3 min at 121°C for methyl methacrylate-embedded semithin sections (2 μm) of tonsil. For better immunostaining of CD3, they pretreated slides with 0.1% trypsin at 37°C for 20 min, followed by a heating AR protocol in the pressure cooker.

*Formalin-Fixed, Acid-decalcified Tissues*

AR-IHC is also suitable for formalin-fixed, acid decalcified tissues *(95–98)*.

*Immunofluorescence*

D'Ambra-Cabry et al. *(99)* reported the use of AR as a potential option in retrospective studies of skin diseases such as bullous pemphigoid and eosinophilic dermatitis by direct immunofluorescence, based on a comparative study of fresh frozen tissue and archival paraffin-embedded tissue. Fariss et al. *(100)* reported excellent immunolocalization of the tissue inhibitor of metalloproteinases-3 (TIMP-3) in formalin-paraffin sections of retina/choroid of adult human eyes using AR with sodium phosphate/citrate buffer, pH 3.5, and immunofluorescence. However, in spite of these successes, two major issues must be solved before AR immunofluorescence can be performed routinely on archival tissues: 1) the sensitivity obtained by AR immunofluorescence on fixed paraffin sections is lower than that of frozen section; and 2) possible altered AR immunofluorescence staining patterns may be found. For example, Al-Rifai et al. *(101)* demonstrated spurious intercellular staining within the epidermis, which was not found in frozen section, when combining heating AR and 0.3% trypsin digestion. There are preliminary indications that improved results may be obtained by combining AR with low antibody concentration and a signal amplification system *(102)*.

In Situ *Hybridization*

Following earlier studies concerning application of high-temperature pretreatment for *in situ* hybridization (ISH) *(103–105)*, Oliver et al. *(106)* conducted a quantitative comparison of several pretreatment regimens, including microwave heating, autoclave heating at 100°C or above, conventional heating at 90°C, and proteinase K treatment for ISH. The sensitivity of these pretreatments was evaluated by densitometric analysis of the ISH signal on autoradiographs, based on the same experimental conditions. The strongest ISH signal was achieved by the microwave heating RNA retrieval procedure, which showed identical results for frozen and paraffin sections. Enzymatic digestion yielded 50% intensity compared with frozen sections. Both the autoclave and heating at 90°C yielded a weaker signal, around 10% compared with the frozen section. In this study the protocol of microwave heating pretreatment involved immersing sections in 10 m*M* citrate buffer, pH 6.0, and microwaving at full power in an 800-W microwave oven for three 5-min periods, with washes in diethylpyrocarbonate (DEPC)-treated water.

*Extraction of Protein*

Ikeda et al. *(107)* developed a protocol for extraction of diagnostically useful proteins from formalin-fixed, paraffin-embedded tissue sections. By heating the deparaffinized tissue sections at 100°C for 20 min before incubation at 60°C for 2 h, in RIPA solution, 121.5 μg of protein may be extracted from 5 mm$^2$ × 50 μm-thick tissue. They have successfully detected a variety of membrane-bound, cytosolic and nuclear proteins in archival paraffin-embedded tissues to obtain valuable information for molecular biology.

## NONSPECIFIC STAINING AFTER AR TREATMENT

Most studies have reported satisfactory results without false positivity for the majority of antibodies tested *(14,16–19,21,25,51,55,58,60–62,75)*. Recently, Baas et al. *(108)* reported that a potential false-positive result might be obtained after extreme AR treatment when using the monoclonal antibody DO7 (anti-p53), and there have been sporadic reports of unexpected staining problems with other antibodies. However, further investigation is needed, correlating AR findings with both frozen section studies and molecular assays. The current consensus is that the immunolocalization of p53 in paraffin sections after AR treatment is comparable with the pattern obtained in frozen sections, as evidenced by a number of studies demonstrating that AR pretreatment significantly increased the intensity of immunostaining using the antibody DO-7 but did not alter the pattern of immunoreactivity for p53 *(109–111)*. Binks et al. *(109)* compared the staining of DO-7 obtained with and without AR, comparing paraffin sections and direct smears of specimens from pulmonary non-small cell carcinomas; they were not able to find evidence of false positivity.

Nevertheless, some studies have reported spurious nonspecific staining using a broad panel of antibodies, particularly with different heating conditions and/or different AR solutions or pH values. For example, weak positive nuclear staining was found when using a low-pH AR solution for monoclonal antibody to MIB1 *(79)*. Sebenik and Wieczorek *(112)* attributed nonspecific nuclear staining after AR to the secondary antibody, suggesting that retitration of the second antibody may be necessary if nonspecific staining occurs when using AR-IHC. By contrast, Wieczorek et al. *(113)* concluded that the problem resides with the primary antibody, based on a study employing three AR solutions, 1% zinc sulfate (pH 4.9), 0.01 *M* citrate buffer (pH 6.0), and 0.01 *M* Tris buffer (pH 9.0), with careful control groups to exclude potential false-positive staining caused by endogenous biotin or electrostatic binding of immunoglobulins. These authors concluded that nuclear positivity is "nonspecific" and is easily eliminated by retitration of primary antibody to a higher dilution. In another study, focal keratin staining was observed in archival paraffin sections of malignant melanoma and plasma cell tumors when using heating AR, possibly resulting from detection of low levels of keratins in these tumors *(114)*.

One potential cause of increased background after AR heating may be the unmasking of endogenous biotin, which does account for some positive staining in frozen tissue sections, particularly for liver, kidney, adrenal cortex, and thyroid tissues *(115,116)*. In this instance, the cytoplasmic biotin reaction appears as a fine granular staining pattern in the cytoplasm. Kashima et al. *(116)* found a high incidence (93/208) of this form of positive cytoplasmic biotin staining in thyroid lesions, especially in papillary carcinoma. This effect can be avoided by using a routine blocking procedure or using a detection system free of avidin-biotin complex. It is important to keep in mind that any background staining that may be observed in frozen tissue sections may also be found in microwave heated archival paraffin sections; the use of appropriate controls is helpful in identifying this problem.

Care must be taken when testing new antibodies using different AR protocols, as emphasized recently by Swanson *(73)*. Indeed variable patterns of immunostaining have been reported when using different AR protocols *(117)*. The incorporation of frozen sections into a test battery, employed to select an optimal AR protocol, is extremely helpful in identifying any false positivity or unexpected change of staining pattern, as exemplified by the monoclonal antibody to thrombospondin (TSP) immuno-localization in prostate and bladder mentioned above.

## DETACHED TISSUE SECTIONS AFTER AR HEATING PROCESS

Pretreated slides, either commercially charged or poly-L-lysine coated, are recommended prior to heating treatment in order to protect tissue sections from detachment. We found that thinner tissue sections adhere more firmly than thicker sections but that different types of tissue vary in adhesiveness to glass slides. In our experience, tissues with more fat detach more readily. Baking the paraffin sections in a 60°C oven overnight appears to increase adhesion. In a few difficult cases, the use of heating at lower temperature (90°C) for extended heating times may reduce tissue damage and detachment, while achieving a similar AR result.

To improve adhesion, a postadhesive method was designed by Pateraki and Kontogeorgos *(118),* for stored paraffin sections that were already mounted on regular plain slides (without treatment for adhesion), but required heating by AR pretreatment. The reported protocol is as follows: place paraffin sections in an oven at 58°C for 3 h. After two to three dips in acetone, immerse the slides in a Coplin jar containing Vectabond glue (Vector, Burlingame, CA) for 30 min, cover the jar, follow with three washes in distilled water, and dry in oven overnight. To ensure safety, the process of using acetone and Vectabond glue should be done under a hood. We tried this postadhesion method in our laboratory recently and achieved satisfactory results. In our experience, for some difficult cases, commercially charged slides may be additionally coated with chemicals such as poly-L-lysine, 3-aminopropyltriethoxysilane (APES), or albumin with gelatin in order to enhance adhesion of tissue sections. Mote et al. *(119)* summarized their experience for maximal tissue retention by using positively charged slides coated with Mayer's albumin adhesive. To reduce background staining due to APES-coated slides when using the immunogold-silver staining method, Krenacs and Krenacs *(120)* found that washing slides in a buffer containing 0.1% detergent after heating treatment may yield clean background and sharp contrast for both paraffin- or resin-embedded tissue sections.

## COUNTERSTAIN AFTER AR-IHC

Hematoxylin used for counterstaining after AR-IHC staining is satisfactory in most situations, except when an AR solution at low pH is used. After heating of the tissue section in low-pH solution, nuclear staining for hematoxylin may be compromised. Since a low-pH AR solution is preferable for some antigens, it is necessary to find a way to correct this acid-induced defect of counterstaining. Recently, we found that dropping 1$N$ NaOH solution on slides for 10 min may improve hematoxylin counterstaining following low-pH AR: otherwise, other nuclear stains should be employed.

## INCONSISTENT RESULTS OF AR-IHC

Although inconsistency is becoming less of a problem in larger laboratories, where many slides are processed routinely every day, irregular results may be expected in both clinical and research laboratories, when technicians or research staff are not familiar with the AR process. Possible causes include the following factors:

1. The AR treatment itself is inconsistent, due to such factors as variable heating temperatures resulting from using different microwave ovens, nonuniform Coplin jars or other containers of different sizes, variability in the total numbers of containers placed in the oven, location of Coplin jars at different sites in the oven, or allowing drying of the slides during the heating process due to loss of AR solution during boiling.
2. The AR solution contains a mixture of chemicals that may undergo unanticipated reactions, a problem that occurs when heating different buffer solutions together at high temperature, or combining various metal salt solutions, particularly calcium salts.
3. The immunohistochemical staining procedure is performed incorrectly. One particular pitfall is that following successful AR, it is frequently necessary to retitrate primary and secondary antibodies to higher dilutions reflective of the greater availability of antigen (effectively gives increased sensitivity). Sections that previously were "negative" may now stain positively.
4. A decrease in the intensity of IHC may occur in cut sections on slides stored for protracted periods. This phenomenon is not well understood and is not uniform or predictable for different antigens or different storage tissues or conditions. We have compared our own experience for common antibodies such as p53, and p27 with that in the literature *(121–124)*. We performed IHC staining of 23 cases of bladder cancer showing demonstrable *p53* gene mutations by molecular analysis, using stored slides and freshly cut paraffin sections. Of the 23 cases, p53-positive immunostaining was found in 9/23 (39%) and 18/23 (78%) of stored and nonstored slides, respectively, when AR was not used. When AR was employed on additional slides from these cases, p53-positive staining was significantly increased, to 18/23 (78%) and 19/23 (83%) of stored and nonstored slides, respectively (Stein et al., unpublished data).

   In short, AR "levels the playing field" for storage conditions, just as it does for fixation conditions. Similar conclusions have been documented in the literature *(121,123)*. For example, Grabau et al. *(121)* carefully compared the influence of tumor tissue paraffin sections stored at −80°C, 4°C, or 20°C, or unmounted at 4°C for 3 years using eight antibodies for IHC staining. They concluded that the intensity of IHC was decreased with increasing storage temperature but that the immunoreactivity may be restored by an efficient AR protocol. In our hands, monoclonal antibody to p27 (clone DCS-72.F6; Neomarkers, Fremont, CA) requires fresh-cut tissue sections to achieve the strongest intensity of staining even with our regular AR protocol *(125)*. Preliminary studies suggest that a combination of a heating AR protocol with signal amplification may be needed to reach a satisfactory intensity of staining for p27.

5. Fixatives are different. Several formalin substitutes have been used in pathology, in attempts to preserve antigenicity better for IHC. An AR protocol that is optimal for routinely formalin-fixed, paraffin-embedded tissue sections may be invalid for tissues exposed to other fixatives. Prento and Lyon *(126)* compared the preformance of six commercial fixatives, proposed as formalin substitutes, with that of formalin and concluded that the best immunostaining was achieved by combining formalin fixation with AR, and further, that none of the six commercial formalin substitutes was adequate for preservation of morphology and antigenicity in a broad sense. Zhang et al. *(127)* compared the efficiency of AR-IHC between tissues fixed in formalin or in Lillie's fixative (formalin-alcohol-acetic acid). They demonstrated that 14 of the 15 antibodies tested showed decreased immunoreactivity in Lillie-fixed tissue, particularly for nuclear immunostaining. Multitissue blocks made of alcohol-stored tissues should not be expected to perform in a similar fashion to formalin-fixed tissues and may yield different IHC results from those previously documented for routine formalin-paraffin sections *(50,51)*. Immunostaining for p53 following microwave heating of alcohol-fixed tissue sections has been reported to produce false-positive or unexpected staining patterns in some tissue components *(128)*. Note that improperly fixed tissues, partially exposed to formalin, due to large size or inadequate formalin solution, are in essence "fixed" in alcohol during later stages of processing; not surprisingly, such tissues show irregular results by AR.

## LOW-TEMPERATURE HEATING METHOD

As previously described, the heating condition is the most important factor that influences the effectiveness of the AR technique. Studies in our laboratories with multiple antigen/antibody combinations repeatedly confirmed the existence of a reverse correlation between heating temperature ($T$) and heating time ($t$), or, to state it another way, the effectiveness of AR heating treatment equals the product of $T \times t$. In our laboratory, very few antigens, such as p16 (unpublished data), collagen type III, yield optimal results at lower temperatures. This also is the experience of others *(129–132)*. Igarashi et al. *(30)* found that overnight heating at a lower temperature (60°C) was better for muscle actin (HHF35) and smooth-muscle actin (CCG 7). Elias and Margiotta *(130)* reported on a low-temperature AR method using a water bath at 80°C for 2 h in citrate buffer pH 6.0, claiming an effectiveness equivalent to the standard method for antibodies to steroid hormone receptor. Recently, Koopal and colleagues *(131)* reported that low-temperature (80°C) overnight heating using Tris-HCl buffer, pH 9, with 16 antibodies produced satisfactory results. Other investigators have reported that, for ER, superior results are achieved by low-temperature heating (60°C overnight) with acetic acid at pH 7–8, or 0.2 *M* boric acid at pH 7 *(132)*. It has been proposed, and is generally agreed, that lower temperatures may have advantages for better preservation of morphology and may reduce the risk of detachment of sections from the slides *(129–132)*. In any event, if temperatures are reduced, then the heating time must equally be extended to achieve satisfactory results, based on the principle of $T \times t$.

## COMBINING AR AND OTHER "UNMASKING" METHODS

Although the heating AR technique enhances the staining of many antibodies for routinely processed paraffin sections, some antigens, particularly cell surface markers, are still not readily demonstrable in archival tissues. If successful immunostaining is not obtained using regular heating methods, lower or higher temperatures and extended heating periods may be employed, or AR may be combined with enzymatic digestion or signal amplification methods. Malisius et al. *(133)* reported combining the use of heating AR and a tyramide amplification method for immunostaining of CD2, CD3, CD4, and CD5 in archival paraffin sections. Subsequently, Butmarc et al. *(134)* adopted a combined AR-IHC protocol with biotinylated tyramine enhancement for monoclonal antibody to CD5 (Leu-1, clone L17F12; Becton Dickinson, San Jose, CA) and obtained satisfactory results in 75% of cases of chronic lymphocytic leukemia (CLL), 86% of mantle cell lymphomas, and 100% of T-cell lymphomas. Successful staining of paraffin sections for CD5 has now been repeated by several authors *(61,135)*.

Kawai and Osamura *(136)* studied the efficiency of AR and tyramide signal amplification on archival praffin sections and demonstrated that although amplification alone may enhance the intensity for some antibodies, it may also be significantly improved by the AR heating method. For some antibodies, such as MIB1, amplification alone will not suffice but must be preceded by AR. Other strategies remain to be explored, exemplified by the sequential use of reducing agents (mercaptoethanol or sodium hydrosulfite) followed by the AR heating procedure, an approach that restored vimentin reactivity in some spindle cell tumors that otherwise were falsely negative *(137)*.

## ACKNOWLEDGMENTS

Parts of this chapter were adapted with permission from refs. *51* and *62*.

## REFERENCES

1. Coons, A. H., Creech, H. J., and Jones, R. N. (1941) Immunological properties of an antibody containing a fluorescent group. *Proc. Soc. Exp. Biol. Med.* **47**, 200–202.
2. Taylor, C. R. and Kledzik, G. (1981) Immunohistologic techniques in surgical pathology. A spectrum of new special strains. *Hum. Pathol.* **12**, 590–596.
3. Taylor, C. R. and Burns. J. (1974) The demonstration of plasma cells and other immuno-globulin containing cells in formalin-fixed, paraffin-embedded tissues using peroxidase labelled antibody. *J. Clin. Pathol.* **27**, 14–20.
4. Taylor, C. R. (1979) Immunohistologic studies of lymphomas: new methodology yields new information and poses new problems. *J. Histochem. Cytochem.* **27**, 1189–1191.
5. Taylor, C. R. (1980) Immunohistologic studies of lymphoma: past, present and future. *J. Histochem. Cytochem.* **28**, 777–787.
6. Pinkus, G. S. (1982) Diagnostic immunocytochemistry of paraffin-embedded tissues. *Hum. Pathol.* **13**, 411–415.
7. Elias, J. M. (1990) *Immunohistopathology: A Practical Approach to Diagnosis.* 1st ed. ASCP Press, Chicago, pp. 1–9.
8. Bhan, A. K. (1995) *Immunoperoxidase*, 2nd ed. Raven, New York, pp. 711–723.
9. DeLellis, R. A. (1988) *Advances in Immunohistochemistry.* Raven, New York, pp. 1–45.
10. Leong, A. S.-Y., Milios, J., and Duncis, C. G. (1988) Antigen preservation in microwave-irradiated tissues: a comparison with formaldehyde fixation. *J. Pathol.* **156**, 275–282.
11. Fraenkel-Conrat, H., Brandon, B. A., and Olcott, H. S. (1947) The reaction of formaldehyde with proteins. IV. Participation of indole groups. *J. Biol. Chem.* **168**, 99–118.
12. Fraenkel-Conrat, H. and Olcott, H. S. (1948) Reaction of formaldehyde with proteins. VI.

Cross-linking of amino groups with phenol, imidazole, or indole groups. *J. Biol. Chem.* **174**, 827–843.

13. Fraenkel-Conrat, H. and Olcott, H. S. (1948) The reaction of formaldehyde with proteins. V. Cross-linking between amino and primary amide or guanidyl groups. *J. Am. Chem. Soc.* **70**, 2673–2684.

14. Shi, S. R., Key, M. E., and Kalra, K. L. (1991) Antigen retrieval in formalin-fixed, paraffin-embedded tissues: an enhancement method for immunohistochemical staining based on microwave oven heating of tissue sections. *J. Histochem. Cytochem.* **39**, 741–748.

15. Shi, S. R., Cote, C., Kalra, K. L., Taylor, C. R., and Tandon, A. K. (1992) A technique for retrieving antigens in formalin-fixed, routinely acid-decalcified, celloidin-embedded human temporal bone sections for immunohistochemistry. *J. Histochem. Cytochem.* **40**, 787–792.

16. Suurmeijer, A. J. and Boon, M. E. (1993) Notes on the application of microwaves for antigen retrieval in paraffin and plastic tissue sections. *Eur. J. Morphol.* **31**, 144–150.

17. Cattoretti, G., Pileri, S., Parravicini, C., et al. (1993) Antigen unmasking on formalin-fixed, paraffin-embedded tissue sections. *J. Pathol.* **171**, 83–98.

18. Gown, A. M., de Wever, N., and Battifora, H. (1993) Microwave-based antigenic unmasking. A revolutionary new technique for routine immunohistochemistry. *Appl. Immunohistochem.* **1**, 256–266.

19. Leong, A. S.-Y. and Milios, J. (1993) An assessment of the efficacy of the microwave antigen-retrieval procedure on a range of tissue antigens. *Appl. Immunohistochem.* **1**, 267–274.

20. Swanson, P. E. (1993) Methodologic standardization in immunohistochemistry. A doorway opens [Editorial]. *Appl. Immunohistochem.* **1**, 229–231.

21. Dookhan, D. B., Kovatich, A. J., and Miettinen, M. (1993) Non-enzymatic antigen retrieval in immunohistochemistry. Comparison between different antigen retrieval modalities and proteolytic digestion. *Appl. Immunohistochem.* **1**, 149–155.

22. Taylor, C. R., Shi, S. R., Chaiwun, B., Young, L., Imam, S. A., and Cote, R. J. (1994) Strategies for improving the immunohistochemical staining of various intranuclear prognostic markers in formalin-paraffin sections: androgen receptor, estrogen receptor, progesterone receptor, p53 protein, proliferating cell nuclear antigen, and Ki-67 antigen revealed by antigen retrieval techniques. [See comments]. *Hum. Pathol.* **25**, 263–270.

23. Cuevas, E. C., Bateman, A. C., Wilkins, B. S., et al. (1994) Microwave antigen retrieval in immunocytochemistry: a study of 80 antibodies. *J. Clin. Pathol.* **47**, 448–452.

24. Gu, J., Forte, M., Xenachis, C., Tarazona, N., Windsor, J., and Santoian, E. C. (1994) Immunohistochemical demonstration of PCNA and Ki 67 positivities in stimulated myochariocytes indicates dividing potential for cardiac muscle cells in adult heart. *Cell Vision* **1**, 91–92.

25. Brown, R. W. and Chirala, R. (1995) Utility of microwave-citrate antigen retrieval in diagnostic immunohistochemistry. *Mod. Pathol.* **8**, 515–520.

26. Cote, R. J., Shi, S.-R., and Taylor, C. R. (1998) Development and standardization of antigen retrieval (AR) for immunohistochemistry (IHC): current issues. *J. Histochem. Cytochem.* **46**, A13.

27. Elias, J. M. (1998) Technical considerations in immunocytochemistry. *Cell Vision* **5**, 35–36.

28. Shin, R.-W., Iwaki, T., Kitamoto, T., and Tateishi, J. (1991) Hydrated autoclave pretreatment enhances TAU immunoreactivity in formalin-fixed normal and Alzheimer's disease brain tissues. *Lab. Invest.* **64**, 693–702.

29. Bankfalvi, A., Navabi, H., Bier, B., Bocker, W., Jasani, B., and Schmid, K. W. (1994) Wet autoclave pretreatment for antigen retrieval in diagnostic immunohistochemistry. *J. Pathol.* **174**, 223–228.

30. Igarashi, H., Sugimura, H., Maruyama, K., et al. (1994) Alteration of immunoreactivity

by hydrated autoclaving, microwave treatment, and simple heating of paraffin-embedded tissue sections. *APMIS* **102**, 295–307.

31. Umemura, S., Kawai, K., Osamura, R. Y., and Tsutsumi, Y. (1995) Antigen retrieval for bcl-2 protein in formalin-fixed, paraffin-embedded sections. *Pathol. Int.* **45**, 103–107.

32. Pons, C., Costa, I., von Schilling, B., Matias-Guiu, X., and Prat, J. (1995) Antigen retrieval by wet autoclaving for p53 immunostaining. *Appl. Immunohistochem.* **3**, 265–267.

33. Norton, A. J., Jordan, S., and Yeomans, P. (1994) Brief, high-temperature heat denaturation (pressure cooking): a simple and effective method of antigen retrieval for routinely processed tissues. *J. Pathol.* **173**, 371–379.

34. Miller, R. T. and Estran, C. (1995) Heat-induced epitope retrieval with a pressure cooker—suggestions for optimal use. *Appl. Immunohistochem.* **3**, 190–193.

35. Pasha, T., Montone, K. T., and Tomaszewski, J. E. (1995) Nuclear antigen retrieval utilizing steam heat. *Lab. Invest.* **72**, 167A.

36. Pertschuk, L. P., Kim, Y. D., Axiotis, C. A., et al. (1994) Estrogen receptor immunocytochemistry: the promise and the perils. *J. Cell Biochem. Suppl.* **19**, 134–137.

37. Taylor, C. R., Shi, S. R., Chen, C., Young, L., Yang, C., and Cote, R. J. (1996) Comparative study of antigen retrieval heating methods: microwave, microwave and pressure cooker, autoclave, and steamer. *Biotech. Histochem.* **71**, 263–270.

38. Greenwell, A., Foley, J. F., and Maronpot, R. R. (1991) An enhancement method for immunohistochemical staining of proliferating cell nuclear antigen in archival rodent tissues. *Cancer Lett.* **59**, 251–256.

39. Suurmeijer, A. J. (1994) Optimizing immunohistochemistry in diagnostic tumor pathology with antigen retrieval. *Eur. J. Morphol.* **32**, 325–330.

40. Siitonen, S. M., Kallioniemi, O. P., and Isola, J. J. (1993) Proliferating cell nuclear antigen immunohistochemistry using monoclonal antibody 19A2 and a new antigen retrieval technique has prognostic impact in archival paraffin-embedded node-negative breast cancer. *Am. J. Pathol.* **142**, 1081–1089.

41. Pavelic, Z. P., Portugal, L. G., Gootee, M. J., et al. (1993) Retrieval of p53 protein in paraffin-embedded head and neck tumor tissues. *Arch. Otolaryngol. Head Neck Surg.* **119**, 1206–1209.

42. Merz, H., Rickers, O., Schrimel, S., Orscheschek, K., and Feller, A. C. (1993) Constant detection of surface and cytoplasmic immunoglobulin heavy and light chain expression in formalin-fixed and paraffin-embedded material. *J. Pathol.* **170**, 257–264.

43. Lucassen, P. J., Ravid, R., Gonatas, N. K., and Swaab, D. F. (1993) Activation of the human supraoptic and paraventricular nucleus neurons with aging and in Alzheimer's disease as judged from increasing size of the Golgi apparatus. *Brain Res.* **632**, 105–113.

44. Evers, P. and Uylings, H. B. (1994) Effects of microwave pretreatment on immunocytochemical staining of vibratome sections and tissue blocks of human cerebral cortex stored in formaldehyde fixative for long periods. *J. Neurosci. Methods* **55**, 163–172.

45. Spires, S. E., Banks, E. R., Davey, D. D., Jennings, C. D., Wood, D. P., Jr., and Cibull, M. L. (1994) Proliferating cell nuclear antigen in prostatic adenocarcinoma: correlation with established prognostic indicators. *Urology* **43**, 660–666.

46. Shin, H. J. C., Shin, D. M., and Ro, J. Y. (1994) Optimization of proliferating cell nuclear antigen immunohistochemical staining by microwave heating in zinc sulfate solution. *Mod. Pathol.* **7**, 242–248.

47. Mintze, K., Macon, N., Gould, K. E., and Sandusky, G. E. (1995) Optimization of proliferating cell nuclear antigen (PCNA) immunohistochemical staining: a comparison of methods using three commercial antibodies, various fixation times, and antigen retrieval solution. *J. Histotechnol.* **18**, 25–30.

48. Gerasimov, G., Bronstein, M., Troshina, K., et al. (1995) Nuclear p53 immunoreactivity

in papillary thyroid cancers is associated with two established indicators of poor prognosis. *Exp. Mol. Pathol.* **62**, 52–62.

49. Stieber, A., Mourelatos, Z., and Gonatas, N. K. (1996) In Alzheimer's disease the Golgi apparatus of apopulation of neurons without neurofibrillary tangles is fragmented and atrophic. *Am. J. Pathol.* **148**, 415–426.

50. Momose, H., Mehta, P., and Battifora, H. (1993) Antigen retrieval by microwave irradiation in lead thiocyanate. *Appl. Immunohistochem.* **1**, 77–82.

51. Shi, S.-R., Cote, R. J., Young, L. L., and Taylor, C. R. (1997) Antigen retrieval immunohistochemistry: practice and development. *J. Histotechnol.* **20**, 145–154.

52. Cattoretti, G., Becker, M. H. G., Key, G., et al. (1992) Monoclonal antibodies against recombinant parts of the Ki-67 antigen (MIB 1 and MIB 3) detect proliferating cells in microwave-processed formalin-fixed paraffin sections. *J. Pathol.* **168**, 357–363.

53. Swanson, P. E. (1994) Microwave antigen retrieval in citrate buffer. *Lab. Med.* **25**, 520–522.

54. Hopwood, D. (1994) Epitope retrieval—survey and prospect. *Eur. J. Morphol.* **32**, 317–324.

55. Shi, S.-R., Gu, J., Kalra, K. L., Chen, T., Cote, R. J., and Taylor, C. R. (1995) Antigen retrieval technique: a novel approach to immunohistochemistry on routinely processed tissue sections. *Cell Vision* **2**, 6–22.

56. Boon, M. E. and Kok, L. P. (1995) Breakthrough in pathology due to antigen retrieval. *Mal. J. Med. Lab. Sci.* **12**, 1–9.

57. Cattoretti, G. and Suurmeijer, A. J. H. (1995) Antigen unmasking on formalin-fixed paraffin-embedded tissues using microwaves: a review. *Adv. Anat. Pathol.* **2**, 2–9.

58. Taylor, C. R., Shi, S.-R., and Cote, R. J. (1996) Antigen retrieval for immunohistochemistry. Status and need for greater standardization. *Appl. Immunohistochem.* **4**, 144–166.

59. Werner, M., Von Wasielewski, R., and Komminoth, P. (1996) Antigen retrieval, signal amplification and intensification in immunohistochemistry. *Histochem. Cell Biol.* **105**, 253–260.

60. Shi, S. R., Cote, R. J., and Taylor, C. R. (1997) Antigen retrieval immunohistochemistry: past, present, and future. *J. Histochem. Cytochem.* **45**, 327–343.

61. Shi, Y., Li, G.-D., and Liu, W.-P. (1997) Recent advances of the antigen retrieval technique. Linchuang Yu Shiyan Binglixue Zazhi (*J. Clin. Exp. Pathol.*). **13**, 265–267.

62. Shi, S. R., Cote, R. J., and Taylor, C. R. (1998) Antigen retrieval immunohistochemistry used for routinely processed celloidin-embedded human temporal bone sections: standardization and development. *Auris Nasus Larynx* **25**, 425–443.

63. McNicol, A. M. and Richmond, J. A. (1998) Optimizing immunohistochemistry: antigen retrieval and signal amplification. *Histopathology* **32**, 97–103.

64. Evers, P., Uylings, H. B., and Suurmeijer, A. J. (1998) Antigen retrieval in formaldehyde-fixed human brain tissue. *Methods* **15**, 133–140.

65. Shi, S. R., Cote, R. J., Young, L., Imam, S. A., and Taylor, C. R. (1996) Use of pH 9.5 Tris-HCl buffer containing 5% urea for antigen retrieval immunohistochemistry. *Biotech. Histochem.* **71**, 190–196.

66. Shi, S. R., Chaiwun, B., Young, L., Cote, R. J., and Taylor, C. R. (1993) Antigen retrieval technique utilizing citrate buffer or urea solution for immunohistochemical demonstration of androgen receptor in formalin- fixed paraffin sections. *J. Histochem. Cytochem.* **41**, 1599–1604.

67. Shi, S. R., Chaiwun, B., Young, L., Imam, A., Cote, R. J., and Taylor, C. R. (1994) Antigen retrieval using pH 3.5 glycine-HCl buffer or urea solution for immunohistochemical localization of Ki-67. *Biotech. Histochem.* **69**, 213–215.

68. Imam, S. A., Young, L., Chaiwun, B., and Taylor, C. R. (1995) Comparison of two microwave based antigen-retrieval solutions in unmasking epitopes in formalin-fixed tissue for immunostaining. *Anticancer Res.* **15**, 1153–1158.

69. Hazelbag, H. M., van den Broek, L. J., van Dorst, E. B., Offerhaus, G. J., Fleuren, G. J., and Hogendoorn, P. C. (1995) Immunostaining of chain-specific keratins on formalin-fixed, paraffin-embedded tissues: a comparison of various antigen retrieval systems using microwave heating and proteolytic pre-treatments. *J. Histochem. Cytochem.* **43**, 429–437.

70. Kwaspen, F., Smedts, F., Blom, J., et al. (1995) Periodic acid as a nonenzymatic enhancement technique for the detection of cytokeratin immunoreactivity in routinely processed carcinomas. *Appl. Immunohistochem.* **3**, 54–63.

71. Pileri, S. A., Roncador, G., Ceccarelli, C., et al. (1997) Antigen retrieval techniques in immunohistochemistry: comparison of different methods. *J. Pathol.* **183**, 116–123.

72. van den Berg, F. M., Baas, I. O., Polak, M. M., and Offerhaus, G. J. A. (1993) Detection of p53 overexpression in routinely paraffin-embedded tissue of human carcinomas using a novel target unmasking fluid. *Am. J. Pathol.* **142**, 381–385.

73. Swanson, P. E. (1997) HIERanarchy: the state of the art in immunohistochemistry. *Am. J. Clin. Pathol.* **107**, 139–140.

74. Taylor, C. R. and Cote, R. J. (1994) *Immunomicroscopy: A Diagnostic Tool for the Surgical Pathologist*, 2nd ed. W. B. Saunders, Philadelphia.

75. Taylor, C. R. (1994) The current role of immunohistochemistry in diagnostic pathology. *Adv. Pathol. Lab. Med.* **7**, 59–105.

76. Katoh, A. and Breier, S. (1994) Nonspecific antigen retrieval solutions. *J. Histotechnol.* **17**, 378.

77. O'Reilly, P. E., Raab, S. S., Niemann, T. H., Rodgers, J. R., and Robinson, R. A. (1997) p53, proliferating cell nuclear antigen, and Ki-67 expression in extrauterine leiomyosarcomas. *Mod. Pathol.* **10**, 91–97.

78. Shi, S.-R., Cote, R. J., Chaiwun, B., et al. (1998) Standardization of immunohistochemistry based on antigen retrieval technique for routine formalin-fixed tissue sections. *Appl. Immunohistochem.* **6**, 89–96.

79. Shi, S. R., Imam, S. A., Young, L., Cote, R. J., and Taylor, C. R. (1995) Antigen retrieval immunohistochemistry under the influence of pH monoclonal antibodies. *J. Histochem. Cytochem.* **43**, 193–201.

80. Evers, P. and Uylings, H. B. (1994) Microwave-stimulated antigen retrieval is pH and temperature dependent. *J. Histochem. Cytochem.* **42**, 1555–1563.

81. Evers, P. and Uylings, H. B. (1997) An optimal antigen retrieval method suitable for different antibodies on human brain tissue stored for several years in formaldehyde fixative. *J. Neurosci. Methods.* **72**, 197–207.

82. Taylor, C. R., Shi, S.-R., Chaiwun, B., et al. (1994) Correspondence. Standardization and reproducibility in diagnostic immunohistochemistry. *Hum. Pathol.* **25**, 1107–1109.

83. Jones, M. D., Banks, P. M., and Caron, B. L. (1981) Transition metal salts as adjuncts to formalin for tissue fixation. *Lab. Invest.* **44**, 32A.

84. Shi, S. R., Tandon, A. K., Cote, C., and Kalra, K. L. (1992) S-100 protein in human inner ear: use of a novel immunohistochemical technique on routinely processed, celloidin-embedded human temporal bone sections. *Laryngoscope* **102**, 734–738.

85. Shi, S. R., Tandon, A. K., Haussmann, R. R., Kalra, K. L., and Taylor, C. R. (1993) Immunohistochemical study of intermediate filament proteins on routinely processed, celloidin-embedded human temporal bone sections by using a new technique for antigen retrieval. *Acta Otolaryngol.* (Stockh.) **113**, 48–54.

86. Shi, S. R., Cote, R. J., Yang, C., et al. (1996) Development of an optimal protocol for antigen retrieval: a 'test battery' approach exemplified with reference to the staining of retinoblastoma protein (pRB) in formalin-fixed paraffin sections. *J. Pathol.* **179**, 347–352.

87. Grossfeld, G. D., Shi, S. R., Ginsberg, D. A., et al. (1996) Immunohistochemical detection of thrombospondin-1 in formalin-fixed, paraffin-embedded tissue. *J. Histochem. Cytochem.* **44**, 761–766.

88. Lan, H. Y., Hutchinson, P., Tesch, G. H., Mu, W., and Atkins, R. C. (1996) A novel method of microwave treatment for detection of cytoplasmic and nuclear antigens by flow cytometry. *J. Immunol. Methods* **190**, 1–10.

89. Leers, M. P., Schutte, B., Theunissen, P. H., Raemaekers, F. C., and Nap, M. (1999) Heat pretreatment increases resolution in DNA flow cytometry of paraffin-embedded tumor tissue. *Cytometry* **35**, 260–266.

90. Overton, W. R., Catalano, E., and McCoy, J. P. (1996) Method to make paraffin-embedded breast and lymph tissue mimic fresh tissue in DNA analysis. *Cytometry* **26**, 166–171.

91. Redkar, A. A. and Krishan, A. (1999) Flow cytometric analysis of estrogen, progesterone receptor expression and DNA content in formalin-fixed, paraffin-embedded human breast tumors. *Cytometry* **38**, 61–69.

92. Boon, M. E., Beck, S., and Kok, L. P. (1995) Semiautomatic PAPNET analysis of proliferating (MiB-1-positive) cells in cervical cytology and histology. *Diagn. Cytopathol.* **13**, 423–428.

93. Brorson, S. H. (1998) The combination of high-accelerator epoxy resin and antigen retrieval to obtain more intense immunolabeling on epoxy sections than on LR-White sections for large proteins. *Micron* **29**, 89–95.

94. Hand, N. M. and Church, R. J. (1998) Superheating using pressure cooking: its use and application in unmasking antigens embedded in methyl methacrylate. *J. Histotechnol.* **21**, 231–236.

95. Barou, O., Laroche, N., Palle, S., Alexandre, C., and Lafage-Proust, M. H. (1997) Pre-osteoblastic proliferation assessed with BrdU in undecalcified, epon-embedded adult rat trabecular bone. *J. Histochem. Cytochem.* **45**, 1189–1195.

96. Blythe, D., Hand, N. M., Jackson, P., Barrans, S. L., Bradbury, R. D., and Jack, A. S. (1997) Use of methyl methacrylate resin for embedding bone marrow trephine biopsy specimens. *J. Clin. Pathol.* **50**, 45–49.

97. Krenacs, T. and Rosendaal, M. (1998) Connexin43 gap junctions in normal, regenerating, and cultured mouse bone marrow and in human leukemias: their possible involvement in blood formation [See comments]. *Am. J. Pathol.* **152**, 993–1004.

98. Sormunen, R. and Leong, A. S.-Y. (1998) Microwave-stimulated antigen retrieval for immunohistology and immunoelectron microscopy of resin-embedded sections. *Appl. Immunohistochem.* **6**, 234–237.

99. D'Ambra-Cabry, K., Deng, D. H., Flynn, K. L., Magee, K. L., and Deng, J. S. (1995) Antigen retrieval in immunofluorescent testing of bullous pemphigoid [See comments]. *Am. J. Dermatopathol.* **17**, 560–563.

100. Fariss, R. N., Apte, S. S., Olsen, B. R., Iwata, K., and Milam, A. H. (1997) Tissue inhibitor of metalloproteinases-3 is a component of Bruch's membrane of the eye. *Am. J. Pathol.* **150**, 323–328.

101. Al-Rifai, I., Kanitakis, J., Faure, M., and Claudy, A. (1997) Immunofluorescence diagnosis of bullous dermatoses on formalin-fixed tissue sections after antigen retrieval [Letter; comment]. *Am. J. Dermatopathol.* **19**, 103–105.

102. Merz, H., Malisius, R., Mannweiler, S., et al. (1995) ImmunoMax. A maximized immunohistochemical method for the retrieval and enhancement of hidden antigens. *Lab. Invest.* **73**, 149–156.

103. Lan, H. Y., Mu, W., Ng, Y. Y., Nilolic-Paterson, D. J., and Atkins, R. C (1996) A simple, reliable, and sensitive method for nonradioactive in situ hybridization: use of microwave heating to improve hybridization efficiency and preserve tissue morphology. *J. Histochem. Cytochem.* **44**, 281–287.

104. McMahon, J. and McQuaid, S. (1996) The use of microwave irradiation as a pretreatment to in situ hybridization for the detection of measles virus and chicken anaemia virus in formalin-fixed paraffin-embedded tissue. *Histochem. J.* **28**, 157–164.

105. Sibony, M., Commo, F., Callard, P., and Gasc, J.-M. (1995) Enhancement of mRNA in situ hybridization signal by microwave heating. *Lab. Invest.* **73**, 586–591.

106. Oliver, K. R., Heavens, R. P., and Sirinathsinghji, D. J. S. (1997) Quantitative comparison of pretreatment regimens used to sensitive in situ hybridization using oligonucleotide probes on paraffin-embedded brain tissue. *J. Histochem. Cytochem.* **45**, 1707–1713.

107. Ikeda, K., Monden, T., Kanoh, T., et al. (1998) Extraction and analysis of diagnostically useful proteins from formalin-fixed, paraffin-embedded tissue sections. *J. Histochem. Cytochem.* **46**, 397–404.

108. Baas, I. O., van den Berg, F. M., Mulder, J.-W. R., et al. (1996) Potential false-positive results with antigen enhancement for immunohistochemistry of the p53 gene product in colorectal neoplasms. *J. Pathol.* **178**, 264–267.

109. Binks, S., Clelland, C. A., Ronan, J., and Bell, J. (1997) p53 gene product expression in resected non-small cell carcinoma of the lung, with studies of concurrent cytological preparations and microwave antigen retrieval. *J. Clin. Pathol.* **50**, 320–323.

110. Daidone, M. G., Benini, E., Rao, S., Pilotti, S., and Silvestrini, R. (1998) Fixation time and microwave oven irradiation affect immunocytochemical p53 detection in formalin-fixed paraffin sections. *Appl. Immunohistochem.* **6**, 140–144.

111. McCluggage, G., McBride, H., Maxwell, P., and Bharucha, H. (1997) Immunohistochemical detection of p53 and bcl-2 proteins in neoplastic and non-neoplastic endocervical glandular lesions. *Int. J. Gynecol. Pathol.* **16**, 22–27.

112. Sebenik, M. and Wieczorek, R. (1995) Nonspecific nuclear staining (NNS) after antigen retrieval (AR). *Lab. Invest.* **72**, 168A.

113. Wieczorek, R., Stover, R., and Sebenik, M. (1997) Nonspecific nuclear immunoreactivity after antigen retrieval using acidic and basic solutions. *J. Histotechnol.* **20**, 139–143.

114. Guiter, G. E., Kwan, P. W., and DeLellis, R. A. (1995) Unwanted tissue immunoreactivities following microwave antigen retrieval: a critical analysis. *Lab. Invest.* **72**, 165A.

115. Bussolati, G., Gugliotta, P., Volante, M., Pace, M., and Papotti, M. (1997) Retrieved endogenous biotin: a novel marker and a potential pitfall in diagnostic immunohistochemistry [See comments]. *Histopathology* **31**, 400–407.

116. Kashima, K., Yokoyama, S., Daa, T., Nakayama, I., Nickerson, P. A., and Noguchi, S. (1997) Cytoplasmic biotin-like activity interferes with immunohistochemical analysis of thyroid lesions: a comparison of antigen retrieval methods. *Mod. Pathol.* **10**, 515–519.

117. Mighell, A. J., Robinson, P. A., and Hume, W. J. (1995) Patterns of immunoreactivity to an anti-fibronectin polyclonal antibody in formalin-fixed, paraffin-embedded oral tissues are dependent on methods of antigen retrieval. *J. Histochem. Cytochem.* **43**, 1107–1114.

118. Pateraki, M. and Kontogeorgos, G. (1997) Postadhesive technique for archival paraffin sections. *Biotech. Histochem.* **72**, 168–170.

119. Mote, P. A., Leary, J. A., and Clarke, C. L. (1998) Immunohistochemical detection of progesterone receptors in archival breast cancer. *Biotech. Histochem.* **73**, 117–127.

120. Krenacs, T. and Krenacs, L. (1999) Immunogold-silver staining (IGSS) for single and multiple antigen detection in archived tissues following antigen retrieval. *Appl. Immunohistochem. Mol. Morphol.* **7**, 93–94.

121. Grabau, K. A., Nielsen, O., Hansen, S., et al. (1998) Influence of storage temperature and high-temperature antigen retrieval buffers on results of immunohistochemical staining in sections stored for long periods. *Appl. Immunohistochem.* **6**, 209–213.

122. Jacobs, T. W., Prioleau, J. E., Stillman, I. E., and Schnitt, S. J. (1996) Loss of tumor marker-immunostaining intensity on stored paraffin slides of breast cancer. *J. Natl. Cancer Inst.* **88**, 1054–1059.

123. Kato, J., Sakamaki, S., and Niitsu, Y. (1995) More on p53 antigen loss in stored paraffin slides. *N. Engl. J. Med.* **333**, 1507–1508.

124. Prioleau, J. and Schnitt, S. I. (1995) p53 antigen loss in stored paraffin slides. *N. Engl. J. Med.* **332**, 1521–1522.

125. Cote, R. J., Shi, Y., Groshen, S., et al. (1998) Association of p27[kip1] levels with recurrence and survival in patients with stage C prostate carcinoma. *J. Natl. Cancer Inst.* **90**, 916–920.

126. Prento, P. and Lyon, H. (1997) Commercial formalin substitutes for histopathology. *Biotech. Histochem.* **72**, 273–282.

127. Zhang, P. J., Wang, H., Wrona, E. L., et al. (1998) Effects of tissue fixatives on antigen preservation for immunohistochemistry: a comparative study of microwave antigen retrieval on Lillie fixative and neutral buffered formalin. *J. Histotechnol.* **21**, 101–106.

128. Allison, R. T. and Best, T. (1998) p53, PCNA and Ki-67 expression in oral squamous-cell carcinomas—the vagaries of fixation and microwave enhancement of immunocytochemistry. *J. Oral Pathol. Med.* **27**, 434–440.

129. Carson, N. E., Gu, J., and Ianuzzo, C. D. (1998) Detection of myosin heavy chain in skeletal muscles using monoclonal antibodies on formalin fixed, paraffin embedded tissue sections. *J. Histotechnol.* **21**, 19–24.

130. Elias, J. M. and Margiotta, M. (1997) Low temperature antigen restoration of steroid hormone receptor proteins in routine paraffin sections. *J. Histotechnol.* **20**, 155–158.

131. Koopal, S. A., Coma, M. I., Tiebosch, A. T. M. G., and Suurmeijer, A. J. H. (1998) Low-temperature heating overnight in Tris-HCl buffer pH 9 is a good alternative for antigen retrieval in formalin-fixed paraffin-embedded tissue. *Appl. Immunohistochem.* **6**, 228–233.

132. Peston, D. and Shousha, S. (1998) Low temperature heat mediated antigen retrieval for the demonstration of oestrogen and progesterone receptors in formlin-fixed paraffin sections. *J. Pathol.* **186**, A21.

133. Malisius, R., Merz, H., Heinz, B., Gafumbegete, E., Koch, B. U., and Feller, A. C. (1997) Constant detection of CD2, CD3, and CD5 in fixed and paraffin-embedded tissue using the peroxidase-mediated deposition of biotin-tyramide. *J. Histochem. Cytochem.* **45**, 1665–1672.

134. Butmarc, J. R., Koureau, H. P., Levi, E., and Kadin, M. E. (1998) Improved detection of CD5 epitope in formalin-fixed paraffin-embedded sections of benign and neoplastic lymphoid tissues by using biotinylated tyramine enhancement after antigen retrieval. *Am. J. Clin. Pathol.* **109**, 682–688.

135. Dorfman, D. M. and Shahsafaei, A. (1997) Usefulness of a new CD5 antibody for the diagnosis of T-cell and B-cell lymphoproliferative disorders in paraffin sections. *Mod. Pathol.* **10**, 859–863.

136. Kawai, K. and Osamura, R. Y. (1998) Epitope retrieval vs. amplification techniques in diagnostic immunohistochemistry. *J. Histochem. Cytochem.* **46**, A14.

137. Moll, B. and Jelveh, Z. (1998) Antigen recovery for vimentin immunostaining by reducing agents for two cases of spindle cell tumors. *J. Histotechnol.* **21**, 45–48.

138. Shi, S.-R., Gu, J., Kalra, K. L., Chen, T., Cote, R. J., and Taylor, C. R. (1997) Antigen retrieval technique: a novel approach to immunohistochemistry on routinely processed tissue sections. In: Go, J. (ed.). *Analytical Morphology: Theory, Application, and Protocols*, Eaton Publishing, Natick, MA, pp. 1–40.

139. Tacha, D. E. and Chen, T. (1994) A modified antigen retrieval method. A calibration technique for microwave ovens. *J. Histotechnol.* **17**, 365–366.

140. Lan, H. Y., Mu, W., Nikolic-Patterson, D. J., and Atkins, C. A. (1995) A novel, simple, reliable, and sensitive method for multiple immunoenzyme staining: use of microwave oven heating to block antibody crossreactivity and retrieve antigens. *J. Histochem. Cytochem.* **43**, 97–102.

141. Stirling, J. W. and Graff, P. S. (1995) Antigen unmasking for immunoelectron microscopy: labeling is improved by treating with sodium ethoxide or sodium metaperiodate, then heating on retrieval medium. *J. Histochem. Cytochem.* **43**, 115–123.
142. Wilson, D. F., Jiang, D.-J., Pierce, A. M., and Wiebkin, O. W. (1996) Antigen retrieval for electron microscopy using a microwave technique for epithelial and basal lamina antigens. *Appl. Immunohistochem.* **4**, 66–71.
143. Strater, J., Gunthert, A. R., Bruderlein, S., and Moller, P. (1995) Microwave irradiation of paraffin-embedded tissue sensitizes the TUNEL method for in situ detection of apoptotic cells. *Histochemistry* **103**, 157–160.
144. Lucassen, P. J., Labat-Moleur, F., Negoescu, A., and van Lookeren Campagne, M. (2000) Microwave-enhanced in situ end-labeling of apoptotic cells in tissue sections; pitfalls and possibilities. In: Shi, S.-R., Gu, J., and Taylor, C. R., eds., Antigen Retrieval Techniques: Immunohistochemistry and Molecular Morphology. Eaton, Natick, MA, pp. 71–91.
145. Harkins, L. E. and Grizzle, W. E. (1995) Advanced techniques in immunohistochemistry, Workshop #35, National Society for Histotechnology Annual Symposium 1995, Buffalo, New York, October 7–13, 1995.
146. Shi, S.-R., Cote, R. J., and Taylor, C. R. (1999) Standardization and further development of antigen retrieval immunohistochemistry: strategies and future goals. *J. Histotechnol.* **22**, 177–192.

# PROTOCOLS

## ANTIGEN RETRIEVAL (AR) TECHNIQUE

### MATERIALS AND REAGENTS

1. Microwave oven. Various kinds of domestic cooking microwave oven with an output power around 1000 W have been widely applied. In our hands, the use of a Sharp model R-4A46 (900 W, 2450 MHz) with a multiple-sequence power setting may provide a satisfactory way to run some of the protocols discussed below. Also available are laboratory microwave ovens that have been designed with controlled temperature regulators that can measure temperature accurately during the heating process, such as the H2550 Laboratory Microwave Processor (EBS, Agawam, MA).
2. Other heating equipment. Autoclaves used for sterilization allow superheating at 120°C. A domestic pressure cooker with an operating pressure of 103 kPa/15 psi can also be used for superheating. A plastic steamer designed by Ventana Biotek Systems (Newport Beach, CA) may be applied for convenient use with the autostainer from the same manufacturer. Hotplates and water baths are available for conventional heating methods at boiling or other desired temperatures. BioGenex (San Ramon, CA) have recently made available a range of automated AR instruments and immunostainers that incorporate their own heating devices and computer-controlled protocols. In our experience these new devices are simple and effective.
3. Slide container. Coplin jars made of plastic material are commonly used for AR heating sections. Any other larger plastic container may be employed for slides and the AR solution. Notice that a loose cover is important to avoid increased pressure during boiling.
4. A plastic pressure cooker (Nordicware, MN) is available that can be set in a microwave oven to reach a "superheating" condition.
5. AR solutions. See below.

## METHODS

### Requirements for Routinely Formalin-Fixed, Paraffin-Embedded Tissue Sections Prior to AR Treatment

1. Mount the tissue sections on slides coated with either poly-L-lysine or APES, or charged slides provided by Fisher Scientific (Pittsburgh, PA), dry the slides at 60°C for at least 1 h in order to make the tissue sections adhere to the slides. Overnight incubation of the slides produces optimal adhesion.
2. Deparaffinize by using the routine procedure: histoclear or xyline, rehydration by graded alcohols, and blocking of endogenous peroxidase by an $H_2O_2$-methanol solution.
3. Rinse the slides in distilled water, 3 changes for 15 min.

### Microwave Heating Method

1. For more details, see ref. *14.* Place slides in plastic Coplin jars containing an AR solution, such as distilled water, buffer solution, metal salt solution, urea, pH 3.5 glycine-HCl, and so forth.
2. Cover the jars with loose-fitting screw caps and heat in the microwave oven for either 5 or 10 min. The 10-min heating time is divided into two 5-min cycles with an interval of 1 min between cycles to check on the fluid level in the jars. If necessary, add more AR solution after the first 5 min to avoid drying the tissue sections.
3. After heating, remove the Coplin jars from the oven and allow them to cool for 15 min.
4. Then rinse the slides in distilled water twice and in PBS for 5 min, ready for IHC staining.

### Laboratory Microwave Processor

For more details on this device (model H2550, EBS), see ref. *138.*

1. This microwave processor displays temperature readouts. Its power output is regulated by a temperature feedback mechanism and a timer, and it is capable of monitoring both temperature and heating time.
2. Fix a Coplin jar on a thick capboard or plastic plate and set it in the center portion of the turntable, filling the jar with distilled water or tap water. Set the temperature probe into the Coplin jar through a hole of the cap, to measure the temperature.
3. Set all test jars around the central probe jar, as close as possible.
4. Turn on the microwave processor. Set the temperature and the time at temperature as required.
5. Turn on the turntable, making sure that the jars are not moved by the probe and the table.
6. Start the microwave heating; the timer is automatically controlled.
7. For heating at 100°C, the heating time should be divided into 5-min cycles as mentioned above.
8. This processor is particularly useful for the test battery approach (see below).

### Calibration Technique for Microwave Oven

For details, see ref. *139.*

1. Microwave at high power (800 W) for 2–3 min until the solution comes to a rapid boil; then turn off the oven. Note the exact time it took for the solution to boil.
2. Set oven power on 3–4 (30–40% power or 250 + 50 W) or on defrost (low power). Heat for 7–10 min. (If using a 500-W oven, set power level at 5–6 or 50–60%, i.e., 250 W). Adjust the setting so oven cycles on and off every 20–30 s and the solution boils about 5–10 s each cycle.
3. The following formula can be used to determine the power setting: $S = 250/P \times 10$, where $S$ is the microwave power setting for AR, and $P$ is the output power of the individual microwave oven. For example, if a microwave oven output power is 800 W, then the power

setting for AR (*S*) is: $S = 250/800 \times 10 = 3.1$. Therefore, it should be set at 3 and heated for 7–10 min.

## Autoclave Heating Method

For details, see refs. *28* and *29*.

1. Set the slides in Coplin jars or other kinds of containers filled with AR solutions. Fix the covers in place with a special tape designed for the autoclave.
2. Set the jars with slides in the center portion of the autoclave.
3. Tightly close the door of the autoclave as required by the instructions.
4. Set the temperature at 120°C for 10 min.
5. Allow a cool-down period of 20–30 min, after heating.
6. For IHC staining, follow the same procedure as with microwave heating.

## Pressure Cooking Method

For details, see ref. *33*.

1. Fill domestic cooker with an operating pressure of 103 kPa/15 psi one-third with AR solution and heat by a hotplate to boiling.
2. Suspend slides in metal slide racks, place quickly into the AR solution, and tightly replace the pressure lid.
3. Start the timer when the pressure indicator valve reaches the maximum (around 4 min). The optimal period of pressurized boiling is 1–2 min.
4. After heating, depressurize the cooker and cool under running water; then remove the lid. Add cold tap water to replace the hot AR solution. A 15–20 min cooling time may be required, as in the microwave heating method.

## Microwave + Plastic Pressure Cooker

For details, see refs. *36* and *37*.

1. Fill three plastic staining jars, containing as many as 25 slides, with AR solution and place in a plastic pressure cooker (Nordicware, MN). Add about 600 mL of distilled water to the pressure cooker to reach one-half of the volume, making sure that all three jars stand in the water at a stable position.
2. Place the plastic pressure cooker containing three staining jars with tissue slides, as described above, in the microwave oven (a Sharp carousel oven, model R-4A46 with multiple-sequence cooking, which can switch from one power level setting to another automatically). Set the oven, at maximum power (900 W, 2450 MHz) for 15 min to boil the water, followed by a 40% power setting for another 15 min (simmer) to maintain boiling.
3. Cool down for 15–20 min, followed by the procedure as for the regular microwave AR method.

## Steam Heating AR

For details, see refs. *35* and *37*.

1. Use Steam HIER (Heat-Induced Epitope Retrieval) provided by BioTek (Ventura, CA) for convenient use with the BioTek autostainer.
2. Set the slides in the TechMate slide holder in the regular face-to-face orientation, maintaining the capillary gap.
3. Fill the steamer with distilled water to the top line, and heat it to boiling point. Then turn the dial to 30 min for the next step.
4. Add 10 mL to each of the 10-well trays in the gray tile provided by BioTek, and place the slide holder on the gray tile with the tips of the slide pairs in the AR solution located in

the 10-well trays. Then place the gray tile with the slide holder into the center of the steam chamber.
5. Place the whole steam chamber base on the steamer, while boiling vigorously for 20 min.
6. After cooling the heated slides for 15 min, follow the remaining procedure for IHC as in the microwave heating method.

## Test Battery to Develop an Optimal AR Protocol for Certain Antibodies Tested

For details, see refs. *58, 60, 86,* and *138.*

1. Number 10 slides 1–10 and use them for three pH values of the AR solution and three different heating conditions, as follows:

| AR solution | pH 1–2 | pH 6–7 | pH 8–10 |
|---|---|---|---|
| Super-high temperature | #1 | #4 | #7 |
| High temperature | #2 | #5 | #8 |
| Mid-high temperature | #3 | #6 | #9 |

A tenth slide stands as a non-AR control. For convenience of practice, we use citrate buffer at pH 6 to represent the middle-pH range, Tris-HCl buffer of pH 1.5 to represent low-pH range, and Tris-HCl buffer of pH 10 or EDTA-NaOH solution of pH 8 to represent the higher range of pH.
2. Super-high temperature: place slides in the microwave oven at 100°C for 5 min × 4, or heat in an autoclave at 120°C for 10 min. High temperature: microwave at 100°C for 10 min as in the original protocol. Mid-high temperature: place in a temperature-controlled microwave oven or water bath at 90°C for 10 min. If possible, a H2550 microwave processor may be used to maintain the heating conditions accurately, as noted above.
3. The use of an autostainer is recommended to perform IHC staining.
4. Evaluate the intensity for all 10 slides to identify the optimal protocol for the antibody under test.
5. Additional studies, including different AR solutions, or different heating methods may be used if necessary.
6. The test battery method can also be performed in two steps: first step, testing three AR solutions at different pH values as listed above, with one regular temperature (100°C for 10 min) to find the optimal pH value of AR solution; second step, testing optimal heating condition based on the established pH value.

## AR Multiple Immunoenzyme Staining

This method is used on PLP- or formalin-fixed tissue sections *(140).*

1. Preincubate with 10% fetal calf serum and 10% normal goat serum for 10 min.
2. Drain the sections and then incubate with the primary antibody, for the required time, followed by a PBS wash.
3. Incubate with secondary antibody, followed by another PBS wash.
4. Incubate with label (such as APAAP, or others) for 30 min and develop with the Fast Blue BB base (Sigma, St. Louis, MO) substrate in the presence of 2 m$M$ levamisole (for APAAP).
5. Then microwave the sections in citrate buffer, pH 6.0, for 5 min, twice, as routine AR treatment in order to block antibody crossreactivity completely between two rounds of immunostaining procedure.
6. Then wash the sections in PBS, prior to adding a second different primary antibody and repeating **steps 1–4**. If using peroxidase label, a blocking step of incubation in 0.3% $H_2O_2$ in methanol for 20 min may be added after **step 1**.

7. If a third or fourth primary antibody is required for multiple immunolabeling, **step 5** (microwave heating procedure) should be repeated between each of the immunostaining procedures.

For paraffin sections, the same procedure (i.e., **step 5** microwave heating) is used between each immunostaining sequence, except that: 1) microwave heating is used prior to the first primary antibody staining as in AR (i.e., before **step 1**); and 2) as some surface antigens may not tolerate the two 5-min microwave heating periods, this method may be limited to nuclear and cytoplasmic antigens.

## AR Immunoelectron Microscopy

For details, see ref. *141*.

1. Cut routinely processed epoxy resin-embedded tissue block, flatten with chloroform vapor, and collect on Formvar-coated nickel grids.
2. Etch the grids with either a saturated aqueous solution of sodium metaperiodate for 1 h or 10% fresh saturated solution of sodium ethoxide (prepared overnight) diluted with anhydrous ethanol for 2 min.
3. Heat the grids in a preheated microwave oven to the boiling point; heat the sections at 95–100°C by a hotplate for 10 min. (Citrate buffer pH 6.0 is recommended as the AR solution.)
4. After heating, allow the grids to cool for 15 min before immunostaining.

Recently, Wilson et al. *(142)* studied AR-immunoelectron microscopy (IEM) based on a comparison between routinely processed TAAB resin- and LR White (LRW) resin-embedded ultrathin sections of both human and rat tissues with the use of antibodies to collagen types I, III, IV, and VI, lamin, and pan-cytokeratin. They found that AR enhanced the detection of types III, IV, and VI collagens and pan-cytokeratin for LRW resin and the detection of type IV collagen for routine TAAB resin-embedded sections. Their protocol was as follows:

1. Each grid bearing an ultrathin section was inverted onto a 0.5-mL solution of 0.01 *M* citrate buffer in a small plastic vial cap. Three such caps, each containing one grid, were placed in a glass Petri dish and heated in an 700-W Sharp domestic microwave oven for 5 min at high power (70–80°C at the end of the cycle).
2. After heating, the grids were washed in PBS for 5 min. For TAAB resin-embedded sections, an additional wash in PBS containing 50 m*M* $NH_4Cl$ and 3% sucrose for 1 h is important to reduce the background staining.
3. The immunogold staining procedure is then performed, using 1% BSA and 1% Tween 20 and PBS for wash and diluent of reagents. To develop more effective protocols for AR-IEM, Wilson's group is examining the pH of the buffer and the use of Tris-HCl buffer based on the test battery concept.

## Microwave Heating for Enhancement of mRNA In Situ Hybridization

For details, see refs. *103* and *105*.

1. After deparaffinization as described above, rinse the sections in 0.85% NaCl for 5 min.
2. Then immerse the slides in 600 mL sodium citrate buffer (0.01 *M*) at pH 6.0 and heat at full power in the microwave oven for 7 min to reach the boiling point and another 5 min with a short interruption to check the liquid level. If necessary, add more distilled water to restore the initial level of buffer solution, as in the microwave AR method. After heating, allow the sections to remain in the same solution for 20 min to cool.
3. Follow with the *in situ* hybridization (ISH) procedure.

## Microwave Heating for the TdT-Mediated Nick End-Labeling Technique (TUNEL)

For details, see ref. *143*.

1. For paraffin sections for enhancement of TUNEL.
2. Tissue sections that are immersed in 10 m*M* citrate buffer, pH 6.0, and heated by microwave for 1 min at 750 W show enhanced sensitivity with the TUNEL method.

Recent studies demonstrated that the microwave heating method plays a key role in enhancement of TUNEL outcome for various tissues fixed in formalin, when an optimal protocol is developed based on the test battery principle *(144)*. Generally speaking, a citrate buffer of pH 3.0 may achieve good TUNEL results for archival formalin-fixed tissue sections of brain and thyroid tissues. However, each individual tissue should be tested to establish an optimal protocol, since a variety of factors such as different kinds and sizes of tissues, autolysis prior to fixation, conditions of fixation, and so on, may influence the optimal pretreatment protocol as well as the concentration of reagents for the TUNEL method *(144)*.

## Nonheating AR Method for Routinely Formalin-Fixed, Acid-Decalcified, Celloidin-Embedded Tissue Sections

For details, see refs. *15, 84,* and *85*.

1. Preparation of AR solution. Saturated sodium hydroxide (NaOH)-methanol solution is made of 50–100 g of NaOH mixed with 500 mL methanol in a brown-colored bottle by vigorous shaking. After standing for 1–2 weeks, the supernant can be used diluted 1:3 in methanol. (This is the NaOH-methanol solution.)
2. Wash the celloidin-embedded tissue sections in distilled water for 10 min and mount on either 0.1% poly-L-lysine (Sigma) or aqueous mounting media-coated slides by pressing the section down with filter paper, and trimming the tissue along the edges of the slides.
3. Place a few drops of 0.1% Poly-L-lysine on the slide to cover the whole section and dry briefly in an oven. Pay attention to controlling the optimal drying time of the sections in the oven. Do not dry excessively.
4. Immerse the slides in a freshly prepared AR solution of NaOH-methanol, diluted in 1:3, for 30 min. Rinse the slides in 100% methanol for 15 min × 2, 70% methanol 15 min × 2, PBS 15 min × 2, 0.3% Triton X-100, for 10 min, and PBS 15 min. The slides are then ready for immunohistochemical staining.
5. For thicker (50 μm) celloidin tissue sections, a floating method may be used to avoid mounting the sections on the glass slides.
   The NaOH-methanol AR solution has also been used for enhancement of immunostaining on acid-decalcified, formalin-fixed paraffin sections for antibodies such as keratin AE1, AE3, vimentin, glial fibrillery acidic protein, desmin, muscle-specific actin, S-100, NSE, κ, λ, bcl-2, L-26, and T-cell (UCHL) *(145)*.
6. A combined AR heating and nonheating method may be used to improve some antibodies that give poor results using the nonheating method alone. In this combined method, the heating-induced AR is performed before immersing the slides in the NaOH-methanol solution *(146)*.

# 15

# Tyramide Amplification in Immunohistochemistry

## Naoko Sanno, MD, PhD, Akira Teramoto, MD, DMSc, and R. Yoshiyuki Osamura, MD, PhD

## INTRODUCTION

Immunohistochemistry is an important technique in both investigative and diagnostic pathology, for correlating the localization of antigens and tissue morphology. Since the advent of immunocytochemistry in the late 1940s, significant progress has been made in methods for labeling, visualizing, and detecting cell constituents. Two techniques have been widely accepted for the demonstration of protein: the peroxidase anti-peroxidase (PAP) *(1)* technique and the avidin-biotin-complex (ABC) *(2)* method. The PAP technique uses peroxidase as the labeling agent, and ABC is based on the noncovalent binding between avidin and biotin. The production of highly specific polyclonal and monoclonal antibodies improves sensitivity as well as specificity of immunocytochemical reactions and allows visualization of specific proteins. More recently, heat-induced epitope retrieval *(3,4)* has become one of the most effective techniques for enhancing the sensitivity of immunohistochemistry. This method significantly improves the number of antigens that can be demonstrated in routine formalin-fixed paraffin sections.

Another amplification system called catalyzed reporter deposition (CARD) or the tyramide amplification technique (TAT) has been reported recently. In 1989, Bobrow et al. *(5)* described a novel application of horseradish peroxidase (HRP) as an analyte-dependent reporter enzyme to catalyze the deposition of additional peroxidase molecules on the surface in a solid phase immunoassay. With the use of a biotinylated tyramide amplification step, this technique produced about a 200-fold increase in sensitivity in solid phase assay *(5)*. CARD has been successfully incorporated into immunohistochemical and *in situ* hybridization techniques in many fields of investigation.

## PRINCIPLE

This method relies on the ability of HRP to catalyze the dimerization of biotinylated tyramine (tyramide), followed by the deposition of a large number of avidin-biotin-peroxidase complexes or peroxidase-labeled streptoavidin molecules on the complex **(Fig. 1)**. The deposition of immune complex oxidizes biotinylated tyramine in the presence of hydrogen peroxide to produce radical species that react with electron

From: *Morphology Methods: Cell and Molecular Biology Techniques*
Edited by: R. V. Lloyd © Humana Press, Totowa, NJ

## Schematics

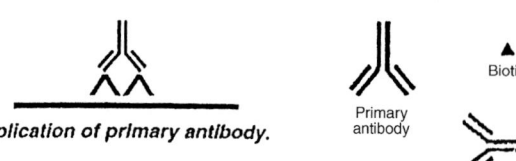

**Application of primary antibody.**

Primary
antibody

Biotin

Streptoavidin-biotin
HRP complex

Streptoavidin
HRP complex

HRP
enzyme

Link antibody

Antigen

Biotinyl tyramide
complex

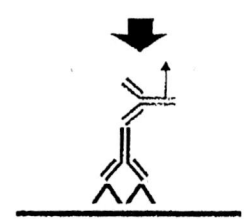

**Application of biotinylated link
antibody.**

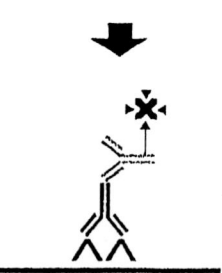

**Application of streptavidin-biotin HRP
complex.**

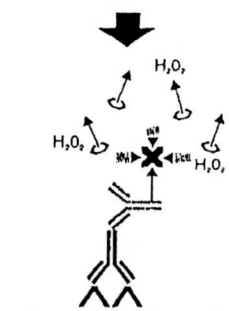

**Application of amplification reagent
(biotinyl tyramide).**

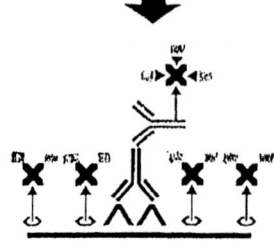

**Application of streptavidin/HRP.**

moieties (e.g., tyrosine, tryptophans, and others) of protein molecules present in the vicinity of the HRP label *(5)*. The net effect is that many peroxidase molecules surround a single HRP label.

## PREPARATION OF BIOTINYLATED TYRAMINE

Biotinylated tyramine is produced by adding 100 mg of sulphosuccinimidyl 6-(biotinamido) hexanoate (sulpho-NHS-LC Biotin; Pierce & Warriner, Chester, UK) to 40 mL of 50 m$M$ borate buffer (pH 8.0). Subsequently, 30 mg of tyramine hydrochloride is added (Sigma), following the method of Kerstens et al. *(6)*. The solution is stirred overnight at room temperature, filtered and then diluted in 0.05 $M$ Tris-HCl buffer (pH 7.6) containing 0.03% hydrogen peroxide. Biotinylated tyramide is available as part of a catalyzed signal amplification (CSA) system from DAKO (Carpinteria, CA) *(7)*.

## METHODOLOGY OF CARD IN IMMUNOHISTOCHEMISTRY

In general, 10% formalin or 4% paraformaldehyde-fixed paraffin-embedded tissues are suitable for CARD. Thin sections (4–6 µm) are cut and mounted on silane-coated slides or poly-L-lysine-coated slides. Prior to staining, the tissue slides must be deparaffinized to remove embedding media and then rehydrated.

A potential problem with the increased sensitivity afforded by this system is the possibility of increased background staining due to endogenous avidin binding activity or endogenous peroxidase activity. Adequate blocking of endogenous biotin is necessary. Incubation with 3% hydrogen peroxide for 5 min is used for this purpose. Then specimens are incubated for 5 min with a protein block to suppress nonspecific binding of subsequent reagents, followed by incubation with an appropriately characterized and diluted primary antibody. This is followed by sequential 15-min incubations with biotinylated antibody, streptavidin-biotin-peroxidase complex, biotinyl tyramide (amplification reagent), and streptavidin-peroxidase. Staining is completed by a 1–5-minute incubation with 3,3′-diaminobenzidine tetrahydrochloride (DAB), which results in a brown-colored precipitate at the antigen site (see **Protocol**).

Appropriate control experiments are required for proper interpretation of specific staining. The specificity of the method is assessed by several controls:

1. Replacement of the polyclonal primary antibodies with normal rabbit serum.
2. Replacement of the monoclonal primary antibody with mouse immunoglobulin.
3. Omission of the streptoavidin-HRP.
4. Omission of the biotinyl tyramide.
5. Preincubation of the primary antibody with excess specific antigen (immunohistochemical absorption test).

## ADVANTAGES OF CARD IN ENDOCRINE PATHOLOGY

This technique has been recognized to be effective, and many studies have been published showing its advantage for scientific investigations. The technique is especially

◄ **Fig. 1.** Principle of catalyzed reporter deposition (CARD) system. Biotinylated secondary antibody against primary antibody is reacted to the streptavidin-biotin-complex, biotinyl tyramide (amplification reagent), and then streptavidin peroxidase.

**Table 1**
**Comparison of the Immunohistochemical Sensitivities for Detecting Follicle-Stimulating Hormone-β (FSH-β) by the CARD System, the Indirect Immuno-peroxidase Method, and the ABC Method Using Graded Dilution of Primary Antibody**[a]

| Method | Dilution of anti-FSH-β mouse monoclonal antibody 1 | | | | | | | | | |
|---|---|---|---|---|---|---|---|---|---|---|
| | $1 \times 10^7$ | $1 \times 10^6$ | $5 \times 10^5$ | $1 \times 10^5$ | $5 \times 10^4$ | $1 \times 10^4$ | $5 \times 10^3$ | $1 \times 10^3$ | $5 \times 10^2$ | $1 \times 10^2$ |
| CSA | – | ++ | + | +++ | +++ | HBG | HBG | HBG | HBG | HBT |
| ID | – | – | – | – | – | – | – | + | + | ++ |
| ABC | – | – | – | – | – | ++ | ++ | ++ | ++ | ++ |

[a]CARD, catalyzed reporter deposition; ID, indirect immunoperoxidase method; ABC, avidin-biotin complex method; HBG, high background. Grading: +, weakly positive; ++, positive, +++, strongly positive.

applicable to endocrine tissues, to demonstrate hormones and peptides and to correlate morphology with function for understanding mechanisms regulating hormonal production. CARD has become very useful to increase sensitivity for immunodetection in endocrine and other areas of pathology.

The authors have previously applied this method to tissues from 50 cases of clinically nonfunctioning pituitary adenomas without evidence of endocrinologic signs of hormone secretion *(8)*. The primary antibodies used were the β subunit of anti-follicle-stimulating hormone (FSH-β), the β-subunit of luteinizing hormone (LH-β), and the α subunit of LH (αSU) mouse monoclonal antibodies (UCB-Bioproducts, France). When CARD was applied to normal human pituitary gland, each of the gonadotropin subunits was positive even when the antibodies were diluted 1:10⁷, which is 1000-fold more than the standard indirect immunoperoxidase method and 100-fold more than the standard avidin-biotin complex method **(Table 1; Fig. 2)**. Using CARD, 35 of 50 cases of nonfunctioning adenomas were immunopositive for FSH-β and/or LH-β. In contrast, all were negative using the regular indirect immunohistochemical method **(Figs. 3 and 4)**. Although these subunits detected by CARD may not result in known biologic effects, this highly sensitive detection system may provide new information about the nature of hormone production in these tumors as well as insight into their histogenetic origin.

The advantage of CARD in investigative immunohistochemistry has been clearly shown. Studies are currently being done on its utility in a diagnostic setting. von Wasielwski et al. *(9)* used this technique with 85 different antibodies on routinely fixed, paraffin-embedded tissues using a simple protocol. They found a 5–50-fold (maximum 500-fold) increase in sensitivity compared with conventional immunohistochemistry with most of the antibodies tested. In their study, it was pointed out that the optimal dilution of antibodies varied with the different antibodies. Although this mechanism is obscure, the differences in tissue and composition of the surrounding tissue may partly affect the results. Thus, detailed studies are necessary to determine the optimal concentration of antibody for each.

The concentration of tyramide probably influences the final results *(3,10)*. Mengel et al. *(11)* tested the relationship of tyramide concentration and primary antibodies in 133 different dilution combinations. They reported that the highest concentrations of both tyramide and antibody paradoxically resulted in a weaker staining with nonspecific

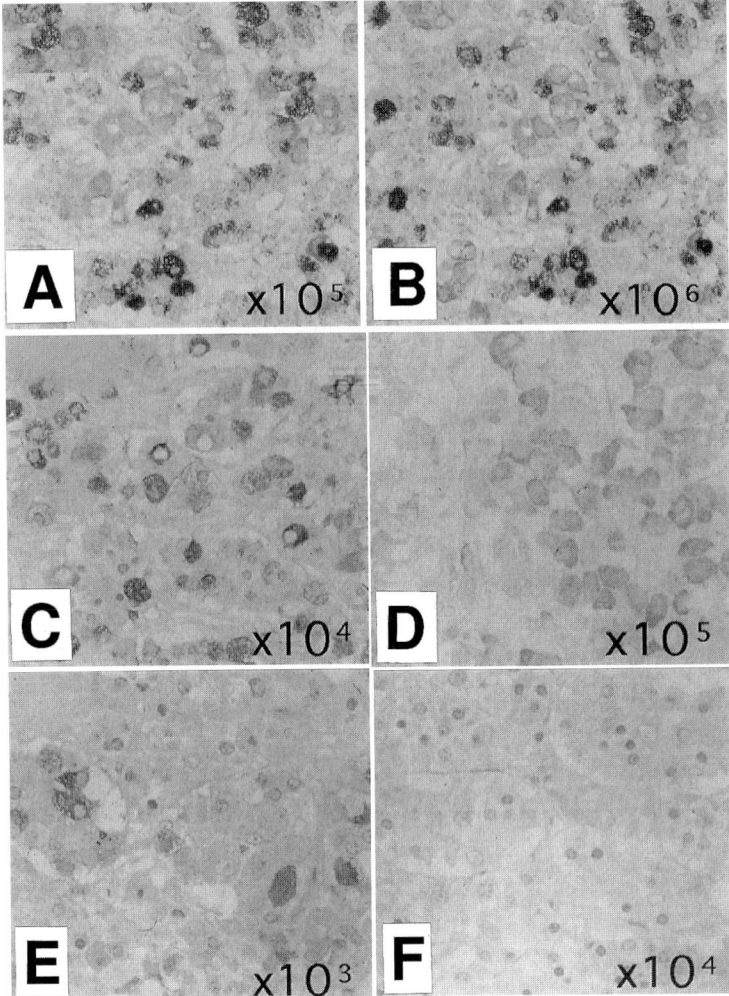

Fig. 2. (**A,B**) The catalyzed reporter deposition system for normal pituitary gland using anti-FSH-β antibody diluted 1:100,000. Immunoreactivity for FSH-β is clearly demonstrated. (**C,D**) Avidin-biotin complex (ABC) method using anti-FSH-β antibody diluted 1:100,000. Immunoreactivity for FSH-β was not detected. (**E,F**) Indirect immunoperoxidase method and original magnification ×400. The sensitivity of CARD revealed a 1000-fold increase compared with the indirect immunoperoxidase method and a 100-fold increase over the standard ABC method.

coloring. In routine laboratory work, commercially available antibody and biotinylated tyramine are used. With these commercial reagents, it is recommended to start with a 10–50-fold higher dilution of the antibody than used regularly.

A paradoxic effect in which the positive cells decrease in intensity compared with standard staining is sometimes observed. It has been reported that at rather high antibody dilutions, some of the markers show a decrease in the proportion of positive cells, although the signal was stronger compared with standard staining *(10)*. One must consider this effect when quantitation of positive cells is important in the evaluation of proliferation markers, for example.

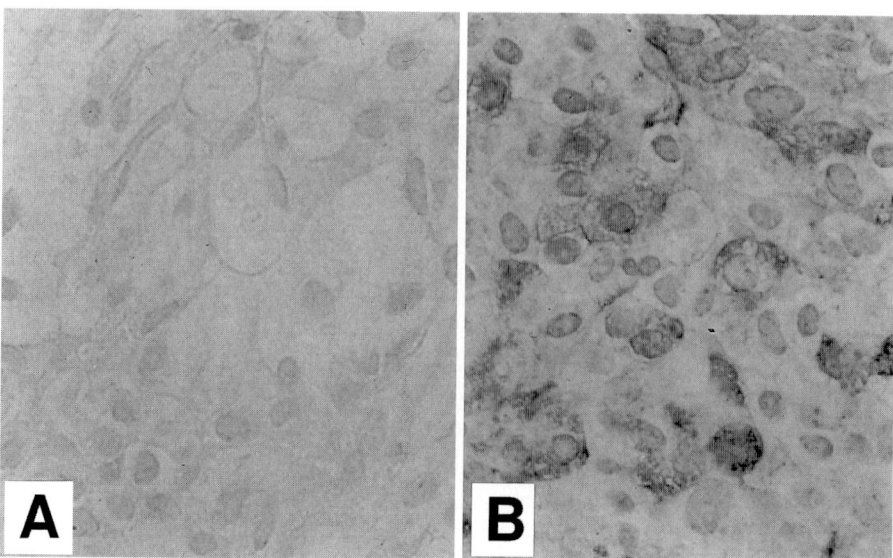

**Fig. 3.** Results of the catalyzed signal amplification system in the analysis of nonfunctioning adenomas. (**A**) By standard ABC method on the same adenoma tissue, immunoreactivity for FSH-β is observed only in scattered cells (case #28). (**B**) Using catalyzed signal amplification, immunoreactivity for FSH-β is seen in the cytoplasm of the adenoma cells.

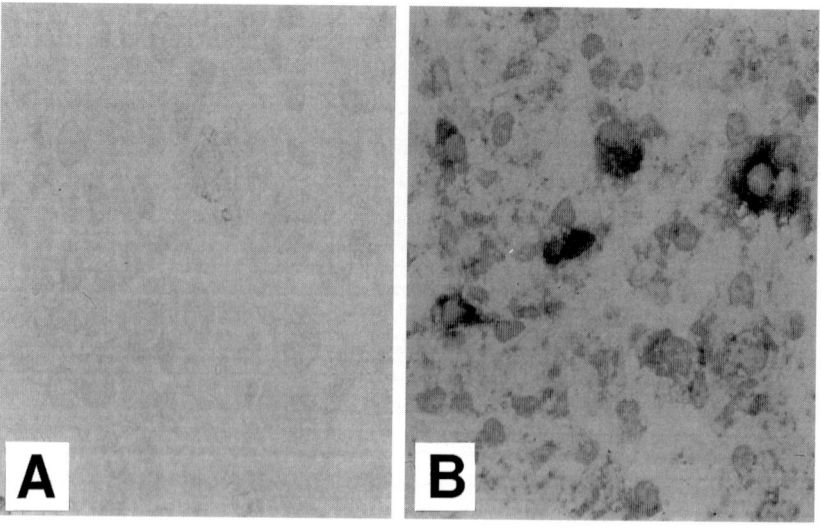

**Fig. 4.** (**A**) Immunohistochemical staining for the α-subunit in a nonfunctioning pituitary adenoma. Immunoreactivity for the α-subunit is not observed by the ABC method. (**B**) Positive immunoreactivity is clearly observed in the cytoplasm of the adenoma cells after catalyzed signal amplification. Original magnification ×600.

The fixation and preservation of tissues may also influence the results. Four percent paraformaldehyde or 10% paraffin-embedded tissues usually work well *(9,10)*.

Another advantage of CARD for immunohistochemistry is that lower concentrations of antibodies can be used. In studies with antibodies available in limited amounts, this may be beneficial.

## CARD WITH OTHER TECHNIQUES

It is possible to use CARD in combination with other techniques for signal amplification. Heat-induced antigen retrieval has been increasingly used for immunostaining with many primary antibodies *(4,12)*. This target retrieval can be done using a pressure cooker or microwave. The tissue section is heated prior to staining in 0.01 *M* citrate buffer, pH 6.0. Charged slides should be used with target retrieval to avoid detachment of tissues from slides. The combination of heat-induced retrieval and CARD has been successfully used in the detection of many antigens, including estrogen receptor and adrenomedulin *(13)*.

Kohler et al. *(14)* described the combination of the CARD protocol with the use of the small Nanogold probes, followed by silver intensification. Sections are processed as for the CARD-DAB method up to the rinsing steps after incubation with biotinylated tyramide. Then sections are incubated with streptavidin conjugated with 1.4 nm Nanogold diluted 1:60 in TMGS buffer for 60 min at 20°C. Then sections are treated with secondary antibody. They noted that this combination has provided improved, highly sensitive, and intense immunolabeling by both light and electron microscopy.

## CARD IN DOUBLE IMMUNOSTAINING

The application of CARD in double immunostaining of two different antigens has been described. Hunyady et al. *(15)* demonstrated amplified fluorescent staining for the first antigen using CARD and subsequent conventional staining for a second antigen in the same host species, i.e., pairs of mouse monoclonal antibodies. Teramoto et al. *(16)* also recently described double staining with two mouse monoclonal antibodies on paraffin sections. The first antibody was visualized with DAB (brown) or 3-amino-9-ethylcarbazole (AEC; brownish red) using the CARD method, and then the second antibody was visualized with 5-bromo-4-chloro-3-indolyl phosphate/nitroblue tetrazolium (BCIP/NBT; purple blue) using the alkaline-phosphatase anti-alkaline phosphatase (APAAP) method. Cross reactivity was not detected between the first and second reactions. It has been suggested that the first primary antibody is covered with deposit and is not detectable by a secondary antibody without amplification.

## CARD IN IMMUNOELECTRON MICROSCOPY

Applications of CARD to immunoelectron microscopic studies have been described *(17–19)*. It has been successfully applied for the evaluation of amylase, heat-shock protein 70, and insulin in rat pancreatic tissues.

## CARD *IN SITU* HYBRIDIZATION

The CARD technique when applied to *in situ* hybridization increases the sensitivity of nonradioactive *in situ* hybridization. Several authors have reported the use of CARD for detection of DNA sequences and mRNA *(20–23)*. Although recent advances in evaluation of nonradioactive ISH led to greater detection sensitivity, sometimes a more sensitive detection method for low copy number of gene products may be required.

In this procedure we use antisense and sense probes that are also employed for conventional ISH. The probes are labeled with biotin 11dUTP (Boehringer Mannheim) by the terminal deoxyribonucleotidyl transferase reaction. Target retrieval can be performed by heating the frozen tissue sections in 10 mmol/L citric acid (pH 6.0) in a microwave oven for 5 min (up to 95°C) and digested with 1 mg/mL proteinase K at 23°C for 10 min. The retrieval and digestion differ from tissues and expected signals.

Sections are then treated with 0.2 *N* HCl for 20 min followed by incubation with 0.25% (vol/vol) acetic anhydride in triethanolamine for 10 min. To reduce endogenous peroxidase background, slides are immersed in 3% $H_2O_2$ in methanol for 30 min and covered with prehybridization buffer for 1 h at room temperature. Thereafter, the sections can be hybridized with the probe at 42°C for 18 h. After stringent washing for 10 min at 42°C, slides are subject to amplification with reagents including primary streptoavidin HRP (1:400) for 15 min, biotinyl-tyramide solution for 15 min, and secondary streptavidin HRP for 15 min, with fresh 1× Tris-buffered saline/Tween (TBST) washing after each step. The reaction product is visualized by developing the slides in DAB chromogen/$H_2O_2$ solution for 3–5 min.

Using this amplification method, a positive reaction has been seen with antibodies previously described as unsuitable for paraffin section immunostaining. Although the CARD method involves multiple steps and is time consuming, it is a more sensitive and specific method for the detection of proteins by immunostaining in many laboratory investigations. The CARD system is one of the most powerful amplification methods in immunochemistry developed to date.

## REFERENCES

1. Sternberger, L. A., Hardy, P. H. Jr., Cuculis, J. J., and Meyer, H. G. (1970) The unlabeled antibody enzyme method of immunocytochemistry. Preparation and properties of soluble antigen-antibody complex (horseradish peroxidase-antiperoxidase) and its use in identification of spirochetes. *J. Histochem. Cytochem.* **18**, 315–333.
2. Hsu, S. M., Raine, L., and Fanger, H. (1981) Use of avidin-biotin-peroxidase complex (ABC) in immunoperoxidase techniques: a comparison between ABC and unlabeled antibody (PAP) procedures. *J. Histochem. Cytochem.* **29**, 577–581.
3. Merz, H., Malisius, R., Mannweiler, S., et al. (1995) ImmunoMax. A maximized immunohistochemical method for the retrieval and enhancement of hidden antigens. *Lab. Invest.* **73**, 149–156.
4. Shi, S. R., Key, M. E., and Kalra, K. L. (1991) Antigen retrieval in formalin-fixed, paraffin-embedded tissues: an enhancement method for immunohistochemical staining based on microwave oven heating of tissue sections. *J. Histochem. Cytochem.* **39**, 741–748.
5. Bobrow, M. N., Harris, T. D., Shaughnessy, K. K., and Litt, G. J. (1989) Catalyzed reporter deposition, a novel method of signal amplification. Application to immunoassays. *J. Immunol. Methods* **125**, 279–285.
6. Kerstens, M. J., Poddighe, P. J., and Hanselaar, A. G. J. M. (1995) A novel in situ hybridization amplification method based on the deposition of biotinylated tyramine. *J. Histochem. Cytochem.* **43**, 347–352.

7. Dako Corporation. (1998) Instructions for DAKO catalyzed signal amplification (CSA) system, peroxidase for mouse primary antibodies. Dako Corporation, Carpinteria, CA.

8. Sanno, N., Teramoto, A., Sugiyama, M., Itoh, Y., and Osamura, R. Y. (1996) Application of catalyzed signal amplification in immunodetection of gonadotropin subunits in clinically nonfunctioning pituitary adenomas. *Am. J. Clin. Pathol.* **106**, 16–21.

9. von Wasielewski, R., Mengel, M., Gignac, S., Wilkens, L., Werner, M., and Georgii A. (1997) Tyramine amplification technique in routine immunohistochemistry *J. Histochem. Cytochem.* **45**, 1455–1459.

10. King, G., Payne, S., Walker, F., and Murray, G. I. (1997) A highly sensitive detection method for immunohistochemistry using biotinylated tyramine. *J. Pathol.* **183**, 237–241.

11. Mengel, M., Werner, M., and von Wasielewski. (1999) Concentration dependent and adverse effects in immunohistochemistry using the tyramine amplification technique. *Histochem. J.* **31**, 195–200.

12. Kaufmann, O., Baume, H., and Dietel, H. (1998) Detection of oestrogen receptors in non-invasive and invasive transitional cell carcinomas of the urinary bladder using both conventional immunohistochemistry and the thyramide staining amplification (TSA) technique. *J. Pathol.* **186**, 165–168.

13. Tajima, A., Osamura, R. Y., Takekoshi, S., et al. (1999) Distribution of adrenomedullin (AM), proadrenomedullin N-terminal 20 peptide, and AM mRNA in the rat gastric mucosa by immunocytochemistry and in situ hybridization. *Histochem. Cell Biol.* **112**, 139–146.

14. Kohler, A., Lauritzen, B., and Van Noorden, J. F. C. (2000) Signal amplification in immunohistochemistry at the light microscopic level using biotinylated tyramide and nanogold-silver staining. *J. Histochem. Cytochem.* **48**, 933–942.

15. Hunyady, B., Krempels, K., Harta, G., and Meqey, E. (1996) Immunohistochemical signal amplification by catalyzed reporter deposition and its application in double immunostaining. *J. Histochem. Cytochem.* **44**, 1353–1362.

16. Teramoto, N., Szekely, L., Pokrovskaja, K., et al. (1998) Simultaneous detection of two independent antigens by double staining with two mouse monoclonal antibodies. *J. Virol. Methods* **73**, 89–97.

17. Mayer, G. and Bendayan, M. (1997) Biotinyl-tyramide: a novel approach for electron microscopic immunocytochemistry. *J. Histochem. Cytochem.* **45**, 1449–1454.

18. Schmidt, B. F., Chao, J., Zhu, Z., DeBiasio, R. L., and Fisher, G. (1997) Signal amplification in the detection of single-copy DNA and RNA by enzyme-catalyzed deposition (CARD) of the novel fluorescent reporter substrate tyramide. *J. Histochem. Cytochem.* **45**, 365–373.

19. Matsuno, A., Sanno, N., Tahara, S., et al. (1999) Silent somatotroph adenoma, detected by catalyzed signal amplification and non-radioisotopic in situ hybridization. *Endocr. J.* **46 (suppl.)**, S81–84.

20. Sanno, N. and Osamura, R. Y. (1998) Catalyzed reporter deposition method for amplifying endocrine products. *Endocr. Pathol.* **9**, 195–199.

21. von Gijlswijk, R. P. and Zijlmans, H. J. (1997) Fluorochrome-labeled tyramides: use in immunohistochemistry and fluorescence in situ hybridization. *J. Histochem. Cytochem.* **45**, 375–382.

22. Koji, T., Kanemitsu, Y., Hoshino, A., and Nakane, P. K. (1997) A novel amplification method of nonradioactive in situ hybridization signal for specific RNA with biotinylated tyramine. *Acta Histochem. Cytochem.* **30**, 401–406.

23. Oka, H., Jin, L., Scheithauer, B. W., et al. (1999) Growth hormone-releasing hormone receptor (GHRH-R) mRNA expression in human pituitary adenomas: a study by catalyzed reporter deposition in situ hybridization (CARD-ISH). *Endocr. Pathol.* **10**, 27–36.

# PROTOCOLS

## CARD IMMUNOHISTOCHEMISTRY

1. Place 3% hydrogen peroxide on the tissue specimen. Incubate for 5 min and rinse with buffer (0.05 $M$ Tris-HCl, pH 7.6, containing 0.3 $M$ NaCl and 0.1% Tween 20).
2. Protein block: serum-free protein in phosphate-buffered saline (PBS) with 0.015 $M$ sodium azide. Incubate for 5 min. Do not rinse off protein block.
3. Primary antibody: mouse monoclonal or rabbit polyclonal. Incubate for 15 min and rinse with buffer.
4. Link antibody: biotinylated rabbit anti-mouse (for mouse monoclonal primary antibody) or mouse anti-rabbit (for rabbit polyclonal primary antibody) immunoglobulins in Tris-HCl buffer containing carrier protein and 0.015 $M$ sodium azide. Incubate for 15 min and rinse with buffer.
5. Streptoavidin-biotin complex: horseradish peroxidase labeled. Incubate for 15 min and rinse with buffer.
6. Amplification reagent: biotinyl tyramide and hydrogen peroxide. Incubate for 15 min and rinse with buffer.
7. Streptoavidin-peroxidase. Incubate for 15 min and rinse with buffer.
8. Substrate-chromogen solution: DAB. Incubate for 1–5 min and rinse with DW.
9. Hematoxylin counterstain. Rinse with DW.
10. Mounting: use aqueous-based mounting medium.

# 16

# Application of Immunohistochemistry in the Diagnosis of Lymphoid Lesions

### Paul J. Kurtin, MD

## INTRODUCTION

The value of phenotypic studies to aid in the diagnosis and classification of reactive and neoplastic disorders of lymphocytes and myeloid cells has been established. In general, immunophenotyping has five main goals. First, it can be used to distinguish reactive from neoplastic lymphoid proliferations in lymph nodes and in extranodal sites. Second, it can help to determine that a poorly differentiated malignant neoplasm is a lymphoma. Third, once a neoplasm has been determined to be a lymphoma or leukemia, the marker studies can establish the lineage: B-cell, T-cell, NK cell, myeloid, or histiocytic. These distinctions are clinically important and form the basis for lymphoma classification by the World Health Organization classification of neoplastic diseases of lymphoid tissues *(1)*. Fourth, phenotypic data can be used to help classify lymphomas because some lymphomas express a characteristic constellation of antigens. Finally, phenotypic data can be used to help distinguish Hodgkin's disease from non-Hodgkin's lymphomas. Because many different phenotypic modalities are available (flow cytometry, paraffin section immunohistochemistry, frozen section immunohistochemistry, molecular genetics studies) there is often confusion about the best modality to employ to solve specific differential diagnostic problems. The techniques are complementary, and none is perfect for every application. In this chapter, the phenotypic characterization of malignant lymphomas and leukemias by immunohistochemistry is discussed.

## HANDLING TISSUES SUSPECTED OF BEING INVOLVED BY HEMATOLYMPHOID NEOPLASMS AND GENERAL APPROACH TO THE DIFFERENTIAL DIAGNOSIS

Optimal utilization of the available phenotypic modalities starts with ensuring that the tissue is preserved in the proper form to maximize the data that can be obtained from the sample. Not every case will require every possible ancillary study for an adequate evaluation, but one cannot always predict which study will be necessary in an individual case at the time that tissue is submitted to the pathology laboratory.

From: *Morphology Methods: Cell and Molecular Biology Techniques*
Edited by: R. V. Lloyd © Humana Press, Totowa, NJ

Therefore, development of a protocol for tissue handling when hematolymphoid disorders are suspected is suggested. In general, coordination among the clinician, the surgeon, and the pathologist should occur so that all tissue is received unfixed. Because the first priority is always excellent morphology, at least one piece of tissue should be fixed in B5 or another mercury-based fixative. If mercury disposal consideratons make this impossible, zinc formalin is an alternative to B5. B5 not only produces excellent morphology, it is also a superior fixative for preserving antigenicity of many surface and cytoplasmic proteins that can be detected by immunohistochemistry *(2)*. The next priority is to freeze a portion of the sample. A sample protocol can be found in Chapter 13. The snap frozen tissue can be used for both frozen section immunoperoxidase stains and molecular genetics studies. Using frozen section immunohistochemistry, one can test for every marker that can be tested for by flow cytometry and the immunoarchitecture of the process is preserved, a decided advantage of immunohistochemistry over flow cytometry. Although a piece of tissue approximately $5 \times 5 \times 2$ mm is usually optimal for frozen section immunohistochemistry, the cells in even the tiniest of endoscopic biopsy specimens or needle biopsy specimens can be successfully phenotyped by this technique. In selected cases, particularly those in which the diagnosis of acute leukemia or other blastic neoplasms is suspected, a specimen should be submitted for cytogenetic testing.

The immunohistochemical techniques that are applied in general surgical pathology are applicable to hematopathology as well. Good fixation, utilization of antigen retrieval strategies optimized for each antigen that is tested for, selection of optimal primary antibodies, utilization of a sensitive immunohistochemical method, proper titering of all primary and secondary antibodies, using diluents containing a detergent to inhibit nonspecific antibody binding, proper buffer pH, appropriate utilization of positive and negative controls, and establishing a quality control program are all necessary for successful phenotyping of hematolymphoid cells by immunohistochemistry. The principles of immunohistochemistry technique are discussed in Chapter 13.

From this point on, the approach to an individual case (including the phenotyping studies to be performed) is based on the clinical history and the morphology of the process. Because diagnosis and classification of hematolymphoid neoplasms is primarily based on morphology, diagnostic accuracy and efficiency and the most cost effective application of immunohistochemistry hinge on morphology to drive the choice of phenotyping modality and the panel of antibodies utilized in each case. In some cases paraffin section immunohistochemistry is the best tool to resolve the differential diagnosis (e.g., separating Hodgkin's lymphoma from non-Hodgkin's lymphoma, distinguishing large cell lymphoma from other malignancies, and determining B-cell or T-cell lineage on morphologically typical large cell lymphomas). In other cases (e.g., separating interfollicular lymphoid hyperplasia from T-cell lymphoma or using a panel of T-cell and B-cell lineage-associated antigens to classify a low-grade B-cell lymphoma), frozen section immunohistochemistry is the optimal phenotyping modality. The panels of antibodies are selected to solve specific differential diagnostic problems as discussed in the remainder of the chapter. Comprehensive lists of antigens to which antibodies are readily available and that are useful for phenotyping lymphomas and leukemias are presented in **Tables 1** and **2**.

**Table 1**
**Frozen Section Immunohistochemistry: Useful Markers of Hematolymphoid Cells**

| | |
|---|---|
| Routine B-cell-associated antigens | Routine T-cell-associated antigens |
| CD20 | CD2 |
| CD23 | CD3 |
| CD10 | CD5 |
| κ light chains | CD7 |
| λ light chains | |
| Other B-cell-associated antigens | Other T-cell-associated antigens |
| CD19 | CD1 |
| CD21 | CD4 |
| CD22 | CD8 |
| IgG | TCR-$\alpha\beta$ |
| IgA | TCR-$\gamma\delta$ |
| IgM | CD56 |
| IgD | CD103 |
| CD103 (hairy cell leukemia) | |
| Myeloid lineage-associated antigens | Antigens for special applications |
| CD15 | CD45 |
| CD33 | CD30 |
| CD13 | bcl-2 |
| CD14 | Terminal deoxynucleotidyl transferase |
| c-kit | CD34 |

## MORPHOLOGY-BASED IMMUNOPHENOTYPING STRATEGIES

### *Malignant Lymphoma vs Reactive Lymphoid Hyperplasia*

The usual morphologic patterns that raise the differential diagnosis of lymphoma vs reactive lymphoid hyperplasia include the following: 1) the distinction of follicular lymphoid hyperplasia from follicular lymphoma; 2) the distinction between prominent mantle zones and mantle cell lymphoma growing in a mantle zone pattern; and 3) the distinction between paracortical hyperplasia and T-cell lymphoma, B-cell lymphoma, and Hodgkin's lymphoma. Three types of phenotypic criteria can be used to distinguish malignant lymphoma from reactive lymphoid hyperplasia: clonality, expression of an aberrant phenotype, and expression of proteins that reflect a lymphoma-associated genetic abnormality.

### *Immunoglobulin Light Chain Restriction*

With few exceptions, morphologically atypical lymphoid infiltrates that can be demonstrated to be clonal by immunohistochemistry can be safely considered to be malignant lymphoma *(3–5)*. By immunohistochemistry, clonality is established by demonstrating κ or λ immunoglobulin light chain restriction in a cell population. Therefore, the criterion of clonality as an indicator of malignancy can only be applied to B-cell lymphomas. Using immunoperoxidase studies on frozen sections, 80–85% of all B-cell lymphomas can be shown to be clonal by demonstrating light chain restriction

**Table 2**
**Paraffin Section Immunohistochemistry: Useful Markers of**
**Hematolymphoid Cells**[a]

| B-cell-associated antigens | T-cell associated antigens |
|---|---|
| **CD20 (L-26)** | **CD3** (polyclonal anti-ε chain) |
| CD79a | CD5 |
| CD10 | CD45R0 (UCHL-1) |
| CD23 | CD43 |
| CD45RA (4KB5) | CD4 |
| κ light chains | CD8 |
| λ light chains | βF1 |
| Immunoglobulin heavy chains | CD1a |
| Myeloid lineage antigens | Special application antigens |
| CD34 (blasts) | CD45 |
| Myeloperoxidase | CD15 |
| Lysozyme | CD30 |
| CD68 (KP-1 and PGM-1) | Terminal deoxynucleotidyl transferase |
| CD15 | bcl-2 |
| Factor VIII | Cyclin D1 (bcl-1) |
| CD61 | p80/ALK-1 |
| Hemoglobin | Fascin |
| NK and cytotoxic lymphocyte antigens | Dendritic cell antigens |
| CD56 | CD21 |
| CD57 | CD35 |
| Tia-1 | CD1a |
| Granzyme B | S-100 |
| | Fascin |

[a]Antigens listed in bold face type are the first line markers used to distinguish B- and T-cell phenotype in non-Hodgkin's lymphomas. All others are selectively applied on a case-by-case basis based on the morphologic differential diagnosis.

**(Fig. 1)** *(3,6–9)*. The remaining 15–20% lack sufficient surface or cytoplasmic immuno-globulin to be detected by frozen section immunohistochemistry. Using the immunoper-oxidase technique in paraffin sections of well-fixed tissue with antigen retrieval or protease digestion, properly titered primary antibodies, and a sensitive detection system, κ or λ immunoglobulin light chain restriction can be demonstrated in approximately 70% of B-cell lymphomas **(Fig. 2)** *(2,10)*.

There are several pitfalls in staining frozen or paraffin sections for immunoglobulin. Here are a few of which to be aware. Because immunoglobulin is ubiquitous in serum and plasma, background staining of interstitial immunoglobulin can obscure cell-related immunoglobulin. Certain extranodal sites, such as skin and tissues in which the lymphoid infiltrates are associated with collagen sclerosis, also present a consistent problem with background staining because the antiimmunoglobulin antibodies have a tendency to bind to collagen. These problems are unavoidable but can be minimized by optimally titered antisera and by using detergents such as Tween 20 in the immunohistochemistry buffers and antibody diluents.

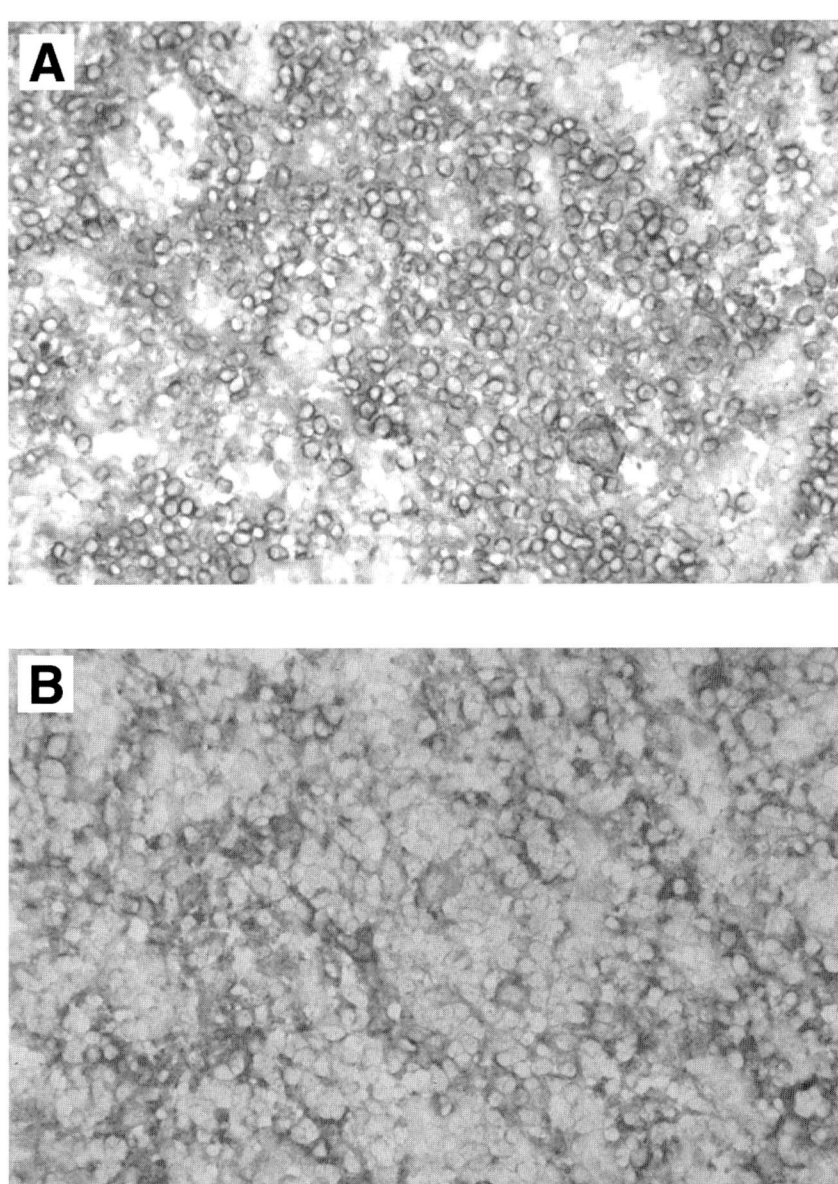

**Fig. 1.** Immunoperoxidase stains for κ (**A**) and λ (**B**) immunoglobulin light chains on frozen sections of a case of B-cell small lymphocytic lymphoma. Note the distinct membrane staining for κ light chains in the small lymphocytes and only hazy background staining for λ light chains within the intermixed histiocytes. This is a monoclonal staining pattern for κ and supports a diagnosis of malignant lymphoma of B-cell lineage. (*See* **Color Plate 6** following page 208.)

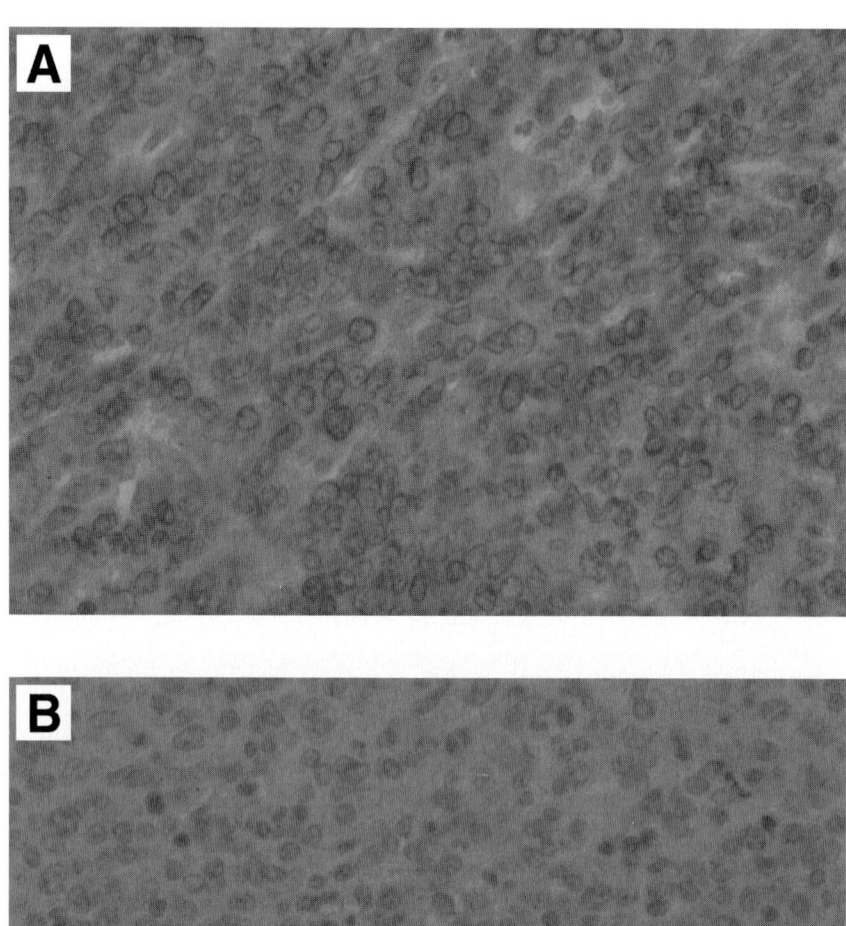

**Fig. 2.** Immunoperoxidase stains for κ (**A**) and λ (**B**) immunoglobulin light chains on paraffin sections of a case of mantle cell lymphoma. Note the distinct, perinuclear cytoplasmic staining for κ, but not for λ light chains. (*See* **Color Plate 6** following page 208.)

To determine light chain restriction reliably, κ and λ antisera must be "matched" to one another to ensure equal sensitivity in detecting immunoglobulin. This does not mean equal dilutions of the anti-κ and anti-λ antisera, rather, the dilutions of the anti-κ and anti-λ antisera should be adjusted so that adjacent sections of positive control tissue stain with equal intensity for κ and λ immunoglobulin light chains. For paraffin sections, properly titered κ and λ antisera will demonstrate polyclonal staining of mantle zone lymphocytes, not just plasma cells.

Similar areas of κ- and λ-stained histologic sections must be compared with each other. In assessing light chain restriction, demonstrating that the cell population in question lacks staining for one light chain is as important as demonstrating positivity for the other light chain. It is also important to pay attention to the quality of the staining. In frozen sections, lymphocytes will have membrane immunoglobulin staining. By contrast, in a properly stained paraffin sections, the most reliable staining pattern for κ and λ light chains is cytoplasmic (often perinuclear) staining. "Membrane" staining of lymphocytes in paraffin sections should be suspected as representing artifactual background positivity even if it looks "monoclonal."

Plasma cells and neoplasms with abundant plasma cells often do not stain well for immunoglobulin in frozen sections. Paraffin section immunoperoxidase stains are usually better for showing abundant cytoplasmic immunoglobulin in these instances.

Absence of immunoglobulin from a population of cells thought to be a B-cell lymphoma is nondiagnostic. It does not denote a neoplastic process nor does it exclude one.

### Aberrant Phenotypes

The second way by which the diagnosis of malignant lymphoma can be supported in a morphologically difficult case is to demonstrate that the atypical lymphocytes have an "aberrant" phenotype. Phenotypic aberrancy is a characteristic of cell populations, not individual cells. It refers to expression of a constellation of antigens by the cells that is not expected to be present in large populations of normal reactive lymphocytes. Usually, this can mean either loss of an expected antigen or gain of an unexpected antigen by the cells. It should be stressed that small populations of normal lymphocytes can express unusual constellations of antigens, particularly in very young children, in immune-deficient individuals, and in individuals with autoimmune disorders. This does not constitute phenotypic aberrancy but is a reflection of immune system immaturity or dysfunction rather than neoplasia.

The most common example of phenotypic "aberrancy" is expression of CD5 by the neoplastic B-lymphocytes of chronic lymphocytic leukemia, small lymphocytic lymphoma, or mantle cell lymphoma *(3,11–13)*. Most cases of these types of B-cell neoplasms will show coexpression of CD5 with pan B-cell markers such as CD19, CD20, or CD22 **(Fig. 3)**. This property of these lymphomas can be demonstrated most reliably in frozen section immunoperoxidase stains or by flow cytometry *(14)*. Recently, a monoclonal antibody to CD5 that identifies the antigen in paraffin sections has become commercially available *(15–18)*. Optimal fixation and immunohistochemistry technique are critical if the paraffin-reactive anti-CD5 antibody is to be used reliably because the intensity of CD5 expression by neoplastic B-cells is considerably weaker than the intensity of CD5 expression by normal T-cells. Because large numbers of CD5-positive

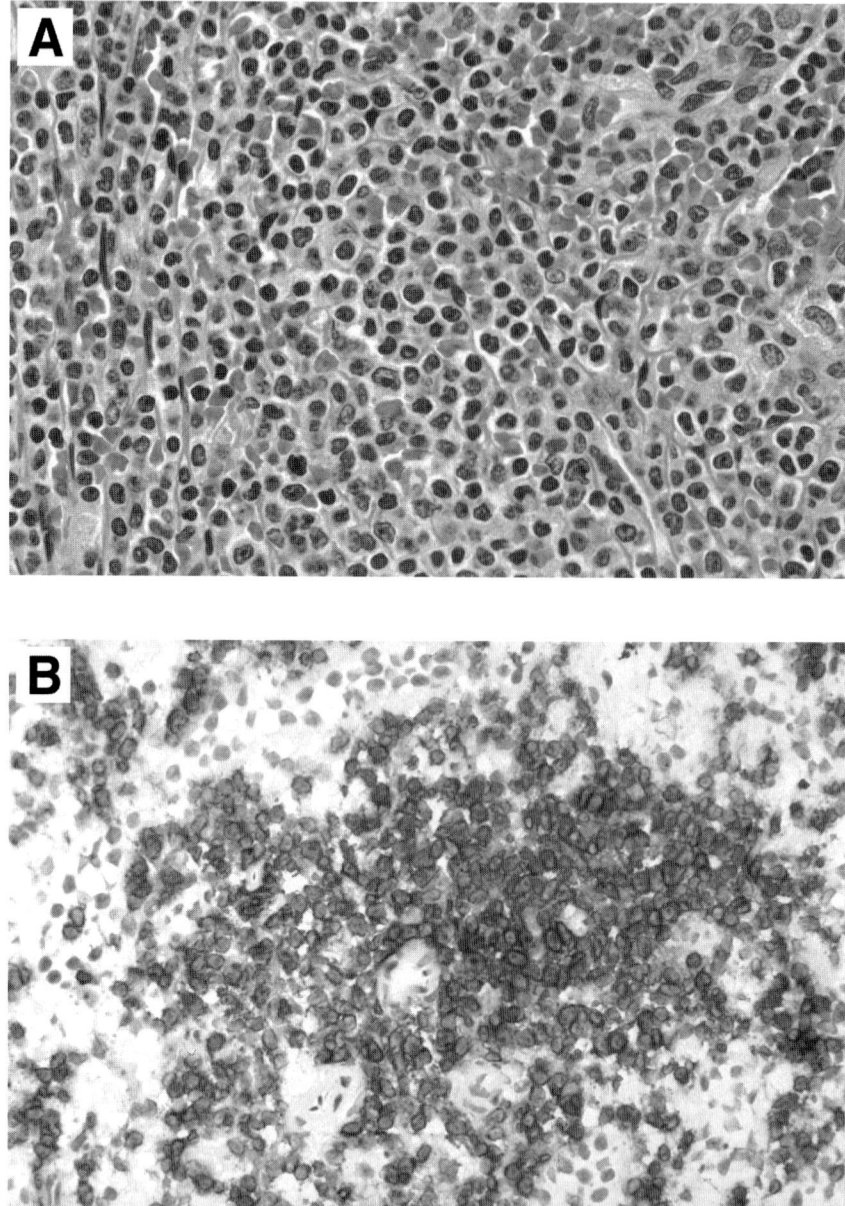

**Fig. 3.** B-cell small lymphocytic lymphoma. (**A**) Hematoxylin and eosin-stained section demonstrating the typical cytologic features of this neoplasm. Frozen section immunoperoxidase stains for (**B**) CD20, (**C**) CD5, (**D**) CD3, and (**E**) CD23 are also illustrated here. The neoplastic cells are CD20-positive B-cells that express the T-cell lineage-associated antigen CD5 but that are negative for CD3. Thus they exhibit an "aberrant phenotype." In addition, they are positive for CD23. Thus, they have a CD20, CD5, and CD23 positive phenotype that is characteristic for B-cell small lymphocytic lymphoma (see also **Table 7**). (*See* **Color Plate 6** following page 208.)

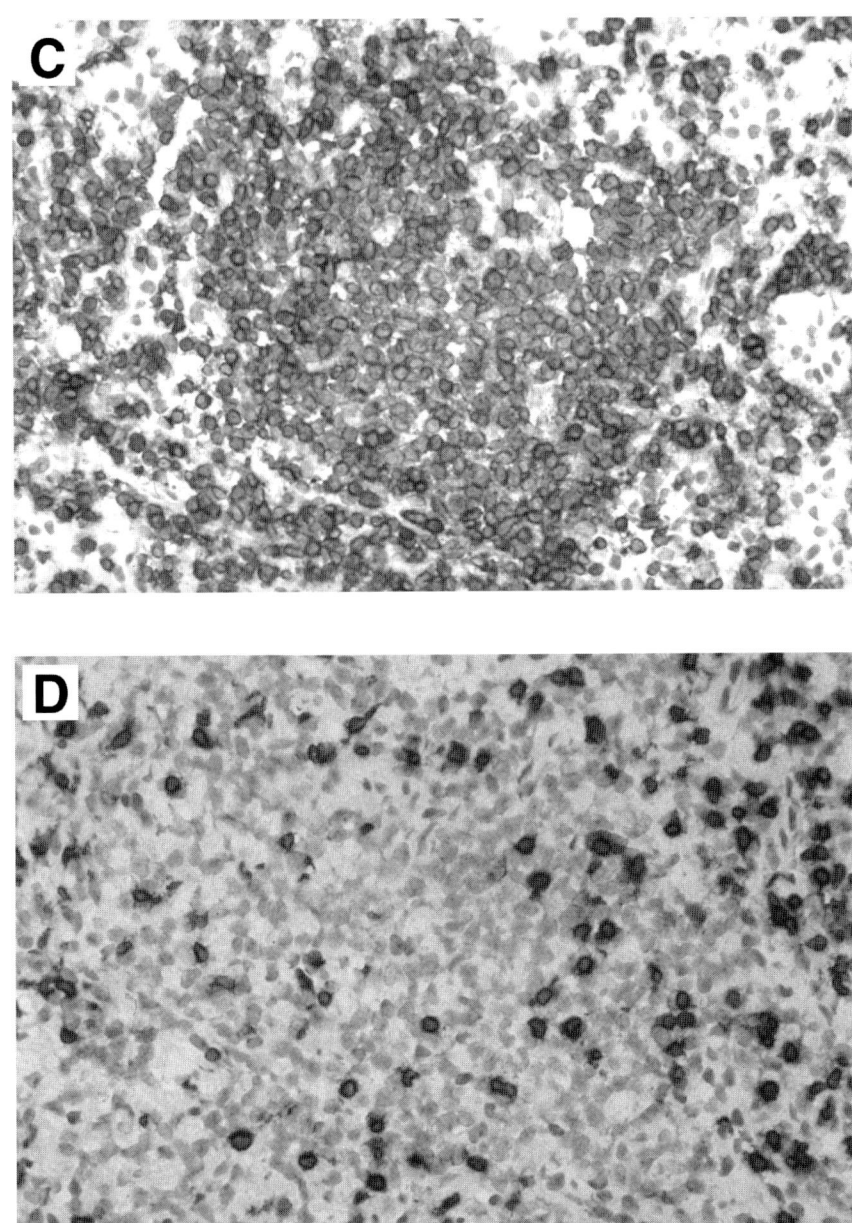

**Fig. 3.** *Continued*

B-cells are not observed in reactive lymphoid hyperplasia, demonstration of CD5 by a large population of B-cells supports the diagnosis of a B-cell lineage malignant lymphoma.

Another example of an "aberrant" phenotype in B-cell lymphomas is CD43 coexpression with B-cell antigens *(2,3,19–22).* Most normal reactive populations of B-lymphocytes do not express CD43. Conversely, many cases of B-cell small lymphocytic lymphoma, chronic lymphocytic leukemia, and mantle cell lymphoma and a subset of

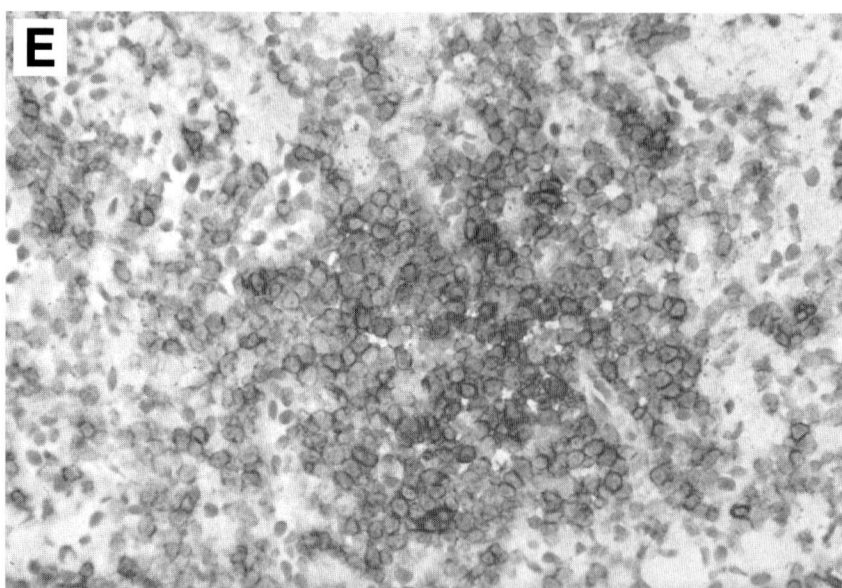

**Fig. 3.** *Continued*

other B-cell lymphomas will express CD43. CD43 can be demonstrated in both frozen sections and in paraffin sections. In paraffin sections, CD43 is best preserved with B5 fixation, but can be unmasked in paraffin sections by using EDTA antigen retrieval. Its expression is weaker on neoplastic B-lymphocytes than on normal T-cells.

Because there is no phenotypically demonstrable clonal marker for T-lymphocytes, one of the most useful applications of immunophenotyping is to demonstrate an aberrant phenotype in morphologically atypical proliferations of T-lymphocytes *(3,23–27)*. An aberrant phenotype will support a diagnosis of T-cell lymphoma, but a normal phenotype does not exclude lymphoma. More than 70% of peripheral T-cell lymphomas have aberrant phenotypes, and the presence of an aberrant phenotype correlates closely with the ability to demonstrate clonal T-cell antigen receptor gene rearrangements using molecular genetics techniques *(28,29)*. In T-cell lymphomas, phenotypic aberrancy usually manifests by loss of one of the pan-T-cell antigens that is expected to be present on post-thymic T-lymphocytes: CD2, CD3, CD5, and/or CD7 **(Fig. 4)**. The antigen most frequently lost from neoplastic T-cells is CD7, followed in decreasing frequency by CD5, CD3, and CD2. Other less common manifestations of phenotypic aberrancy in T-cell populations are either absence of staining for both CD4 and CD8 or presence of staining for both CD4 and CD8.

There are anatomic site- and disease-specific exceptions to the criteria for phenotypic aberrancy. Reactive T-cell populations in the skin, such as those that occur in chronic dermatitis and lichen planus as well as in other conditions, will frequently lack expression of CD7 *(30–32)*. Therefore, absence of CD7 from a T-cell population in the skin does not constitute evidence in favor of the diagnosis of cutaneous T-cell lymphoma even though most cases of cutaneous T-cell lymphoma lack CD7 expression *(33–35)*. The normal thymus *(36)* and the T-cell infiltrates in thymomas *(37,38)* contain substantial numbers of normal dual-expressing CD4- and CD8-positive T-cells and T-cells that

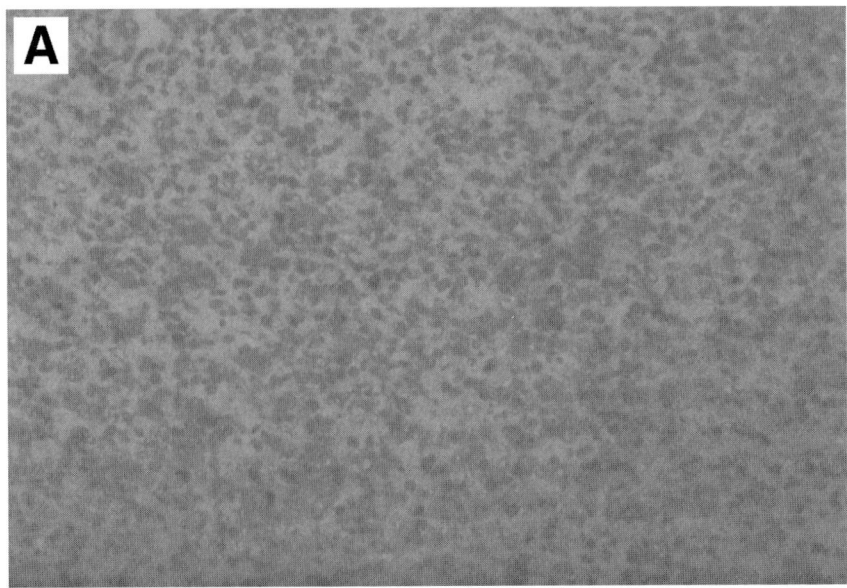

**Fig. 4.** Angioimmunoblastic T-cell lymphoma stained in frozen sections for (**A**) CD2 and (**B**) CD7. The neoplastic T-cells are positive for CD2 (as well as for CD3 and CD5; not illustrated), but they are negative for CD7. This is an "aberrant phenotype" because the T-cells are negative for an antigen (CD7) that is normally expressed on nonneoplastic T-cells. These findings would support a diagnosis of a peripheral T-cell lymphoma. (*See* **Color Plate 6** following page 208.)

are negative for both CD4 and CD8. Therefore, dual CD4 and CD8 positivity or double CD4 and CD8 negativity should not be used as a criterion for lymphoma in the thymus or in lymphocytes within mediastinal masses until normal thymus and thymoma are considered in the differential diagnosis. As more blood and tissue specimens are studied from patients with viral infections and immunodeficiencies (particularly the acquired immunodeficiency syndrome [AIDS]), it has become clear that phenotypically unusual peripheral blood- and tissue-based T-cell populations can be expanded in these conditions. Finally, almost all the published data on phenotypic aberrancy in peripheral T-cell lymphomas come from immunoperoxidase studies performed on frozen tissue sections in which the immunoarchitecture of the process can be assessed. It is not clear that flow cytometry gives similar results and it is not clear what the criteria for T-cell phenotypic aberrancy should be when flow cytometry is employed as the phenotyping modality. Therefore, supporting the diagnosis of a T-cell lymphoma on the basis of phenotypic aberrancy is only valid in frozen tissue sections where direct correlation between phenotype and morphology can be made and should be done only after carefully considering the clinical situation, the anatomic site from which the specimen is obtained, and the morphology of the process.

### Demonstrating Protein Products of Oncogenes or Abnormally Expressed Genes

One final approach to supporting a diagnosis of malignant lymphoma appears to be evolving with the advent of antibodies that recognize protein products of oncogenes or abnormally translocated genes in certain malignant lymphomas. Three types of malignant lymphoma have characteristic chromosomal translocations that result in the overexpression of genes with the production of protein products that are characteristically not expressed in the normal cell counterparts of the neoplasm. These include bcl-2 overexpression as the result of the t(14;18) in follicular lymphomas, cyclin D1 (bcl-1) overexpression as the result of the t(11;14) in mantle cell lymphomas, and p80/ALK-1 expression as the result of the t(2;5) in CD30-positive anaplastic large cell lymphomas. bcl-2, cyclin D1, and p80/ALK-1 can all be recognized by paraffin section immunohistochemistry using commercially available antibodies.

The best established application of immunohistologic demonstration of an oncogene product to support a diagnosis of malignant lymphoma is showing bcl-2 protein expression in the intrafollicular lymphocytes of follicular lymphomas (39–43). In normal and reactive lymph nodes stained with antibodies to bcl-2, the germinal center B-lymphocytes are negative for bcl-2. There is positivity for bcl-2 in mantle zone B-cells and a subset of inter- and intrafollicular T-cells. By contrast, 80% of follicular lymphomas show bcl-2 expression in the neoplastic follicular center cells **(Fig. 5)**. Thus, faced with the morphologic differential diagnosis of follicular lymphoma vs follicular hyperplasia, demonstrating bcl-2 positivity in the intrafollicular B-cells strongly supports a diagnosis of follicular lymphoma. On the other hand, absence of bcl-2 staining of intrafollicular lymphocytes, although characteristic of follicular hyperplasia, does not exclude follicular lymphoma because 20% of follicular lymphomas are bcl-2 negative. Recently, it has been suggested that normal monocytoid B-lymphocytes in reactive lymph nodes do not express bcl-2. By contrast, neoplastic monocytoid B-cells in nodal marginal zone B-cell lymphomas are bcl-2 positive. Therefore, demonstrating bcl-2 expression by

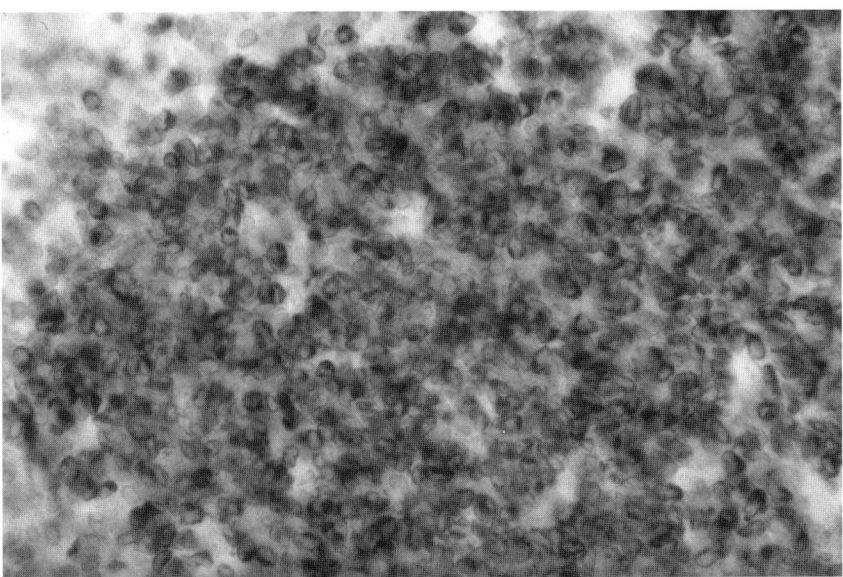

**Fig. 5.** Follicular lymphoma stained for bcl-2 in paraffin sections. Note that the intrafollicular lymphocytes strongly express bcl-2. This finding supports the diagnosis of follicular lymphoma. (*See* **Color Plate 6** following page 208.)

monocytoid B-lymphocytes should suggest the possibility of nodal marginal zone B-cell lymphoma. Finally, cells from many other types of lymphomas will express bcl-2 in the absence of the t(14;18). Therefore, the diagnostic utility of bcl-2 immunohisto-chemistry is specifically restricted to distinguishing follicular lymphoma from follicular hyperplasia and monocytoid B-cell hyperplasia from nodal marginal zone B-cell lymph-oma. It has no utility in distinguishing diffusely growing lymphomas from reactive processes or from one another *(44)*.

Antibodies to cyclin D1 that react in routinely fixed and processed paraffin sections can be very useful reagents for the evaluation of mantle cell lymphomas *(45–49)*. In a number of studies of normal tissues, there are few to no normal nodal, splenic, bone marrow, or gastrointestinal lymphocytes that are cyclin D1 positive. In contrast, cells from a very high percentage of cases of mantle cell lymphomas of all types (mantle zone pattern, diffuse pattern, monocytoid cytology, pleomorphic cytology, and blastic cytology) will exhibit *nuclear* cyclin D1 positivity as a result of the t(11;14) **(Fig. 6)**. This translocation moves the cyclin D1 gene from chromosome 11 into the regulatory control of the immunoglobulin heavy chain gene locus on chromosome 14 and causes overexpression of cyclin D1. Successful staining for cyclin D1 in paraffin sections of mantle cell lymphoma can be quite a technical challenge. Utilization of various antigen retrieval strategies (particularly EDTA pretreatment), a cocktail of two or more different monoclonal anti-cyclin D1 antibody preparations, overnight incubation of the tissue sections with primary antibody at 4°C, and a sensitive detection system will all help to ensure successful staining. With an optimized method, one can expect to demonstrate cyclin D1 immunoreactivity in 80–90% of mantle cell lymphomas *(2,49)*. Practically

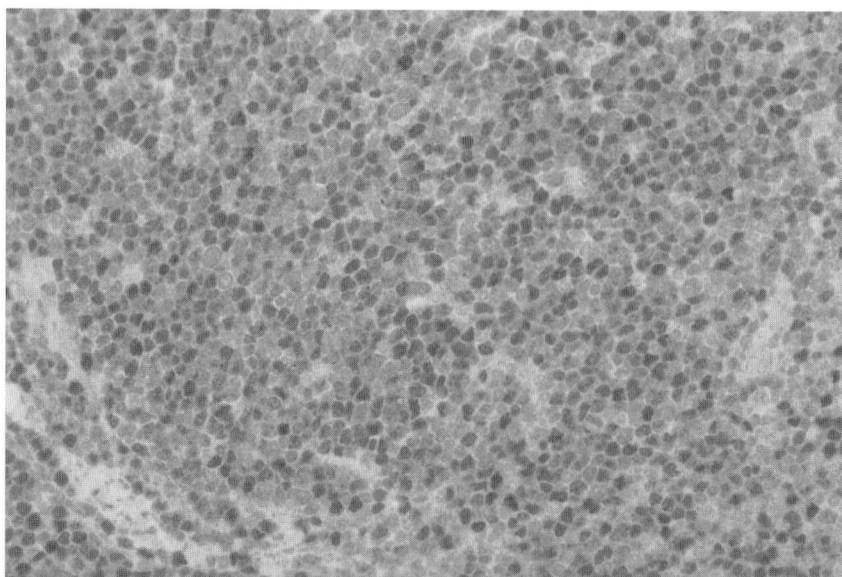

**Fig. 6.** Mantle cell lymphoma stained for cyclin D1 in paraffin sections. The neoplastic lymphocytes exhibit nuclear cyclin D1 positivity. (*See* **Color Plate 7** following page 208.)

speaking, if one finds a substantial number of small lymphocytes in any tissue site that has nuclear staining for cyclin D1, one can support a diagnosis of malignant lymphoma, specifically mantle cell type. Cyclin D1 overexpression also occurs in a subset of cases of multiple myeloma *(50,51)* and is also observed in some adenocarcinomas and squamous cell carcinomas.

Finally, there is no normal lymphoid cell population expressing the p80 protein that can be demonstrated in some cases of CD30-positive anaplastic large cell lymphoma *(52–58)*. p80 is an 80-kDa chimeric protein that is produced as a result of the t(2;5) as portions of the nucleophosmin gene from chromosome 5 and the anaplastic lymphoma kinase gene (ALK-1) from chromosome 2 are deleted, and the remaining portions are fused. The protein is antigenically distinct and is recognized by antibodies to p80. Anaplastic lymphoma kinase can also be detected immunohistochemically. Because the protein recognized by the anti-p80 antibodies is only present in the presence of the t(2;5), demonstrating nuclear and cytoplasmic staining for p80 (**Fig. 7** and **Color Plate 7** following page 208) is tantamount to recognizing the t(2;5) and is strong supporting evidence in favor of a diagnosis of CD30-positive anaplastic large cell lymphoma or one of its variants (small cell, monomorphous, lymphohistiocytic, and so forth). Likewise, anaplastic lymphoma kinase is not identifiable in normal lymphocytes, and demonstrating nuclear and cytoplasmic staining for this protein can also be used to support a diagnosis of CD30-positive anaplastic large cell lymphoma. Other chromosome partners can be translocated to the anaplastic lymphoma kinase locus on chromosome 2p and can result in overexpression of anaplastic lymphoma kinase. In these cases, cytoplasmic, but not nuclear, ALK-1 positivity is observed.

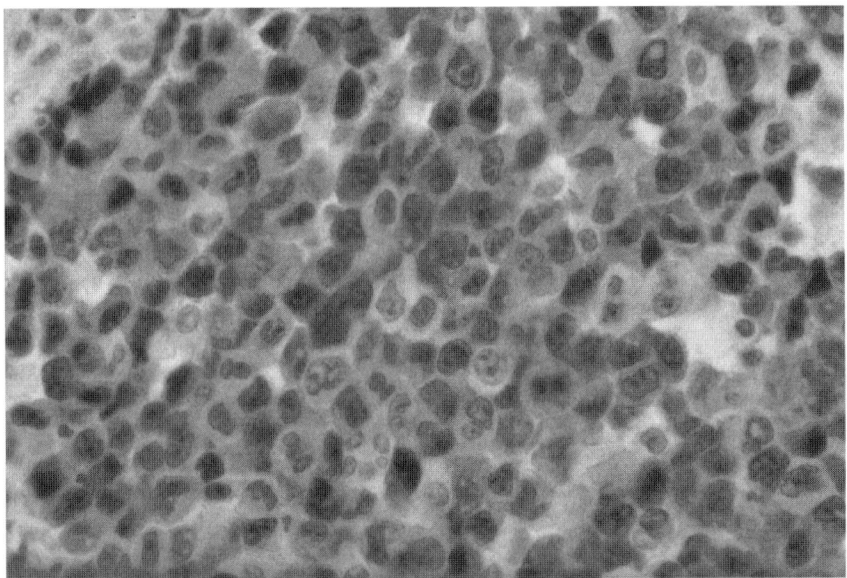

**Fig. 7 (Color Plate 7).** CD30-positive anaplastic large cell lymphoma stained for p80 in paraffin sections. The neoplastic cells exhibit both cytoplasmic and nuclear staining for p80.

## Distinguishing Malignant Lymphomas from Other Neoplasms

One of the most useful applications of immunophenotyping is to determine the nature of high-grade undifferentiated malignant neoplasms. Usually, the differential diagnosis of these tumors revolves around large cell processes (malignant lymphoma, high-grade carcinoma, malignant melanoma, seminoma), small blue cell tumors in children (lymphoma, Ewing's sarcoma, rhabdomyosarcoma, neuroblastoma), and blastic neoplasms (B- and T-lymphocyte precursor lymphoblastic lymphomas/leukemias, granulocytic sarcomas, blastoid variants of mantle cell lymphoma, and Burkitt's lymphoma). By using a panel of monoclonal and/or polyclonal antibodies, one can generally identify the tumor type by paraffin section immunohistochemistry (**Table 3–5**), but in some cases, the lymphoid and myeloid lineage malignancies may require snap frozen tissue or flow cytometry for optimal analysis. Other chapters of this book focus on neoplasms other than lymphomas and leukemias. Because alveolar rhabdomyosarcomas, Ewing's sarcomas, and peripheral neuroectodermal tumors can have characteristic chromosomal translocations, routine cytogenetics or molecular genetics techniques (polymerase chain reaction in particular) can supplement the analysis of these neoplasms.

## Phenotyping Malignant Lymphomas

### Assessing Lymphocyte Lineage

Once a neoplasm has been determined to be malignant lymphoma by histologic criteria or by using the panel of antibodies described above, precise classification of the lymphoma in many instances requires determining T-cell, B-cell, or natural killer (NK) cell lineage. In most cases this can be readily accomplished using paraffin or

**Table 3**
**Paraffin Section Immunoperoxidase Approach to Undifferentiated Large Cell Neoplasms**

| Tumor Type | CD45 | Keratin | S-100/HMB45 | PLAP[a] |
|---|---|---|---|---|
| Lymphoma | +[b] | − | − | − |
| Carcinoma | − | + | ± | ± |
| Melanoma | − | − | + | − |
| Seminoma | − | ±[c] | − | + |

[a]PLAP, placental alkaline phosphatase.

[b]Some cases of CD30-positive anaplastic large cell lymphoma are CD45 negative. Utilization of T- and B-cell lineage-specific antibodies, and antibodies to CD30 can help to identify these cases.

[c]Some seminomas are positive for keratin using antibodies to low molecular weight keratin, such as Cam 5.2. in a dot-like or wispy-appearing filamentous pattern.

**Table 4**
**Paraffin Section Immunoperoxidase Approach to Small Blue Cell Tumors**

| Tumor type | CD45 | CD99 | Neurofilaments | Synaptophysin | Desmin | Keratin |
|---|---|---|---|---|---|---|
| Lymphoma | + | +[a] | − | − | − | − |
| Ewing sarcoma | − | + | − | ± | − | − |
| PNET[b] | − | + | − | + | − | − |
| Neuroblastoma | − | − | + | + | − | − |
| Rhabdomyosarcoma | − | − | − | − | + | − |
| Small cell carcinoma | − | − | − | ± | − | + |

[a]T-precursor lymphoblastic lymphoma in particular.

[b]Peripheral neuroectodermal tumor.

**Table 5**
**Paraffin Section Immunoperoxidase Approach to Blastic Hematolymphoid Neoplasms**

| Disorder | TdT | CD34 | CD43 | MPO/Lys | CD3 | CD79a | CD20 | Cyclin D1 |
|---|---|---|---|---|---|---|---|---|
| T-lymphoblastic | + | − | + | − | + | − | − | − |
| B-lymphoblastic | + | + | ± | − | − | + | ± | − |
| Myeloid/monocytic | ± | + | + | + | − | − | − | − |
| Burkitt's lymphoma | − | − | + | − | − | + | + | − |
| Blastoid mantle cell | − | − | + | − | − | + | + | + |

frozen section immunohistochemistry. Because many B-cell, T-cell, and NK-cell lineage antigens can now be reliably detected in paraffin sections, a very small panel of antibodies employed in paraffin sections will give accurate information about cell lineage in >95% of cases of malignant lymphomas. The most useful B-cell lineage antigens are CD20 (L-26) *(59–63)* and CD79a *(64).* CD20 is expressed by a high percentage of cases of B-cell lymphomas **(Fig. 8)** and almost all of the remainder will be positive for CD79a. Notable exceptions are the rare lymphomas that express

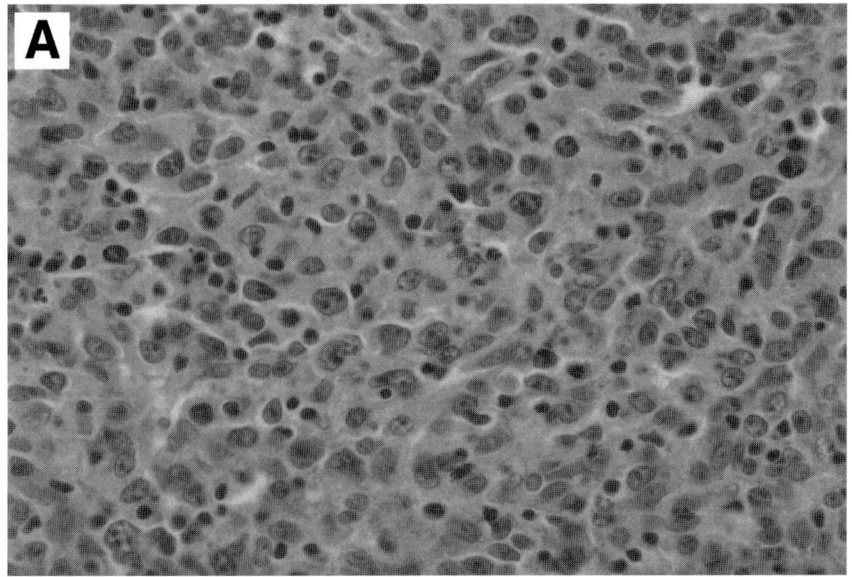

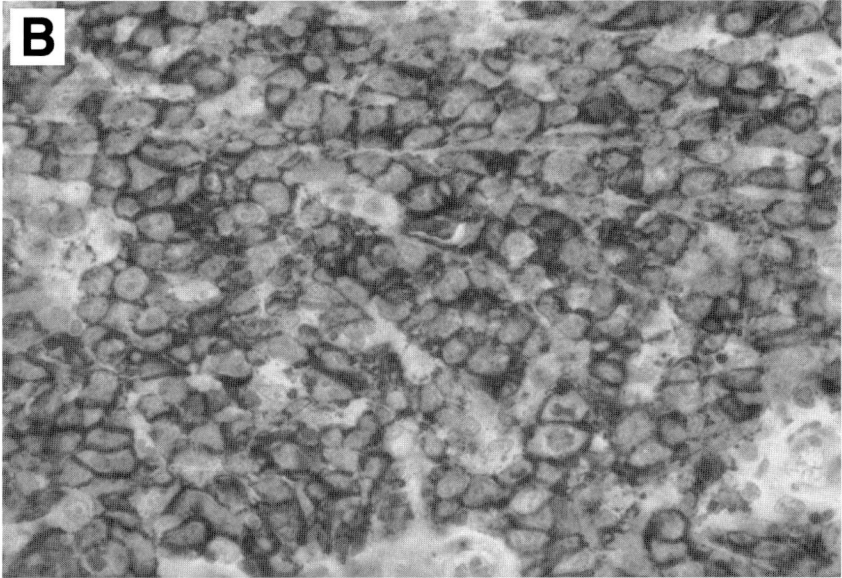

**Fig. 8.** Diffuse large B-cell lymphoma involving a lymph node. (**A**) Hematoxylin and eosin-stained section and (**B**) immunoperoxidase stain for CD20 performed on a paraffin section of the tumor. The morphologic features including effacement of the lymph node architecture and the monomorphous population of large atypical lymphoid cells are diagnostic of malignant lymphoma. The strong staining of the tumor cells for CD20 indicates that the lymphoma is of the B-cell lineage. (*See* **Color Plate 7** following page 208.)

immunoglobulin light or heavy chains as the only markers of B-cell lineage. They usually arise in immunosuppressed hosts. T-cell lymphomas can almost all (75%) be reliably identified by antibodies to CD3 (polyclonal) *(62,63,65–67)* **(Fig. 9)**, and the remainder will be positive for CD45RO (UCHL-1) *(62,65,68)*. Because antibodies to CD45RO lack absolute T-cell lineage specificity, and because NK lymphocytes can express the CD3ε epitope recognized by the paraffin-reactive polyclonal CD3 antibodies, βF1 is another excellent marker for T-cell lineage in paraffin sections *(69–72)*. Therefore, by using antibodies to CD20, CD79a, CD3, CD45RO, and the T-cell receptor β chain (βF1), almost all cases of lymphoma can be reliably determined to be of the B- or T-cell type in paraffin sections.

If frozen section immunoperoxidase stains are used to determine cell lineage, a limited panel will also accurately help to determine phenotype. A typical frozen section immunohistochemistry panel might include antibodies to the B-cell lineage-associated antigens CD19, CD20, CD22, CD10, CD23, and κ and λ light chains and to the T-cell lineage-associated antigens CD2, CD3, CD5, and CD7. This panel will successfully determine B- or T-cell lineage in almost all lymphoma cases.

Recognizing NK cells by phenotypic characteristics can be problematic because they have phenotypes that overlap with cytotoxic T-cells and granulocytes. They are separable by immunophenotype using frozen tissue, as illustrated in **Table 6.** NK cell and cytotoxic T-cell antigens that can be assessed in paraffin sections include CD3ε (polyclonal anti-CD3 recognizes the CD3ε chain, which is expressed in both NK and T-cells) *(73–75)*, CD56 *(76,77)*, CD8, Tia-1 *(78)*, and granzyme B *(79)*. In addition, some NK cell neoplasms (e.g., nasal type NK/T cell lymphoma) exhibit strong, uniform Epstein-Barr virus (EBV) expression using *in situ* hybridization techniques with probes that recognize small EBV-encoded RNAs (EBER) *(80)*. Because NK cells do not rearrange T-cell receptor genes, an additional diagnostic criterion for NK cell malignancies is to demonstrate absence of clonal T-cell receptor gene rearrangements *(76)*.

## Classifying Lymphomas Based on Phenotypic Characteristics

A detailed discussion of the diagnostic criteria for lymphoma classification is beyond the scope of this chapter. An excellent synopsis of the morphologic and phenotypic criteria by which lymphomas are classified can be found in the description of the Revised European American Lymphoma Classification *(81)*. Although lymphomas should be classified by a combined morphologic and phenotypic approach, certain lymphomas of B-cell lineage express characteristic constellations of antigens.

As a neoplasm of immature B-cells, the cells of B-lymphocyte precursor lymphoblastic lymphoma/leukemia virtually always express terminal deoxynucleotidyl transferase and are usually positive for CD34. Most cases express CD19 and CD10, and there is variable expression of CD20 and CD22 (cytoplasmic and/or surface) *(82–86)*. Examples of B-precursor lymphoblastic lymphomas are almost always surface immunoglobulin light chain negative. A subset of cases formerly referred to as "pre-B-cell phenotype" is positive for cytoplasmic μ-immunoglobulin heavy chain and does not express surface immunoglobulin heavy or light chains *(87)*.

The phenotypic "fingerprints" of B cell small lymphocytic lymphoma/chronic lymphocytic leukemia *(3,13,81,88–91)* **(Fig. 3)**, mantle cell lymphoma *(11,12,45,46,48,81,88, 92–94)*, splenic marginal zone B-cell lymphoma *(81,95–99)*, marginal zone B-cell

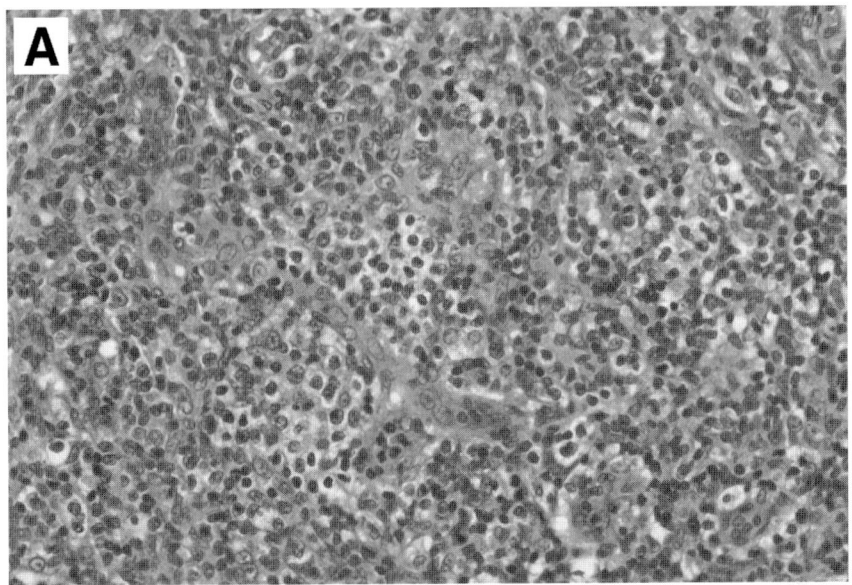

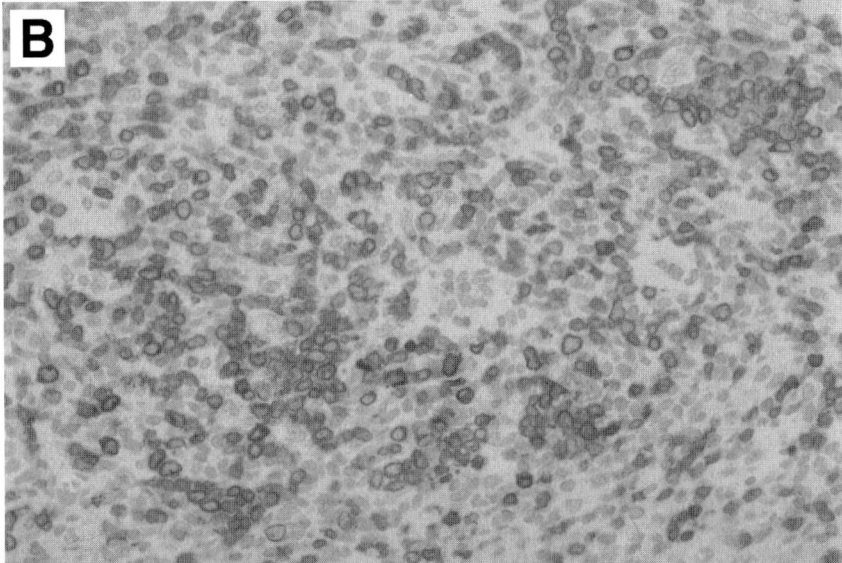

**Fig. 9.** Angioimmunoblastic T-cell lymphoma involving a lymph node. (**A**) Hematoxylin and eosin-stained section and (**B**) immunoperoxidase stain for CD3 performed on a paraffin section of the tumor. The morphologic features including architectural effacement of the lymph node and the clustered atypical cells with clear cytoplasm are diagnostic of malignant lymphoma. The strong cytoplasmic staining of the cells for CD3 indicates that the lymphoma is of the T-cell lineage. (*See* **Color Plate 7** following page 208.)

**Table 6**
**Phenotypic Characteristics of NK Cells and Cytotoxic T-Cells**

| Antigen | NK Cells | Cytotoxic T-cells |
|---|---|---|
| CD2 | + | + |
| CD3 | − | + |
| CD5 | − | + |
| CD7 | + | + |
| CD8 | − | + |
| CD16 | + | − |
| Tia-1/granzyme B | + | + |
| Antigen recognition receptor | KIR[a] | αβ or γδ |
| MHC restriction | Class I | Class II |
| Function | Kill target cells | TCR and ADCC[b] |

[a]Killing inhibitory receptor.
[b]T-cell receptor and antibody-dependent cell-mediated cytotoxicity.

**Table 7**
**Immunophenotypic Features of Small B-Cell Lymphoproliferative Disorders:**
**Frozen Section and Paraffin Section Approach**

| | SIg | Cyclin D1 | CD20 | CD23 | CD10 | CD5 | CD3 |
|---|---|---|---|---|---|---|---|
| Small lymphocytic/chronic lymphocytic leukemia | Monoclonal | − | + | + | − | + | − |
| Lymphoplasmacytic | Monoclonal cIg in plasma cells | | ± | ± | ± | ± | − |
| Mantle cell[a] | Monoclonal | + | + | − | − | + | − |
| Marginal zone: splenic, nodal, and extranodal | Monoclonal | − | + | − | − | − | − |
| Hairy cell leukemia[b] | Monoclonal | − | + | − | − | − | − |
| Follicular lymphoma | Monoclonal | − | + | ± | + | − | − |

[a]Cells in most cases of mantle cell lymphoma (>80%) exhibit nuclear cyclin D1 staining.
[b]In addition to the above, the cells of hairy cell leukemia characteristically and brightly co-express CD22 and CD11c and are positive for CD103 (frozen sections). By paraffin section immunohistochemistry, hairy cells are also positive for tartrate-resistant acid phosphatase and DBA.44.

lymphoma of the mucosa-associated lymphoid tissue (MALT) type *(2,11,81,100),* and follicular lymphomas *(2,3,12,13, 81,91,101)* are presented in **Table 7.** In the past, frozen section immunohistochemistry or flow cytometry was necessary to make these distinctions *(102).* Using frozen section immunoperoxidase stains and antibodies to CD19, CD20, CD10, CD23, CD5, and CD3, these low-grade lymphomas can be separated from one another. Recently, antibodies to CD5 *(15,16,18,20),* CD23 *(2,103–105),* CD10 *(2,20,106–109)* **(Fig. 10)**, and cyclin D1 *(2,45–49)* **(Fig. 6)** that perform well in fixed, paraffin sections have become available, and they can also be used to resolve the differential diagnosis of the small B-cell lymphomas. Lymphoplasmacytic lymphomas do not have a consistent phenotype *(110–114).* Some express CD5 and CD23, similar to B-cell

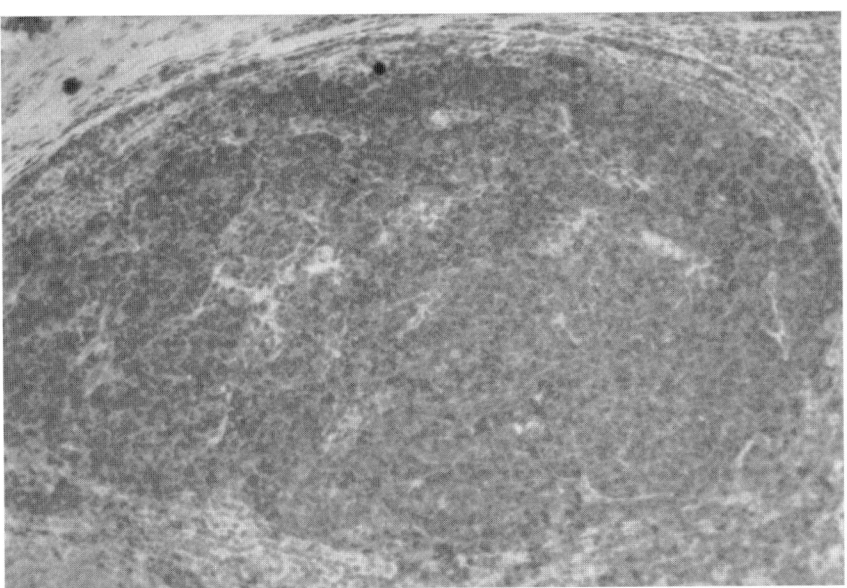

**Fig. 10.** Immunoperoxidase stain for CD10 performed on a paraffin section of a follicular lymphoma. Note the strong cytoplasmic staining for CD10 in the neoplastic follicular center cells. (*See* **Color Plate 7** following page 208.)

small lymphocytic lymphoma and chronic lymphocytic leukemia. Some express CD5 without CD23, similar to mantle cell lymphoma, some are positive for CD10, and some cases express neither CD5 nor CD23. However, a consistent defining feature of lymphoplasmacytic lymphoma is plasma cells within the lesion that have a monoclonal staining pattern for κ or λ immunoglobulin light chains. Paraffin section immunohistochemistry is best suited for this application.

Large B-cell lymphomas are almost always positive for CD19 and CD20 *(13,91,106,115)* (**Fig. 8**), a sizable subset expresses CD10 *(13,91,106,108,115–118)* and/ or bcl-2 *(44,108,117–119)*, and rare examples are positive for CD5 *(117,120)*. Surface immunoglobulin expression occurs at a lower frequency in large B-cell lymphomas than in the lower grade B-cell lymphomas, particularly those that arise in the mediastinum *(121)*. Up to 40% are immunoglobulin negative. Practically speaking, large B-cell lymphomas are sufficiently phenotyped in paraffin sections utilizing antibodies to CD20.

The high-grade histologic features of Burkitt's lymphoma (intermediate cell size, high mitotic rate, and intermixed tingible body macrophages) can render cases of Burkitt's lymphoma difficult to distinguish from B-precursor or T-precursor lymphoblastic lymphoma/leukemia by morphology alone. However, Burkitt's lymphomas are uniformly terminal deoxynucleotidyl transferase and CD34 negative, and they almost always express surface immunoglobulin. In addition, they are usually positive for CD10 together with pan-B antigens, including CD19, CD20, and CD22 *(13,81,91,122)*.

In rare cases, it is difficult by morphology alone to make the clinically important distinction between multiple myeloma and either low-grade lymphomas with extensive plasmacytic differentiation (lymphoplasmacytic lymphoma, MALT-type lymphoma) or large B-cell lymphomas with plasmacytoid features (formerly termed immunoblastic

**Table 8**
**Immunophenotypic Differences Between Multiple Myeloma and Non-Hodgkin's Lymphomas with Plasmacytic Differentiation: Paraffin Sections**[a]

| Marker | Multiple myeloma | Lymphoma |
|---|---|---|
| CD45 | Negative to weak | Strong positive |
| CD79a | Positive | Positive |
| CD20 | Negative to weak | Strong positive |
| CD138 (Syndecan) | Positive | Negative |
| Immunoglobulin heavy chain isotype | Any except IgM | IgM |
| Epithelial membrane antigen | Positive | Negative to weak |

[a]This distinction cannot always be made based solely on morphologic and phenotypic criteria; see text.

lymphoma). This problem can sometimes be resolved by immunophenotype, as illustrated in **Table 8** *(123)*. However, there is phenotypic overlap between plasma cell neoplasms and B-cell lymphomas, so in difficult cases it is prudent to defer to the clinical presentation rather than to rely solely on morphology and phenotype. Serum or urine paraproteins, particularly of the IgA or IgG isotype, lytic bone lesions, renal failure, and hypercalcemia strongly favor multiple myeloma over lymphoma.

The phenotypes of the different T-cell lymphoma types are not nearly as homogeneous as those seen in specific B-cell lymphomas. In general, the neoplastic cells in cutaneous T-cell lymphoma (mycosis fungoides) *(33–35)*, peripheral T-cell lymphoma not otherwise characterized *(3,23–27)*, and angioimmunoblastic T-cell lymphoma *(81,124–127)* are phenotypically indistinguishable. In each there is usually an aberrant T-cell phenotype, as discussed above, and most examples are tumors of CD4-positive helper lymphocytes that express the αβ form of the T-cell antigen receptor. In addition, angioimmunoblastic T-cell lymphomas often contain EBV-positive, nonneoplastic T- or B-lymphocytes *(128)* and disorganized meshworks of CD23-positive follicular dendritic cells admixed with the neoplastic T-cells *(129,130)* (**Fig. 11**).

A subset of unusual extranodal T-cell lymphomas has phenotypic characteristics of cytolytic T-cells *(78)*. These cases include subcutaneous panniculitis-like T-cell lymphomas *(131,132)*, hepatosplenic γδ and αβ T-cell lymphomas *(78,133)*, and enteropathy-type T-cell lymphomas *(78,79,134)*. They are recognized as neoplasms of cytolytic T-cells because they express CD3 (membrane CD3, not only cytoplasmic epsilon chain), the common framework antigens of the αβ or γδ T-cell receptors and cytolytic granule proteins, such as Tia-1 (**Fig. 12**) and granzyme B. In addition, most cases of subcutaneous panniculitis like T-cell lymphoma are CD8 positive, and most cases of enteropathy-type T-cell lymphomas are CD4 and CD8 negative. Normal T-cells that express the γδ form of the T-cell antigen receptor are negative for CD5, CD4, and CD8, and the hepatosplenic γδ-T-cell lymphomas usually recapitulate this phenotype.

Cases of T-cell precursor lymphoblastic lymphoma/leukemia are characterized by an immature phenotype corresponding to stages in intrathymic T-cell development (**Table 9**) *(36,135–137)*. They uniformly express terminal deoxynucleotidyl transferase, an enzyme that can be detected by paraffin section immunohistochemistry. Unlike B-lymphocyte precursor lymphoblastic lymphoma/leukemia, they are usually negative for

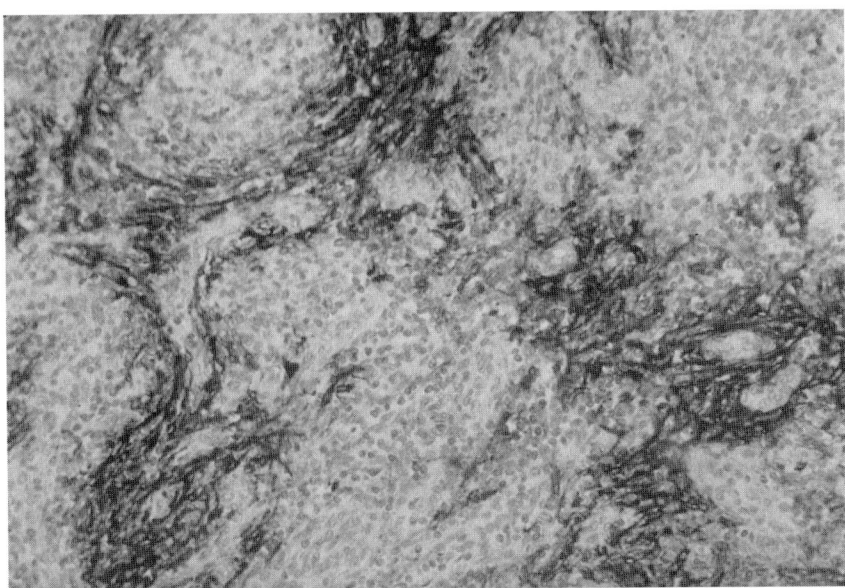

**Fig. 11.** Immunoperoxidase stain for CD23 performed on a paraffin section of an angioimmunoblastic T-cell lymphoma. Same case as **Fig. 9.** Note the strong CD23-positive follicular dendritic cell meshworks that are characteristically observed in this type of T-cell lymphoma. (*See* **Color Plate 7** following page 208.)

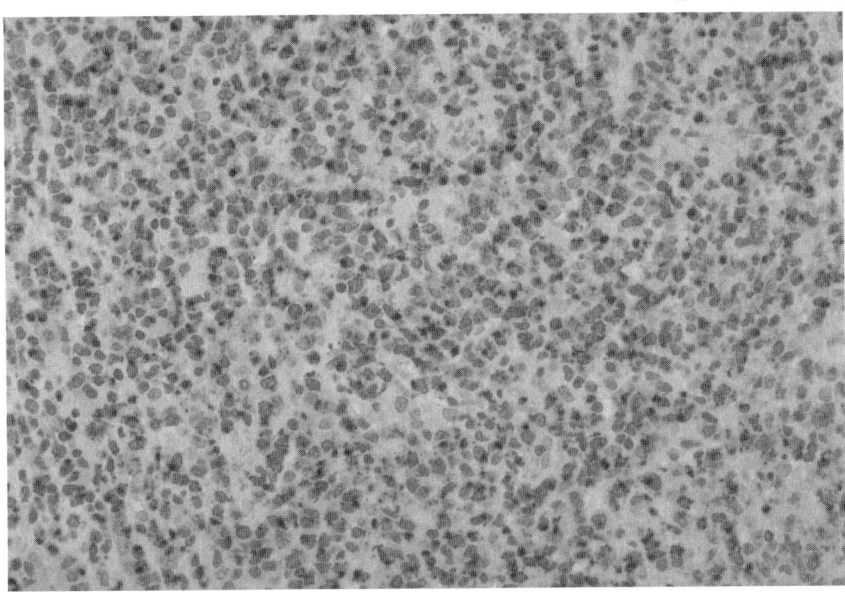

**Fig. 12.** Immunoperoxidase stain performed on a paraffin section for the cytolytic granule protein Tia-1 in a case of NK/T-cell lymphoma of the nasal type. The neoplastic cells exhibit coarse granular cytoplasmic positivity, characteristic of the pattern of staining that is observed for this antigen. (*See* **Color Plate 7** following page 208.)

**Table 9**
**Phenotypes of T-Lymphoblastic Lymphoma/Leukemia**

| Normal Counterpart | CD1 | CD2 | CD7 | CD3 | CD4 | CD8 | TdT |
|---|---|---|---|---|---|---|---|
| Immature thymocyte[a] | − | + | + | − | − | − | + |
| Cortical thymocyte | + | + | + | +c | + | + | + |
| Medullary thymocyte | − | + | + | +s | ±[b] | ±[b] | ± |

cCytoplasmic CD3 expression.
sSurface CD3 expression.
[a]CD2, CD7, CD4, and TdT can all be expressed by myeloid lineage acute leukemias.
[b]At the medullary thymocyte stage, CD4 and CD8 expression are mutually exclusive.

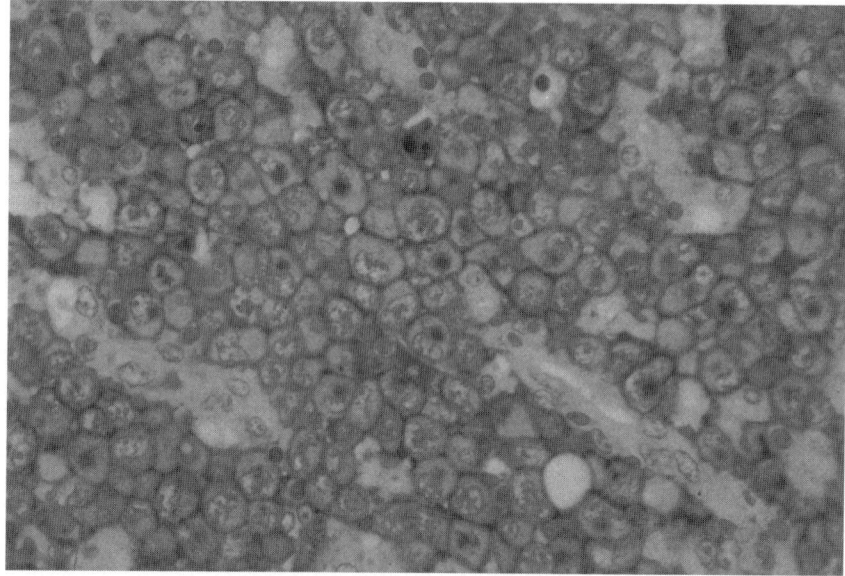

**Fig. 13.** Immunoperoxidase stain for CD30 performed on a paraffin section of a CD30-positive anaplastic large cell lymphoma. Note the strong, uniform staining of the neoplastic cells. The membrane and dot-like cytoplasmic staining pattern is characteristic for this marker.

CD34. The other T-cell lineage antigens expressed by lymphoblastic lymphoma/leukemia are best demonstrated in frozen sections.

CD30-positive anaplastic large cell lymphoma deserves special mention because of its unique morphology and phenotypic features *(138–144)*. Immunohistochemistry is required for the diagnosis of CD30-positive anaplastic large cell lymphoma, because the expression of CD30 by the neoplastic cells in part defines the entity. Furthermore, CD30-positive anaplastic large cell lymphomas share morphologic features with both Hodgkin's lymphoma and metastatic malignancies, and these must be distinguished from one another. Typically, the cells of CD30-positive anaplastic large cell lymphoma are strongly and uniformly positive for CD30 (**Fig. 13** and **Color Plate 7** following page 208), CD45 (in 70% of cases), epithelial membrane antigen (in up to 70% of cases), and HLA-DR. In paraffin sections, T-cell lineage-associated markers such as

**Table 10**
**Distinction of CD30-Positive Anaplastic Large Cell Lymphoma from Hodgkin's Lymphoma Based on Immunophenotype and Genetic Studies**

| Marker | CD30-positive anaplastic large cell lymphoma (%) | Classical Hodgkin's lymphoma (%) |
|---|---|---|
| CD30 | + (strong, all cells) | + (variable, cell subset) |
| CD45 | + | – |
| CD15 | ± (15%) | + (80%) |
| CD43 | + (70%) | – |
| CD45RO | + (50%) | – |
| CD3 (paraffin) | + (50%) | – |
| EMA | + | – |
| p80/ALK-1 | + | – |
| Clonal T-cell receptor gene rearrangements | + (70%) | – |

CD3, CD45RO, and CD43 are expressed in at least 70% of cases. Frozen section immunohistochemistry can be used to assess the tumor for other T-cell lineage antigens such as CD2, CD5, and CD7 in cases that fail to exhibit a T-cell phenotype by paraffin section immunohistochemistry. Recently, it has been shown that a substantial subset of CD30-positive anaplastic large cell lymphomas expresses cytolytic granule proteins such as Tia-1, granzyme B, and perforin *(141)*. The remainder of the cases expresses neither T-cell nor B-cell lineage markers. The recent World Health Organization classification of hematolymphoid malignancies does not recognize B-cell lineage CD30-positive anaplastic large cell lymphoma as a distinct disorder. The characteristic phenotype of CD30-positive anaplastic large cell lymphoma and its distinction from Hodgkin's disease based on immunohistochemistry are listed in Table 10.

## Distinguishing Hodgkin's Lymphoma from Non-Hodgkin's Lymphomas

The more the phenotype and molecular characteristics of Hodgkin's lymphomas are studied, the closer the resemblance of Hodgkin's lymphoma to non-Hodgkin's lymphomas becomes. Nonetheless, Hodgkin's lymphoma is morphologically and clinically distinctive, and its separation from non-Hodgkin's lymphomas remains critical to patient management. Although morphology remains the premier criterion by which Hodgkin's lymphoma is distinguished from non-Hodgkin's lymphomas, phenotypic data can be a useful supplement to the morphology (**Table 11**).

As Hodgkin's lymphoma is currently understood, it can be broadly separated into the nodular lymphocyte predominance type and all other types, now generically termed *classical Hodgkin's lymphoma (1)*. This distinction is based on traditional morphologic criteria and on the phenotypic differences between the L and H variants of Reed-Sternberg cells of lymphocyte predominance type and the Hodgkin cells of classical Hodgkin's lymphoma.

CD15 and CD30 are frequently expressed by the Hodgkin's cells in classical types of Hodgkin's disease (**Fig. 14**), whereas leukocyte common antigen (CD45) is only rarely expressed in paraffin section immunohistochemical studies *(145–152)*. A highly

**Table 11**
**Distinction of Classical Hodgkin's Lymphoma from Non-Hodgkin's Lymphoma**
**(NHL) Based on the Phenotype of the Large Cells**

| Marker | Hodgkin's lymphoma (classical) | Non-Hodgkin's lymphoma |
|---|---|---|
| CD45 | Negative | Positive |
| CD15 | Positive (80%) | Negative (90%) |
| CD30 | Positive (90+%) | Variable |
| CD20 | Often negative if positive, cell subset, variable intensity | Positive in most B-cell lineage NHL: strong, uniform staining |
| CD3 | Usually negative | Positive in most T-cell lineage NHLs |
| Fascin | Positive | Negative to weak |

sensitive marker for Reed-Sternberg cells and variants from cases of classical Hodgkin's lymphoma is fascin **(Fig. 14)**, an actin bundling protein *(153)*. Hodgkin's cells are also positive for activation antigens, including HLA-DR, interleukin-2 receptors (CD25), the transferrin receptor (CD71) *(150,154)*, and CD95 (APO-1/Fas) *(155)*.

Reed-Sternberg cells of the classical types of Hodgkin's lymphoma exhibit variable expression from case to case of pan-B- and pan-T-cell markers. There have been consistent reports of subpopulations of Reed-Sternberg cells positive for CD19, CD20, CD22, and CD75 in up to 80% of the classical types of Hodgkin's lymphoma *(156–161)*. T-cell and cytolytic lymphocyte granule-associated antigens are expressed by a small percentage of cases of Hodgkin's lymphoma of the nodular sclerosis and mixed cellularity types *(157,162–165)*. CD2, CD3, CD4, and the T-cell antigen receptor framework antigens are the markers that have been most frequently studied in this regard. CD45RO and CD43 are almost never expressed by Hodgkin's cells. Finally, Reed-Sternberg cells have only rarely been considered positive for antigens expressed on histiocytic or myelomonocytic cells, except for CD15.

These observations can be exploited diagnostically to separate cases of classical Hodgkin's lymphoma from non-Hodgkin's lymphomas *(152)*. Paraffin section immuno-histochemistry is preferred for this application. In general, if the large cells of a lymphoma express CD15 and CD30 and are negative for CD45 (this is a critical negative result), the diagnosis of Hodgkin's lymphoma of one of the classical types is supported. On the other hand, if the large cells of a lymphoma are positive for CD45 (95% of non-Hodgkin's lymphomas express CD45) and positive or negative for either CD15 or CD30, the diagnosis of non-Hodgkin's lymphoma is supported. Expression of CD20 in cells of classical Hodgkin's lymphoma, when observed, tends to be variable; some cells are strongly positive, others are negative, and still others exhibit weak staining. Conversely, the cells of non-Hodgkin's lymphomas of a B-cell lineage, including T-cell-rich B-cell lymphomas, generally show strong and uniform staining for CD20. T-cell lineage antigen expression by Hodgkin's cells (CD3, CD45R0, or CD43) is distinctly uncommon in paraffin sections and usually can be taken as evidence against the diagnosis of Hodgkin's lymphoma.

In cases of nodular lymphocyte predominance Hodgkin's lymphoma, the L and H variants of Reed-Sternberg cells have a distinctive and consistent phenotype, indicative of B-cell lineage. They are positive for pan B-cell markers, such as CD20 and CD22 in

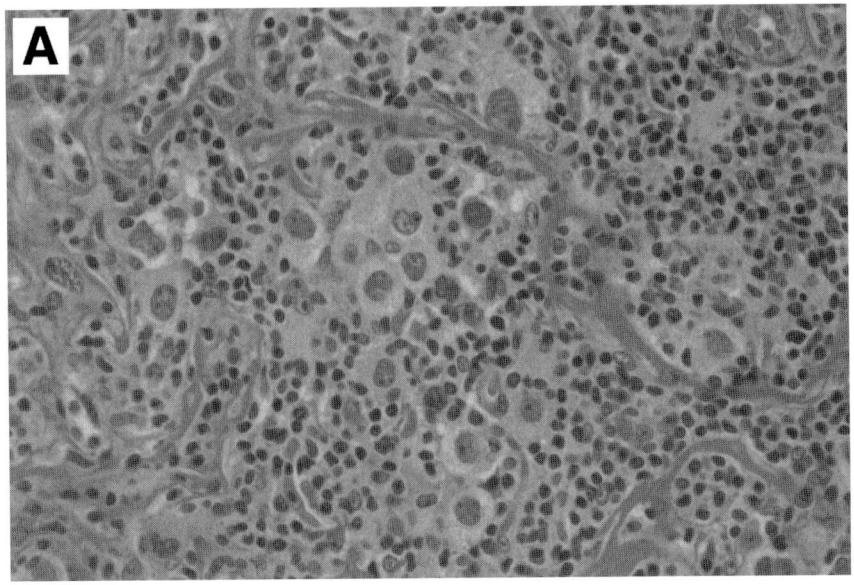

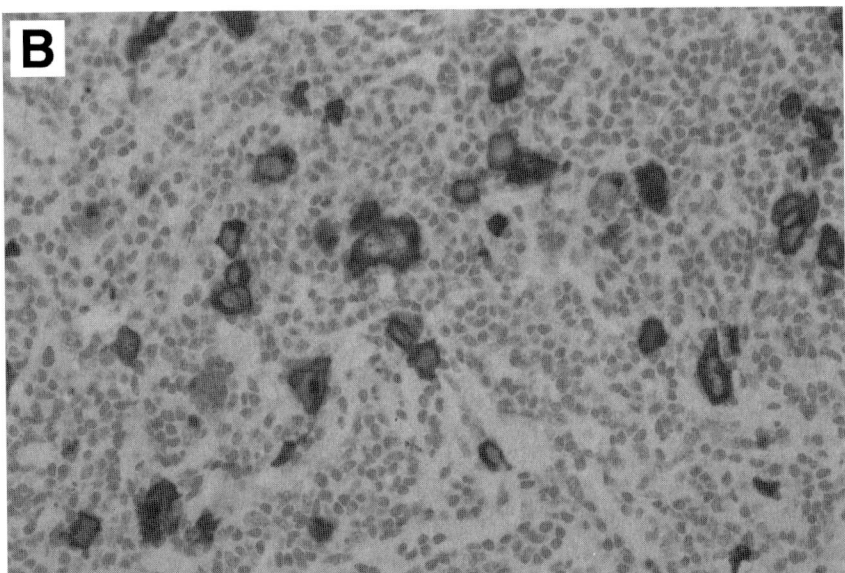

**Fig. 14.** Hodgkin's lymphoma, nodular sclerosis type involving a lymph node. (**A**) Hematoxylin and eosin-stained section demonstrating the large neoplastic cells admixed with small lymphocytes, eosinophils, plasma cells, and macrophages. Immunoperoxidase stains for (**B**) CD15, (**C**) CD30, and (**D**) fascin performed on paraffin sections. The neoplastic cells are positive for all three antigens. Note that the stains for CD15 and CD30 decorate the tumor cells in a membrane and punctate perinuclear staining pattern, whereas fascin positivity is diffusely present in the cytoplasm. (*See* **Color Plate 8** following page 208.)

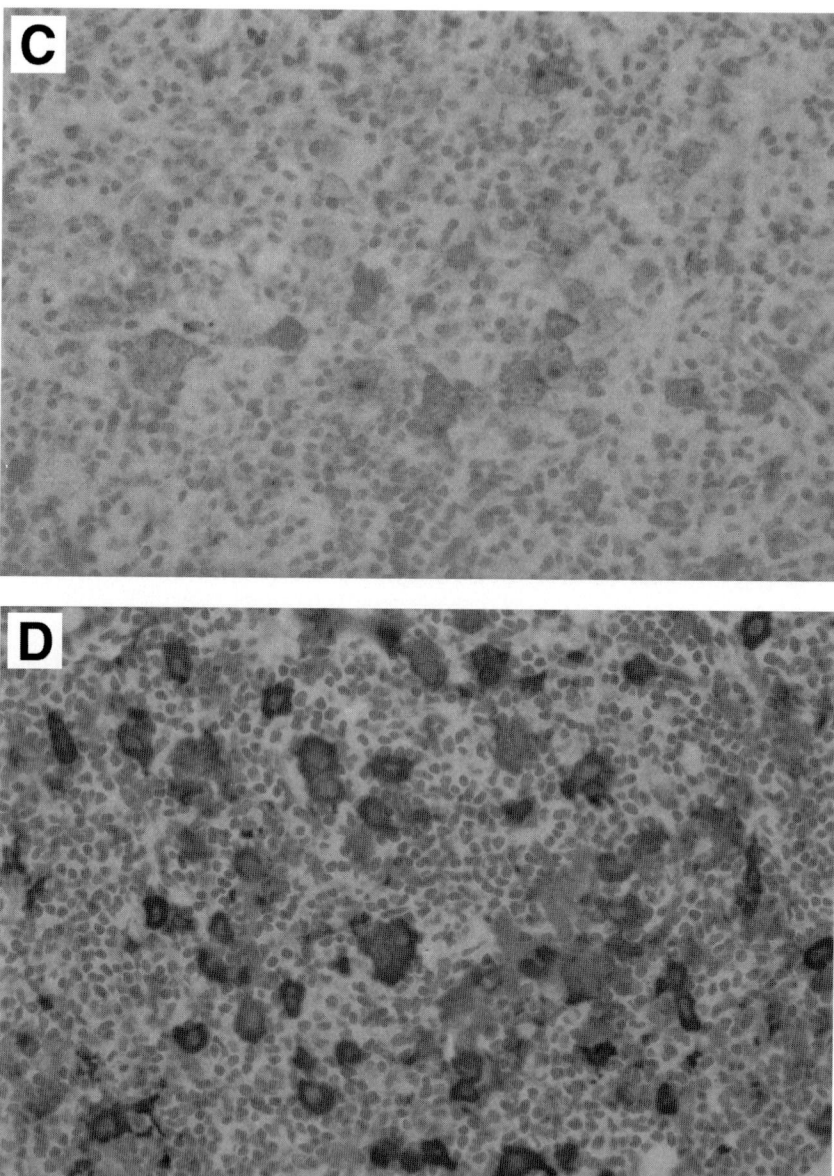

**Fig. 14.** *Continued*

frozen sections and CD20 in paraffin sections *(166–173)* (**Fig. 15** and **Color Plate 8** following page 208). L and H variants of Reed-Sternberg cells consistently stain for J-chain, a polypeptide synthesized exclusively by B-lymphocytes *(174,175)*, and it has been demonstrated *(159,160)*, that L and H variants express CD79a, a component of the B-cell-specific phosphoprotein heterodimer associated with the B-cell antigen receptor complex. In contrast to other types of Hodgkin's lymphoma, the L and H variants of Reed-Sternberg cells are frequently positive for CD45; they lack staining for CD15, and they have variable staining for CD30. In a subset of cases, they express epithelial membrane antigen *(166,169)*.

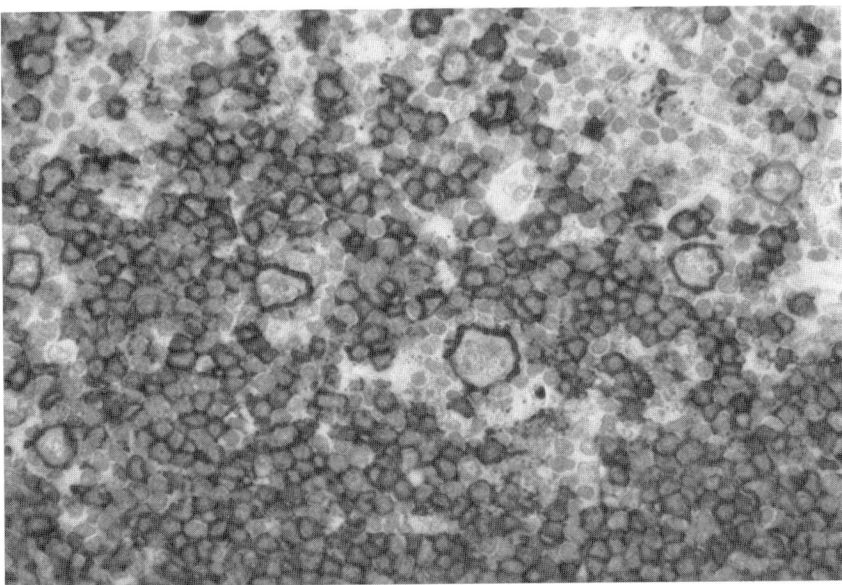

**Fig. 15 (Color Plate 8).** Hodgkin's lymphoma, nodular lymphocyte predominance type, involving a lymph node. Immunoperoxidase stain performed on paraffin sections for CD20. The large L and H variants of Reed-Sternberg cells, characteristically observed in this type of Hodgkin's lymphoma, are positive for CD20. They are admixed with numerous CD20-positive nonneoplastic small lymphocytes, but the cells in the immediate vicinity of the large cells are negative for CD20. They are CD3- and CD57-positive T-cells (not illustrated).

In most cases of nodular lymphocyte predominance Hodgkin's lymphoma, most of the small lymphocytes accompanying the Reed-Sternberg variants are polyclonal B-lymphocytes. Most express membrane IgM and IgD similar to mantle zone lymphocytes, and they are mixed with numerous follicular dendritic reticulum cells positive for CD21 and CD35 *(173–175)*. Characteristically, the small lymphocytes that are immediately adjacent to the L and H variants ("rosetting" around the L and H variants) have a distinctive phenotype. They are CD2, CD3, CD4, and CD57 positive and contain strong nuclear bcl-6 positivity, a feature unique to lymphocyte perdominance Hodgkin's lymphoma *(176–178)*.

## Myeloid and Histiocytic Lineage Malignancies

Diagnosis, classification, selection of therapy, and assessment of prognosis of the acute myeloid leukemias is optimally accomplished by a combination of morphology of Wright-Giemsa-stained blood and bone marrow aspirate smears, cytochemistry, flow cytometry, cytogenetics, and in some instances, molecular genetic testing *(1,83,102, 179–185)*. However, there are times when fresh tissue is not available for these ancillary studies; in these instances immunohistochemistry performed on frozen or paraffin sections can at least aid in the distinction between myeloid and lymphoid cell lineages and distinguish T-precursor from B-precursor types of lymphoblastic leukemia *(186–193)*. The phenotypic characteristics of T and B-lymphocyte precursor lymphoblastic lymphoma/leukemia are presented in **Tables 5** and **9 (Fig. 16)**. A strategy employing

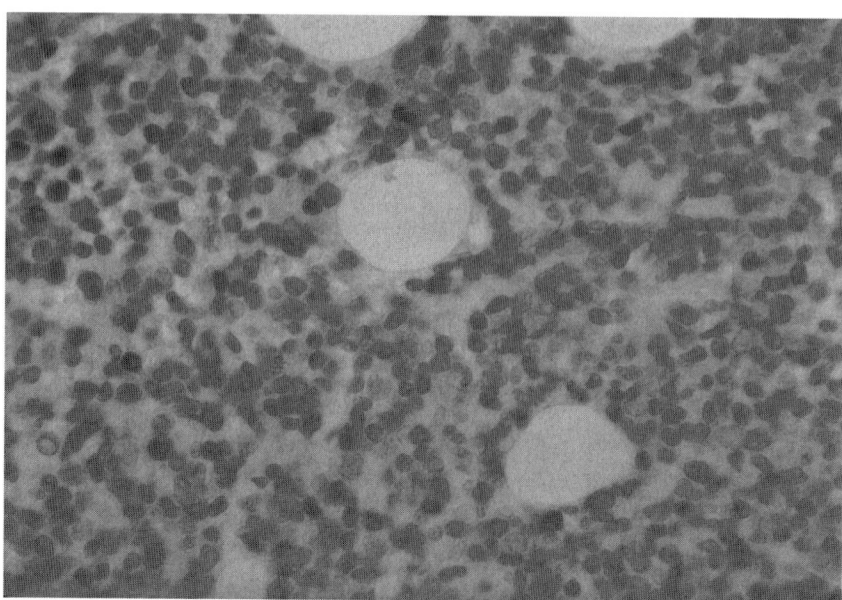

**Fig. 16.** Acute lymphoblastic leukemia, B-lymphocyte precursor type. Immunoperoxidase stain for terminal deoxynucleotidyl transferase (TdT) performed in paraffin sections. The leukemic blasts exhibit strong nuclear staining for TdT. This result is observed in almost all cases of T-lymphocyte precursor and B-lymphocyte precursor acute lymphoblastic leukemia. In this case the neoplastic cells expressed the B-cell lineage-associated antigens CD19 and CD10 (not illustrated). (*See* **Color Plate 8** following page 208.)

**Table 12**
**Paraffin Section Immunoperoxidase Characterization of Acute Myelogenous Leukemia by FAB Type**[a]

| FAB type | MPO | Lysozyme | CD68 | FVIII/CD61 | Hemoglobin | CD34/TdT |
|----------|-----|----------|------|------------|------------|----------|
| M1/M0 | + | + | − | − | − | + |
| M2 | + | + | − | − | − | ± |
| M3 | + | + | − | − | − | − |
| M4 | ± | + | ± | − | − | ± |
| M5 | − | + | + | − | − | ± |
| M6 | − | − | − | − | + | − |
| M7 | − | − | − | + | − | − |

[a]FAB, French-American-British classification; MPO, myeloperoxidase; CD68, PG-M1; FVIII, factor VIII-related antigen (von Willebrand factor); TdT, terminal deoxynucleotidyl transferase.

paraffin section immunohistochemistry to distinguish among the types of acute myelogenous leukemia defined by the French-American-British (FAB) classification scheme is presented in **Table 12 (Fig. 17)**. Although most cases of acute myelogenous leukemia will have a pattern of antigen expression that is characteristic for each FAB type, overlap in antigen expression between the different acute myelogenous leukemia types occurs. In addition, proper recognition of myeloblastic leukemia without differentiation (M0), promyelocytic leukemia (M3), myelomonoblastic leukemia with (M4eo) and

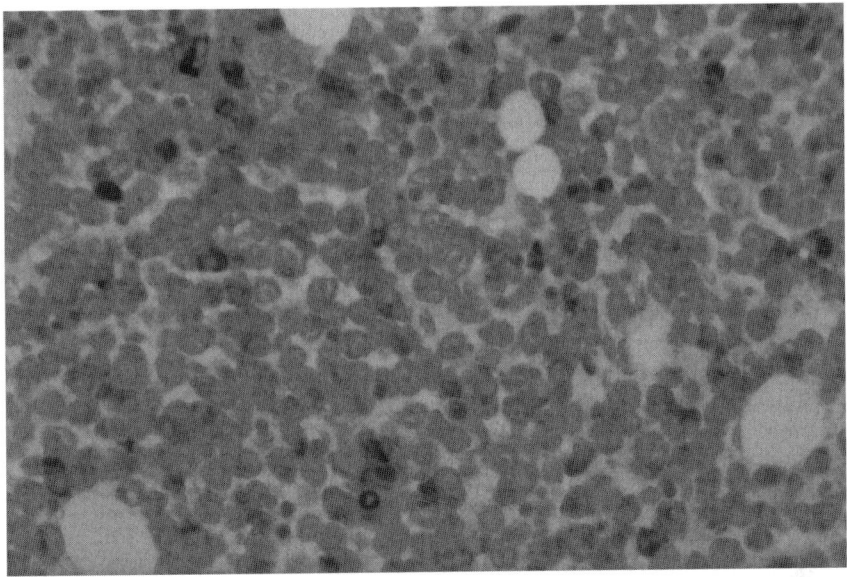

**Fig. 17.** Acute myelogenous leukemia, FAB subtype M1. Immunoperoxidase stain for myelo-peroxidase. The blasts are variably positive for myeloperoxidase (see also **Table 12**). (*See* **Color Plate 8** following page 208.)

without eosinophils (M4), and monoblastic leukemia (M5) requires evaluation of Wright-Giemsa-strained blood and bone marrow aspirates, cytochemical stains, particularly peroxidase and α-naphthyl butyrate esterase, and cytogenetic analysis *(181)*. Finally, therapeutic strategies employing monoclonal antibodies directed against CD33 and other myeloid lineage antigens now require phenotypic analysis of acute leukemias by flow cytometry. Therefore, phenotyping by immunohistochemistry has limited usefulness in the assessment of acute leukemias.

Neoplasms of differentiated monocytic/macrophage lineage are a complex group of tumors of phagocytic histiocytes and antigen presenting cells. They are distinguished from one another on the basis of clinical, morphologic, and phenotypic criteria. In general, benign and malignant cells of a monocyte-macrophage lineage are positive for CD14, CD68 *(194–196)*, and lysozyme *(197)*. They also often express HLA-DR antigens.

Tumors of the antigen-presenting cells include Langerhans' cell histiocytosis, follicular dendritic cell sarcoma, interdigitating reticulum cell sarcoma, and unspecified dendritic cell neoplasms *(198)*. Except for Langerhans' cell histiocytosis, these neoplasms are very rare. Most tumors of the antigen-presenting cells will express fascin *(153)*, and then they can be distinguished from one another by their pattern of expression of other antigens as illustrated in **Table 13** and by morphology. CD1a expression by neoplastic cells is unusual outside of Langerhans' cell histiocytosis **(Fig. 18)** and T-lymphocyte precursor lymphoblastic lymphoma *(199)*. Therefore, CD1a has very restricted diagnostic specificity. Cells in follicular dendritic reticulum cell sarcomas will usually express CD21, CD23, and/or CD35 *(200–203)*. Interdigitating reticulum cells are often positive for S-100 protein but have few other distinguishing phenotypic characteristics *(198,203–205)*. Because the dentritic cell tumors usually have a spindle cell morphology and because the antigens expressed by these tumors are not strictly

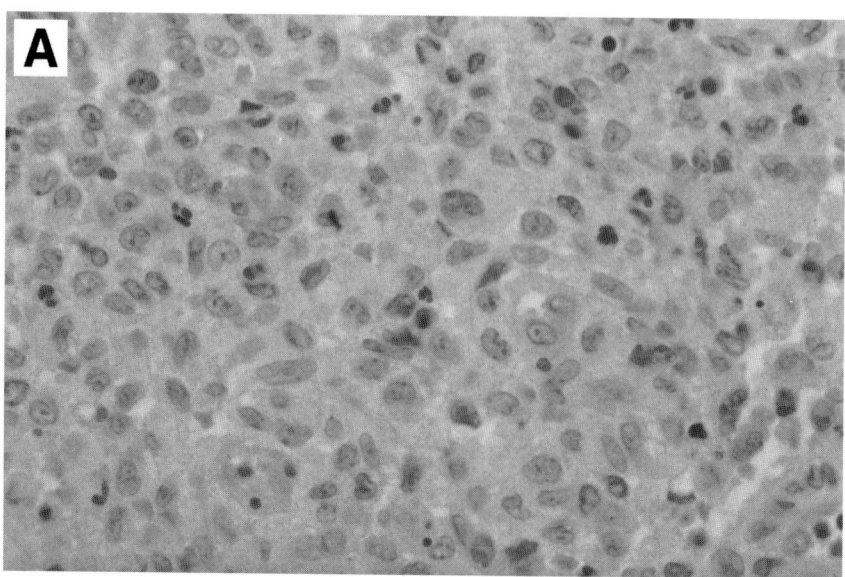

**Fig. 18.** Langerhans' cell histiocytosis involving the parotid gland. (**A**) Hematoxylin and eosin stain exhibits the typical cytologic features of this disorder. The architecture of the salivary gland is effaced by a monomorphous population of large cells with grooved and irregular nuclei and abundant pale eosinophilic cytoplasm. (**B**) Immunoperoxidase stain for CD1a performed in paraffin sections. The tumor cells are positive. (*See* **Color Plate 8** following page 208.)

**Table 13**
**Phenotypic Characteristics of Antigen-Presenting Cell Tumors**

| Tumor type | Fascin | CD1a | S-100 | CD21 | CD23 | CD35 |
|---|---|---|---|---|---|---|
| Langerhans' cell histiocytosis | + | + | + | − | − | − |
| Follicular dendritic cell sarcoma | + | − | − | + | + | + |
| Interdigitating reticulum cell sarcoma | + | − | + | − | − | − |

lineage specific, care must be taken to exclude other spindle cell sarcomas and sarcomatoid carcinomas when considering the diagnosis of a dendritic cell tumor.

Most cases formerly diagnosed as malignant histiocytosis have been shown to be neoplasms of other cell lineages (e.g., CD30-positive anaplastic large cell lymphoma and others) *(206,207)* or reactive processes characterized by prominent hemophagocytosis *(198).* The criteria for diagnosis of a true histiocytic malignancy (now termed histiocytic sarcoma) include demonstrating expression of histiocytic lineage antigens by the neoplastic cells and demonstrating absence of antigens expressed by T-cells, B-cells, melanocytes, and epithelial cells *(207–210).* Because histiocytes do not rearrange the T-cell antigen receptor genes or the immunoglobulin heavy and light chain genes, sometimes this also requires application of molecular genetics techniques to demonstrate absence of clonal rearrangements of these genes. In general, neoplastic histiocytes will express CD45, CD45RO, CD45RA, CD43, lysozyme, CD68, and CD14 in varying combinations (**Fig. 19**).

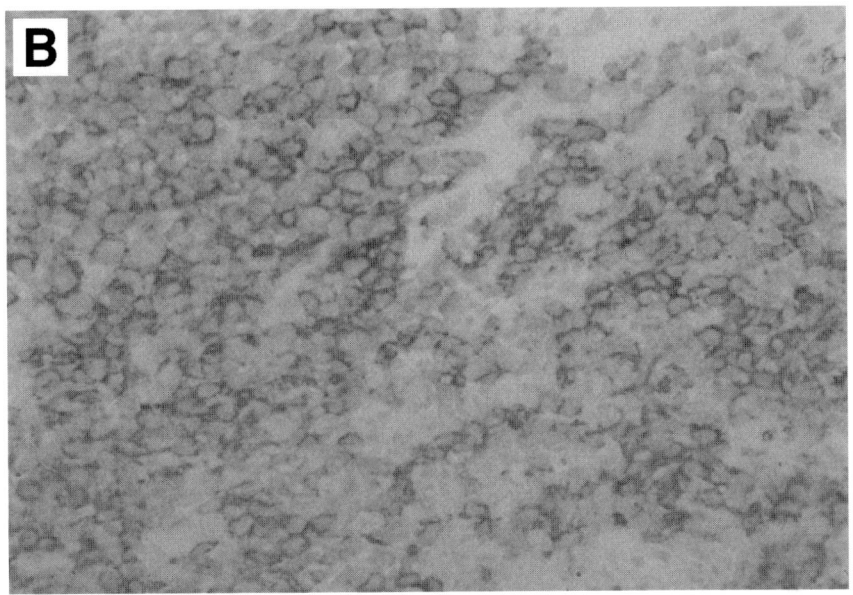

**Fig. 18.** *Continued*

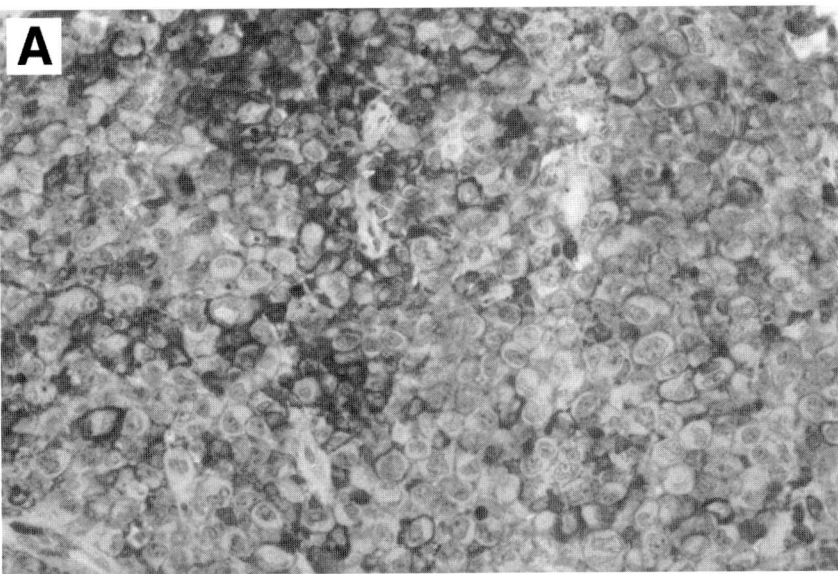

**Fig. 19.** Histiocytic sarcoma. Immunoperoxidase stains for (**A**) CD45 and (**B**) CD68 performed in paraffin sections and for (**C**) CD13 performed on frozen sections of the tumor. The neoplastic cells are positive for all three of these antigens and negative for a large number of T-cell and B-cell lineage-associated antigens. In conjunction with the morphology (not illustrated), the findings indicate that the neoplastic cells are of a histiocytic lineage. (*See* **Color Plate 8** following page 208.)

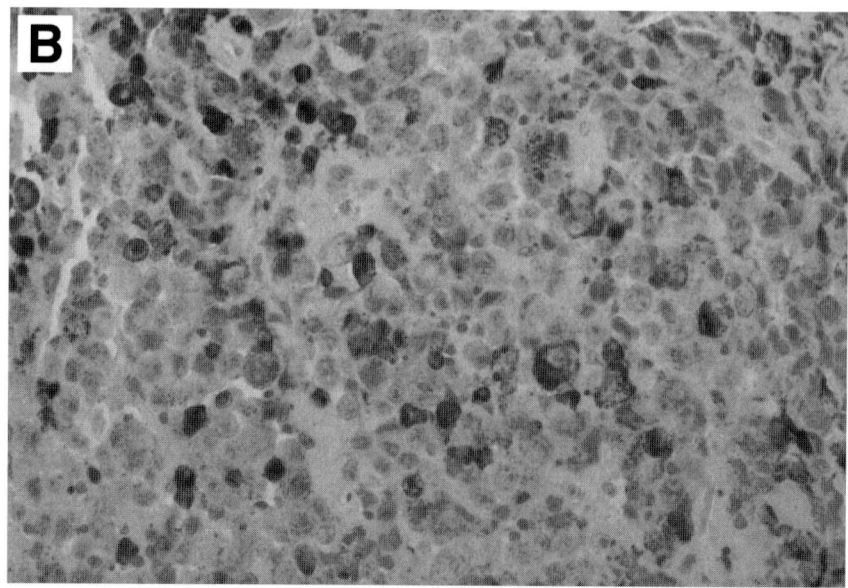

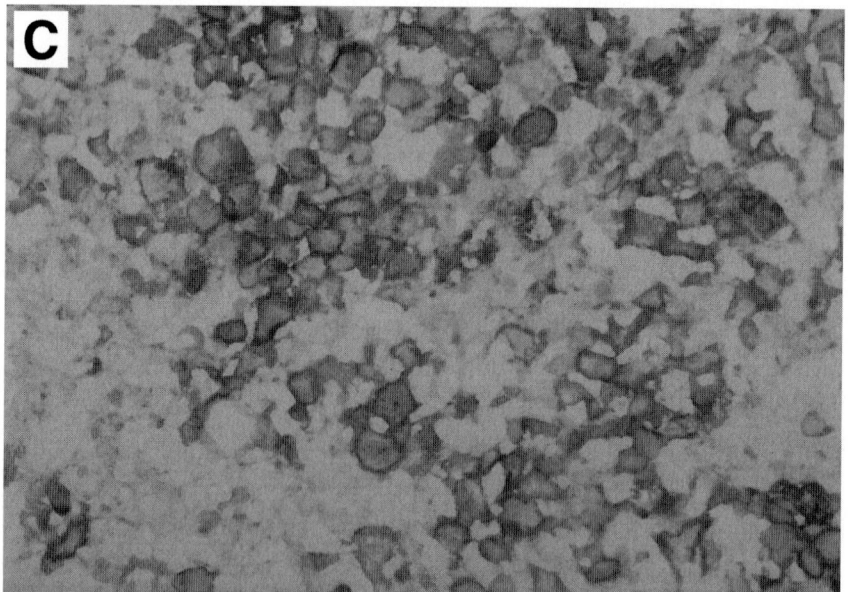

**Fig. 19.** *Continued*

## REFERENCES

1. Harris, N. L., Jaffe, E. S., Diebold, J., et al. (2000) The World Health Organization classification of hematological malignancies report of the Clinical Advisory Committee Meeting, Airlie House, Virginia, November 1997. *Mod. Pathol.* **13**, 193–207.
2. Kurtin, P. J., Hobday, K. S., Ziesmer, S., and Caron, B. L. (1999) Demonstration of

distinct antigenic profiles of small B-cell lymphomas by paraffin section immunohistochemistry. *Am. J. Clin. Pathol.* **112**, 319–329.

3. Picker, L. J., Weiss, L. M., Medeiros, L. J., Wood, G. S., and Warnke, R. A. (1987) Immunophenotypic criteria for the diagnosis of non-Hodgkin's lymphoma. *Am. J. Pathol.* **128**, 181–201.

4. Tbakhi, A., Edinger, M., Myles, J., Pohlman, B., and Tubbs, R. R. (1996) Flow cytometric immunophenotyping of non-Hodgkin's lymphomas and related disorders. *Cytometry* **25**, 113–124.

5. Witzig, T. E., Banks, P. M., Stenson, M. J., et al. (1990) Rapid immunotyping of B-cell non-Hodgkin's lymphomas by flow cytometry. A comparison with the standard frozen-section method. *Am. J. Clin. Pathol.* **94**, 280–286.

6. Harris, N., Poppema, S., and Data, R. (1982) Demonstration of immunoglobulin in malignant lymphomas, use of an immunoperoxidase technic on frozen sections. *Am. J. Clin. Pathol.* **78**, 14–21.

7. Stein, H., Bonk, A., Tolksdorf, G., Lennert, K., Rodt, H., and Gerdes, J. (1980) Immunohistologic analysis of the organization of normal lymphoid tissue and non-Hodgkin's lymphomas. *J. Histochem. Cytochem.* **28**, 746–760.

8. Tubbs, R., Sheibani, K., Weiss, R., Sebek, B., and Deodhar, S. (1981) Tissue immunomicroscopic evaluation of monoclonality in B-cell lymphomas. *Am. J. Clin. Pathol.* **76**, 24–28.

9. Strauchen, J. A. and Mandeli, J. P. (1991) Immunoglobulin expression in B-cell lymphoma. Immunohistochemical study of 345 cases. *Am. J. Clin. Pathol.* **95**, 692–695.

10. Ashton-Key, M., Jessup, E., and Isaacson, P. G. (1996) Immunoglobulin light chain staining in paraffin-embedded tissue using a heat mediated epitope retrieval method. *Histopathology* **29**, 525–531.

11. Zukerberg, L. R., Medeiros, L. J., Ferry, J. A., and Harris, N. L. (1993) Diffuse low-grade B-cell lymphomas. Four clinically distinct subtypes defined by a combination of morphologic and immunophenotypic features. *Am. J. Clin. Pathol.* **100**, 373–385.

12. Harris, N. L., Nadler, L. M., and Bhan, A. K. (1984) Immunohistologic characterization of two malignant lymphomas of germinal center type (centroblastic/centrocytic and centrocytic) with monoclonal antibodies. Follicular and diffuse lymphomas of small-cleaved-cell type are related but distinct entities. *Am. J. Pathol.* **117**, 262–272.

13. Stein, H., Lennert, K., Feller, A. C., and Mason, D. Y. (1984) Immunohistological analysis of human lymphoma: correlation of histological and immunological categories. *Adv. Cancer Res.* **42**, 67–147.

14. Jennings, C. D. and Foon, K. A. (1997) Recent advances in flow cytometry: application to the diagnosis of hematologic malignancy. *Blood* **90**, 2863–2892.

15. Kaufmann, O., Flath, B., Spath-Schwalbe, E., Possinger, K., and Dietel, M. (1997) Immunohistochemical detection of CD5 with monoclonal antibody 4C7 on paraffin sections. *Am. J. Clin. Pathol.* **108**, 669–673.

16. Butmarc, J. R., Kourea, H. P., Levi, E., and Kadin, M. E. (1998) Improved detection of CD5 epitope in formalin-fixed paraffin-embedded sections of benign and neoplastic lymphoid tissues by using biotinylated tyramine enhancement after antigen retrieval. *Am. J. Clin. Pathol.* **109**, 682–688.

17. Butmarc, J., Kourea, H. P., and Kadin, M. E. (1997) CD5 immunostaining of lymphoid neoplasms in paraffin sections. *Am. J. Clin. Pathol.* **107**, 496–497.

18. Dorfman, D. M. and Shahsafaei, A. (1997) Usefulness of a new CD5 antibody for the diagnosis of T-cell and B-cell lymphoproliferative disorders in paraffin sections. *Mod. Pathol.* **10**, 859–863.

19. Contos, M. J., Kornstein, M. J., Innes, D. J., and Ben-Ezra, J. (1992) The utility of CD20

and CD43 in subclassification of low-grade B-cell lymphoma on paraffin sections. *Mod. Pathol.* **5**, 631–633.

20. de Leon, E. D., Alkan, S., Huang, J. C., and Hsi, E. D. (1998) Usefulness of an immunohisto-chemical panel in paraffin-embedded tissues for the differentiation of B-cell non-Hodgkin's lymphomas of small lymphocytes. *Mod. Pathol.* **11**, 1046–1051.

21. Lai, R., Weiss, L. M., Chang, K. L., and Arber, D. A. (1999) Frequency of CD43 expression in non-Hodgkin lymphoma. A survey of 742 cases and further characterization of rare CD43+ follicular lymphomas. *Am. J. Clin. Pathol.* **111**, 488–494.

22. Ngan, B. Y., Picker, L. J., Medeiros, L. J., and Warnke, R. A. (1989) Immunophenotypic diagnosis of non-Hodgkin's lymphoma in paraffin sections. Co-expression of L60 (Leu-22) and L26 antigens correlates with malignant histologic findings. *Am. J. Clin. Pathol.* **91**, 579–583.

23. Weiss, L. M., Crabtree, G. S., Rouse, R. V., and Warnke, R. A. (1985) Morphologic and immunologic characterization of 50 peripheral T-cell lymphomas. *Am. J. Pathol.* **118**, 316–324.

24. Borowiz, M. J., Reichert, T. A., Brynes, R. K., et al. (1986) The phenotypic diversity of peripheral T-cell lymphomas: the Southeastern Cancer Study Group experience. *Hum. Pathol.* **17**, 567–574.

25. Grogan, T. M., Fielder, K., Rangel, C., et al. (1985) Peripheral T-cell lymphoma: aggressive disease with heterogeneous immunotypes. *Am. J. Clin. Pathol.* **83**, 279–288.

26. Hastrup, N., Ralfkiaer, E., and Pallesen, G. (1989) Aberrant phenotypes in peripheral T-cell lymphomas. *J. Clin. Pathol.* **42**, 398–402.

27. Nasu, K., Said, J., Vonderheid, E., Olerud, J., Sako, D., and Kadin, M. (1985) Immunopathology of cutaneous T-cell lymphomas. *Am. J. Pathol.* **119**, 436–447.

28. Henni, T., Gaulard, P., Divine, M., et al. (1988) Comparison of genetic probe with immunophenotype analysis in lymphoproliferative disorders: a study of 87 cases. *Blood* **72**, 1937–1943.

29. Kurtin, P., Lust, J., and Thibodeau, S. (1991) Correlation of aberrant T-cell phenotype (ATCP) with clonal T-cell receptor gene rearrangements in peripheral T-cell lymphomas. *Mod. Pathol.* **4**, 76a.

30. Rijlaarsdam, J. U., Boorsma, D. M., de Haan, P., Sampat, S., and Willemze, R. (1990) The significance of Leu 8 negative T cells in lymphoid skin infiltrates: malignant transformation, selective homing or T-cell activation? *Br. J. Dermatol.* **123**, 587–593.

31. Rijlaarsdam, J. U., Scheffer, E., Meijer, C. J., and Willemze, R. (1992) Cutaneous pseudo-T-cell lymphomas. A clinicopathologic study of 20 patients. *Cancer* **69**, 717–724.

32. Smoller, B., Bishop, K., and Glusac, E. (1995) Lymphocyte antigen abnormalities in inflammatory dermatoses. *Appl. Immunohistochem.* **3**, 127–131.

33. Smoller, B., Bishop, K., Glusac, E., Kim, Y., Bhargava, V., and Warnke, R. (1995) Reassessment of lymphocyte immunophenotyping in the diagnosis of patch and plaque stage lesions of mycosis fungoides. *Appl. Immunohistochem.* **3**, 32–36.

34. Ralfkiaer, E., Wantzin, G. L., Mason, D. Y., Hou-Jensen, K., Stein, H., and Thomsen, K. (1985) Phenotypic characterization of lymphocyte subsets in mycosis fungoides. Comparison with large plaque parapsoriasis and benign chronic dermatoses. *Am. J. Clin. Pathol.* **84**, 610–619.

35. van der Putte, S. C., Toonstra, J., van Wichen, D. F., van Unnik, J. A., and van Vloten, W. A. (1988) Aberrant immunophenotypes in mycosis fungoides. *Arch. Dermatol.* **124**, 373–380.

36. Reinherz, E. L., Kung, P. C., Goldstein, G., Levey, R. H., and Schlossman, S. F. (1980) Discrete stages of human intrathymic differentiation: analysis of normal thymocytes and leukemic lymphoblasts of T-cell lineage. *Proc. Natl. Acad. Sci. USA* **77**, 1588–1592.

37. Mokhtar, N., Hsu, S. M., Lad, R. P., Haynes, B. F., and Jaffe, E. S. (1984) Thymoma: lymphoid and epithelial components mirror the phenotype of normal thymus. *Hum. Pathol.* **15**, 378–384.

38. Sato, Y., Watanabe, S., Mukai, K., et al. (1986) An immunohistochemical study of thymic epithelial tumors. II. Lymphoid component. *Am. J. Surg. Pathol.* **10**, 862–870.

39. Zutter, M., Hockenbery, D., Silverman, G. A., and Korsmeyer, S. J. (1991) Immunolocalization of the Bcl-2 protein within hematopoietic neoplasms. *Blood* **78**, 1062–1068.

40. Pezzella, F., Tse, A. G., Cordell, J. L., Pulford, K. A., Gatter, K. C., and Mason, D. Y. (1990) Expression of the bcl-2 oncogene protein is not specific for the 14;18 chromosomal translocation. *Am. J. Pathol.* **137**, 225–232.

41. Ngan, B. Y., Chen-Levy, Z., Weiss, L. M., Warnke, R. A., and Cleary, M. L. (1988) Expression in non-Hodgkin's lymphoma of the bcl-2 protein associated with the t(14;18) chromosomal translocation. *N. Engl. J. Med.* **318**, 1638–1644.

42. Krajewski, S., Bodrug, S., Gascoyne, R., Berean, K., Krajewska, M., and Reed, J. C. (1994) Immunohistochemical analysis of Mcl-1 and Bcl-2 proteins in normal and neoplastic lymph nodes. *Am. J. Pathol.* **145**, 515–525.

43. Gaulard, P., d'Agay, M. F., Peuchmaur, M., et al. (1992) Expression of the bcl-2 gene product in follicular lymphoma. *Am. J. Pathol.* **140**, 1089–1095.

44. Lai, R., Arber, D. A., Chang, K. L., Wilson, C. S., and Weiss, L. M. (1998) Frequency of bcl-2 expression in non-Hodgkin's lymphoma: a study of 778 cases with comparison of marginal zone lymphoma and monocytoid B-cell hyperplasia. *Mod. Pathol.* **11**, 864–869.

45. Zukerberg, L. R., Yang, W. I., Arnold, A., and Harris, N. L. (1995) Cyclin D1 expression in non-Hodgkin's lymphomas. Detection by immunohistochemistry. *Am. J. Clin. Pathol.* **103**, 756–760.

46. Yang, W. I., Zukerberg, L. R., Motokura, T., Arnold, A., and Harris, N. L. (1994) Cyclin D1 (Bcl-1, PRAD1) protein expression in low-grade B-cell lymphomas and reactive hyperplasia. *Am. J. Pathol.* **145**, 86–96.

47. Swerdlow, S. H., Zukerberg, L. R., Yang, W. I., Harris, N. L., and Williams, M. E. (1996) The morphologic spectrum of non-Hodgkin's lymphomas with BCL1/cyclin D1 gene rearrangements. *Am. J. Surg. Pathol.* **20**, 627–640.

48. Swerdlow, S. H., Yang, W. I., Zukerberg, L. R., Harris, N. L., Arnold, A., and Williams, M. E. (1995) Expression of cyclin D1 protein in centrocytic/mantle cell lymphomas with and without rearrangement of the BCL1/cyclin D1 gene. *Hum. Pathol.* **26**, 999–1004.

49. Vasef, M. A., Medeiros, L. J., Koo, C., McCourty, A., and Brynes, R. K. (1997) Cyclin D1 immunohistochemical staining is useful in distinguishing mantle cell lymphoma from other low-grade B-cell neoplasms in bone marrow. *Am. J. Clin. Pathol.* **108**, 302–307.

50. Hoyer, J. D., Hanson, C. A., Fonseca, R., Greipp, P. R., Dewald, G. W., and Kurtin, P. J. (2000) The (11;14)(q13;q32) translocation in multiple myeloma. A morphologic and immunohistochemical study. *Am. J. Clin. Pathol.* **113**, 831–837.

51. Vasef, M. A., Medeiros, L. J., Yospur, L. S., Sun, N. C., McCourty, A., and Brynes, R. K. (1997) Cyclin D1 protein in multiple myeloma and plasmacytoma: an immunohistochemical study using fixed, paraffin-embedded tissue sections. *Mod. Pathol.* **10**, 927–932.

52. Benharroch, D., Meguerian-Bedoyan, Z., Lamant, L., et al. (1998) ALK-positive lymphoma: a single disease with a broad spectrum of morphology. *Blood* **91**, 2076–2084.

53. Cataldo, K. A., Jalal, S. M., Law, M. E., et al. (1999) Detection of t(2;5) in anaplastic large cell lymphoma: comparison of immunohistochemical studies, FISH, and RT-PCR in paraffin-embedded tissue. *Am. J. Surg. Pathol.* **23**, 1386–1392.

54. Falini, B., Bigerna, B., Fizzotti, M., et al. (1998) ALK expression defines a distinct group of T/null lymphomas with a wide morphological spectrum. *Am. J. Pathol.* **153**, 875–886.

55. Pittaluga, S., Wiodarska, I., Pulford, K., et al. (1997) The monoclonal antibody ALK1 identifies a distinct morphological subtype of anaplastic large cell lymphoma associated with 2p23/ALK rearrangements. *Am. J. Pathol.* **151**, 343–351.

56. Pulford, K., Lamant, L., Morris, S. W., et al. (1997) Detection of anaplastic lymphoma kinase (ALK) and nucleolar protein nucleophosmin (NPM)-ALK proteins in normal and neoplastic cells with the monoclonal antibody ALK1. *Blood* **89**, 1394–1404.

57. Shiota, M., Fujimoto, J., Takenaga, M., et al. (1994) Diagnosis of t(2;5)(p23;q35)-associated Ki-1 lymphoma with immunohistochemistry. *Blood* **84**, 3648–3652.

58. Simonitsch, I., Panzer-Gruemayer, E. R., Ghali, D. W., et al. (1996) NPM/ALK gene fusion transcripts identify a distinct subgroup of null type Ki-1 positive anaplastic large cell lymphomas. *Br. J. Haematol.* **92**, 866–871.

59. Cartun, R. W., Coles, F. B., and Pastuszak, W. T. (1987) Utilization of monoclonal antibody L26 in the identification and confirmation of B-cell lymphomas. A sensitive and specific marker applicable to formalin-and B5-fixed, paraffin-embedded tissues. *Am. J. Pathol.* **129**, 415–421.

60. Mason, D. Y., Comans-Bitter, M. W., Cordell, J. L., Verhoeven, M. A., and van Dongen, J. J. (1990) Antibody L26 recognizes an intracellular epitope on the B-cell-associated CD20 antigen. *Am. J. Pathol.* **136**, 1215–1222.

61. Norton, A. J. and Isaacson, P. G. (1987) Monoclonal antibody L26: an antibody that is reactive with normal and neoplastic B lymphocytes in routinely fixed and paraffin wax embedded tissues. *J. Clin. Pathol.* **40**, 1405–1412.

62. Kurtin, P. J. and Roche, P. C. (1993) Immunoperoxidase staining of non-Hodgkin's lymphomas for T-cell lineage associated antigens in paraffin sections. Comparison of the performance characteristics of four commercially available antibody preparations. *Am. J. Surg. Pathol.* **17**, 898–904.

63. Chadburn, A. and Knowles D. M. (1994) Paraffin-resistant antigens detectable by antibodies L26 and polyclonal CD3 predict the B- or T-cell lineage of 95% of diffuse aggressive non-Hodgkin's lymphomas. *Am. J. Clin. Pathol.* **102**, 284–291.

64. Mason, D. Y., Cordell, J. L., Brown, M. H., et al. (1995) CD79a: a novel marker for B-cell neoplasms in routinely processed tissue samples. *Blood* **86**, 1453–1459.

65. Cabecadas, J. M. and Isaacson, P. G. (1991) Phenotyping of T-cell lymphomas in paraffin sections—which antibodies? *Histopathology* **19**, 419–424.

66. Mason, D. Y., Cordell, J., Brown, M., et al. (1989) Detection of T cells in paraffin wax embedded tissue using antibodies against a peptide sequence from the CD3 antigen. *J. Clin. Pathol.* **42**, 1194–1200.

67. Anderson, C., Rezuke, W. N., Kosciol, C. M., Pastuszak, W. T., and Cartun, R. W. (1991) Methods in pathology. Identification of T-cell lymphomas in paraffin-embedded tissues using polyclonal anti-CD3 antibody: comparison with frozen section immunophenotyping and genotypic analysis. *Mod. Pathol.* **4**, 358–362.

68. Norton, A. J., Ramsay, A. D., Smith, S. H., Beverley, P. C., and Isaacson, P. G. (1986) Monoclonal antibody (UCHL1) that recognizes normal and neoplastic T cells in routinely fixed tissues. *J. Clin. Pathol.* **39**, 399–405.

69. Macon, W. R. and Salhany, K. E. (1998) T-cell subset analysis of peripheral T-cell lymphomas by paraffin section immunohistology and correlation of CD4/CD8 results with flow cytometry. *Am. J. Clin. Pathol.* **109**, 610–617.

70. Said, J. W., Shintaku, I. P., Parekh, K., and Pinkus, G. S. (1990) Specific phenotyping of T-cell proliferations in formalin-fixed paraffin-embedded tissues. Use of antibodies to the T-cell receptor beta F1. *Am. J. Clin. Pathol.* **93**, 382–386.

71. Chan, W. C., Borowitz, M. J., Hammami, A., Wu, Y. J., and Ip S. J. (1988) T-cell receptor antibodies in the immunohistochemical studies of normal and malignant lymphoid cells. *Cancer* **62**, 2118–2124.

72. Ng, C. S., Chan, J. K., Hui, P. K., Chan, W. C., and Lo, S. T. (1988) Application of a T cell receptor antibody beta F1 for immunophenotypic analysis of malignant lymphomas. *Am. J. Pathol.* **132**, 365–371.

73. Ohno, T., Yamaguchi, M., Oka, K., Miwa, H., Kita, K., and Shirakawa, S. (1995) Frequent expression of CD3 epsilon in CD3 (Leu 4)-negative nasal T-cell lymphomas. *Leukemia* **9**, 44–52.

74. Ohsawa, M., Nakatsuka, S., Kanno, H., et al. (1999) Immunophenotypic and genotypic characterization of nasal lymphoma with polymorphic reticulosis morphology. *Int. J. Cancer* **81**, 865–870.

75. Suzumiya, J., Takeshita, M., Kimura, N., et al. (1994) Expression of adult and fetal natural killer cell markers in sinonasal lymphomas. *Blood* **83**, 2255–2260.

76. Robertson, M. J. and Ritz, J. (1990) Biology and clinical relevance of human natural killer cells. *Blood* **76**, 2421–2438.

77. Lanier, L. L., Phillips, J. H., Hackett, J., Jr., Tutt, M., and Kumar, V. (1986) Natural killer cells: definition of a cell type rather than a function. *J. Immunol.* **137**, 2735–2739.

78. Felgar, R. E., Macon, W. R., Kinney, M. C., Roberts, S., Pasha, T., and Salhany, K. E. (1997) TIA-1 expression in lymphoid neoplasms. Identification of subsets with cytotoxic T lymphocyte or natural killer cell differentiation. *Am. J. Pathol.* **150**, 1893–1900.

79. de Bruin, P. C., Kummer, J. A., van der Valk, P., et al. (1994) Granzyme B-expressing peripheral T-cell lymphomas: neoplastic equivalents of activated cytotoxic T cells with preference for mucosa-associated lymphoid tissue localization. *Blood* **84**, 3785–3791.

80. Jaffe, E. S., Chan, J. K., Su, I. J., Frizzera, G., Mori, S., Feller, A. C., and Ho, F. C. (1996) Report of the Workshop on Nasal and Related Extranodal Angiocentric T/Natural Killer Cell Lymphomas. Definitions, differential diagnosis, and epidemiology. *Am. J. Surg. Pathol.* **20**, 103–111.

81. Harris, N. L., Jaffe, E. S., Stein, H., et al. (1994) A revised European-American classification of lymphoid neoplasms: a proposal from the International Lymphoma Study Group [See comments]. *Blood* **84**, 1361–1392.

82. Sallan, S. E., Ritz, J., Pesando, J., et al. (1980) Cell surface antigens: prognostic implications in childhood acute lymphoblastic leukemia. *Blood* **55**, 395–402.

83. Pui, C. H., Behm, F. G., and Crist, W. M. (1993) Clinical and biologic relevance of immunologic marker studies in childhood acute lymphoblastic leukemia. *Blood* **82**, 343–362.

84. Khalidi, H. S., Chang, K. L., Medeiros, L. J., et al. (1999) Acute lymphoblastic leukemia. Survey of immunophenotype, French-American-British classification, frequency of myeloid antigen expression, and karyotypic abnormalities in 210 pediatric and adult cases. *Am. J. Clin. Pathol.* **111**, 467–476.

85. Borowitz, M. J. (1990) Immunologic markers in childhood acute lymphoblastic leukemia. *Hematol. Oncol. Clin. North Am.* **4**, 743–765.

86. Borowitz, M. J., Shuster, J. J., Civin, C. I., et al. (1990) Prognostic significance of CD34 expression in childhood B-precursor acute lymphocytic leukemia: a Pediatric Oncology Group study. *J. Clin. Oncol.* **8**, 1389–1398.

87. Vogler, L. B., Crist, W. M., Bockman, D. E., Pearl, E. R., Lawton, A. R., and Cooper, M. D. (1978) Pre-B-cell leukemia. A new phenotype of childhood lymphoblastic leukemia. *N. Engl. J. Med.* **298**, 872–878.

88. Kilo, M. N. and Dorfman, D. M. (1996) The utility of flow cytometric immunophenotypic analysis in the distinction of small lymphocytic lymphoma/chronic lymphocytic leukemia from mantle cell lymphoma. *Am. J. Clin. Pathol.* **105**, 451–457.

89. Harris, N. L. and Bhan, A. K. (1985) B-cell neoplasms of the lymphocytic, lymphoplasmacytoid, and plasma cell types: immunohistologic analysis and clinical correlation. *Hum. Pathol.* **16**, 829–837.

90. Matutes, E. and Catavosky, D. (1994) The value of scoring systems for the diagnosis of biphenotypic leukemia and mature B-cell disorders. *Leuk. Lymphoma* **13**, 11–14.

91. Weisenburger, D., Harrington, J., and Armitage, J. (1990) B-cell neoplasia: a conceptual understanding based on the normal humoral immune system. *Pathol. Annu.* **25**, 99–115.

92. Dorfman, D. M. and Pinkus, G. S. (1994) Distinction between small lymphocytic and mantle cell lymphoma by immunoreactivity for CD23. *Mod. Pathol.* **7**, 326–331.

93. Argatoff, L. H., Connors, J. M., Klasa, R. J., Horsman, D. E., and Gascoyne, R. D. (1997) Mantle cell lymphoma: a clinicopathologic study of 80 cases. *Blood* **89**, 2067–2078.

94. Pittaluga, S., Wlodarska, I., Stul, M. S., et al. (1995) Mantle cell lymphoma: a clinicopathological study of 55 cases. *Histopathology* **26**, 17–24.

95. Dierlamm, J., Pittaluga, S., Wlodarksa, I., et al. (1996) Marginal zone B-cell lymphomas of different sites share similar cytogenetic and morphologic features. *Blood* **87**, 299–307.

96. Hammer, R. D., Glick, A. D., Greer, J. P., Collins, R. D., and Cousar, J. B. (1996) Splenic marginal zone lymphoma. A distinct B-cell neoplasm. *Am. J. Surg. Pathol.* **20**, 613–626.

97. Mollejo, M., Menarguez, J., Lloret, E., et al. (1995) Splenic marginal zone lymphoma: a distinctive type of low-grade B-cell lymphoma. A clinicopathological study of 13 cases. *Am. J. Surg. Pathol.* **19**, 1146–1157.

98. Schmid, C., Kirkham, N., Diss, T., and Isaacson, P. G. (1992) Splenic marginal zone cell lymphoma. *Am. J. Surg. Pathol.* **16**, 455–466.

99. Savilo, E., Campo, E., Mollejo, M., et al. (1998) Absence of cyclin D1 protein expression in splenic marginal zone lymphoma. *Mod. Pathol.* **11**, 601–606.

100. de Wolf-Peeters, C., Pittaluga, S., Dierlamm, J., Wlodarska, I., and Van Den Berghe, H. (1997) Marginal zone B-cell lymphomas including musosa-associated lymphoid tissue type lymphoma (MALT), monocytoid B-cell lymphoma and splenic marginal zone cell lymphoma and their relation to the reactive marginal zone. *Leuk. Lymphoma* **26**, 467–478.

101. Almasri, N. M., Iturraspe, J. A., and Braylan, R. C. (1998) CD10 expression in follicular lymphoma and large cell lymphoma is different from that of reactive lymph node follicles. *Arch. Pathol. Lab. Med.* **122**, 539–544.

102. Carey, J. and Hanson, C. (1994) Flow cytometric analysis of leukemia and lymphoma, in Flow Cytometry and Clinical Diagnosis (Keren, D., Hanson, C., and Hurtubise, P., eds.) American Society of Clinical Pathologists, Chicago IL, pp. 197–308.

103. Kumar, S., Green, G. A., Teruya-Feldstein, J., Raffeld, M., and Jaffe, E. S. (1996) Use of CD23 (BU38) on paraffin sections in the diagnosis of small lymphocytic lymphoma and mantle cell lymphoma. *Mod. Pathol.* **9**, 925–929.

104. Singh, N. and Wright, D. H. (1997) The value of immunohistochemistry on paraffin wax embedded tissue sections in the differentiation of small lymphocytic and mantle cell lymphomas. *J. Clin. Pathol.* **50**, 16–21.

105. Murray, P. G., Janmohamed, R. M., and Crocker, J. (1991) CD23 expression in non-Hodgkin lymphoma: immunohistochemical demonstration using the antibody BU38 on paraffin sections. *J. Pathol.* **165**, 125–128.

106. Doggett, R. S., Wood, G. S., and Horning, S. (1984) The immunologic characterization of 95 nodal and extranodal diffuse large cell lymphomas in 89 patients. *Am. J. Pathol.* **115**, 245–252.

107. Pezzella, F., Munson, P. J., Miller, K. D., Goldstone, A. H., and Gatter, K. C. (2000) The diagnosis of low-grade peripheral B-cell neoplasms in bone marrow trephines. *Br. J. Haematol.* **108**, 369–376.

108. McIntosh, G. G., Lodge, A. J., Watson, P., et al. (1999) NCL-CD10-270: a new monoclonal antibody recognizing CD10 in paraffin-embedded tissue. *Am. J. Pathol.* **154**, 77–82.

109. Kaufmann, O., Flath, B., Spath-Schwalbe, E., Possinger, K., and Dietel, M. (1999) Immuno-histochemical detection of CD10 with monoclonal antibody 56C6 on paraffin sections. *Am. J. Clin. Pathol.* **111**, 117–122.

110. Pangalis, G. A., Angelopoulou, M. K., Vassilakopoulos, T. P., Siakantaris, M. P., and Kittas, C. (1999) B-chronic lymphocytic leukemia, small lymphocytic lymphoma, and lymphoplasmacytic lymphoma, including Waldenstrom's macroglobulinemia: a clinical, morphologic, and biologic spectrum of similar disorders. *Semin. Hematol.* **36**, 104–114.

111. Feiner, H. D., Rizk, C. C., Finfer, M. D., et al. (1990) IgM monoclonal gammopathy/ Waldenstrom's macroglobulinemia: a morphological and immunophenotypic study of the bone marrow. *Mod. Pathol.* **3**, 348–356.

112. San Miguel, J. F., Caballero, M. D., Gonzalez, M., Zola, H., and Lopez, Borrasca A. (1986) Immunological phenotype of neoplasms involving the B cell in the last step of differentiation. *Br. J. Haematol.* **62**, 75–83.

113. Martinsson, U., Sundstrom, C., and Glimelius, B. (1989) Immunophenotype analysis of B-CLL lymphoma and immunocytoma. *APMIS* **97**, 1025–1032.

114. Shapiro, J. L., Miller, M. L., Pohlman, B., Mascha, E., and Fishleder, A. J. (1999) CD5-B-cell lymphoproliferative disorders presenting in blood and bone marrow. A clinicopathologic study of 40 patients. *Am. J. Clin. Pathol.* **111**, 477–487.

115. Nakamine, H., Bagin, R. G., Vose, J. M., et al. (1993) Prognostic significance of clinical and pathologic features in diffuse large B-cell lymphoma. *Cancer* **71**, 3130–3137.

116. Dogan, A., Bagdi, E., Munson, P., and Isaacson, P. G. (2000) CD10 and BCL-6 expression in paraffin sections of normal lymphoid tissue and B-cell lymphomas. *Am. J. Surg. Pathol.* **24**, 846–852.

117. Harada, S., Suzuki, R., Uehira, K., et al. (1999) Molecular and immunological dissection of diffuse large B cell lymphoma: CD5+, and CD5– with CD10+ groups may constitute clinically relevant subtypes. *Leukemia* **13**, 1441–1447.

118. Fang, J. M., Finn, W. G., Hussong, J. W., Goolsby, C. L., Cubbon, A. R., and Variakojis, D. (1999) CD10 antigen expression correlates with the t(14;18)(q32;q21) major breakpoint region in diffuse large B-cell lymphoma. *Mod. Pathol.* **12**, 295–300.

119. Gascoyne, R. D., Adomat, S. A., Krajewski, S., et al. (1997) Prognostic significance of Bcl-2 protein expression and Bcl-2 gene rearrangement in diffuse aggressive non-Hodgkin's lymphoma. *Blood* **90**, 244–251.

120. Matolcsy, A., Chadburn, A., and Knowles, D. M. (1995) De novo CD5-positive and Richter's syndrome-associated diffuse large B cell lymphomas are genotypically distinct. *Am. J. Pathol.* **147**, 207–216.

121. Moller, P., Moldenhauer, G., Momburg, F., et al. (1987) Mediastinal lymphoma of clear cell type is a tumor corresponding to terminal steps of B cell differentiation. *Blood* **69**, 1087–1095.

122. Garcia, C. F., Weiss, L. M., and Warnke, R. A. (1986) Small noncleaved cell lymphoma: an immunophenotypic study of 18 cases and comparison with large cell lymphoma. *Hum. Pathol.* **17**, 454–461.

123. Strickler, J. G., Audeh, M. W., Copenhaver, C. M., and Warnke, R. A. (1988) Immunophenotypic differences between plasmacytoma/multiple myeloma and immunoblastic lymphoma. *Cancer* **61**, 1782–1786.

124. Feller, A. C., Griesser, H., Schilling, C. V., et al. (1988) Clonal gene rearrangement patterns correlate with immunophenotype and clinical parameters in patients with angio-immunoblastic lymphadenopathy. *Am. J. Pathol.* **133**, 549–556.

125. Namikawa, R., Suchi, T., Ueda, R., et al. (1987) Phenotyping of proliferating lymphocytes in angioimmunoblastic lymphadenopathy and related lesions by the double immunoenzymatic staining technique. *Am. J. Pathol.* **127**, 279–287.

126. Kaneko, Y., Maseki, N., Sakurai, M., et al. (1988) Characteristic karyotypic pattern in T-cell lymphoproliferative disorders with reactive angioimmunoblastic lymphadenopathy with dysproteinemia features. *Blood* **72**, 413–421.

127. Watanabe, S., Sato, Y., Shimoyama, M., Minato, K., and Shimosato, Y. (1986) Immunoblastic lymphadenopathy, angioimmunoblastic lymphadenopathy, and IBL-like T-cell lymphoma. A spectrum of T-cell neoplasia. *Cancer* **58**, 2224–2232.

128. Pallesen, G., Hamilton-Dutoit, S. J., and Zhou, X. (1993) The association of Epstein-Barr virus (EBV) with T cell lymphoproliferations and Hodgkin's disease: two new developments in the EBV field. *Adv. Cancer Res.* **62**, 179–239.

129. Leung, C. Y., Ho, F. C., Srivastava, G., Loke, S. L., Liu, Y. T., and Chan, A. C. (1993) Usefulness of follicular dendritic cell pattern in classification of peripheral T-cell lymphomas. *Histopathology* **23**, 433–437.

130. Patsouris, E., Noel, H., and Lennert, K. (1989) Angioimmunoblastic lymphadenopathy—type of T-cell lymphoma with a high content of epithelioid cells. Histopathology and comparison with lymphoepithelioid cell lymphoma. *Am. J. Surg. Pathol.* **13**, 262–275.

131. Salhany, K. E., Macon, W. R., Choi, J. K., et al. (1998) Subcutaneous panniculitis-like-T-cell lymphoma: clinicopathologic, immunophenotypic, and genotypic analysis of alpha/beta and gamma/delta subtypes. *Am. J. Surg. Pathol.* **22**, 881–893.

132. Kumar, S., Krenacs, L., Medeiros, J., et al. (1998) Subcutaneous panniculitic T-cell lymphoma is a tumor of cytotoxic T lymphocytes. *Hum. Pathol.* **29**, 397–403.

133. Cooke, C. B., Krenacs, L., Stetler-Stevenson, M., et al. (1996) Hepatosplenic T-cell lymphoma: a distinct clinicopathologic entity of cytotoxic gamma delta T-cell origin. *Blood* **88**, 4265–4274.

134. Bagdi, E., Diss, T. C., Munson, P., and Isaacson, P. G. (1999) Mucosal intra-epithelial lymphocytes in enteropathy-associated T-cell lymphoma, ulcerative jejunitis, and refractory celiac disease constitute a neoplastic population. *Blood* **94**, 260–264.

135. Feller, A. C., Parwaresch, M. R., Stein, H., Ziegler, A., Herbst, H., and Lennert, K. (1986) Immunophenotyping of T-lymphoblastic lymphoma/leukemia: correlation with normal T-cell maturation. *Leuk. Res.* **10**, 1025–1031.

136. Sheibani, K., Nathwani, B. N., Winberg, C. D., et al. (1987) Antigenically defined subgroups of lymphoblastic lymphoma. Relationship to clinical presentation and biologic behavior. *Cancer* **60**, 183–190.

137. Weiss, L. M., Bindl, J. M., Picozzi, V. J., Link, M. P., and Warnke, R. A. (1986) Lymphoblastic lymphoma: an immunophenotype study of 26 cases with comparison to T cell acute lymphoblastic leukemia. *Blood* **67**, 474–478.

138. Delsol, G., Al Saati, T., Gatter, K. C., et al. (1988) Coexpression of epithelial membrane antigen (EMA), Ki-1, and interleukin-2 receptor by anaplastic large cell lymphomas. Diagnostic value in so-called malignant histiocytosis. *Am. J. Pathol.* **130**, 59–70.

139. Chott, A., Kaserer, K., Augustin, I., et al. (1990) Ki-1-positive large cell lymphoma. A clinicopathologic study of 41 cases. *Am. J. Surg. Pathol.* **14**, 439–448.

140. Filippa, D. A., Ladanyi, M., Wollner, N., et al. (1996) CD30 (Ki-1)-positive malignant lymphomas: clinical, immunophenotypic, histologic, and genetic characteristics and differences with Hodgkin's disease. *Blood* **87**, 2905–2917.

141. Krenacs, L., Wellmann, A., Sorbara, L., et al. (1997) Cytotoxic cell antigen expression in anaplastic large cell lymphomas of T- and null-cell type and Hodgkin's disease: evidence for distinct cellular origin. *Blood* **89**, 980–989.

142. Pileri, S., Bocchia, M., Baroni, C. D., et al. (1994) Anaplastic large cell lymphoma (CD30 +/Ki-1+): results of a prospective clinico-pathological study of 69 cases. *Br. J. Haematol.* **86**, 513–523.

143. Pileri, S. A., Piccaluga, A., Poggi, S., et al. (1995) Anaplastic large cell lymphoma: update of findings. *Leuk. Lymphoma* **18**, 17–25.

144. Tilly, H., Gaulard, P., Lepage, E., et al. (1997) Primary anaplastic large-cell lymphoma in adults: clinical presentation, immunophenotype, and outcome. *Blood* **90**, 3727–3734.
145. Hall, P. A., D'Ardenne, A. J., and Stansfield, A. G. (1988) Paraffin section immunohisto-chemistry. II. Hodgkin's disease and large cell anaplastic (Ki-1) lymphoma. *Histopathology* **13**, 161–169.
146. Hall, P. A. and D'Ardenne, A. J. (1987) Value of CD15 immunostaining in diagnosing Hodgkin's disease: a review of published literature. *J. Clin. Pathol.* **40**, 1298–1304.
147. Hsu, S. M. and Jaffe, E. S. (1984) Leu M1 and peanut agglutinin stain the neoplastic cells of Hodgkin's disease. *Am. J. Clin. Pathol.* **82**, 29–32.
148. Hsu, S. M., Yang, K., and Jaffe, E. S. (1985) Phenotypic expression of Hodgkin's and Reed-Sternberg cells in Hodgkin's disease. *Am. J. Pathol.* **118**, 209–217.
149. Schwarting, R., Gerdes, J., Dürkop, H., Falini, B., Pileri, S., and Stein, H. (1989) Ber-H2: a new anti-Ki-1 (CD30) monoclonal antibody directed at a formol-resistant epitope. *Blood* **74**, 1678–1689.
150. Stein, H., Gerdes, J., Schwab, U., et al. (1982) Identification of Hodgkin and Sternberg-Reed cells as a unique cell type derived from a newly-detected small-cell population. *Int. J. Cancer* **30**, 445–459.
151. Arber, D. A. and Weiss, L. M. (1993) CD15: a review. *Appl. Immunohistochem.* **1**, 17–30.
152. Chittal, S. M., Caveriviere, P., Schwarting, R., et al. (1988) Monoclonal antibodies in the diagnosis of Hodgkin's disease. The search for a rational panel. *Am. J. Surg. Pathol.* **12**, 9–21.
153. Pinkus, G. S., Pinkus, J. L., Langhoff, E., et al. (1997) Fascin, a sensitive new marker for Reed-Sternberg cells of Hodgkin's disease. Evidence for a dendritic or B cell derivation? *Am. J. Pathol.* **150**, 543–562.
154. Stein, H., Mason, D. Y., Gerdes, J., et al. (1985) The expression of the Hodgkin's disease associated antigen Ki-1 in reactive and neoplastic lymphoid tissue: evidence that Reed-Sternberg cells and histiocytic malignancies are derived from activated lymphoid cells. *Blood* **66**, 848–858.
155. Nguyen, P. L., Harris, N. L., Ritz, J., and Robertson, M. J. (1996) Expression of CD95 antigen and bcl-2 protein in non-Hodgkin's lymphomas and Hodgkin's disease. *Am. J. Pathol.* **148**, 847–853.
156. Casey, T., Olson, S. J., Cousar, J. B., and Collins, R. D. (1989) Immunophenotypes of Reed-Sternberg cells: a study of 19 cases of Hodgkin's disease in plastic-embedded sections. *Blood* **74**, 2624–2628.
157. Falini, B., Stein, H., Pileri, S., et al. (1987) Expression of lymphoid-associated antigens on Hodgkin's and Reed-Sternberg cells of Hodgkin's disease. *Histopathology* **11**, 1129–1242.
158. Kadin, M. E., Muramoto, L., and Said, J. (1988) Expression of T-cell antigens on Reed-Sternberg cells in a subset of patients with nodular sclerosing and mixed cellularity Hodgkin's disease. *Am. J. Pathol.* **130**, 345–353.
159. Korkolopoulou, P., Cordell, J., Jones, M., et al. (1994) The expression of the B-cell marker Mb-1 (CD79a) in Hodgkin's disease. *Histopathology* **24**, 511–515.
160. Kuzu, I., Delsol, G., Jones, M., Gatter, K. C., and Mason, D. Y. (1993) Expression of the Ig associated heterodimer (Mb-1 and B29) in Hodgkin's disease. *Histopathology* **22**, 141–144.
161. Schmid, C., Pan, L., Diss, T., and Isaacson, P. G. (1991) Expression of B-cell antigens by Hodgkin's and Reed-Sternberg cells. *Am. J. Pathol.* **139**, 701–707.
162. Abdulaziz, Z., Mason, D. Y., Stein, H., Gatter, K. C., and Nash, J. R. G. (1984) An immunohistochemical study of the cellular constituents of Hodgkin's disease using an monoclonal antibody panel. *Histopathology* **8**, 1–25.
163. Agnarsson, B. A. and Kadin, M. E. (1989) The immunophenotype of Reed-Sternberg cells. A study of 50 cases of Hodgkin's disease using fixed frozen tissues. *Cancer* **63**, 2083–2087.

164. Angel, C. A., Warford, A., Campbell, A. C., Pringle, J. H., and Lauder, I. (1987) The immunohistology of Hodgkin's disease—Reed-Sternberg cells and their variants. *J. Pathol.* **153**, 21–30.

165. Cibull, M. L., Stein, H., Gatter, K. C., and Mason, D. Y. (1989) The expression of the CD3 antigen in Hodgkin's disease. *Histopathology* **15**, 597–605.

166. Bishop, P. W., Harris, M., Smith, A. P., and Elsam, K. J. (1991) Immunophenotypic study of lymphocyte predominance Hodgkin's disease. *Histopathology* **18**, 19–24.

167. Coles, R. B., Cartun, R. W., and Pastuszak, W. T. (1988) Hodgkin's disease, lymphocyte-predominant type: immunoreactivity with B-cell antibodies. *Mod. Pathol.* **1**, 274–278.

168. Hansmann, M. L., Wacker, H. H., and Radzun, H. J. (1986) Paragranuloma is a variant of Hodgkin's disease with predominance of B-cells. *Virchows Arch. Pathol. Anat.* **409**, 171–181.

169. Nicholas, D. S., Harris, S., and Wright, D. H. (1990) Lymphocyte predominance Hodgkin's disease—an immunohistochemical study. *Histopathology* **16**, 157–165.

170. Pinkus, G. S. and Said, J. W. (1985) Hodgkin's disease, lymphocyte predominance type, nodular—a distinct entity? Unique staining profile for L & H variants of Reed-Sternberg cells defined by monoclonal antibodies to leukocyte common antigen, granulocyte-specific antigen, and B-cell specific antigen. *Am. J. Pathol.* **133**, 211–217.

171. Pinkus, G. S. and Said, J. W. (1988) Hodgkin's disease, lymphocyte predominance type, nodular—further evidence for a B cell derivation. L & H variants of Reed-Sternberg cells express L26, a pan B cell marker. *Am. J. Pathol.* **133**, 211–217.

172. Poppema, S. (1992) Lymphocyte-predominance Hodgkin's disease. *Int. Rev. Exp. Pathol.* **33**, 53–79.

173. Poppema, S., Timens, W., and Visser, L. (1985) Nodular lymphocyte predominance type of Hodgkin's disease is a B cell lymphoma. *Adv. Exp. Med. Biol.* **186**, 963–969.

174. Poppema, S. (1980) The diversity of the immunohistological staining pattern of Sternberg-Reed cells. *J. Histochem. Cytochem.* **28**, 788–791.

175. Stein, H., Hansmann, M. L., Lennert, K., Brandtzaeg, P., Gatter, K. C., and Mason, D. Y. (1986) Reed-Sternberg and Hodgkin cells in lymphocyte-predominant Hodgkin's disease of nodular subtype contain J chain. *Am. J. Clin. Pathol.* **86**, 292–297.

176. Falini, B., Bigerna, B., Pasqualucci, L., et al. (1996) Distinctive expression pattern of the BCL-6 protein in nodular lymphocyte predominance Hodgkin's disease. *Blood* **87**, 465–471.

177. Hansmann, M. L., Fellbaum, C., Hui, P. K., and Zwingers, T. (1988) Correlation of content of B cells and Leu7-positive cells with subtype and stage in lymphocyte predominance type Hodgkin's disease. *J. Cancer Res. Clin. Oncol.* **114**, 405–410.

178. Poppema, S. (1989) The nature of the lymphocytes surrounding Reed-Sternberg cells in nodular lymphocyte predominance and in other types of Hodgkin's disease. *Am. J. Pathol.* **135**, 351–357.

179. Bene, M. C., Castoldi, G., Knapp, W., et al. (1995) Proposals for the immunological classification of acute leukemias. European Group for the Immunological Characterization of Leukemias (EGIL). *Leukemia* **9**, 1783–1786.

180. Bradstock, K. F. (1993) The diagnostic and prognostic value of immunophenotyping in acute leukemia. *Pathology* **25**, 367–374.

181. Brunning, R. and McKenna, R. (1994) Tumors of the bone marrow, in *Atlas of Tumor Pathology*, vol. 9 (Rosai, J., ed.) Armed Forces Institute of Pathology, Washington, DC.

182. Hanson, C. A., Gajl-Peczalska, K. J., Parkin, J. L., and Brunning, R. D. (1987) Immunophenotyping of acute myeloid leukemia using monoclonal antibodies and the alkaline phosphatase-antialkaline phosphatase technique. *Blood* **70**, 83–89.

183. Khalidi, H. S., Medeiros, L. J., Chang, K. L., Brynes, R. K., Slovak, M. L., and Arber, D. A. (1998) The immunophenotype of adult acute myeloid leukemia: high frequency of

lymphoid antigen expression and comparison of immunophenotype, French-American-British classification, and karyotypic abnormalities. *Am. J. Clin. Pathol.* **109**, 211–220.

184. Pui, C. H., Raimondi, S. C., Head, D. R., et al. (1991) Characterization of childhood acute leukemia with multiple myeloid and lymphoid markers at diagnosis and at relapse. *Blood* **78**, 1327–1337.

185. Wang, J. C., Beauregard, P., Soamboonsrup, P., and Neame, P. B. (1995) Monoclonal antibodies in the management of acute leukemia. *Am. J. Hematol.* **50**, 188–199.

186. Chuang, S. S. and Li, C. Y. (1997) Useful panel of antibodies for the classification of acute leukemia by immunohistochemical methods in bone marrow trephine biopsy specimens. *Am. J. Clin. Pathol.* **107**, 410–418.

187. Davey, F. R., Olson, S., Kurec, A. S., Eastman-Abaya, R., Gottlieb, A. J., and Mason, D. Y. (1988) The immunophenotyping of extramedullary myeloid cell tumors in paraffin-embedded tissue sections. *Am. J. Surg. Pathol.* **12**, 699–707.

188. Hanson, C. A., Ross, C. W., and Schnitzer, B. (1992) Anti-CD34 immunoperoxidase staining in paraffin sections of acute leukemia: comparison with flow cytometric immunophenotyping. *Hum. Pathol.* **23**, 26–32.

189. Horny, H. P., Campbell, M., Steinke, B., and Kaiserling, E. (1990) Acute myeloid leukemia: immunohistologic findings in paraffin-embedded bone marrow biopsy specimens. *Hum. Pathol.* **21**, 648–655.

190. Pileri, S. A., Ascani, S., Milani, M., et al. (1999) Acute leukaemia immunophenotyping in bone-marrow routine sections. *Br. J. Haematol.* **105**, 394–401.

191. Pinkus, G. S. and Pinkus, J. L. (1991) Myeloperoxidase: a specific marker for myeloid cells in paraffin sections. *Mod. Pathol.* **4**, 733–741.

192. Toth, B., Wehrmann, M., Kaiserling, E., and Horny, H. P. (1999) Immunophenotyping of acute lymphoblastic leukaemia in routinely processed bone marrow biopsy specimens. *J. Clin. Pathol.* **52**, 688–692.

193. Warnke, R. A., Pulford, K. A., Pallesen, G., et al. (1989) Diagnosis of myelomonocytic and macrophage neoplasms in routinely processed tissue biopsies with monoclonal antibody KP1. *Am. J. Pathol.* **135**, 1089–1095.

194. Falini, B., Flenghi, L., Pileri, S., et al. (1993) PG-M1: a new monoclonal antibody directed against a fixative-resistant epitope on the macrophage-restricted form of the CD68 molecule. *Am. J. Pathol.* **142**, 1359–1372.

195. Micklem, K., Rigney, E., Cordell, J., et al. (1989) A human macrophage-associated antigen (CD68) detected by six different monoclonal antibodies. *Br. J. Haematol.* **73**, 6–11.

196. Pulford, K. A., Rigney, E. M., Micklem, K. J., et al. (1989) KP1: a new monoclonal antibody that detects a monocyte/macrophage associated antigen in routinely processed tissue sections. *J. Clin. Pathol.* **42**, 414–421.

197. Pinkus, G. S. and Said, J. W. (1977) Profile of intracytoplasmic lysozyme in normal tissues, myeloproliferative disorders, hairy cell leukemia, and other pathologic processes. An immunoperoxidase study of paraffin sections and smears. *Am. J. Pathol.* **89**, 351–366.

198. Warnke, R., Weiss, L., Chan, J., Cleary, M., and Dorfman, R. (1995) Tumors of the lymph nodes and spleen, in *Atlas of Tumor Pathology*, vol. 14 (Rosai, J., ed.), Armed Forces Institute of Pathology, Washington, DC.

199. Emile, J. F., Wechsler, J., Brousse, N., et al. (1995) Langerhans' cell histiocytosis. Definitive diagnosis with the use of monoclonal antibody O10 on routinely paraffin-embedded samples. *Am. J. Surg. Pathol.* **19**, 636–641.

200. Chan, J. K., Fletcher, C. D., Nayler, S. J., and Cooper, K. (1997) Follicular dendritic cell sarcoma. Clinicopathologic analysis of 17 cases suggesting a malignant potential higher than currently recognized. *Cancer* **79**, 294–313.

201. Perez-Ordonez, B., Erlandson, R. A., and Rosai, J. (1996) Follicular dendritic cell tumor: report of 13 additional cases of a distinctive entity. *Am. J. Surg. Pathol.* **20**, 944–955.

202. Schriever, F., Freedman, A. S., Freeman, G., et al. (1989) Isolated human follicular dendritic cells display a unique antigenic phenotype. *J. Exp. Med.* **169**, 2043–2058.
203. Weiss, L. M., Berry, G. J., Dorfman, R. F., et al. (1990) Spindle cell neoplasms of lymph nodes of probable reticulum cell lineage. True reticulum cell sarcoma? *Am. J. Surg. Pathol.* **14**, 405–414.
204. Feltkamp, C., van Heerde, P., Feltkamp-Vroom, T., and Koudstaal, J. (1981) A malignant tumor arising from interdigitating cells; light microscopical, ultrastructural, immuno- and enzyme-histochemical characteristics. *Virchows Arch. Pathol. Anat.* **393**, 183–192.
205. Nakamura, S., Hara, K., Suchi, T., et al. (1988) Interdigitating cell sarcoma. A morphologic, immunohistologic, and enzyme-histochemical study. *Cancer* **61**, 562–568.
206. Wilson, M. S., Weiss, L. M., Gatter, K. C., Mason, D. Y., Dorfman, R. F., and Warnke, R. A. (1990) Malignant histiocytosis. A reassessment of cases previously reported in 1975 based on paraffin section immunophenotyping studies. *Cancer* **66**, 530–536.
207. Arai, E., Su, W. P., Roche, P. C., and Li, C. Y. (1993) Cutaneous histiocytic malignancy. Immunohistochemical re-examination of cases previously diagnosed as cutaneous histiocytic lymphoma; malignant histiocytosis. *J. Cutan. Pathol.* **20**, 115–120.
208. Kamel, O. W., Gocke, C. D., Kell, D. L., Cleary, M. L., and Warnke, R. A. (1995) True histiocytic lymphoma: a study of 12 cases based on current definition. *Leuk. Lymphoma* **18**, 81–86.
209. Lauritzen, A. F. and Ralfkaier, E. (1995) Histiocytic sarcomas. *Leuk. Lymphoma* **18**, 73–80.
210. Hsu, S. M., Ho, Y. S., and Hsu, P. L. (1991) Lymphomas of true histiocytic origin. Expression of different phenotypes in so-called true histiocytic lymphoma and malignant histiocytosis. *Am. J. Pathol.* **138**, 1389–1404.

# 17

# Applications of Immunohistochemistry in the Diagnosis of Undifferentiated Tumors

## Mark R. Wick, MD and Lisa A. Cerilli, MD

## INTRODUCTION

Not uncommonly, pathologists are confronted with neoplasms that have few distinguishing microscopic characteristics and, therefore, appear to be "undifferentiated." The differential diagnostic considerations in such cases are many, and special studies are almost always necessary to reach a definite conclusion. Determining the probable site of origin for a metastatic carcinoma and making a diagnostic separation between two differentiated but histologically similar neoplasms represent additional challenges in surgical pathology that require ancillary laboratory analyses. This presentation will address approaches to these and other dilemmas that are particularly suited to resolution by immunodiagnosis. The following discussion is not intended to be exhaustive or all-inclusive, rather, the authors' aim is to provide a framework for the contemporary use of immunohistochemistry in the interpretation of solid tumors.

## RECOMMENDED TECHNIQUES AND REAGENTS FOR THE EVALUATION OF SOLID NEOPLASMS IN SURGICAL PATHOLOGY

An extensive array of antibodies is available to the surgical pathologist to facilitate characterization of tumors without histologically specific features. A highly select panel of immunostains, based on the histopathologic impression of the tumor, can be extremely useful to narrow the diagnostic considerations, if not definitively identify the tumor. The antibodies utilized for this purpose in our laboratory are described briefly below.

### Immunohistologic Detection Techniques

The basic substrates of surgical pathology are still formalin-fixed, paraffin-embedded tissue specimens, and this situation is likely to continue over the foreseeable future. Alternative fixatives have been recommended for use in certain contexts in this field (1) but are generally unnecessary if proper attention is given to routine processing procedures. Formalin exerts negative effects on tissue antigens by causing aldehyde-linkage "masking" of protein determinants, and oxidation in the fixative itself amplifies this problem (2). Although such difficulties were serious a decade ago, the recent

From: *Morphology Methods: Cell and Molecular Biology Techniques*
Edited by: R. V. Lloyd © Humana Press, Totowa, NJ

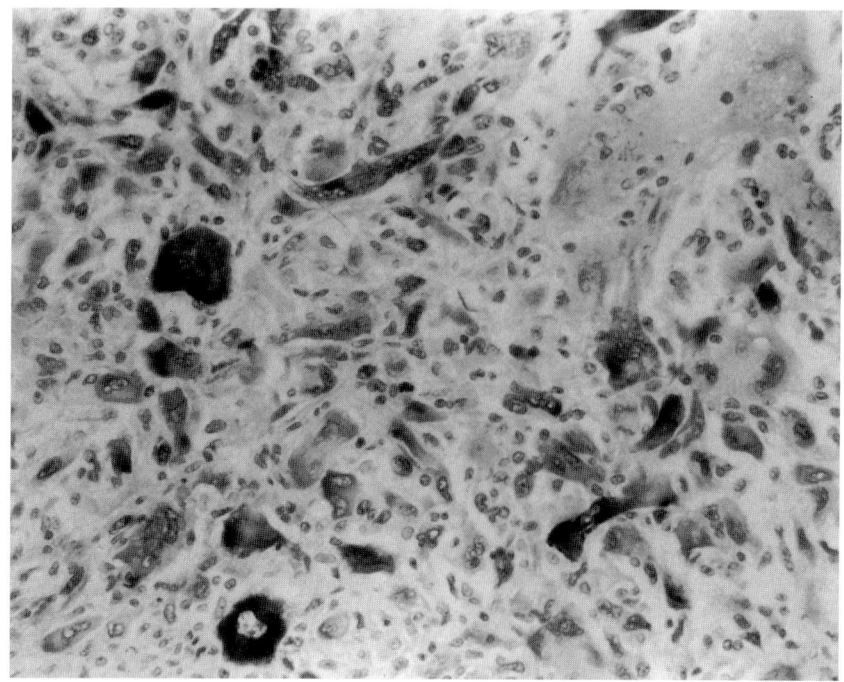

**Fig. 1.** Spindle and pleomorphic cells showing strong keratin labeling that facilitates recognition of sarcomatoid carcinomas. Evaluation of this case disclosed a corresponding renal cell carcinoma with sarcomatoid features.

availability of "epitope retrieval" procedures has done much to lessen the impact that formalin has on the innumoreactivity of clinical specimens *(3)*.

These new techniques, and currently available antibody-enzyme bridge methods, have been reviewed extensively in other reference sources and will not be recounted here *(3–6)*. Nonetheless, it is worthwhile reiterating that select procedures lend themselves to the detection of some antigens and not others *(7,8)*. Moreover, tissues with a high endogenous level of peroxidase (e.g., bone marrow sections) may require a different chromogenic enzyme (alkaline phosphatase, glucose oxidase) to obtain high-quality results *(9,10)*. Each laboratory is encouraged to make such determinations individually, since nuances in techniques at different institutions are well known to affect the outcome.

## Immunohistochemical Reagents

### Intermediate Filaments

Cytokeratins are constituents of the intermediate filaments (IFs) of epithelial cells expressed in various combinations depending on the epithelial type and the degree of differentiation. This class of IFs remains among the most commonly studied determinants in immunohistochemistry. Cytokeratin positivity helps corroborate a diagnosis of sarcomatoid carcinoma **(Fig. 1)** and usually excludes the possibility of sarcoma, malignant lymphoma, or melanoma *(16,17)*. Notable exceptions include synovial sarcoma **(Fig. 2)**, chordoma **(Fig. 3)**, Ewing's sarcoma, and epithelioid sarcoma; some smooth muscle cells may also react.

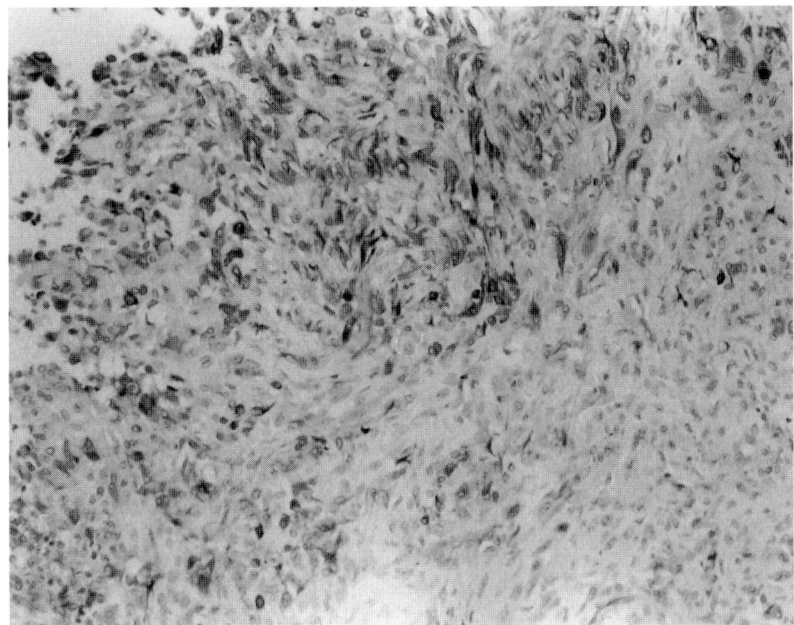

**Fig. 2.** Monophasic spindle cell synovial sarcoma shows unequivocal staining for keratin.

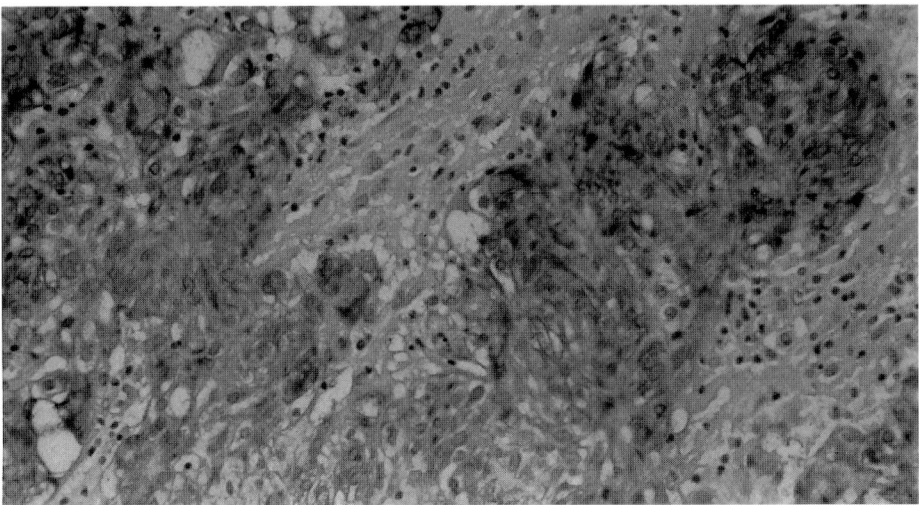

**Fig. 3.** Chordoma stains strongly for cytokeratin.

Monoclonal antibodies are now available to a wide range of keratin proteins (40–67 kDa) *(11)*. To maximize cytokeratin detection, proteolysis (0.4% ficin in phosphate-buffered saline [pH 7.4] for 20 min at room temperature *[12]*) or microwave-mediated epitope retrieval in citrate buffer *(7,8)* is mandatory before application of primary antibodies to rehydrated paraffin sections. Since, in most cases, the question is whether or not *any* cytokeratin is present in a given neoplasm, combinations or "cocktails" of monoclonal antibodies may be prepared to evaluate the widest range of kilodalton

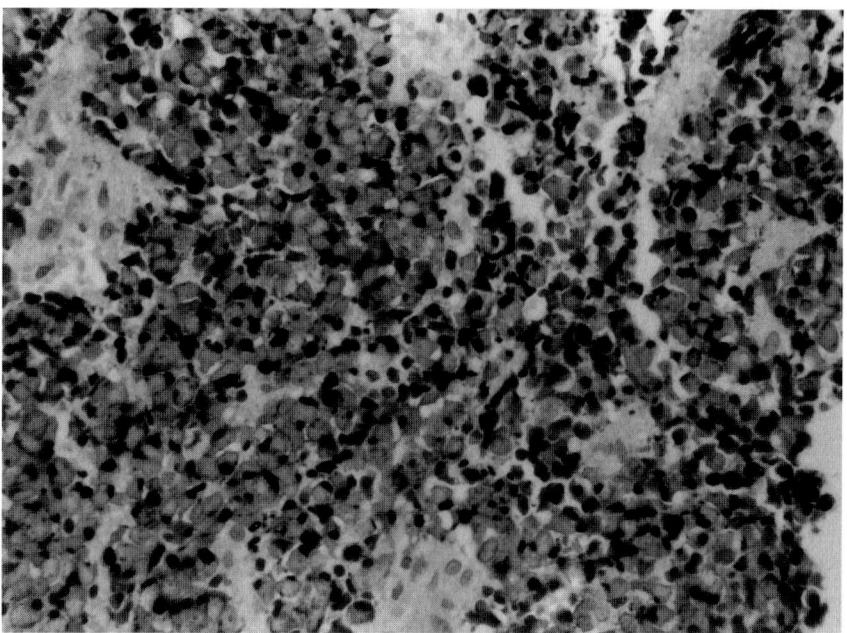

**Fig. 4.** Histologically, Merkel cell carcinoma can easily be misdiagnosed as lymphoma, melanoma, or metastatic small cell carcinoma of the lung. Dot-like cytoplasmic positivity for cytokeratin 20 helps to exclude lymphoma and melanoma. Only rare cases of small cell carcinoma show this pattern of staining.

weights. Keratin "cocktails" are the most useful in the diagnosis of poorly differentiated epithelial tumors *(13)*, but monospecific keratin antibodies also have distinct advantages in selected circumstances. For example, Merkel cell carcinoma of the skin—an example of a small round cell undifferentiated malignancy—regularly expresses keratin 20 **(Fig. 4)**, whereas its differential diagnostic simulators generally do not *(14)*. To prepare keratin antibody "cocktails," each of several monoclonal anticytokeratins is added to the same tube of diluent, in a quantity yielding the optimal titer of that particular antibody. For example, if two antibodies are selected to prepare 4 mL of "cocktail," one with an optimal titer of 1:200 and the other 1:400, 20 μL of the first antibody and 10 μL of the second would be added to the same 4 mL of solution.

Of course, one must be assured that the anticytokeratin preparations being utilized are *specific*. This point is particularly important with respect to the immunodetection of IFs as a group (cytokeratin, vimentin, desmin, glial fibrillary acidic protein [GFAP], and neurofilament protein), since all share a common peptide sequence *(15)*.

Vimentin is an IF that is present in most mesenchymal neoplasms and a variety of other classes of neoplasms (e.g., endometrial carcinoma) *(16–18)*. It is highly valuable as a marker of adequate tissue fixation and processing, in which positive and appropriate vimentin staining provides certainty of the ability of the tissue to react with antibodies in general *(19)*. Desmin is another IF present in mesenchymal lesions, found principally in myogenous tumors such as rhabdomyosarcoma and leiomyosarcoma *(20)* **(Fig. 5)**.

The neural IFs include neurofilament protein and GFAP. The first of these is restricted to neuronal and neuroendocrine cellular proliferations *(21)*; however, it is suboptimally

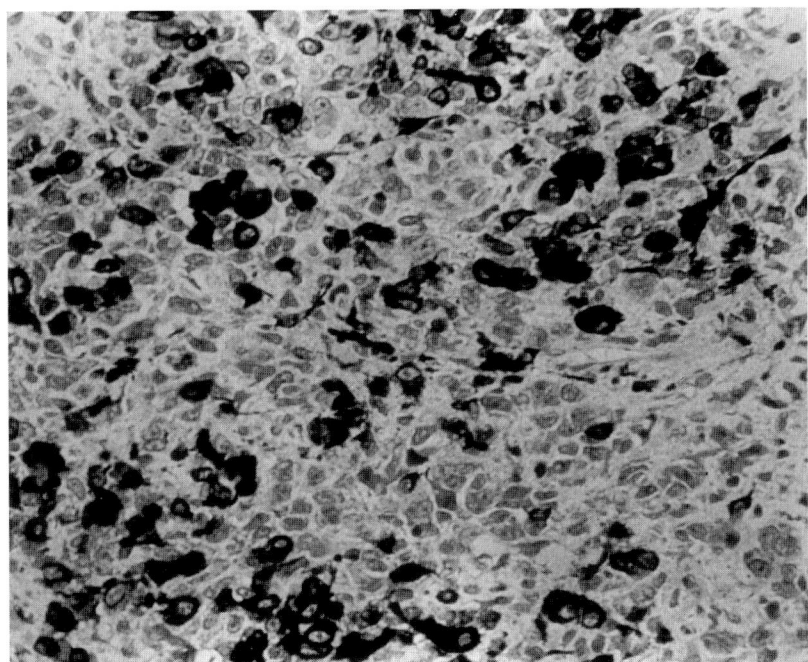

**Fig. 5.** Pleomorphic rhabdomyosarcomas frequently stain for desmin, allowing separation from other pleomorphic sarcomas in adults.

preserved in formalin-fixed specimens. GFAP immunoreactivity characterizes glial neoplasms such as astrocytomas and ependymomas *(22)*. Expression of this reactant appears to be maintained even in poorly differentiated tumors of the nervous system.

*CD45*

CD45 is a surface antigen expressed by viturally all hematolymphoid proliferations, and monoclonal antibodies for this marker are reliably specific *(23)*. However, not all anti-CD45 preparations are identical. Some, such as T29/33, necessitate the use of frozen tissue, whereas others (PD7/26-2B11) are applicable to paraffinized specimens as well *(24)*. The utility of CD45 is enhanced by concomitant staining with panels of antibodies to cytokeratin and S-100 protein. These reagents are helpful in the resolution of such problems as whether or not a polygonal or small cell undifferentiated lesion is a carcinoma, lymphoma, or melanoma.

*Epithelial Membrane Antigen*

Since its description in 1979 *(25)*, epithelial membrane antigen (EMA) has become a widely used determinant in diagnostic immunocytochemistry. EMA represents complex membrane glycoprotein originally isolated from milk fat globules and is unrelated to the keratin family *(25–27)*. EMA reactivity is relatively resistant to the effects of various fixatives and is well preserved in paraffinized specimens.

Providing that only *membrane-based* immunoreactivity is regarded as valid (**Fig. 6**), monoclonal antibodies to this discriminant are useful in determining the epithelial nature of undifferentiated tumors *(28)*. The tissue distribution is largely similar to that

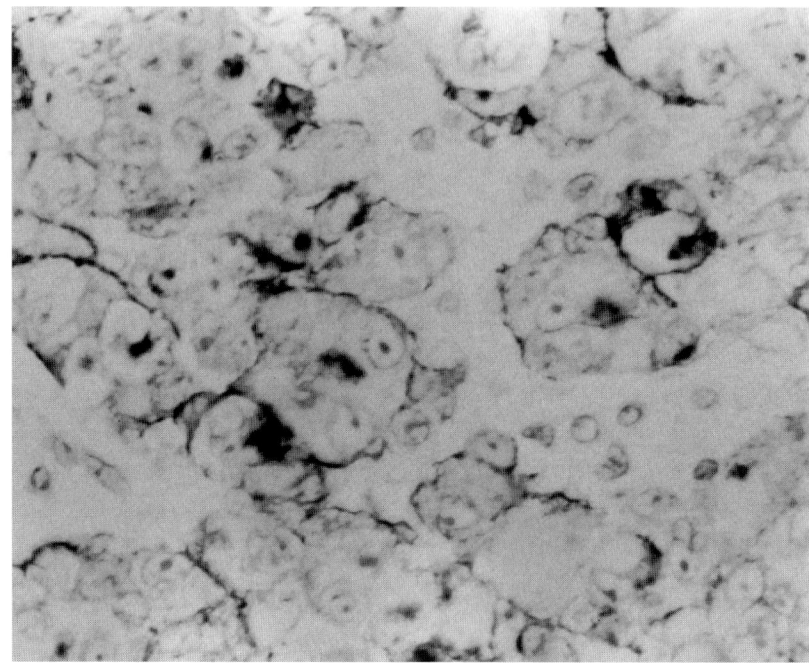

**Fig. 6.** Soft tissue rhabdoid tumor shows common coexpression of vimentin and epithelial markers. This tumor shows membranous straining for epithelial membrane antigen (EMA).

of keratin, with some caveats; in particular, not all epithelia are positive—hepatocellular carcinomas, adrenocortical carcinomas, and malignant germ cell neoplasms are EMA negative *(26,29)*. Also, EMA may be seen outside of nonepithelial lesions, including large cell anaplastic lymphoma, plasmacytoma **(Fig. 7)**, some T-cell lymphomas *(30,31)*, epithelioid sarcoma, and synovial sarcoma; meningioma may also show EMA reactivity *(32)*. The use of supplementary antibody panels that include other epithelial markers obviates potential misclassification.

*MOC-31*

MOC-31 is a 41-kDa glycoprotein that is cell membrane based; it is widely distributed in epithelial cells and tumors in many tissue sites *(33,34)*. Monoclonal antibodies to this determinant have most often been used in the diagnostic distinction between serosal adenocarcinoma and mesothelioma (which typically lacks MOC-31) *(35)*, but they also fail to label a selected group of carcinomas that includes hepatocellular carcinoma, germ cell malignancies, and renal cell carcinoma *(36)*.

*Tumor-Associated Glycoprotein (B72.3)*

The monoclonal antibody known as B72.3 labels a plasmalemmal glycoprotein designated tumor-associated glycoprotein-72 (TAG-72) *(37)*. It was isolated from a human breast carcinoma cell line and appears to be virtually pan-carcinomatous in distribution *(37,38)*. In likeness to MOC-31, however, TAG-72 is characteristically absent in mesotheliomas, adrenocortical carcinomas, hepatocellular carcinomas, renal

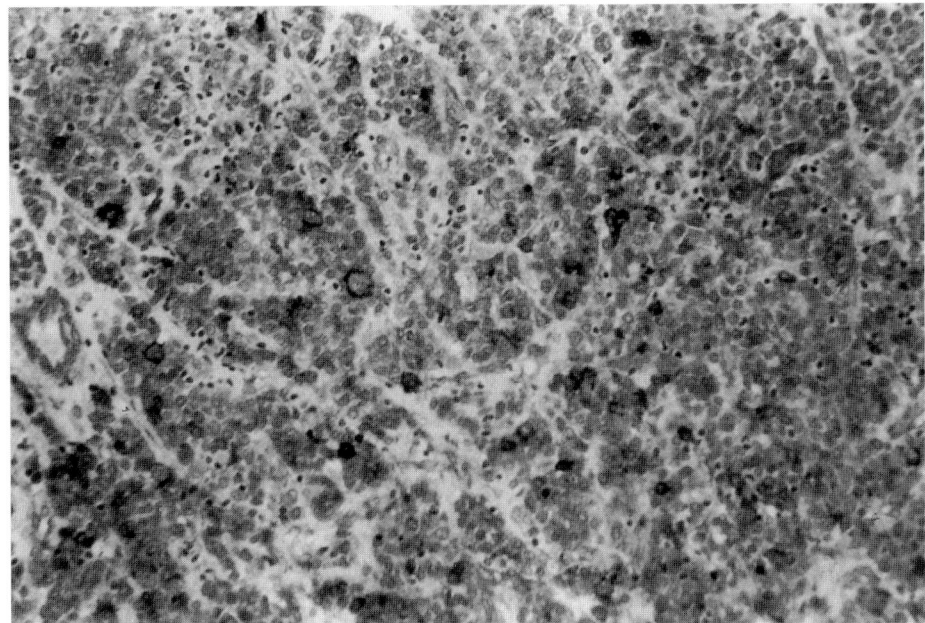

**Fig. 7.** Plasmacytomas frequently show epithelial membrane antigen (EMA; shown) and vimentin positivity, which may lead to misclassification as a carcinoma or sarcoma.

cell carcinoma, nasopharyngeal carcinoma, thyroid carcinomas, and malignant germ cell tumors *(36,38)*.

*Placental Alkaline Phosphatase*

The isoenzyme of alkaline phosphatase that is expressed by the normal placenta (PLAP) is also evident as an oncofetal antigen in some genitourinary, gastrointestinal, and pulmonary carcinomas *(29)*. Moreover, it is nearly universally seen in germ cell tumors *(39–41)*. Immunostains for PLAP therefore have their greatest application in separating germ cell neoplasms from somatic tumors. We have found anti-PLAP to be an extremely useful screening reagent for possible germ cell differentiation in undifferentiated tumors. When positive, stains for more specific oncofetal determinants, e.g., α-fetoprotein (AFP) and human chorionic gonadotrophin, as well keratin subclasses, EMA, and such markers as CD30, should be performed *(41)*.

Anti-PLAP consistently identifies both seminoma **(Fig. 8)** and embryonal carcinoma *(39,40)*. Both of these differ from most other epithelial malignancies in that they generally lack EMA reactivity. Seminoma also lacks keratin reactivity in 80% of cases, whereas embryonal carcinoma is keratin positive. CD30 is helpful in this distinction, as it is present in embryonal carcinoma **(Fig. 9)** and is only rarely if ever seen in seminoma *(40,41a)*. In contrast to germ cell tumors, most PLAP-positive somatic tumors uniformly express EMA and lack CD30 *(29)*. Therefore, a panel that includes cytokeratin, EMA, CD30, and PLAP may be useful in distinguishing among these pleomorphic neoplasms.

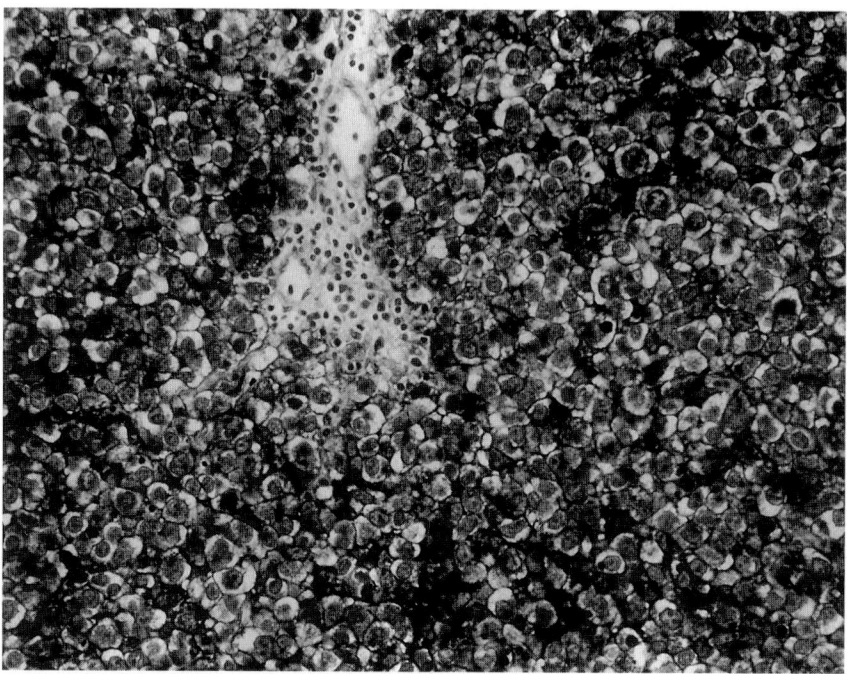

**Fig. 8.** Placental alkaline phosphatase labels this seminoma.

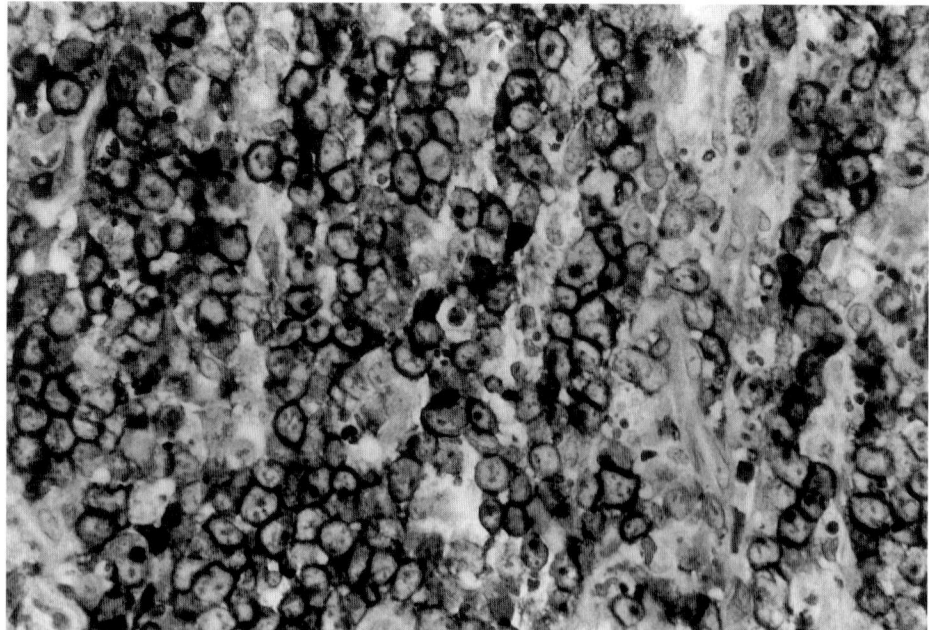

**Fig. 9.** CD30 is frequently positive in embryonal carcinoma (shown) and absent in seminoma, a helpful distinction in keratin-positive germ cell tumors lacking specific features on routine histology.

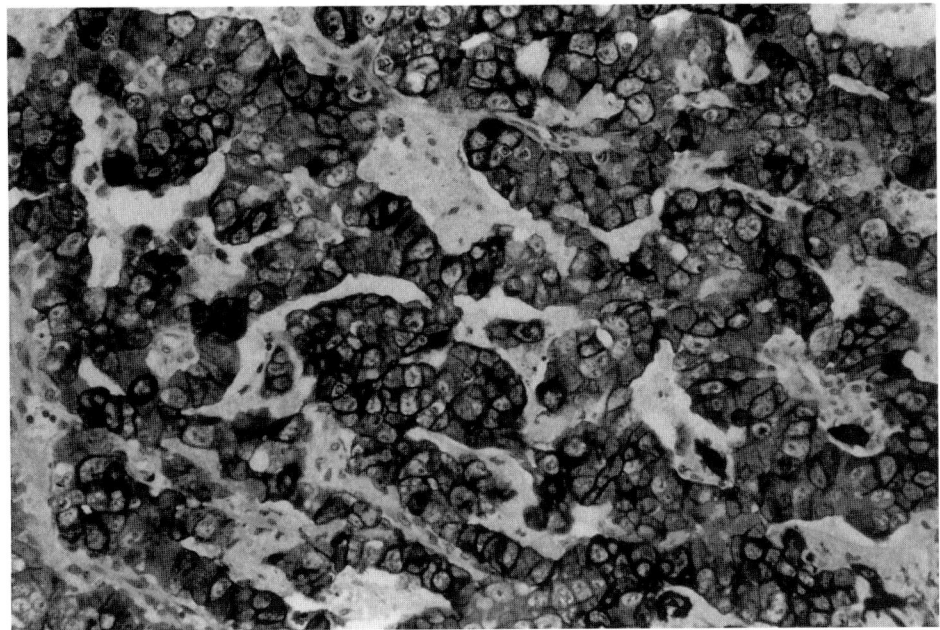

**Fig. 10.** Strong α-fetoprotein reactivity in mixed embryonal-yolk sac carcinoma.

### α-Fetoprotein

AFP is another oncofetal antigen that is relatively restricted in its expression to nonseminomatous germ cell tumors **(Fig. 10)**, hepatocellular carcinomas, and rare examples of somatic malignancies *(42–44)*. Hence, it has its greatest use in immunohisto-chemistry in the characterization of polygonal large cell undifferentiated neoplasms. In an AFP-positive polygonal cell tumor, PLAP positivity is desirable before favoring the diagnosis of germ cell tumor over hepatocellular carcinoma, which is devoid of alkaline phosphatase immunoreactivity *(29)*.

### Carcinoembryonic Antigen

Carcinoembryonic antigen (CEA) has enjoyed its greatest recognition in clinical medicine as a serologic indicator of the growth of colorectal cancer. Immunohistochemi-cally, CEA is strongly expressed in colorectal adenocarcinoma **(Fig. 11)**, but it may be found in other epithelial tumors *(45–47)*. Monoclonal antibodies to CEA represent prototypic epitope-specific probes, which recognize a small portion of a large antigen *(45,47)*. Different carcinomas express a common portion of the CEA molecule but may also produce mutually exclusive epitopes that are tissue restricted. If the latter are "mapped" by immunohistochemical surveys of primary carcinomas, the corresponding antibody reagents can be employed to yield diagnostic information on the possible sources of lesions that present with metastasis. For example, adenocarcinomas of the lung, breast, and gastrointestinal (GI) tract show uniform reactivity with certain epitope-specific anti-CEAs and thereby may narrow the potential anatomic sites for metastatic glandular neoplasms *(45–47)*. In general, monoclonal anti-CEA preparations are prefera-ble to polyclonal heteroantisera, because most of the latter are inferior in specificity and require adsorption with crossreacting antigens.

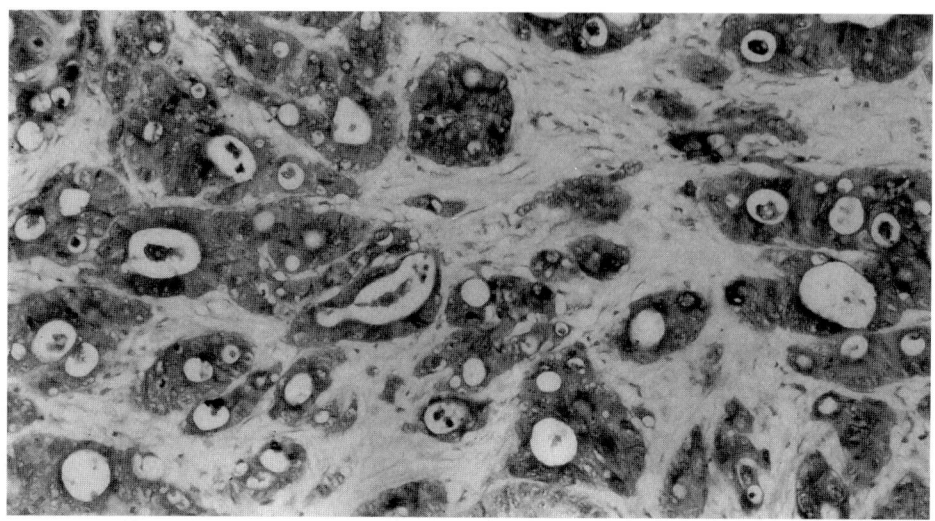

**Fig. 11.** Adenocarcinoma of the colon shows uniform strong carcinoembryonic antigen (CEA) expression.

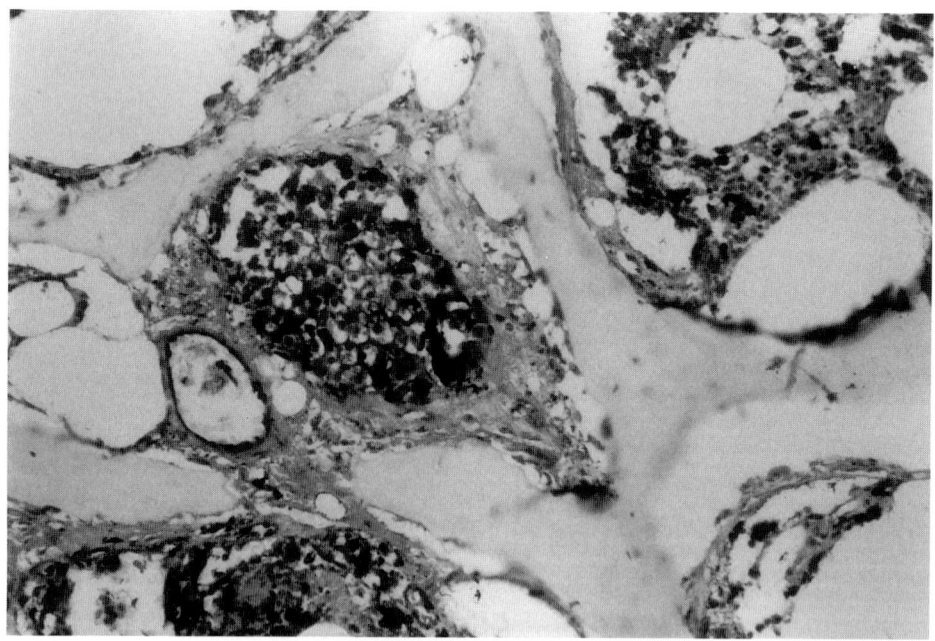

**Fig. 12.** The immunohistochemical demonstration of prostate-specific antigen (PSA) in a bony metastasis is accepted as a reliable marker for adenocarcinoma of prostatic origin.

*Prostate-Specific Antigen*

As its name suggests, prostate-specific antigen (PSA) is largely restricted in its tissue distribution to epithelial cells of the prostate and prostatic adenocarcinomas **(Fig. 12)** *(48–50)*. This cytoplasmic determinant appears to be expressed at an early stage of embryonic development, inasmuch as even poorly differentiated prostatic malignancies

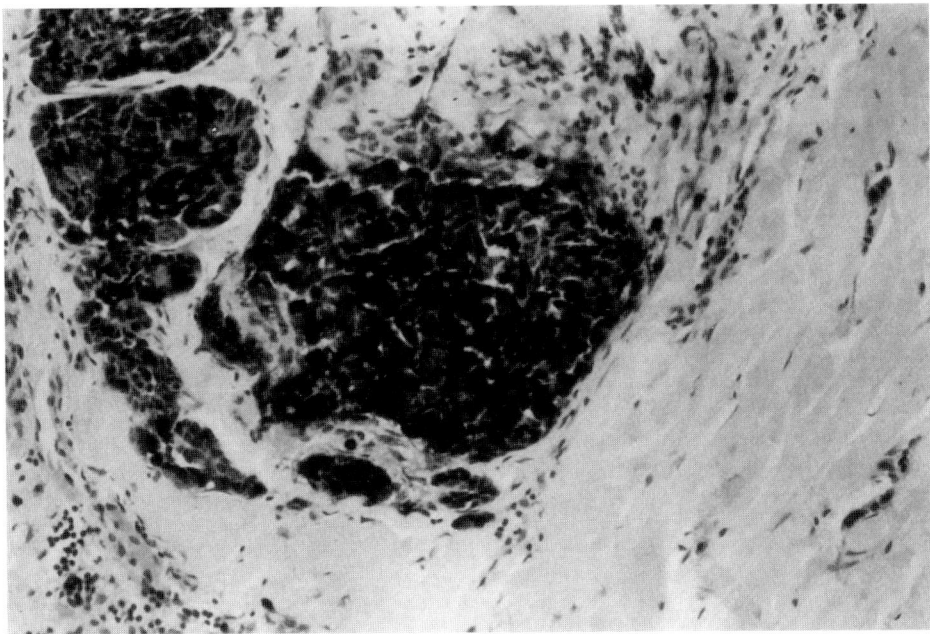

**Fig. 13.** The presence of thyroglobulin reactivity in a metastatic tumor showing a nested pattern helps establish the diagnosis of poorly differentiated thyroid carcinoma.

display its presence. Hence, PSA represents one of exceedingly few tissue-specific markers available in diagnostic immunohistochemistry and has been widely utilized in the recognition of metastatic prostate cancer.

Some reports have impugned the exclusivity of this association and have documented apparent PSA reactivity in selected adenocarcinomas of the bladder *(51)* or in neuroendocrine tumors *(52)*. In our experience, however, such problems are likely to be related to the heteroantiserum reagents that were used in these studies. We have employed a monoclonal anti-PSA in analyses of over 3000 carcinomas from diverse locations; to date, it has demonstrated absolute specificity for prostatic neoplasms in these cases. However, extraprostatic expression of PSA has also been noted in periurethral gland adenocarcinoma in women, rectal carcinoid, and extramammary Paget's disease *(52a)*. These entities can generally be easily addressed using a panel of immunostains that includes PSA, as well as neuroendocrine markers and CEA.

*Thyroglobulin*

Thyroglobulin is likewise restricted in its expression to the follicular thyroid epithelium and related neoplasms *(53,54)*. The greatest utility of this marker is to confirm a thyroid origin for metastatic carcinomas in lymph nodes, lung, or bone. Thyroglobulin reactivity is also helpful in the diagnosis of struma ovarii. Unfortunately, "anaplastic" thyroid cancer seldom demonstrates this determinant immunohistochemically *(55)*, limiting its usefulness in the evaluation of undifferentiated tumors to "solid" follicular carcinomas, including "insular" thyroid carcinoma and insular carcinoma **(Fig. 13)** *(56)*.

*Thyroid Transcription Factor-1*

Thyroid transcription factor-1 (TTF-1) is a DNA-binding transcriptional regulator protein present in normal thyroid follicular cells, subsets of respiratory and alveolar epithelium, and the diencephalon *(57–61)*; TTF-1 binds preferentially to thyroglobulin and thyroperoxidase promoters *(62)*. It is commonly present in the malignant counterparts of these cells, namely, adenocarcinomas of the lung and thyroid *(60,61)*; most anaplastic thyroid carcinomas lack this marker; some authors have found it in 25% of such cases *(62)*, whereas thyroglobulin reactivity is virtually never detected *(55)*. We believe more studies need to be performed to determine the usefulness of this marker in the identification of anaplastic thyroid tumors.

The uniform *lack* of TTF-1 in adenocarcinoma of the breast, however, is quite useful to discriminate primary lung tumors from metastases of a breast tumor. In addition, small cell neuroendocrine carcinomas of the lung and selected extrapulmonary sites label for TTF-1 *(63,64)*. Because of mutually exclusive staining patterns in this group of tumors, it has been touted as a differential diagnostic discriminant in the separation of Merkel cell carcinoma of the skin and metastatic small cell carcinoma of the lung *(65)*. The initial results using TTF-1 seem promising, although more data are needed before we can advocate the use of this antibody for routine clinical use.

*Gross Cystic Disease Fluid Protein-15*

Over the past 20 years, progress has been made in characterizing several proteins manufactured by mammary epithelial cells; in particular, one such soluble product that is found in the fluid contents of "fibrocystic" breasts has been studied extensively *(66,67)*. This determinant is known as gross cystic disease fluid protein-15 (GCDFP-15) and is strongly expressed by cells with apocrine characteristics. GCDFP-15 is present in the tumor cells of approximately 55–70% of breast carcinomas, regardless of histologic subtype *(66,67)* **(Fig. 14)**. Indeed, catalog studies of various metastatic carcinomas indicate that GCDFP-15 is *restricted* to breast tumors, making it a valuable indicator of anatomic origin in the evaluation of metastases of unknown origin. The only caveat is salivary duct carcinoma, with its morphologic similarity to high-grade ductal carcinoma, which also strongly expresses this marker **(Fig. 15)**. GCDFP-15 is much more specific than lactalbumin, BCA225, or CA15-3, proteins that may be observed in a diverse group of unrelated carcinomas in extramammary locations *(68–71)*.

*Estrogen Receptor Protein*

Intuitively, one would expect that estrogen receptor protein (ERP) would be restricted to carcinomas of the breast and mullerian tract. Although those sites certainly account for *most* ERP-positive tumors, lesions such as thyroid carcinoma, transitional cell carcinoma of the bladder, prostatic carcinoma, aggressive angiomyxoma *(71a)*, and even rare examples of gastric, pulmonary, and hepatocellular carcinoma may express ERP *(72–74)*.

*CA-125*

CA-125 is a glycoproteinaceous membrane constituent first identified nearly 20 years ago in ovarian cell lines *(75)*. Early on, it was recognized that a closely similar or identical moiety was expressed by mesothelial cells and mesotheliomas as well *(76)*.

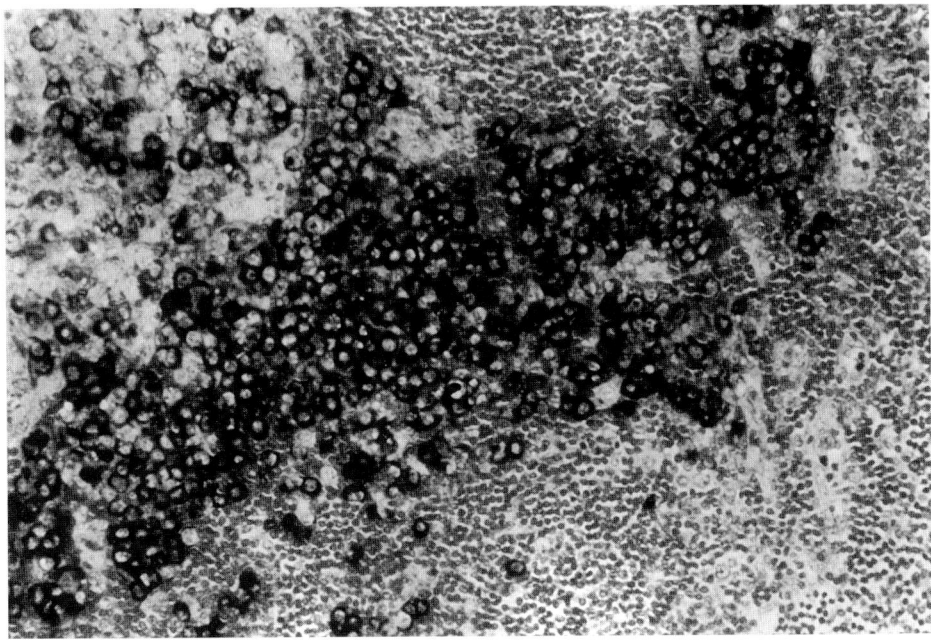

**Fig. 14.** The cells of metastatic lobular carcinoma are readily detected in a lymph node using an antibody to GCDFP-15.

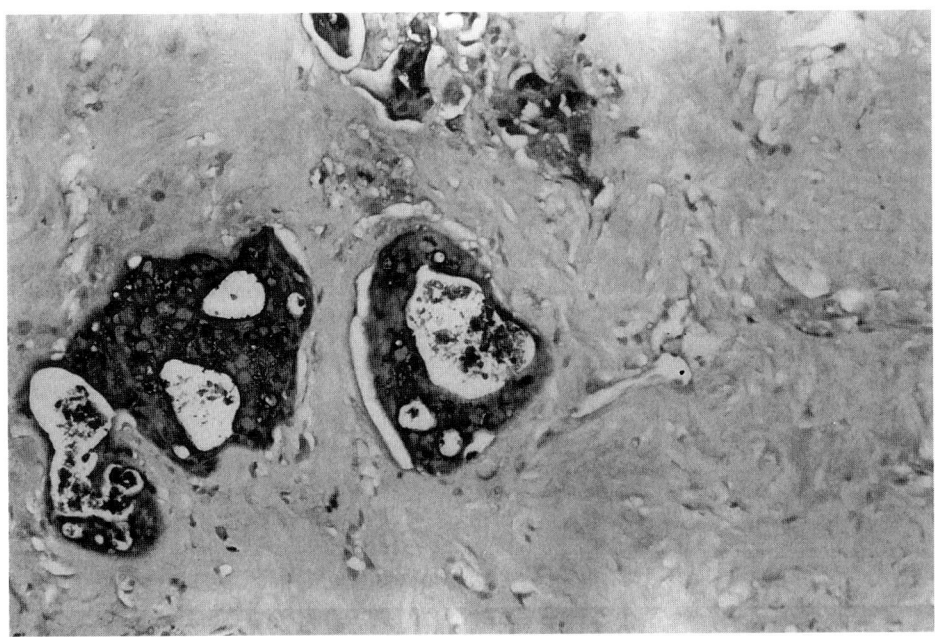

**Fig. 15.** This metastatic adenocarcinoma shows strong GCDFP-15 reactivity. Although typically regarded as a marker for breast cancer, salivary duct carcinoma (SDC) also shows strong reactivity. SDC lacks estrogen receptor and progesterone receptor staining, which may be useful in differential diagnosis.

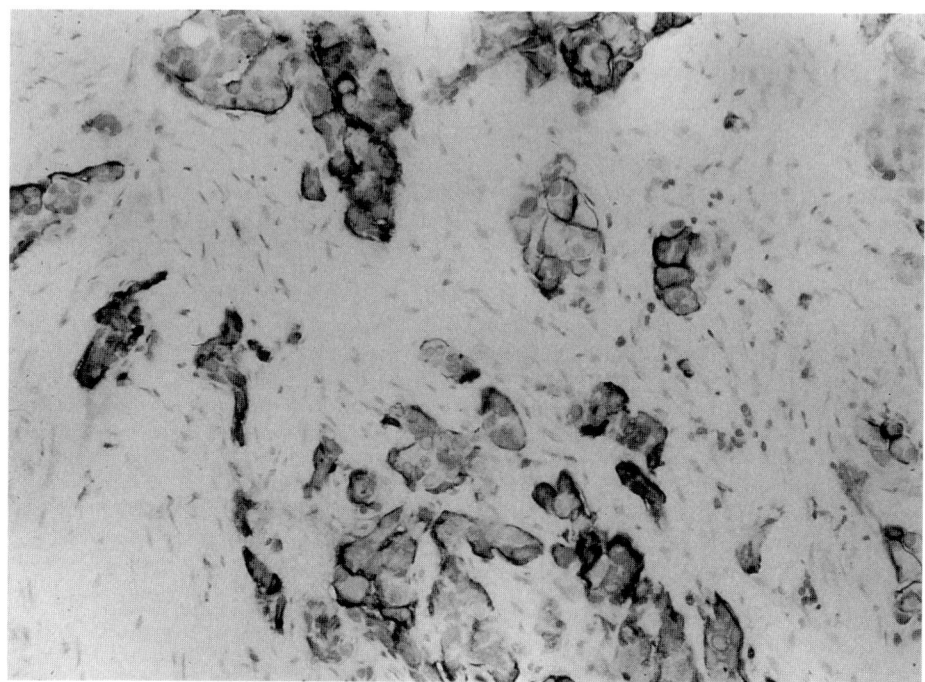

**Fig. 16.** Adenocarcinoma of the ovary frequently shows strong reactivity for CA-125.

Today it is recognized as the marker most strongly associated with epithelial gynecologic tumors **(Fig. 16)**. CA-125 has been detected immunohistochemically, most often in neoplasms of the müllerian tract, but roughly one-half of tumors of the biliary tree and pancreas may also react *(77,78)*. Carcinomas in other sites are uncommonly CA-125 positive. In diagnostic pathology CA-125 plays a role in identifying the primary locations of metastatic carcinoma of unknown origin. The use of CA-125 antibodies is recommended not in a solitary setting but in combination with CEA, GCDF5-15, and vimentin to discriminate among the sites of origin of metastatic carcinoma.

*CA19-9*

CA19-9 is a glycoprotein that is related to the Le[a] blood group antigen, and it is labeled by the monoclonal antibody known as 1116NS19-9 *(79)*. The latter reagent was raised against a human colon carcinoma cell line. Most carcinomas of the GI tract, pancreas, biliary tree, urinary bladder, ovaries, and endometrium manifest CA19-9 reactivity *(80)* **(Fig. 17)**, whereas tumors of other anatomic locations are only sporadically positive. In particular, hepatocellular carcinoma and renal cell carcinoma are consistently negative for this marker *(71,80)*.

*CD15*

CD15 is a hematopoietic differentiation antigen shared by granulocytes, monocytes, Reed-Stemberg cells, and subsets of neoplastic B-cells and T-lymphocytes *(81,82)*.

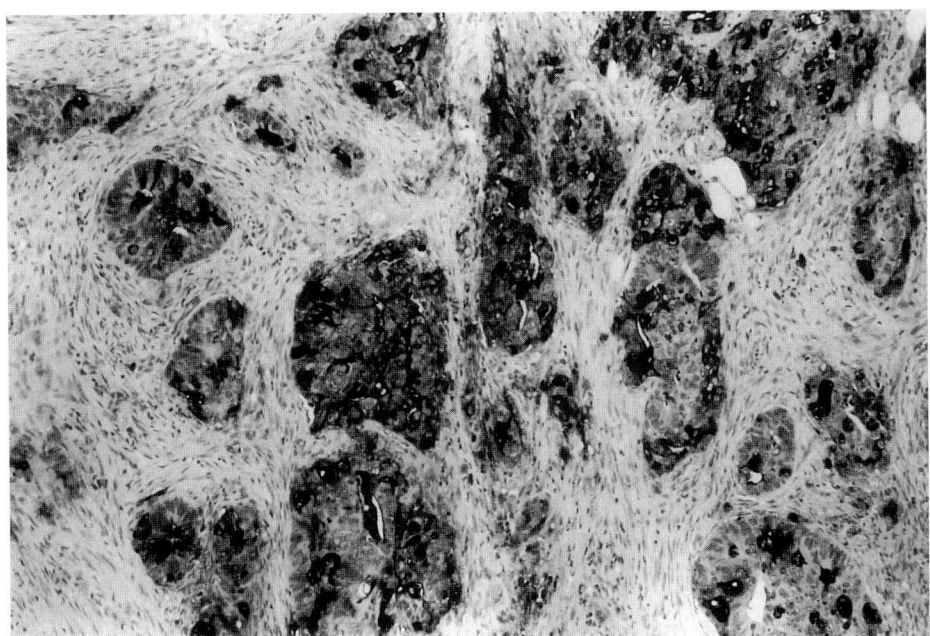

**Fig. 17.** CA19-9 highlights the malignant cells of pancreatic adenocarcinoma. This marker may also be positive in tumors of the gastrointestinal tract, biliary tree, urinary bladder, ovaries, and endometrium.

Several epithelial cell lines (most of which are glandular or neuroendocrine) also expresses CD15 *(81)*, and it may be detected immunohistochemically in their malignant counterparts **(Fig. 18)**. Reactivity for CD15 is applicable to the separation of metastatic adenocarcinomas from histologically similar lesions such as malignant mesotheliomas, since it is lacking in the latter tumors *(83)*. When used in combination with leukocyte common antigen (CD45), CD15 also provides information on the cell lineage of lymphoreticular neoplasms; CD15-reactive tumors in this class predominantly include Hodgkin's disease **(Fig. 19)** and T-cell non-Hodgkin's lymphomas *(82)*.

### Calretinin

Calretinin is a calcium-binding protein that is virtually universally expressed by mesothelial cells and malignant mesotheliomas *(84)*. Calretinin is a recognized useful marker for discriminating mesothelioma of the epithelial type from adenocarcinoma in both paraffin sections and serous effusions *(85,85a–c)*. A limited number of cases of sarcomatoid mesothelioma may also show calretinin expression, separating these lesions from a spectrum of other spindle cell neoplasms of the pleura *(85d)*.

### S-100 Protein

S-100 protein is a calcium flux determinant, expressed by normal melanocytes, Langerhans' histiocytes, cartilaginous cells, adipocytes, Schwann cells, astrocytes, oligodendroglia, ependyma, eccrine sweat glands, reticulum cells, salivary glands, and myoepithelial cells *(86–89)*. It has its greatest use in the diagnosis of solid tumors in

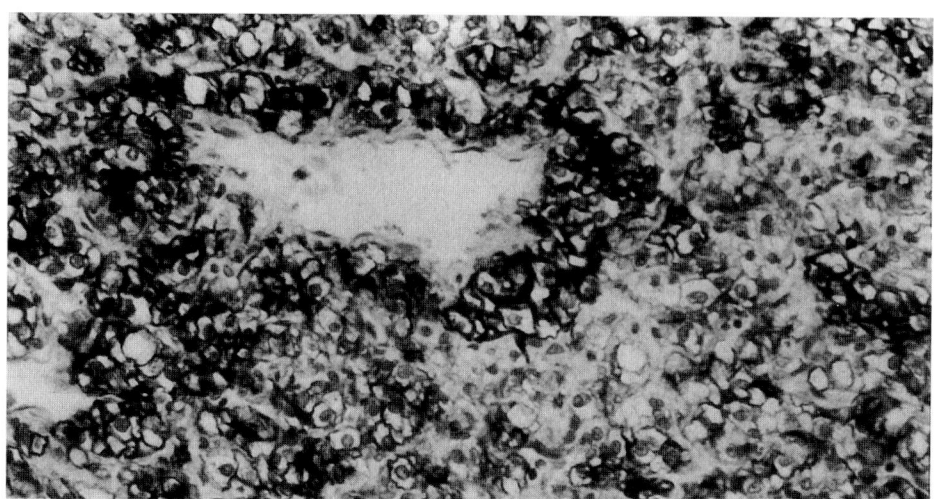

**Fig. 18.** Positive CD15 (Leu-M1) staining is diagnostically useful to distinguish carcinoma from epithelial mesothelioma. It is not useful, however, in sarcomatoid tumors.

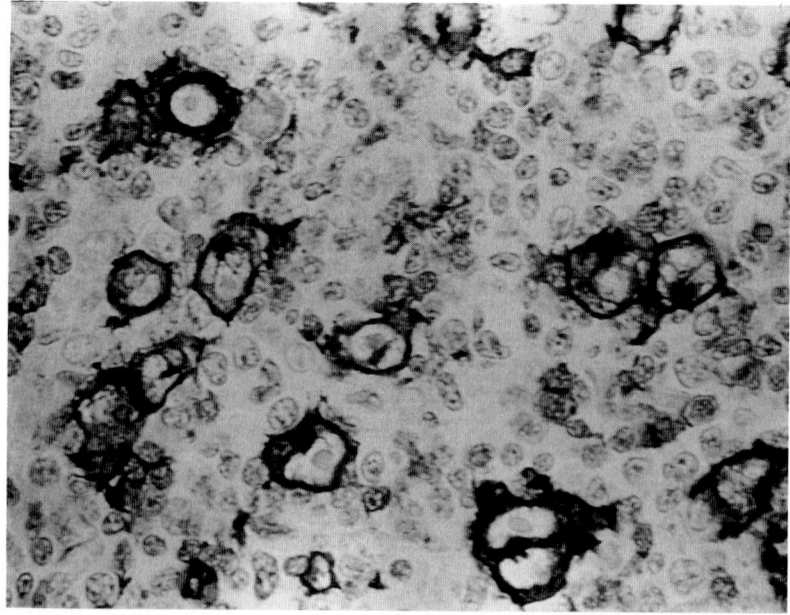

**Fig. 19.** CD15 (Leu-M1) highlights the Reed-Sternberg cells in Hodgkin's disease.

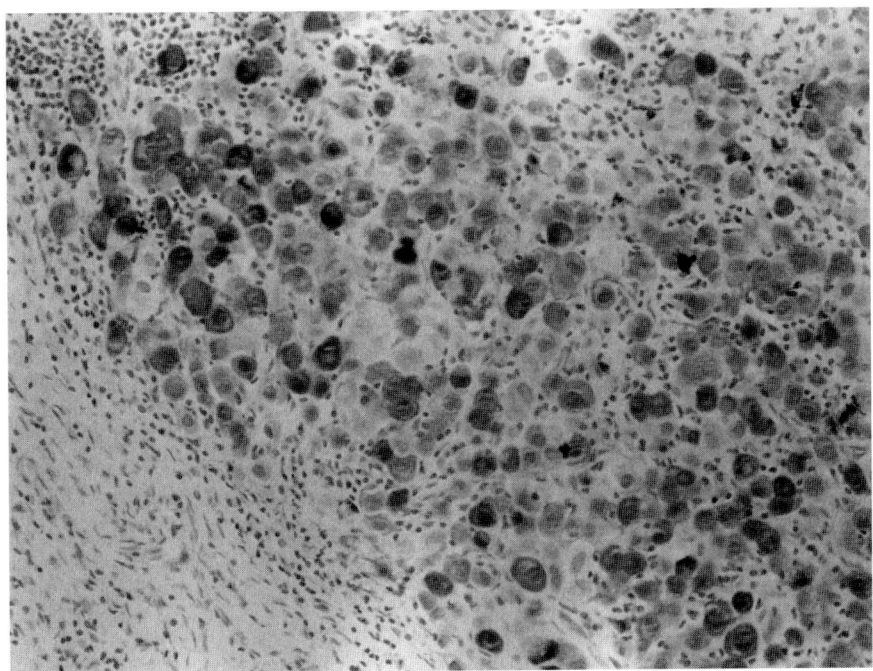

**Fig. 20.** The presence of S-100 protein reactivity is useful to corroborate a diagnosis of metastatic adenocarcinoma of the breast.

the identification of malignant melanoma, clear-cell sarcoma (melanoma of soft parts), glioma, and malignant peripheral nerve sheath tumor.

Both polyclonal and monoclonal antibodies to this protein are available and have been used widely. One caution when interpreting diagnostic immunostains for S-100 protein is the diversity of potentially reactive cell types, as listed above. Hence, such stains should not be used alone in the characterization of any given neoplasm. For example, we have found that a surprisingly large number of carcinomas are invested with a reactive, intratumoral, Langerhans' histiocyte population. The S100 protein reactivity that the latter displays may be misinterpreted as *tumor cell* positivity, resulting in an erroneous diagnosis of malignant melanoma.

Also, it has become evident that certain poorly differentiated epithelial malignancies have the ability to express S-100 protein. Carcinomas of the breast **(Fig. 20)**, genitourinary tract, pancreas, salivary glands, and sweat glands are most notable for this antigenic expression *(86–88)*. Again, failure to include antibodies to cytokeratin and EMA in assessments of such lesions may well result in diagnostic misadventure.

*Other Melanocyte Markers*

Several other monoclonal antibody reagents with selectivity for melanocytic determinants have entered general practice as well. These include HMB-45 *(90,91)*, anti-tyrosinase, MART-1 (Melan-A), and KBA62 *(92)*. Such markers exhibit excellent labeling of formalin-fixed tissues and show absolute or nearly absolute specificity for melanocytes and melanocytic neoplasms **(Fig. 21)**. In our experience and that of others *(90–92)*, HMB-45 and anti-tyrosinase do not label carcinomas or lymphomas; on the

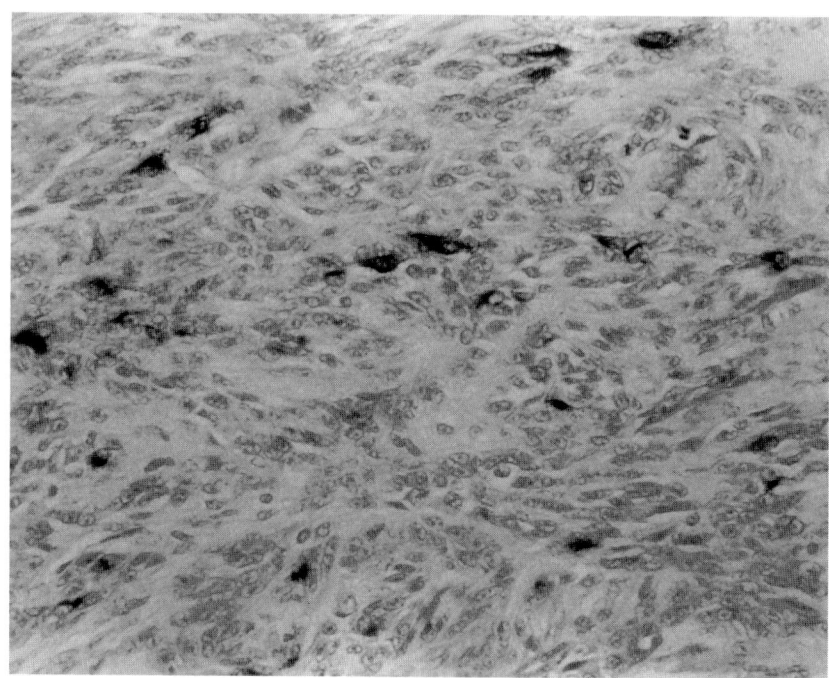

**Fig. 21.** The presence of HMB-45 reactivity in the spindle cell variant of malignant melanoma is rare but helps distinguish this tumor from other spindle cell tumors.

other hand, MART-1 reactivity is seen in most adrenocortical carcinomas, and KBA62 may decorate occasional poorly differentiated squamous and renal cell carcinomas as well as rare rhabdomyosarcomas *(92)*. However, the overall utility of those two antibodies as melanocytic markers is still high.

*Chromogranin A*

Chromogranin A (CgA) is a protein that is indigenous to the matrices of neurosecretory granules. It has a relatively ubiquitous distribution in neuroendocrine tissues, *e.g.*, those of the anterior pituitary gland, thyroid C-cell system, parathyroid glands, paraganglion system, adrenal medulla, and pancreatic islets *(93,94)*. Accordingly, antibodies to CgA are exceedingly specific in the delineation of neuroendocrine differentiation in epithelial tumors. However, because CgA reactivity is directly related to the relative number of cytoplasmic endocrine granules in any given neoplasm, this marker is somewhat insensitive. For example, roughly 30% of small cell neuroendocrine carcinomas and neuroblastomas will exhibit labeling with anti-CgA, and it is distinctly unusual for primitive neuroectodermal tumors to do so *(95)*.

*Synaptophysin*

Synaptophysin is an integral membrane glycoprotein of presynaptic vesicles that is detectable in a wide range of normal and neoplastic neuroendocrine cells **(Fig. 22)** *(95–99)*. Monoclonal antibodies to this marker are widely used in diagnostic surgical pathology and cytopathology with good success. In light of its subcellular associations,

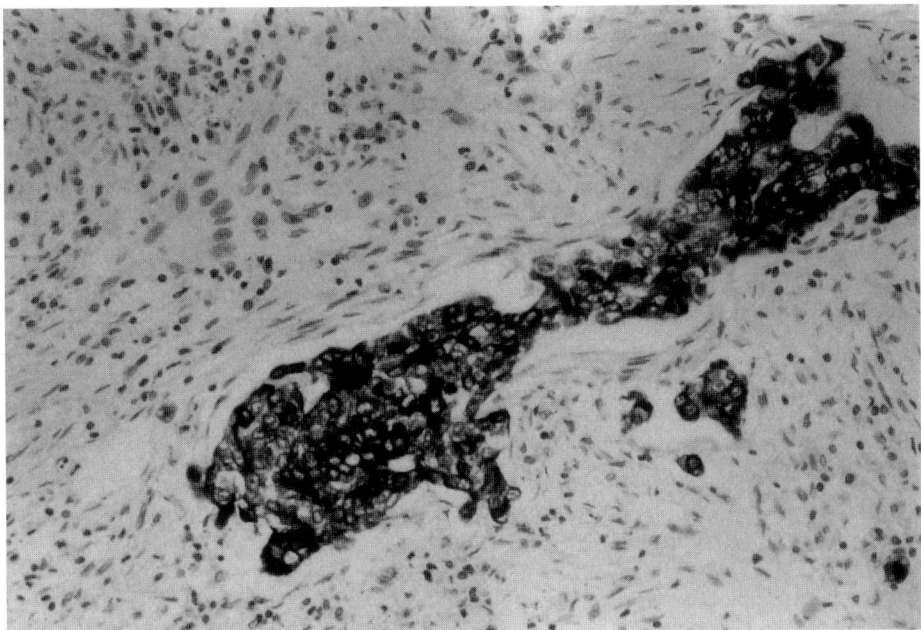

**Fig. 22.** Desmoplastic small cell tumors show a diverse array of differentiation, which often includes strong synaptophysin reactivity.

one might assume that synaptophysin would have a tissue distribution synonymous to that of the chromogranins; however, in practicality, that is not true *(99)*. The vesicles that contain synaptophysin can colocalize with neurofilaments or epithelial filaments, showing expression independent of other neuronal differentiation markers. A sizable proportion of neuroendocrine neoplasms will label for CGA, but not synaptophysin, and the converse of that relationship also applies. Thus, anti-synaptophysin and anti-chromogranin should be conceptualized as complementary reagents.

## CD57

CD57 is recognized by the monoclonal antibody HNK-1 and was initially described as a membrane antigen of hematopoietic cells, namely, natural killer lymphocytes. *(100)*. Subsequent analyses have demonstrated a relative rarity of CD57-positive malignant lymphomas *(101)*, relegating this marker to a minor role in hematopathology; it is most often seen in "large granular cell" lymphoproliferative disorders, and in the Reed-Sternberg-like cells of lymphocyte-predominant Hodgkin's disease (which is probably, in actuality, a B-cell lymphoma) *(101)*. Unexpectedly, HNK-1 was shown to recognize an epitope of myelin-associated glycoprotein and a glycoproteinaceous determinant in neurosecretory granule matrices *(100,102,103)*. Hence, it is most often used today in *solid* tumor pathology, to label schwannian, neuroectodermal, and neuroendocrine proliferations. Because of the variety of cell lineages involved in this group, it should be apparent that results of other immunostains are required to put the significance of CD57 reactivity into proper perspective, in any given case.

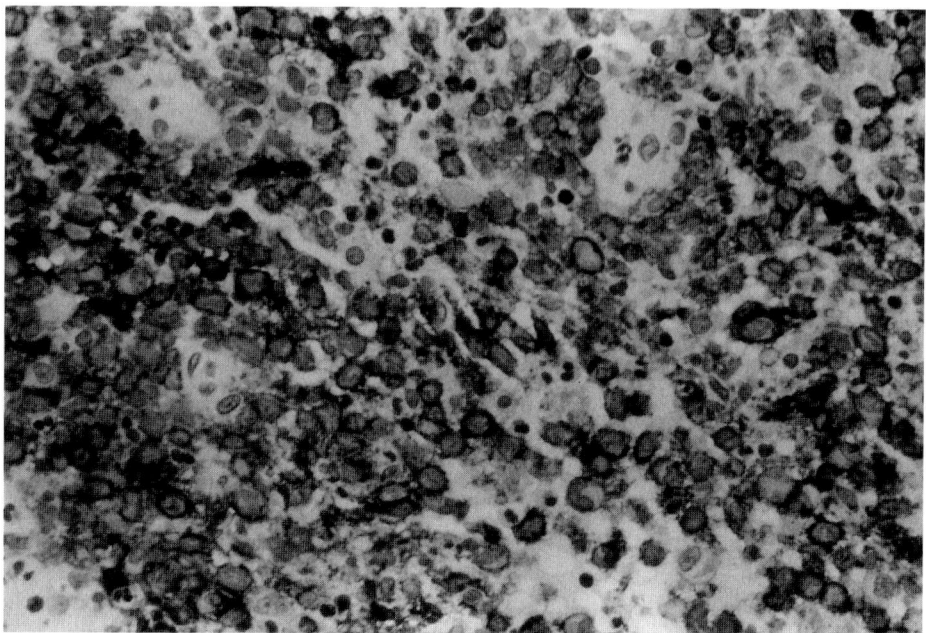

**Fig. 23.** Among small round cell tumors, CD99 shows a characteristic membranous pattern in acute lymphoblastic lymphoma (shown) as well as Ewing's sarcoma.

### CD99 (MIC-2; p30/32 Protein)

The membranocytoplasmic protein known as CD99 in the hematopoietic antigen cluster designation is the same molecule that has been called "MIC-2" or "p30/32 protein" *(101)*. The function of this moiety remains uncertain but is known to be expressed in virtually all primitive neuroectodermal tumors (PNETs) and Ewing's sarcomas *(104)*. The specificity of CD99 antibodies for a neuroectodermal lineage is not absolute, however, because they may also label a minority (<15%) of alveolar rhabdomyosarcomas as well as the overwhelming majority (90%) of lymphoblastic lymphomas **(Fig. 23)** *(101)*. CD99 is observed in up to 20% of neuroendocrine carcinomas in some body sites as well *(105)*. This is an important fact, because it may obscure the difference between NEC and PNET in selected instances; this is particularly true in light of the potential for keratin reactivity that both of these lesions have.

### Muscle-Specific Actin

The molecular family of actin proteins includes some moieties that are confined to cells with muscular differentiation, whereas others are seen in epithelia as well. In the diagnosis of mesenchymal neoplasms, monospecific antibodies to the former are highly desirable. One such reagent, HHF-35, demonstrates an excellent level of specificity and sensitivity for smooth and striated muscular proliferations, including leiomyosarcomas, hemangiopericytomas, and rhabdomyosarcomas *(106,107)*. As such, it is a valuable adjunct to anti-desmin antibodies in the immunohistologic detection of myogenic tumors.

## CD34

CD34, or the human hematopoietic progenitor cell antigen, is recognized by several monoclonal antibodies including My10, QBEND10, and BI-3C5 *(108)*. It is a 110 kDa protein that, as its name suggests, is expressed by embryonic cells of the hematopoietic system. These include lymphoid and myelogenous elements and also endothelial cells *(109)*. Correspondingly, again in the setting of soft tissue tumors, CD34 is a potential indicator of vascular differentiation. It is highly sensitive for endothelial differentiation, regardless of tumor grade, and recognizes >85% of angiosarcomas and Kaposi's sarcomas *(108,109)*. Nevertheless, the specificity of CD34 is a problem, inasmuchas it has been reported in some leiomyosarcomas, peripheral nerve sheath tumors, and epithelioid sarcomas, which could potentially simulate variants of angiosarcoma or hemangioendothelioma *(109)*. In addition, CD34 is so commonly present in dermatofibrosarcoma protuberans (and its variants), spindle cell lipoma, and solitary fibrous tumor that it is regularly used as an adjunct for the diagnosis of those tumors *(110,111)*. Thus, as endothelial markers, antibodies to CD34 are best used in a panel of reagents that is designed to account for these other diagnostic possibilities.

## CD31

The platelet-endothelial cell adhesion molecule-1 (PECAM-1) is also known as CD31. It is a 130-kDa transmembrane glycoprotein shared by vascular lining cells, megakaryocytes, platelets, and selected other hematopoietic elements and is recognized by the monoclonal antibody JC/70A *(108,111)*. This marker is highly restricted to endothelial neoplasms among all tumors of the soft tissue, and its sensitivity is also excellent. In our hands, virtually 100% of angiosarcomas are CD31 positive, regardless of grade or histotype *(112)* (**Fig. 17**), and the same statement applies to hemangioma and hemangioendothelioma variants. It must be acknowledged, however, that Kaposi's sarcoma is labeled more consistently for CD34 than CD31, for unknown reasons.

## *Thrombomodulin*

Thrombomodulin (TMN) is a 75-kDa membrano-cytoplasmic glycoprotein that is distributed among endothelial cells, mesothelial cells, osteoblasts, mononuclear phagocytic cells, and selected epithelia *(113)*. Its physiologic role is to convert thrombin from a coagulant protein to an anticoagulant. TMN may be present in some metastatic carcinomas and most mesotheliomas *(114)*, both of which may be confused with epithelioid angiosarcomas; therefore, it should be interpreted with this caveat in mind and in the context of the results of other vascular markers. Nevertheless, TMN has proved to be a highly *sensitive* indicator of endothelial differentiation, particularly in poorly differentiated vascular malignancies. Kaposi's sarcoma is likewise consistently immunoreactive for this determinant *(115)*. Thus, its inclusion in antibody panels is certainly worthwhile.

## Ulex europaeus I *Agglutinin*

*Ulex europaeus I* (UEAI) agglutinin is not an antibody reagent but instead represents a lectin produced by the gorse plant. It recognizes the Fuc-$\alpha$-1-2-Gal linkage in fucosylated oligosaccharides, which comprise portions of various glycoproteins *(116)*. In particular, the H blood group antigen and carcinoembryonic antigen regularly bind to UEAI, as does a separate fucosylated protein that is expressed by endothelial cells. Biotinylated

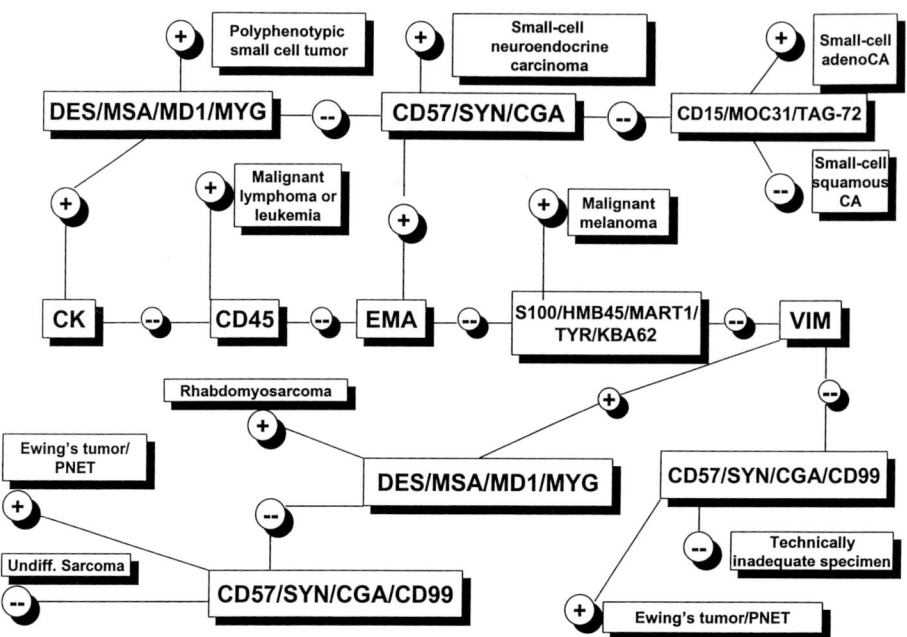

**Fig. 24.** Algorithmic approach to the immunohistochemical diagnosis of small cell tumors.

*Ulex* may be used an a histochemical reagent in surgical pathology, or, alternatively, unlabeled lectin can be employed, with its binding to tissue subsequently detected by application of biotinylated anti-*Ulex* and avidin-biotin-peroxidase complex. Because of the nonspecificity of UEAI for endothelial differentiation, as mentioned, it is absolutely necessary to utilize this lectin as part of a histochemical-immunohistochemical panel *(117)*. For example, epithelioid sarcoma and various metastatic carcinomas may also bind *Ulex*, in addition to vascular neoplasms. However, the extremely high sensitivity of UEAI justifies its continued use as a potential endothelial determinant.

## APPROACHES TO GENERIC CLASSIFICATION PROBLEMS IN SOLID TUMOR PATHOLOGY

### Small Round Cell Neoplasms

Small cell neoplasms that cause potential diagnostic confusion include embryonal rhabdomyosarcoma, Ewing's sarcoma, neuroblastoma, PNET (peripheral neuroepithelioma), malignant lymphoma, and small cell carcinomas **(Fig. 24)**. The anatomic locations and microscopic details of these tumors and other aspects of their clinical presentations have a strong bearing on the relative likelihood of respective diagnoses, and immunohistochemical analyses may be tailored according to such considerations.

### Small Cell Tumors of Soft Tissue

Embryonal rhabdomyosarcomas are typified by their immunoreactivity for desmin and muscle-specific actin. When used in combination, antibodies to these determinants yield virtually absolute sensitivity for myogenic sarcomas and allow for exclusion of other diagnostic possibilities *(18,20,106)*. Similarly, small cell lymphomas exhibit CD45 in a uniform manner, whereas other neoplasms in this category are devoid of reactivity *(23)*, with extremely rare exceptions *(117a)*. Neuroblastomas and neuroepitheliomas express synaptophysin and CD57 *(118)*. Rhabdomyosarcomas also share potential reactivity for CD57, but, excluding rare examples of PNET with divergent differentiation *(119)*, primitive striated muscle sarcomas are distinguished from small cell neurogenic neoplasms by their desmin or actin positivity, as mentioned above.

A particularly diagnostically challenging member of this tumor group is, in fact, Ewing's sarcoma/PNET. Our approach is to make this unqualified diagnosis when a small cell neoplasm of soft tissue is reactive for CD99, with or without vimentin, synaptophysin, and CD57, and in the absence of desmin, actin, S100 protein, and CD45. Some PNETs may also synthesize keratin in an "aberrant" fashion *(118)*.

*Small Cell Carcinomas*

Small cell carcinomas may exhibit neuroendocrine, squamous, or glandular differentiation. However, a determination of cell lineage is often difficult on conventional microscopy of these lesions, prompting the use of discriminating immunostains. CD57, synaptophysin, and CGA are restricted to neuroendocrine neoplasms in this morphologic class of lesions; nevertheless, it should be reemphasized that they are seen in only 30–50% of cases *(120)*. Thus, most of the time, immunohistologic studies will not successfully corroborate a neuroendocrine lineage for small cell carcinoma, even though it is really present, and one could seriously question whether or not such analyses are cost effective. On the other hand, small cell adenocarcinomas usually dislay MOC-31 or TAG-72, whereas small cell squamous tumors do not. All three forms of small cell carcinoma express cytokeratin reactivity, separating them from most small cell sarcomas of soft tissue. This fact is more important than one would assume at first glance, because Eusebi et al. *(120a)* have described small cell carcinomas that were aberrantly labeled for myogenic markers as well as keratin. One should also keep in mind the potential overlap between PNET and small cell neuroendocrine carcinoma, in regard to the possibility of conjoint staining for CD99 and keratin in those neoplasms.

## Spindle Cell Neoplasms

Spindle cell ("sarcomatoid") carcinomas have been described in a diversity of organs, including the skin, aerodigestive tract, kidney, bladder, female genital tract, lung, and other sites, and must be distinguished from melanomas and true sarcomas in such locations *(121)* (**Fig. 25**). In addition, spindle cell mesenchymal malignancies of soft tissue can be difficult to separate from one another diagnostically.

*Spindle Cell Carcinomas*

In mucosal or organ-based locations, any spindle cell tumor that displays reactivity for cytokeratin or EMA can be defined as carcinomatous *(121)*. Most sarcomatoid carcinomas also coexpress vimentin, but the latter protein should be regarded as nonspecific when seen in conjunction with other intermediate filaments.

*Spindle Cell Amelanotic Malignant Melanomas*

Melanomas that are composed exclusively of fusiform cells retain the general immunohistochemical attributes of other malignant melanocytic neoplasms, with one exception. For unknown reasons, spindle cell melanomas react poorly or not at all with

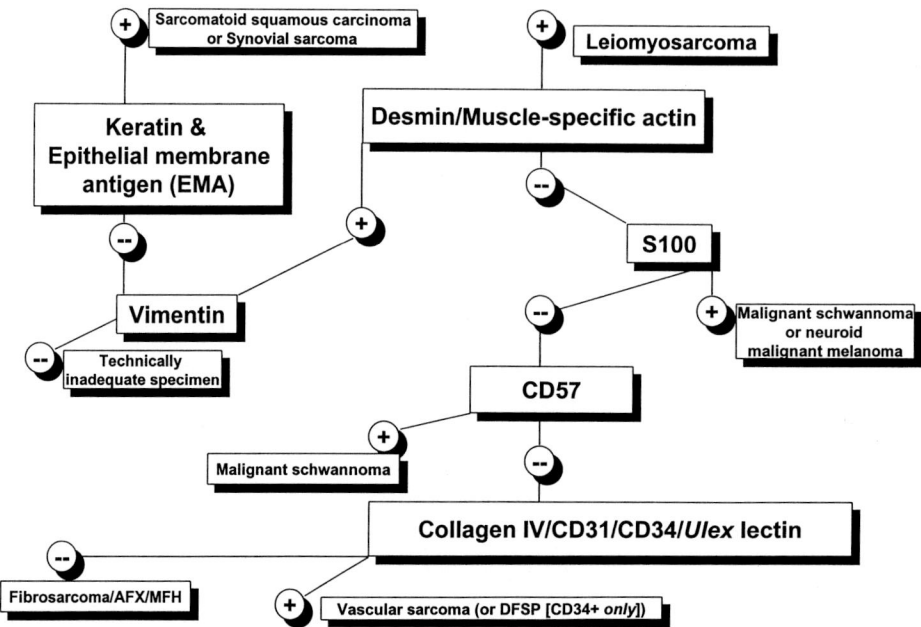

**Fig. 25.** Algorithmic approach to the immunohistochemical diagnosis of malignant spindle cell tumors.

HMB-45, KBA62, MART-1, and antityrosinase, making the conjoint use of antibodies to S100, vimentin, EMA, and cytokeratin paramount in this diagnostic context *(92,122)*. Positivity for either of the last two of these determinants excludes the possibility of malignant melanoma.

*Spindle Cell Sarcomas of Soft Tissue*

The malignant spindle cell neoplasms of soft tissue include fibrosarcoma, leiomyosarcoma, malignant peripheral nerve sheath tumor, monophasic synovial sarcoma, malignant fibrous histiocytoma (MFH), and Kaposi's sarcoma *(18,111)*. Among these, fibrosarcoma is characterized by its reactivity for vimentin, to the exclusion of desmin, actin, S-100 protein, CD57, cytokeratin, and EMA. In addition, it lacks endothelial markers. Leiomyosarcomas are unique in their diffuse reactivity for desmin and muscle-specific actin but may also express S-100 protein and CD57. Malignant peripheral nerve sheath tumors generally demonstrate only the last two of these four determinants. Synovial sarcoma represents the sole lesion in this category that is capable of cytokeratin and EMA synthesis; it may also express CD57 and CD99 *(18)*. The immunohistologic features of Kaposi's sarcomas are controversial. Some investigators have obtained consistent labeling of such neoplasms with UEAI, CD34, or CD31 *(108)*, whereas others *(117)* (including ourselves) have observed this finding to be less than universal. Fortunately, conventional histologic analysis and clinical correlation usually allow for confident identification of this lesion.

## Epithelioid (Large Polygonal Cell) Neoplasms

Metastatic carcinoma and melanoma, large cell malignant lymphoma, "syncytial" Hodgkin's disease, epithelioid sarcoma, clear cell sarcoma, epithelioid angiosarcoma, epithelioid leiomyosarcoma, and epithelioid malignant peripheral nerve sheath tumor are the differential diagnostic possibilities when dealing with polygonal cell malignancies in lymph nodes and soft tissues. The detailed characterization of metastatic carcinomas is addressed in a subsequent section.

### Lymph Nodal Lesions

In lymph nodes, epithelioid neoplasms most commonly represent large cell lymphomas or metastases of visceral carcinomas and malignant melanomas. The separation of these differential diagnostic possibilities principally involves the use of antibodies to CD45, cytokeratin, EMA, and S-100 protein. Carcinomas are reactive for cytokeratin, with or without EMA, whereas only lymphomas exhibit CD45 *(23)*. It is essential to avoid equating CD15 reactivity with the presence of Reed-Sternberg cells, to the exclusion of other immunophenotypic results, since anaplastic carcinomas may also display this determinant *(81,82)*. Similarly, EMA is rarely present in some Hodgkin's and non-Hodgkin's lymphomas, making it unreliable as an exclusive marker for carcinomas *(123)*. Malignant melanomas and some poorly differentiated carcinomas exhibit S100 protein, but only the latter tumors generally express cytokeratin *(86)*.

### Soft Tissue Lesions

An epithelioid malignancy of soft tissue is relatively unlikely to represent metastatic carcinoma or melanoma, unless widespread dissemination by these neoplasms is clinically obvious. Primary non-Hodgkin's lymphomas of soft tissues do exist, but they are quite rare. Indeed, in this tissue compartment, epithelioid sarcoma, epithelioid angiosarcoma, epithelioid malignant Schwannoma, epithelioid leiomyosarcoma, and clear cell sarcoma are the principal differential diagnostic considerations *(18)*.

Epithelioid sarcoma is a polyphenotypic neoplasm that may express vimentin, S-100 protein, and even desmin and UEL affinity, overlapping with the immunohistologic attributes of other mesenchymal tumors *(124–126)*. However, it is unique among polygonal cell sarcomas in its consistent expression of cytokeratin; EMA reactivity is also observed in 75% of epithelioid sarcomas *(124)*. Epithelioid angiosarcoma exhibits vimentin positivity, reactivity for CD31, thrombomodulin, and CD34, and UEA1 binding *(109,111,112,115,117)*. This profile usually includes a lack of keratin positivity, but rare examples of epithelioid angiosarcoma have demonstrated aberrant keratin expression, a potential pitfall in interpretation *(127)*. B72.3 may also unexpectedly label such tumors *(128)*. Clear cell sarcoma is the only one of these lesions that binds HMB-45, KBA62, and anti-tyrosinase; it displays S-100 protein and vimentin as well but lacks cytokeratin, EMA, desmin, CD31, CD34, and UEL affinity *(18,126)*. HMB-45, anti-tyrosinase, and KBA62 are important in distinguishing between clear cell sarcoma and epithelioid malignant peripheral nerve sheath tumors, which are typically S-100 protein positive but nonreactive for more melanocyte-selective markers *(129)*. Lastly, only epithelioid leiomyosarcoma expresses desmin and muscle-specific actin.

Another possible diagnostic entity in this category is the "gastrointestinal stromal tumor," which may assume the form of an undifferentiated polyhedral-cell neoplasm.

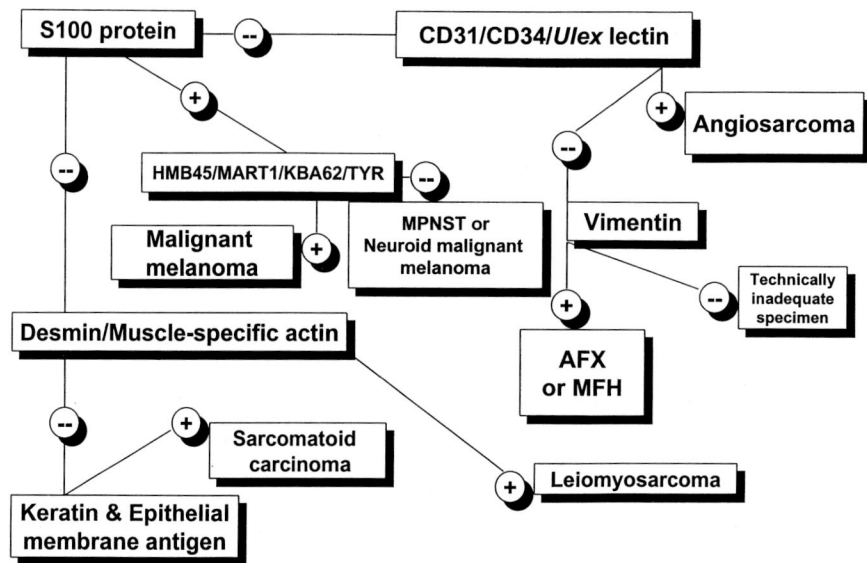

**Fig. 26.** Algorithmic approach to the immunohistochemical diagnosis of malignant pleomorphic tumors.

It is distinctive in its immunoreactivity for vimentin and CD117 (C-*kit* protein). with or without CD34, desmin, muscle-specific actin, or S-100 protein *(130)*.

## Pleomorphic Neoplasms

Tumors that are capable of assuming extremely bizarre, pleomorphic cellular shapes include MFH, pleomorphic liposarcoma, pleomorphic rhabdomyosarcoma, "dedifferentiated" leiomyosarcoma, malignant peripheral nerve sheath tumor, and metastatic carcinoma or melanoma **(Fig. 26)**.

### Metastatic Carcinomas and Melanomas

Metastases of carcinomas and melanomas with a pleomorphic microscopic appearance are delineated by their reactivity with anticytokeratin or with melanocyte-selective markers, respectively. These immunophenotypes are not shared by any other pleomorphic malignancy.

### Pleomorphic Soft Tissue Tumors

Malignant peripheral nerve sheath tumors are unique among pleomorphic soft tissue sarcomas, because they are the only lesions in this class that are capable of diffuse expression of S100 protein, with or without CD57 *(18)*. Pleomorphic rhabdomyosarcomas are globally positive for desmin and muscle-specific actin; in addition, they express myoglobin, Myo-D1, or myogenin, all of which are proteins that are restricted to striated muscle *(18,111,118)*. Of the latter, Myo-D1 and myogenin are strictly limited to the nuclei. Although it is a relatively insensitive marker in other forms of rhabdomyosarcoma, myoglobin assumes some importance among pleomorphic neoplasms, because "dedifferentiated" leiomyosarcomas also exhibit desmin and actin reactivity but are devoid of myoglobin *(18)*. Pleomorphic MFH lacks all antigens except vimentin;

pleomorphic liposarcoma differs from this profile in showing focal S100 protein positivity in signet ring or multivacuolated tumor lipoblasts.

## DIFFERENTIAL IMMUNOHISTOCHEMICAL DIAGNOSIS OF METASTATIC CARCINOMAS

In the present climate of medical cost consciousness and corresponding limitations on the scope of radiographic procedures, the pathologist is often called on to aid in predicting the sources of metastatic non-small cell carcinomas of unknown origin. It is in this realm of inquiry that permutations in the expression of several tissue-restricted determinants achieve their greatest importance.

As mentioned previously, PLAP, CEA, S-100 protein, AFP, EMA, and other tissue-specific determinants are useful in this context. Among the latter, for example, anti-PSA can be employed to detect metastatic prostatic carcinomas, anti-GCDFP-15 is specific for metastatic breast cancers, and anti-thyroglobulin is effective in the recognition of metastatic carcinomas of the thyroid.

Other discriminants listed in the introductory sections of this chapter are found in carcinomas originating in more than one tissue location. Nevertheless, because their distributions are not identical, *combinations* of reactivity can be used to generate probability tables that reflect the likelihood of anatomic origins *(131)*. These, in turn, can be used to construct interpretative algorithms, such as that provided in **Fig. 27**. The "entry" point for interpretation is the box labeled "keratin mixture" at the left center of the figure, with arrows proceeding from that point representing the relative results obtained for each respective marker. As one would expect, "first-tier" immuno-reactants (those with a high degree of individual specificity and predictive value) yield a definitive interpretation early in the algorithm, such as those attending positivity for PSA or thyroglobulin. A seemingly paradoxical "early" entry in the algorithm is that associated with seminoma and adrenocortical carcinoma, because although they are undeniably epithelial in character (as determined by their conventional morphologic features), both fail typically to label with the specified mixture of antikeratin antibodies. The highly unusual nature of that phenomenon among epithelial malignancies in general allows for the "early" segregation of seminoma and adrenocortical carcinoma, which, in turn, can be distinguished from one another by their relative frequencies of reactivity for vimentin and PLAP.

The midportion ("second tier") of the algorithm is devoted to separating two singular epithelial malignancies (mesothelioma and germ cell tumors) from somatic carcinomas that typically do not express specialized differentiation markers. The latter principally include nasopharyngeal carcinoma, renal cell carcinoma, hepatocellular carcinoma, and rare keratin-positive variants of adrenocortical carcinoma. Entry into this segment of the interpretative sequence is determined by positivity or negativity for TAG-72 and MOC-31, both of which label the vast majority of other epithelial malignancies.

Finally, the largest and most complex part of the algorithm (the "third tier") is that seen at the far right of **Fig. 27**, where several different combinations of eight immunodeterminants (GCDFP-15, CEA, S-100 protein, PLAP, CA-125, CA19-9, CK20, and ERP) are listed. This portion of the analysis deals with markers that are relatively nonspecific in and of themselves, and that are therefore shared by tumors of

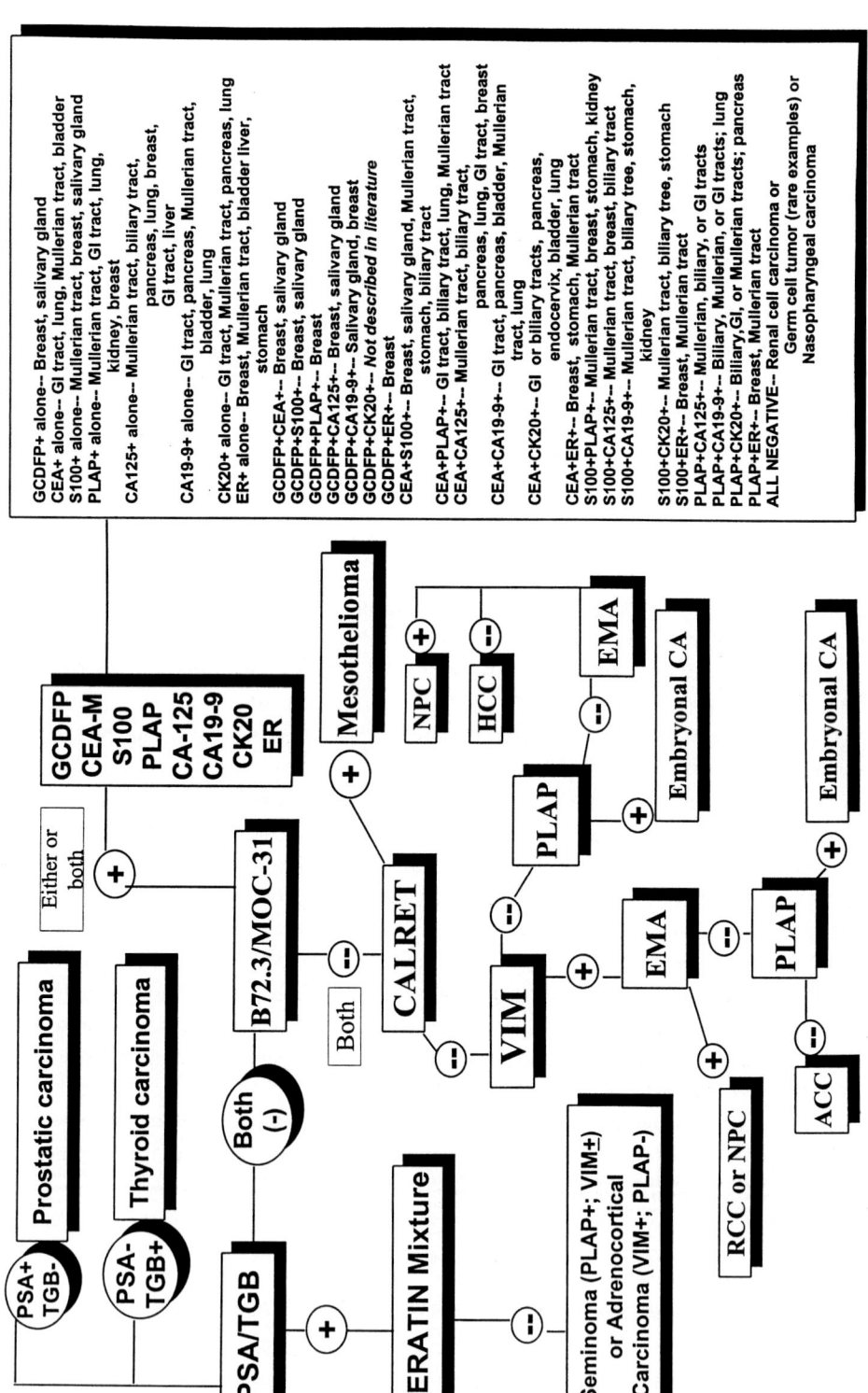

**Fig. 27.** Algorithmic approach to the immunohistochemical diagnosis of malignant epithelial tumors of unknown origin.

several topographic origins. The proffered listings attempt to account for most, if not all, of the combinations of immunoreactivity that may be encountered. In addition, they cite the primary sites that are most likely for each combination of results, in order of frequency. For example, the combination of CEA+/CA19-9+ (lower midportion of tier 3) would suggest a tumor origin in the GI tract, pancreas, bladder, endocervix, and lung, in relative order of likelihood. Although such lists may seem unsatisfying at first glance (because they do not provide only one anatomic choice), integration of clinical information and morphologic details will usually allow for a single diagnostic choice to be made with confidence.

Because the information presented here has been accumulated over a long period, the authors have had the opportunity to undertake validation studies of the algorithm for metastatic carcinoma of unknown origin (MCUO). In those cases in which primary sites were found after initial biopsy of the tumor, either by clinical evaluations or by autopsy, overall correspondence with the pathologic prediction for the site of origin has been 67%. Others have had similar rates of success with immunohistochemical evaluations. This figure may not seem overly impressive, but it should be compared with the accuracy and efficiency of radiographic and other nonmorphologic methods, as outlined in the introduction of this presentation. Moreover, when mistaken predictions occurred in our experience, they most often involved tumors in the same general immunophenotypic class; for example, a tumor felt to be most consistent with a pancreatic carcinoma actually proved to be a lung cancer with a largely superimposable set of immunoreactivity patterns.

Another practical aspect of our algorithmic approach is still early in its evolution but is mentioned her for interest. Oncologists are well aware that only a minority of MCUO cases have definable primary tumors that can be found during life. Accordingly, most patients in this group have been treated with a standard consensus panel of chemotherapeutic agents designed to have activity against a broad spectrum of malignancies. Nevertheless, that approach is probably not an optimal one compared with *specific* available treatments aimed at specific organ sites, judging by the dismal survival results that are realized. Thus, some oncologist colleagues have elected to administer chemotherapy regimens that are dictated by the predictive result of the MCUO algorithm. In other words, if lung carcinoma is the favored interpretation by the pathologist, a set of drugs tailored to that diagnosis would be given to the patient even if no pulmonary mass were evident by clinical evaluation. Very preliminary results of such a process have been promising, showing outcomes that are comparable to those of patients with high-stage tumors of *known* anatomic derivation and somewhat better than those treated with generic MCUO therapy. Obviously, considerable work will be required to decide whether this is ultimately the optimal means of treating MCUO cases.

## IMMUNODIAGNOSIS OF NEUROENDOCRINE CARCINOMAS

The tenets that have just been discussed are most applicable to the diagnosis of poorly differentiated carcinomas that lack neuroendocrine differentiation. Neuroendocrine tumors are more homogeneous, regardless of their anatomic origins; all of them share the potential to express CD57, CGA, synaptophysin, and various neuropeptides and amines. Therefore, immunohistology is not useful in predicting where such tumors arose. (*See also* Chapter 18.)

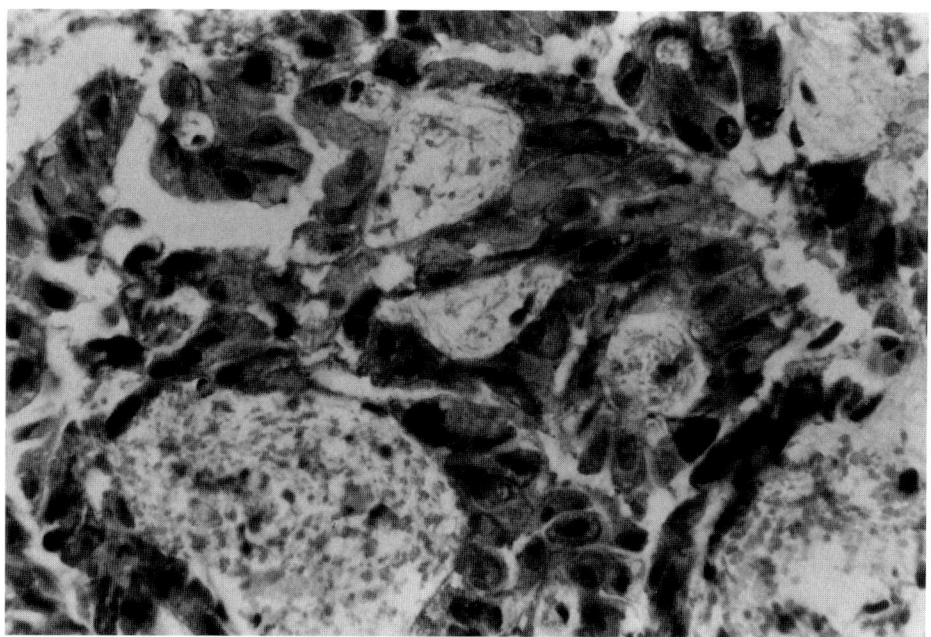

**Fig. 28.** Calretinin shows strong cytoplasmic staining in epithelial mesothelioma.

## OTHER SELECTED DIAGNOSTIC PROBLEMS AMENABLE TO IMMUNOHISTOLOGIC RESOLUTION

Aside from the diagnostic dilemmas that have been addressed up to this point, immunohistochemical analyses may add significant information in the evaluation of other problems in surgical pathology. These are considered briefly in the following sections.

### Malignant Mesothelioma vs Metastatic Carcinoma

For medicolegal and epidemiologic reasons, it is desirable to render a specific and confident diagnosis of malignant mesothelioma. This neoplasm may be mimicked closely by metastatic adenocarcinomas from the breast, lung, ovary, and other sites, which are capable of seeding the pleural or peritoneal surfaces. Determinants that are often seen in these microscopic stimulants are absent in mesothelioma. These include CEA, CD15, CA19-9, MOC-31, TAG-72, PLAP, S-100 protein, TTF-1, BG8 (a blood group antigen-related marker), and GCDFP-15, at least two of which are seen in >95% of adenocarcinomas *(132)*. Conversely, one expects keratin 5/6 and calretinin expression by mesotheliomas **(Fig. 28)** but, with rare exceptions, not in carcinomas. Using this panel approach, one can closely approach absolute specificity in the immunohistochemical diagnosis of mesothelioma, as validated by correlation with ultrastructural characteristics *(133)*.

### Metastatic Carcinoma vs Second Primary Tumor

The tendency for individuals with a one malignancy to develop a second is a well-known phenomenon. Under such circumstances, the surgical pathologist is required to make a decision on the probability of this eventuality, as opposed to recurrence or metastasis of the primary neoplasm. The same tenets that were outlined in an earlier section on metastatic neoplasms can be employed effectively in this context. Whenever possible, the immunophenotypes of both tumors should be studied appropriately; if significant differences are detected, it is virtually certain that two separate primary neoplasms have arisen in the patient under consideration.

## PERSPECTIVES

Throughout this chapter, an integrated multiparametric approach to the immunohistologic diagnosis of poorly differentiated and morphologically ambiguous malignant tumors has been emphasized. This is indeed an effective process, as reflected by its beneficial results in many medical centers. Nonetheless, one should not lose sight of the fact that surgical pathology is still based on skill in morphologic interpretation; one cannot achieve success as a diagnostic immunohistochemist without having a sound foundation in histopathologic analysis. Also, we do not mean to imply that another method of detailed pathologic assessment—electron microscopy—has been relegated to a historical perspective. It is still invaluable in the evaluation of selected neoplasms and often supplies information that is "synergistic" with that of immunohistology.

Also, immunohistochemistry is not itself being outmoded by more "molecular" techniques, which are being increasingly integrated currently into surgical pathology. It is rather anticipated that all these methods will be employed selectively in the future, at the discretion of the pathologist, to provide rapid and comprehensive solutions for problem cases in anatomic pathology.

## REFERENCES

1. Prento, P. and Lyon, H. (1997) Commercial formalin substitutes for histopathology. *Biotech. Histochem.* **72**, 273–282.
2. Arnold, M. M., Srivastava, S., Fredenburgh, J., Stockard, C. R., Myers, R. B., and Grizzle, W. E. (1996) Effects of fixation and tissue processing on immunohistochemical demonstration of specific antigens. *Biotech. Histochem.* **71**, 224–230.
3. Miller, R. T., Swanson, P. E., and Wick, M. R. Fixation and epitope retrieval in diagnostic immunohistochemistry: a concise review with practical considerations. *Appl. Immunohistochem. Mol. Morphol.* **8**, 228–235.
4. Falini, B., and Taylor, C. R. (1983) New developments in immunoperoxidase techniques and their application. *Arch. Pathol. Lab. Med.* **107**, 105–117.
5. Giorno, R. (1984) A comparison of two immunoperoxidase staining methods based on the avidin-biotin interaction. *Diagn. Immunol.* **2**, 161–169.
6. Swanson, P. E. (1988) Foundations of immunohistochemistry: a practical review. *Am. J. Clin. Pathol.* **90**, 333–339.
7. Gown, A. M., DeWever, N., and Battifora, H. (1993) Microwave-based antigenic unmasking: a revolutionary new technique for routine immunohistochemistry. *Appl. Immunohistochem.* **1**, 1256–1266.
8. Cattoretti, G., and Suurmeijer, A. (1995) Antigen unmasking for formalin-fixed paraffin-embedded tissues: a review. *Adv. Anat. Pathol.* **2**, 2–9.
9. DeJong, A. S. H., Van Kessel-Van Mark, M., and Raap, A. K. (1985) Sensitivity of various visualization methods for peroxidase and alkaline phosphatase activity in immunoenzyme histochemistry. *Histochem. J.* **17**, 119–130.

10. Suffin, S. C., Muck, K. B., and Young, J. C. (1979) Improvement of the glucose oxidase immunoenzyme technique: use of a tetrazolium whose formazan is stable without heavy metal chelation. *Am. J. Clin. Pathol.* **71**, 492–496.

11. Miettinen, M. (1993) Keratin immunohistochemistry: update on applications and pitfalls. *Pathol. Annu.* **28**, 113–143.

12. Andrade, R. E., Hagen, K. A., Swanson, P. E., and Wick, M. R. (1988) The use of ficin for proteolysis in immunostaining of paraffin sections. *Am. J. Clin. Pathol.* **90**, 33–39.

13. Battifora, H. (1988) Diagnostic uses of antibodies to keratins. *Prog. Surg. Pathol.* **8**, 1–16.

14. Chan, J. K. C., Suster, S., Wenig, B. M., Tsang, W. Y., Chan, J. B., and Lau, A. L. (1997) Cytokeratin 20 immunoreactivity distinguishes Merkel cell (primary cutaneous neuroendocrine) carcinomas and salivary gland small cell carcinomas from small cell carcinomas of various sites. *Am. J. Surg. Pathol.* **21**, 226–234.

15. Pruss, R. M., Mirsky, R., Raff, M. C., Thrope, R., Dowling, A. J., and Anderton, B. H. (1981) All classes of intermediate filaments share a common antigenic determinant defined by a monoclonal antibody. *Cell* **27**, 419–428.

16. Caselitz, J., Janner, M., Breitbart, E., Weber, K., and Osborn, M. (1983) Malignant melanomas contain only the vimentin type of intermediate filaments. *Virchows Arch. A.* **400**, 43–51.

17. Gustmann, C., Altmannsberger, M., Osborn, M., Griesser, H., and Feller, A. C. (1991) Cytokeratin expression and vimentin content in large cell anaplastic lymphomas and other non-Hodgkin's lymphomas. *Am. J. Pathol.* **138**, 1413–1422.

18. Swanson, P. E., and Wick, M. R. (1995) Soft tissue tumors, in *Diagnostic Immunopathology*, 2nd ed. (Colvin, R., Bhan, A., and McCluskey, R., eds.), Lippincott-Raven, Philadelphia, pp. 599–632.

19. Battifora, H. (1991) Assessment of antigen damage in immunohistochemistry: the vimentin internal control. *Am. J. Clin. Pathol.* **96**, 669–671.

20. Truong, L. D., Rangdaeng, S., Cagle, P. T., Ro, J. Y., Hawkins, H., and Font, R. L. (1990) The diagnostic utility of desmin: a study of 584 cases and review of the literature. *Am. J. Clin. Pathol.* **93**, 305–314.

21. Trojanowski, J. Q., Lee, V. M. Y., and Schlaepfer, W. W. (1984) An immunohistochemical study of human central and peripheral nervous system tumors with monoclonal antibodies against neurofilaments and glial filaments, *Hum. Pathol.* **15**, 248–257.

22. Morrison, C. D. and Prayson, R. A. (2000) Immunohistochemistry in the diagnosis of neoplasms of the central nervous system *Semin. Diagn. Pathol.* **17**, 204–215.

23. Kurtin, P. J. and Pinkus, G. S. (1985) Leukocyte common antigen—a diagnostic discriminant between hematopoietic and nonhematopoietic neoplasms in paraffin sections using monoclonal antibodies: correlation with immunologic studies and ultrastructural localization. *Hum. Pathol.* **16**, 353–365.

24. Wick, M. R. (1988) Monoclonal antibodies to leukocyte common antigen, in *Monoclonal Antibodies in Diagnostic Immunohistochemistry* (Wick, M. R. and Siegal, G. P., eds.), Marcel Dekker, New York, pp. 285–307.

25. Heyderman, E., Steele, K., and Ormerod, M. G. (1979) A new antigen on the epithelial membrane: its immunoperoxidase localization in normal and neoplastic tissue. *J. Clin. Pathol.* **32**, 35–44.

26. Sloane, J. P. and Ormerod, M. G. (1981) Distribution of epithelial membrane antigen in normal and neoplastic tissues and its value in diagnostic pathology, *Cancer* **47**, 178–185.

27. Pinkus, G. S. and Kurtin, P. J. (1985) Epithelial membrane antigen—a diagnostic discriminant in surgical pathology. *Hum. Pathol.* **16**, 929–938.

28. Swanson, P. E., Manivel, J. C., Scheithauer, B. W., and Wick, M. R. (1989) Epithelial membrane antigen in human sarcomas: an immunohistochemical study. *Surg. Pathol.* **2**, 313–322.

29. Wick, M. R., Swanson, P. E., and Manivel, J. C. (1987) Placental-like alkaline phosphatase reactivity in human tumors: an immunohistochemical study of 520 cases. *Hum. Pathol.* **18**, 946–954.
30. Elgin, J., Phillips, J. G., Reddy, V. V., Gibbs, P. O., and Listinsky, C. M. (1999) Hodgkin's and non-Hodgkin's lymphoma: spectrum of morphologic and immunophenotypic overlap. *Ann. Diagn. Pathol.* **3**, 263–275.
31. Salter, D. M., Krajewski, A. S., Miller, E. P., and Dewar, A. E. (1985) Expression of leukocyte common antigen and epithelial membrane antigen in plasmacytic malignancies. *J. Clin. Pathol.* **38**, 843–844.
32. Schnitt, S. J. and Vogel, H. (1986) Meningiomas: diagnostic value of immunoperoxidase staining for epithelial membrane antigen. *Am. J. Surg. Pathol.* **10**, 640–649.
33. Ruitenbeek, C. T., Gouw, A. S. H., and Poppema, S. (1994) Immunocytology of body cavity fluids: MOC-31, a monoclonal antibody discriminating between mesothelial and epithelial cells. *Arch. Pathol. Lab. Med.* **118**, 265–269.
34. Sosolik, R. C., McGaughy, V. R., and DeYoung, B. R. (1997) Anti-MOC31: a potential addition to the pulmonary adenocarcinoma versus mesothelioma immunohistochemistry panel. *Mod. Pathol.* **10**, 716–719.
35. Chenard-Neu, M. P., Kabou, A., Mechine, A., Brolly, F., Orion, B., and Bellocq, J. P. (1998) Immunohistochemistry in the differential diagnosis of mesothelioma and adenocarcinoma: evaluation of 5 new antibodies and 6 traditional antibodies. *Ann. Pathol.* **18**, 460–465.
36. DeYoung, B. R. and Wick, M. R. (2000) Immunohistologic evaluation of metastatic carcinomas of unknown origin: an algorithmic approach. *Semin. Diagn. Pathol.* **17**, 184–193.
37. Thor, A., Ohuchi, N., Szpak, C. A., Johnston, W. W., and Schlom, J. (1986) Distribution of oncofetal antigen tumor-associated glycoprotein-72, defined by monoclonal antibody B72.3. *Cancer Res.* **46**, 3118–3124.
38. Loy, T. S. and Nashelsky, M. B. (1993) Reactivity of B72.3 with adenocarcinomas: an immunohistochemical study of 476 cases. *Cancer* **72**, 2495–2498.
39. Manivel, J. C., Jessurun, J., Wick, M. R., et al. (1987) Placental alkaline phosphatase immunoreactivity in testicular germ cell neoplasms. *Am. J. Surg. Pathol.* **11**, 21–30.
40. Niehans, G. A., Manivel, J. C., Copland, G. T., et al. (1988) Immunohistochemistry of germ cell and trophoblastic neoplasms. *Cancer* **62**, 1113–1123.
41. Suster, S., Moran, C. A., Dominguez-Malagon, H., and Quevedo-Blanco, P. (1998) Germ cell tumors of the mediastinum and testis: a comparative immunohistochemical study of 120 cases. *Hum. Pathol.* **29**, 737–742.
41a. Cheville, J. C., Rao, S., Iczkowski, K. A., Lohse, C. M., and Pankratz, V. S. (2000) Cytokeratin expression in seminoma of the human testis. *Am. J. Clin. Pathol.* **113**, 583–588.
42. Thung, S. N., Gerber, M. A., Sarno, R., and Popper, H. (1979) Distribution of five antigens in hepatocellular carcinoma. *Lab. Invest.* **43**, 101–105.
43. Stiller, D., Bahn, H., and Pressler, H. (1986) Immunohistochemical demonstration of alpha-fetoprotein in testicular germ cell tumors. *Acta. Histochem. Suppl.* **33**, 225–231.
44. Okamoto, T., Hirabayashi, K., and Ishiguro, T. (1993) Immunohistochemical type distinction of alpha-fetoprotein in various alpha-fetoprotein-secreting tumors. *Jpn. J. Cancer Res.* **84**, 360–364.
45. Colcher, D., Hand, P. H., Nuti, M., and Schlom, J. (1983) Differential binding to human mammary and nonmammary tumors of monoclonal antibodies reactive with carcinoembryonic antigen. *Cancer Invest.* **1**, 127–138.
46. Wick, M. R. (1988) Monoclonal antibodies to carcinoembryonic antigen, in *Monoclonal Antibodies in Diagnostic Immunohistochemistry* (Wick, M. R. and Siegal, G. P., eds.), Marcel Dekker, New York, pp. 539–567.

47. Sheahan, K., O'Brien, M. J., Burke, B., et al. (1990) Differential reactivities of carcinoembronic antigen (CEA) and CEA-related monoclonal and polyclonal antibodies in common epithelial malignancies. *Am. J. Clin. Pathol.* **94**, 157–164.

48. Nadji, M., Tabei, S. Z., Castro, A., et al. (1981) Prostatic-specific antigen: an immunohistologic marker for prostatic neoplasms. *Cancer* **48**, 1229–1232.

49. Miller, G. J. (1982) The use of histochemistry and immunohistochemistry in evaluating prostatic neoplasia. *Prog. Surg. Pathol.* **5**, 115–134.

50. Papsidero, L. D., Croghan, G. A., Asirwatham, J., et al. (1985) Immunohistochemical demonstration of prostate-specific antigen in metastases with the use of monoclonal antibody F5. *Am. J. Pathol.* **121**, 451–454.

51. Torenbeek, R., Lagendijk, J. H., Van Diest, P. J., et al. (1998) Value of a panel of antibodies to identify the primary origin of adenocarcinomas presenting as bladder carcinoma. *Histopathology* **32**, 20–27.

52. Bostwick, D. G. (1994) Prostate-specific antigen: current role in diagnostic pathology of prostate cancer. *Am. J. Clin. Pathol.* **102 (4 suppl. 1)**, S31–S37.

52a. Bostwick, D. G. (1994) Prostate specific antigen. Current role in diagnostic pathology of prostate cancer. *Am. J. Clin. Pathol.* **102**, 831–837.

53. Permanetter, W., Nathrath, W. B. J., and Lohrs, U. (1982) Immunohistochemical analysis of thyroglobulin and keratin in benign and malignant thyroid tumors. *Virchows Arch.* **398**, 221–228.

54. DeMicco, C., Ruf, J., Carayon, P., et al. (1987) Immunohistochemical study of thyroglobulin in thyroid carcinomas with monoclonal antibodies. *Cancer* **59**, 471–476.

55. Carcangiu, M. L., Steeper, T., Zampi, G., and Rosai, J. (1985) Anaplastic thyroid carcinoma: a study of 70 cases. *Am. J. Clin. Pathol.* **83**, 135–158.

56. Carcangiu, M. L., Zampi, G., and Rosai, J. (1984) Poorly differentiated ("insular") thyroid carcinoma: a reinterpretation of Langhans' "wuchernde Struma." *Am. J. Surg. Pathol.* **8**, 655–668.

57. Guazzi, S., Price, M., DeFelice, M., et al. (1990) Thyroid nuclear factor-1 (TTF-1) contains a homeodomain and displays a novel DNA binding specificity. *EMBO J.* **9**, 3631–3639.

58. Lazzaro, D., Price, M., DeFelice, D., and DiLauro, R. (1991) The transcription factor TTF-1 is expressed at the onset of thyroid and lung morphogenesis and in restricted regions of the fetal brain. *Development* **113**, 1093–1104.

59. Kelly, S. E., Bachurski, C. J., Burhans, M. S., and Glassor, J. W. (1996) Transcription of the lung-specific surfactant protein C gene is mediated by TTF-1. *J. Biol. Chem.* **27**, 6881–6888.

60. Ordonez, N. G. (2000) Thyroid transcription factor-1 is a marker for lung and thyroid carcinomas. *Adv. Anat. Pathol.* **7**, 123–127.

61. Ordonez, N. G. (2000) Value of thyroid transcription factor-1, E-cadherin, BG8, WT1, and CD44S immunostaining in distinguishing epithelial pleural mesothelioma from pulmonary and nonpulmonary adenocarcinoma. *Am. J. Surg. Pathol.* **24**, 598–606.

62. Katoh, R., Kawaoi, A., Miyagi, E., et al. (2000) Thyroid transcription factor-1 in normal, hyperplastic, and neoplastic follicular thyroid cells examined by immunohistochemistry and nonradioactive in-situ hybridization. *Mod. Pathol.* **13**, 570–576.

63. Kaufmann, O. and Dietel, M. (2000) Expression of thyroid transcription factor-1 in pulmonary and extrapulmonary small cell carcinomas and other neuroendocrine carcinomas of various primary sites. *Histopathology* **36**, 415–420.

64. Agoff, S. N., Lamps, L. W., Philip, A. T., et al. (2000) Thyroid transcription factor-1 is expressed in extrapulmonary small cell carcinomas but not in other extrapulmonary neuroendocrine tumors. *Mod. Pathol.* **13**, 238–242.

65. Hanly, A. J., Elgart, G. W., Jorda, M., Smith, J., and Nadji, M. (2000) Analysis of thyroid

transcription factor-1 and cytokeratin 20 separates Merkel cell carcinoma from small cell carcinoma of lung. *J. Cutan. Pathol.* **27**, 118–120.

66. Wick, M. R., Lillemoe, T. J., Copland, G. T., et al. (1989) Gross cystic disease fluid protein-15 as a marker for breast carcinoma. *Hum. Pathol.* **20**, 281–287.

67. Mazoujian, G., and Haagensen, D. E., Jr. (1990) The immunopathology of gross cystic disease fluid proteins. *Ann. NY Acad. Sci.* **586**, 188–197.

68. Lee, A. K., DeLellis, R. A., Rosen, P. P., et al. (1984) Alpha-lactalbumin as an immunohistochemical marker for metastatic breast carcinomas. *Am. J. Surg. Pathol.* **8**, 93–100.

69. Mesa-Tejada, R., Palakodety, R. B., Leon, J. A., Khatcherian, A. O., and Greaton, C. J. (1988) Immunocytochemical distribution of a breast carcinoma-associated glycoprotein identified by monoclonal antibodies. *Am. J. Pathol.* **130**, 305–314.

70. Loy, T. S., Chapman, R. K., Diaz-Arias, A. A., Bulatao, I. S., and Bickel, J. T. (1991) Distribution of BCA-225 in adenocarcinomas: an immunohistochemical study of 446 cases. *Am. J. Clin. Pathol.* **96**, 326–329.

71. Gatalica, Z. and Miettinen, M. (1994) Distribution of carcinoma antigens CA19-9 and CA15-3; an immunohistochemical study of 400 tumors. *Appl. Immunohistochem.* **2**, 205–211.

71a.McCluggage, W. G., Patternson, A., and Maxweill, P. (2000) Aggressive angiomyxoma of pelvic parts exhibits oestrogen and progesterone receptor positivity. *J. Clin. Pathol.* **53**, 603–605.

72. Yokazaki, H., Takekura, N., Takanashi, A., Tabuchi, J., Haruta, R., and Tahara, E. (1988) Estrogen receptors in gastric adenocarcinoma: a retrospective immunohistochemical analysis. *Virchows Arch. A. (Pathol. Anat.)* **413**, 297–302.

73. Diaz, N. M., Mazoujian, G., and Wick, M. R. (1991) Estrogen-receptor protein in thyroid neoplasms: an immunohistochemical analysis of papillary carcinoma, follicular carcinoma, and follicular adenoma. *Arch. Pathol. Lab. Med.* **115**, 1203–1207.

74. Kaufmann, O., Baume, H., and Dietel, M. (1998) Detection of estrogen receptors in noninvasive and invasive transitional cell carcinomas of the urinary bladder using both conventional immunohistochemistry and the tyramide staining amplification technique. *J. Pathol.* **186**, 165–168.

75. Bast, R. C., Feeney, M., Lazarus, H., et al. (1981) Reactivity of a monoclonal antibody with human ovarian carcinoma. *J. Clin. Invest.* **68**, 1331–1337.

76. Kabawat, S. E., Bast, R. C., Welch, W. R., et al. (1983) Immunopathologic characterization of a monoclonal antibody that recognizes common surface antigens of human ovarian tumors of serous, endometrioid, and clear cell types. *Am. J. Clin. Pathol.* **79**, 98–104.

77. Koelma, I. A., Nap, M., Rodenburg, C. J., and Fleuren, G. J. (1987) The value of tumor marker CA125 in surgical pathology. *Histopathology* **11**, 287–294.

78. Loy, T. S., Quesenberry, J. T., and Sharp, S. C. (1992) Distribution of CA125 in adenocarcinomas: an immunohistochemical study of 481 cases. *Am. J. Clin. Pathol.* **98**, 175–179.

79. Itzkowitz, S. H., Yuan, M., Fukushi, Y., et al. (1988) Immunohistochemical comparison of Le$^a$, monosialosyl Le$^a$ (CA19-9), and disialosyl Le$^a$ antigens in human colorectal and pancreatic tissues. *Cancer Res.* **48**, 3834–3842.

80. Loy, T. S., Sharp, S. C., Andershock, C. J., and Craig, S. B. (1993) Distribution of CA19-9 in adenocarcinomas and transitional cell carcinomas: an immunohistochemical study of 527 cases. *Am. J. Clin. Pathol.* **99**, 726–728.

81. Sheibani, K., Battifora, H., Burke, J. S., and Rappaport, H. (1986) Leu-M1 antigen in human neoplasms: an immunohistologic study of 400 cases. *Am. J. Surg. Pathol.* **10**, 227–236.

82. Arber, D. A. and Weiss, L. (1993) CD15: a review. *Appl. Immunohistochem.* **1**, 17–30.

83. Wick, M. R., Mills, S. E., and Swanson, P. E. (1990) Expression of "myelomonocytic" antigens in mesotheliomas and adenocarcinomas involving the serosal surfaces. *Am. J. Clin. Pathol.* **94**, 18–26.

84. Tos, A. P. and Doglioni, C. (1998) Calretinin: a novel tool for diagnostic immunohistochemistry. *Adv. Anat. Pathol.* **5**, 61–66.

85. Gotzos, V., Wintergerst, E. S., Musy, J. P., Spichtin, H. P., and Genton, C. Y. (1999) Selective distribution of calretinin in adenocarcinomas of the human colon and adjacent tissues. *Am. J. Surg. Pathol.* **23**, 701–711.

85a. Chhieng, D. C., et al. (2000) Calretinin staining pattern aids in the differentiation of mesothelioma from adenocarcinoma in serous effusions. *Cancer* **90**, 194–200.

85b. Kitazume, H., et al. (2000) Cytologic differential diagnosis among reactive mesothelial cells, malignant mesothelioma, and adenocarcinoma: utility of combined E-cadherin and calretinin immunostaining. *Cancer* **90**, 55–60.

85c. Ordonez, N. G. (1998) Value of calretinin immunostaining in differentiating epithelial mesothelioma from lung adenocarcinoma. *Mod. Pathol.* **11**, 929–933.

85d. Atanoos, R. L., Dojcinov, S. D., Webb, R., and Gibbs, A. R. (2000) Antimesothelial markers in sarcomatoid mesothelioma and other spindle cell neoplasms. *Histopathology* **37**, 224–231.

86. Drier, J., Swanson, P. E., Cherwitz, D. L., and Wick, M. R. (1987) S100 protein immunoreactivity in poorly-differentiated carcinomas: immunohistochemical comparison with malignant melanoma. *Arch. Pathol. Lab. Med.* **111**, 447–452.

87. Wick, M. R., Ockner, D. M., Mills, S. E., Ritter, J. H., and Swanson, P. E. (1998) Homologous carcinomas of the breasts, salivary glands, and skin. *Am. J. Clin. Pathol.* **109**, 75–84.

88. Matsushima, S., Mori, M., Adachi, Y., Matsukuma, A., and Sugimachi, K. (1994) S100 protein-positive breast carcinomas: an immunohistochemical study. *J. Surg. Oncol.* **55**, 108–113.

89. Loeffel, S. C., Gillespie, G. Y., Mirmiran, A., Sawhney, D., Askin, F. B., and Siegal, G. P. (1985) Cellular immunolocalization of S100 protein within fixed tissue sections by monoclonal antibodies. *Arch. Pathol. Lab. Med.* **109**, 117–122.

90. Gown, A. M., Vogel, A. M., Hoak, D., Gough, F., and McNutt, M. A. (1986) Monoclonal antibodies specific for melanocytic tumors distinguish subpopulations of melanocytes. *Am. J. Pathol.* **123**, 195–203.

91. Wick, M. R., Swanson, P. E., and Rocamora, A. (1988) Recognition of malignant melanoma by monoclonal antibody HMB-45. *J. Cutan. Pathol.* **15**, 201–207.

92. Kaufmann, O., Koch, S., Burghardt, J., Andring, H., and Dietel, M. (1998) Tyrosinase, Melan-A, and KBA62 as markers for the immunohistochemical identification of metastatic amelanotic melanomas on paraffin sections. *Mod. Pathol.* **11**, 740–746.

93. Lloyd, R. V. and Wilson, B. S. (1983) Specific endocrine tissue marker defined by a monoclonal antibody. *Science* **222**, 628–630.

94. Wilson, B. S. and Lloyd, R. V. (1984) Detection of chromogranin in neuroendocrine cells with a monoclonal antibody. *Am. J. Pathol.* **115**, 458–468.

95. Wick, M. R. (2000) Immunohistology of neuroendocrine and neuroectodermal tumors. *Semin. Diagn. Pathol.* **17**, 194–203.

96. Miettinen, M. (1987) Synaptophysin and neurofilament protein as markers for neuroendocrine tumors. *Arch. Pathol. Lab. Med.* **111**, 813–818.

97. Gould, V. E., Lee, I., Wiedenmann, B., et al. (1986) Synaptophysin: a novel marker for neurons, certain neuroendocrine cells, and their neoplasms. *Hum. Pathol.* **17**, 979–983.

98. Buffa, R., Rindi, G., Sessa, F., et al. (1987) Synaptophysin immunoreactivity and small clear vesicles in neuroendocrine cells and related tumors. *Mol. Cell. Probes* **1**, 367–381.

99. Stirdsberg, M. (1995) The use of chromogranin, synaptophysin, and islet amyloid polypeptide as markers for neuroendocrine tumors. *Ups. J. Med. Sci.* **100**, 169–199.

100. Lipinski, M., Braham, K., Caillaud, J. M., et al. (1983) HNK-1 antibody detects an antigen expressed on neuroectodermal cells. *J. Exp. Med.* **158**, 1775–1780.
101. Chu, P. G., Chang, K. L., Arber, D. A., and Weiss, L. M. (2000) Immunophenotyping of hematopoietic neoplasms. *Semin. Diagn. Pathol.* 17, 236–256.
102. Perentes, E. and Rubinstein, L. J. (1985) Immunohistochemical recognition of human nerve sheath tumors by anti-Leu 7 (HNK-1) monoclonal antibody. *Acta. Neuropathol.* **68**, 319–324.
103. Tsutsumi, Y. (1984) Leu-7 immunoreactivity as a histochemical marker for paraffin-embedded neuroendocrine tumors. *Acta Histochem. Cytochem.* **17**, 15–21.
104. Dehner, L. P. (1993) Primitive neuroectodermal tumor and Ewing's sarcoma. *Am. J. Surg. Pathol.* **17**, 1–13.
105. Lumadue, J. A., Askin, F. B., and Perlman, E. J. (1994) MIC2 analysis of small cell carcinoma. *Am. J. Clin. Pathol.* **102**, 692–694.
106. Tsukada, T., McNutt, M. A., Ross, R., and Gown, A. M. (1987) HHF35, a muscle actin-specific monoclonal antibody. II. Reactivity in normal, reactive, and neoplastic human tissues. *Am. J. Pathol.* **127**, 389–402.
107. Rangdaeng, S. and Truong, L. D. (1991) Comparative immunohistochemical staining for desmin and muscle-specific actin: a study of 576 cases. *Am. J. Clin. Pathol.* **96**, 32–45.
108. Miettinen, M., Lindenmayer, A. E., and Chanbal, A. (1994) Endothelial cell markers CD31, CD34, and BNH9 antibody to H- and Y- antigens: evaluation of their specificity and sensitivity in the diagnosis of vascular tumors and comparison with von Willebrand factor. *Mod. Pathol.* **7**, 82–90.
109. Traweek, S. T., Kandalaft, P. L., Mehta, P., and Battifora, H. (1991) The human hematopoietic progenitor cell antigen (CD34) in vascular neoplasms. *Am. J. Clin. Pathol.* **96**, 25–31.
110. Cohen, P. R., Rapini, R. P., and Farhood, A. I. (1993) Expression of the human hematopoietic progenitor cell antigen CD34 in vascular and spindle cell tumors. *J. Cutan. Pathol.* **20**, 15–20.
111. Suster, S. (2000) Recent advances in the application of immunohistochemical markers for the diagnosis of soft tissue tumors. *Semin. Diagn. Pathol.* 17, 225–235.
112. DeYoung, B. R., Wick, M. R., Fitzgibbon, J. F., et al. (1993) CD31: an immunospecific marker for endothelial differentiation in human neoplasms. *Appl. Immunohistochem.* **1**, 97–100.
113. Karmochkine, M. and Boffa, M. C. (1997) Thrombomodulin: physiology and clinical applications. *Rev. Med. Interne* **18**, 119–125.
114. Ordonez, N. G. (1997) Value of thrombomodulin immunostaining in the diagnosis of mesothelioma. *Histopathology* **31**, 25–30.
115. Appleton, M. A., Attanoos, R. L., and Jasani, B. (1996) Thrombomodulin as a marker of vascular and lymphatic tumors. *Histopathology* **29**, 1531–157.
116. Holthofer, H., Virtanen, I., Karineimi, A. L., et al. (1982) *Ulex europaeus* 1 lectin as a marker for vascular endothelium in human tissues. *Lab. Invest.* **47**, 60–66.
117. Leader, M., Collins, M., Patel, J., and Henry, K. (1986) Staining of factor VIII-related antigen and *Ulex europaeus* 1 (UEA-1) in 230 tumors: an assessment of their specificity for angiosarcoma and Kaposi's sarcoma. *Histopathology* **18**, 1153–1162.
117a. Nandedkar, M. A., Palazzo, J., Abbondanzo, S. L., Lasota, J., and Miettinen, M. (1998) CD45 (leukocyte common antigen) immunoreactivity in metastatic undifferentiated and neuroendocrine carcinoma: a potential diagnostic pitfall. *Mod. Pathol.* **11**, 1204–1210.
118. Devoe, R. and Weidner, N. (2000) Immunohistochemistry of small round-cell tumors. *Semin. Diagn. Pathol.* 17, 216–224.

119. Parham, D. M., Dias, P., Kelly, D. R., Rutledge, J. C., and Houghton, P. (1992) Desmin-positivity in primitive neuroectodermal tumors of childhood. *Am. J. Surg. Pathol.* **16**, 483–492.

120. Guinee, D. G., Jr., Fishback, N. F., Koss, M. N., et al. (1994) The spectrum of immunohistochemical staining of small cell lung carcinoma in specimens from transbronchial and open-lung biopsies. *Am. J. Clin. Pathol.* **102**, 406–414.

120a. Eusebi, V., Damiani, S., Pasquinelli, G., et al. (2000) Small cell neuroendocrine carcinoma with skeletal muscle differentiation. *Am. J. Surg. Pathol.* **24**, 223–230.

121. Wick, M. R. and Swanson, P. E. (1993) "Carcinosarcomas"—current perspectives and an historical review of nosological concepts. *Semin. Diagn. Pathol.* **10**, 118–127.

122. Wick, M. R., Fitzgibbon, J. F., and Swanson, P. E. (1993) Cutaneous sarcomas and sarcomatoid neoplasms of the skin. *Semin. Diagn. Pathol.* **10**, 148–158.

123. Elgin, J., Phillips, J. G., Reddy, V. V., Gibbs, P. O., and Listinsky, C. M. (1999) Hodgkin's and non-Hodgkin's lymphoma: spectrum of morphologic and immunophenotypic overlap. *Ann. Diagn. Pathol.* **3**, 263–275.

124. Manivel, J. C., Wick, M. R., Dehner, L. P., and Sibley, R. K. (1987) Epithelioid sarcoma: an immunohistochemical study. *Am. J. Clin. Pathol.* **87**, 319–326.

125. Schmidt, D. and Harms, D. (1987) Epithelioid sarcoma in children and adolescents: an immunohistochemical study. *Virchows Arch. A* **410**, 423–431.

126. Papas-Corden, P. C., Zarbo, R. J., Gown, A. M., and Crissman, J. D. (1989) Immunohistochemical characterization of synovial, epithelioid, and clear-cell sarcomas. *Surg. Pathol.* **2**, 43–58.

127. Swanson, P. E., Dehner, L. P., Sirgi, K. E., and Wick, M. R. (1994) Cytokeratin-immunoreactivity in malignant tumors of bone and soft tissue. *Appl. Immunohistochem.* **2**, 103–112.

128. Sirgi, K. E., Wick, M. R., and Swanson, P. E. (1993) B72.3 and CD34 immunoreactivity in malignant epithelioid soft tissue tumors: adjuncts in the recognition of endothelial neoplasms. *Am. J. Surg. Pathol.* **17**, 179–185.

129. Wick, M. R., Swanson, P. E., Scheithauer, B. W., and Manivel, J. C. (1987) Malignant peripheral nerve sheath tumor: an immunohistochemical study of 62 cases. *Am. J. Clin. Pathol.* **87**, 425–433.

130. Sarlomo-Rikala, M., Kovatich, A. J., Barusevicius, A., et al. (1998) CD117: a sensitive marker for gastrointestinal stromal tumors that is more specific than CD34. *Mod. Pathol.* **11**, 728–734.

131. Brown, R. W., Campagna, L. B., Dunn, J. K., et al. (1997) Immunohistochemical identification of tumor markers in metastatic adenocarcinoma: a diagnostic adjunct in the determination of primary site. *Am. J. Clin. Pathol.* **107**, 12–19.

132. Moran, C. A., Wick, M. R., and Suster, S. (2000) The role of immunohistochemistry in the diagnosis of malignant mesothelioma. *Semin. Diagn. Pathol.* **17**, 178–183.

133. Wick, M. R., Loy, T., Mills, S. E., et al. (1990) Malignant epithelioid pleural mesothelioma versus peripheral pulmonary adenocarcinoma: a histochemical, ultrastructural, and immunohistochemical study of 103 cases. *Hum. Pathol.* **21**, 759–766.

# 18
## Applications of Immunohistochemistry in the Diagnosis of Endocrine Lesions

### Ricardo V. Lloyd, MD, PhD

## INTRODUCTION

Endocrine cells and tumors can be divided into neuroendocrine and nonneuroendocrine depending on their immunophenotypic and ultrastructural features. Neuroendocrine cells and tumors are members of the diffuse or dispersed neuroendocrine system (DNES), all of which have secretory granules demonstrated by ultrastructural studies. They can also be recognized with broad-spectrum neuroendocrine markers such as chromogranin and synaptophysin. Members of the DNES also produce specific peptides that can serve as target markers for cell and tumor identification.

Nonneuroendocrine cells and tumors include adrenal cortical cells, thyroid follicular cells, and steroid-producing cells of the ovary and testes. Markers such as inhibins are useful for detecting adrenocortical cells and tumors and gonadal tumors, whereas thyroglobulin is relatively specific for thyroid follicular cells.

The use of specific antibodies, also referred to as markers, has provided a great deal of specificity for the characterization of endocrine cells and tumors.

### Endocrine System

The endocrine system can be divided into neuroendocrine and nonneuroendocrine cells and tumors. Most endocrine cells belong to the neuroendocrine category (**Table 1**). The neuroendocrine system is also referred to as the diffuse or dispersed neuroendocrine system (DNES), which consists of endocrine cells that are present in most areas of the body. Although DNES cells are not all derived from the neural crest, as originally proposed, they all share common features, including expression of chromogranin and synaptophysin, and have dense-core granules detected by ultrastructural examination (1–4). The nonneuroendocrine cells consist of the steroid-producing cells of the adrenal cortex, ovary, and testes and the thyroglobulin-producing follicular cells of the thyroid.

There are many broad-spectrum and specific immunohistochemical markers that can be used to characterize cells and tumors of the DNES (**Tables 1** and **2**). Many of these, such as the chromogranin PGP 9.5 and synaptophysin, have been in use for over a decade. Others, such as synaptic proteins and neuroendocrine-specific protein reticulons, have been recently described in the past few years. To determine whether a tumor is

From: *Morphology Methods: Cell and Molecular Biology Techniques*
Edited by: R. V. Lloyd © Humana Press, Totowa, NJ

**Table 1**
**Specific Peptide and Other Markers for Neuroendocrine Cells and Tumors**

| Cell type | Neoplasm(s) | Markers[a] |
|---|---|---|
| Adrenal medulla and paraganglia | Pheochromocytoma, neuroblastoma, ganglioneuroblastoma, ganglioma | Catecholamines, erkephalin, VIP, somatostatin, S-100 protein |
| Gastrointestinal tract | Benign and malignant neuroendocrine tumors | Gastrin, somatostatin, VIP, secretin, insulin, glucagons, pancreatic polypeptide, cholecystokinin |
| Lung | Benign and malignant neuroendocrine tumors, including carcinoids | ACTH, endorphin, calcitonin, secretin, gastrin-releasing peptide |
| Pancreas | Benign and malignant pancreatic endocrine tumors | Insulin, glucagon pancreatic polypeptide, somatostatin, VIP, gastrin, serotonin |
| Parathyroid | Hyperplasia, adenoma, carcinoma | Parathyroid hormone, parathyroid hormone, related peptide |
| Pituitary | Adenoma and carcinoma | ACTH, growth hormone, prolactin, follicle-stimulating hormone, luteinizing hormone, thyroid-stimulating hormone, calcitonin |
| Thyroid C-cell | Medullary thyroid carcinoma | Calcitonin, somatostatin, ACTH |

[a] VIP, vasoactive intestinal polypeptide; ACTH, adrenocorticotropic hormone.

**Table 2**
**Broad-Spectrum Neuroendocrine Markers**

Chromogranin/secretogranin
Synaptophysin
PGP9.5
Leu7
Synaptic proteins
    SNAP-25
    Rab3A
Neural cell adhesion molecules (NCAM) CD57
Peptidylglycine α-amidating monooxygenase
Neuroendocrine-specific proteins (NSP)-reticulons

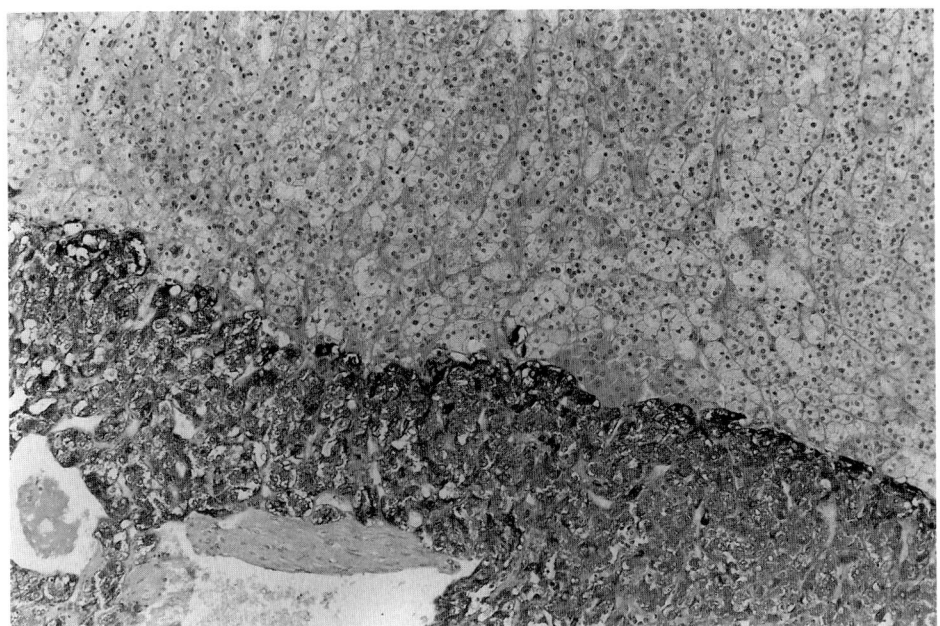

**Fig. 1.** Chromogranin A expression in the adrenal gland. The medullary cells are positive, and the cortical cells are negative. Immunoperoxidase-diaminobenzidine (DAB), original magnification ×200.

neuroendocrine, broad-spectrum markers are used first, followed by more specific markers for specific neuroendocrine organs. This approach is very useful in the characterization of normal endocrine cells and in the diagnosis of endocrine tumors.

## GENERAL (BROAD-SPECTRUM) NEUROENDOCRINE MARKERS

### Chromogranin/Secretogranin

The chromogranin/secretogranin (Cg/Sg) family is composed of several acidic proteins present in the secretory granules of neuroendocrine cells (**Table 2**). The three major Cg/Sg proteins are currently designated CgA, CgB, and SgII. Others include SgIII, SgIV, and SgV (also known as 7β2). The distribution of CgA has been studied extensively in human tumors *(4–7)*. It is present in most neuroendocrine cells and neoplasms. However, most neoplasms with only a few endocrine secretory granules, such as small cell carcinomas of the lung and Merkel's cell carcinomas, do not react strongly with CgA antibodies *(4–7)*. Because of their widespread distribution and high degree of specificity, Cg/Sgs are excellent markers for neuroendocrine cells and neoplasms *(8,9)* (**Figs. 1** and **2**).

Although CgA is a highly specific neuroendocrine marker, it may have limited sensitivity with some tumors. For example, hindgut carcinoids have limited immunoreactivity for CgA, with only 60% of cases positive in a recent series *(10)*. Similarly, pituitary prolactinomas are often negative for CgA. Because hindgut carcinoids and prolactinomas often express CgB, using antibodies against CgB or a cocktail of CgA

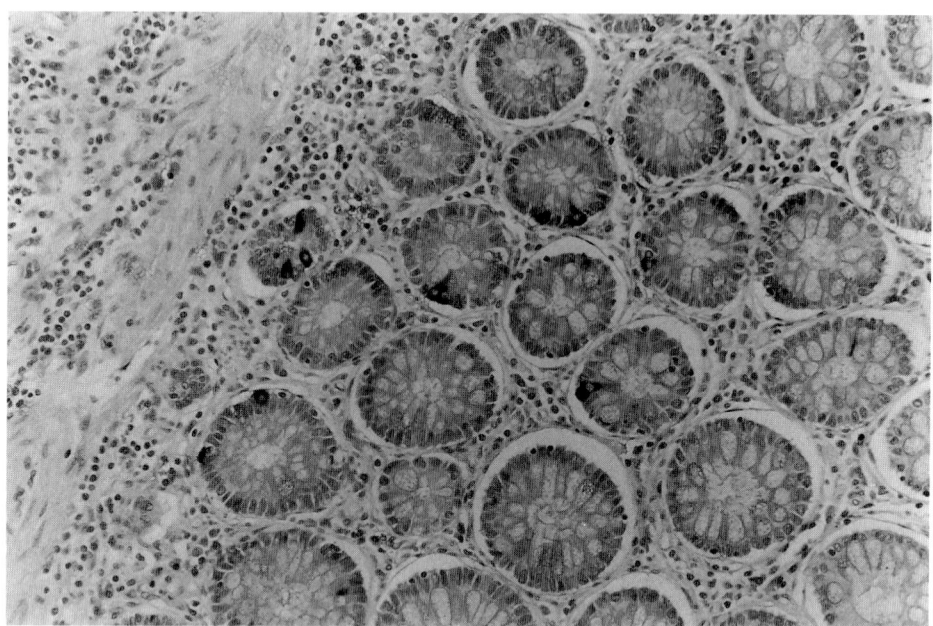

**Fig. 2.** The neuroendocrine cells in the colonic crypts are strongly positive for chromogranin A. Immunoperoxidase-DAB, original magnification ×200.

and -B usually increase the sensitivity for detecting neuroendocrine cells with Cg/ Sg antibodies.

### Synaptophysin

Synaptophysin, a 38-kDa protein molecule, is a component of the membrane of presynaptic vesicles. Gould et al. *(11)* first reported that it was widely distributed in neurons and neuroendocrine cells and their neoplasms. It is another good broad-spectrum neuroendocrine marker. Unlike the Cg/Sg proteins, which are well preserved in formalin-fixed tissues, ethanol fixation provides optimal preservation of the synaptophysin antigen. However, most of the anti-synaptophysin monoclonal antibodies work well in formalin-fixed sections. Synaptophysin is present in vesicles in neuroendocrine cells of the tumors, but immunostaining is present diffusely in the cytoplasm **(Figs. 3** and **4)**.

### Proconvertases

The proconvertases (PCs) are enzymes that process propeptides into active peptides within cells *(12,13)*. Some of these, including PC1/PC3 and PC2, are highly specific for neuroendocrine cells and tumors and can be used as specific neuroendocrine markers. Others, such as PC4, are present in the testes; PC5/6 is more prevalent in the gastrointestinal tract and adrenal.

### Neuron-Specific Enolase

Neuron-specific enolase (NSE), an enzyme also known as γ-enolase, is a highly sensitive but not very specific marker for neuroendocrine cells and tumors. It is

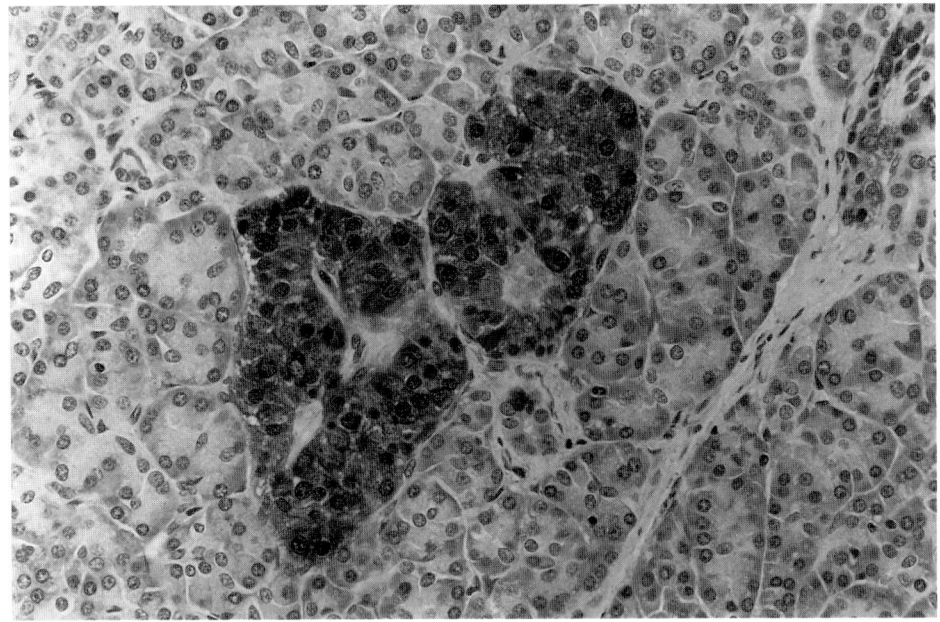

**Fig. 3.** Pancreatic islet showing diffuse staining for the broad-spectrum neuroendocrine marker synaptophysin. Immunoperoxidase-DAB, original magnification ×200.

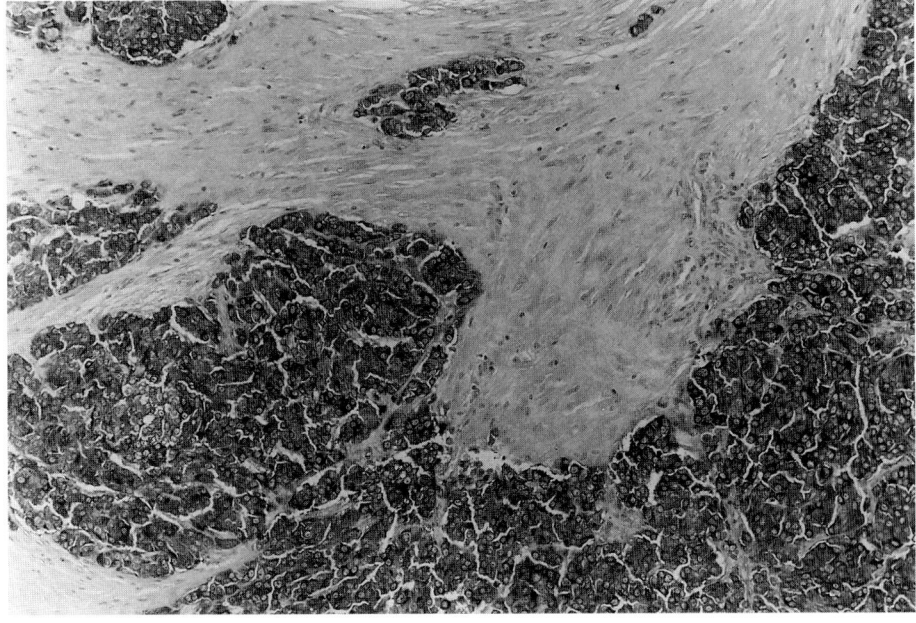

**Fig. 4.** Laryngeal neuroendocrine tumor staining positively for the broad-spectrum neuroendocrine marker synaptophysin. Immunoperoxidase-DAB, original magnification ×300.

commonly found in neurons, peripheral nerves, and neuroendocrine cells *(14,15)*. Some nonneuroendocrine cells and neoplasms also react with antisera against NSE. NSE should be used only with other broad-spectrum markers of neuroendocrine cells in the diagnosis of neuroendocrine tumors because of its relative lack of specificity. It is used only infrequently in our laboratory.

### *Bombesin/Gastrin-Releasing Peptide and Leu7 (HNK-1)*

Bombesin is a tetradecapeptide originally isolated from amphibian skin. It is present in many endocrine cells as well as in central and peripheral neurons *(16)*. Gastrin-releasing peptide (GRP), the proposed mammalian analog of bombesin, has been found in many lung and gastrointestinal endocrine tumors and can be used as a broad-spectrum marker for many endocrine neoplasms *(17)*.

Leu7 (HNK-1), a monoclonal antibody that was produced against a T-cell leukemia cell line, recognizes natural killer cells in blood and lymphoid tissues. It also reacts with small cell carcinomas of the lung as well as with pheochromocytomas and other neuroendocrine neoplasms *(18,19)*.

### *PGP9.5*

PGP9.5 is a soluble protein that was originally isolated from brain. It is a good general marker for neuronal and neuroendocrine tissues *(20,21)*. Interestingly, about half of melanomas stain for PGP9.5, whereas these melanocytic tumors are usually negative for Cg/Sg and for synaptophysin. PGP is a cytoplasmic soluble protein that frequently colocalizes in normal and neoplastic neuroendocrine tissues.

### *Neural Cell Adhesion Molecule*

Neural cell adhesion molecule (NCAM) is a member of the family of membrane-bound glycoproteins present in brain and muscle and other tissues *(22,23)*. It is involved in neuron-neuron and nerve-muscle interactions. The distribution of NCAM in neuroendocrine tissues and tumors was first reported by Jin et al. *(23)*. They found a widespread distribution of NCAM. NCAM is present in about 20% of non-small cell lung carcinomas and is often associated with a poor prognosis *(24)*. Small cell lung carcinomas frequently stain with NCAM antibodies *(25,26)*.

### *Peptidylglycine α-Amidating Monooxygenase*

Amidation is an important step in the maturation of some neuropeptides *(27–30)*. The enzyme peptidylglycine α-amidating monooxygenase (PAM) catalyzes the post-translational modification of many neuropeptides. It consists of two enzymes that convert peptidylglycine substrates into α-amidating products and glyoxylate. The PAM proteins are usually released along with their peptide products during exocytosis, whereas membrane-bound PAM remains associated with the cell *(27)*. Several studies have examined PAM expression in neuroendocrine cells *(30–33)*. A recent study found PAM in all neuroendocrine cell types *(34)*. They found close correlation between PAM expression and at least one of the three principal granin proteins (CgA, CgB, or SgII) *(30)*.

## Synaptic Proteins

A series of proteins involved in neurotransmitter secretion have also been associated with neuroendocrine cells and tumors. These synaptic proteins include SNAP-25 nd Rab3A *(35–42)*.

SNAP-25 was identified as a neuron-specific protein associated with the plasma membrane of the presynaptic nerve terminal in early studies. It was shown to be part of the putative docking complex that is implicated in membrane fusion. A homologous protein, SNAP-23, is expressed ubiquitously in human nonneuronal tissues, including endocrine organs.

Rab3A, a small GTP-binding protein of the rab family, is expressed mainly in neurons and neuroendocrine cells *(39–42)*. Rab3A is thought to be an important control system for exocytosis in neuron and neuroendocrine cells. A recent study showed increased SNAP-25 immunoreactivity in most pituitary prolactin-cell adenomas and in growth hormone-cell adenomas, suggesting that this protein is involved in the mechanism of exocytosis in neoplasms derived from these cell types *(42)*.

## Neuroendocrine-Specific Protein Reticulons

Neuroendocrine-specific protein (NSP)-reticulons are endoplasmic reticulum-associated protein complexes consisting of two closely related protein constituents, NSP-A and NSP-C *(43–46)*. In a recent report, the expression of the NSP-reticulons NSP-A and NSP-C was examined in lung carcinomas. NSP-A and NSP-C were reactive with most carcinoid tumor and small cell lung carcinomas. There was a high concordance between expression of NSP-A and NSP-C in neuroendocrine tumors. These investigators noted that NSP-A was more sensitive than synaptophysin, CgA, Leu7, and neurofilament proteins in detecting neuroendocrine differentiation in non-small cell lung carcinomas *(46)*. They also observed that NSP-A expression showed a stronger correlation with conventional neuroendocrine markers than NCAM *(46)*.

## SPECIFIC MARKERS

### Neuroendocrine Cells and Tumors

Once a neoplasm has been characterized as neuroendocrine by use of the broad-spectrum endocrine markers, antibodies to specific peptide hormones can be used to characterize the lesion better. A summary of the most common peptides and other markers is shown in **Table 2 (Figs. 5** and **6)**. General approaches in the use of these markers should include the following considerations.

1. Very few peptides are restricted to one neuroendocrine cell type or organ. For example, although calcitonin is more commonly produced by the thyroid C-cells, other normal neuroendocrine tissues, including the anterior pituitary gland, also produce calcitonin.

2. Ectopic production of hormones by neoplasms is the rule rather than the exception. Although medullary thyroid carcinomas usually produce calcitonin, hypercalcitonemia may be associated with other tumors such as atypical laryngeal carcinoids, lung carcinoids, or small cell lung carcinomas and others.

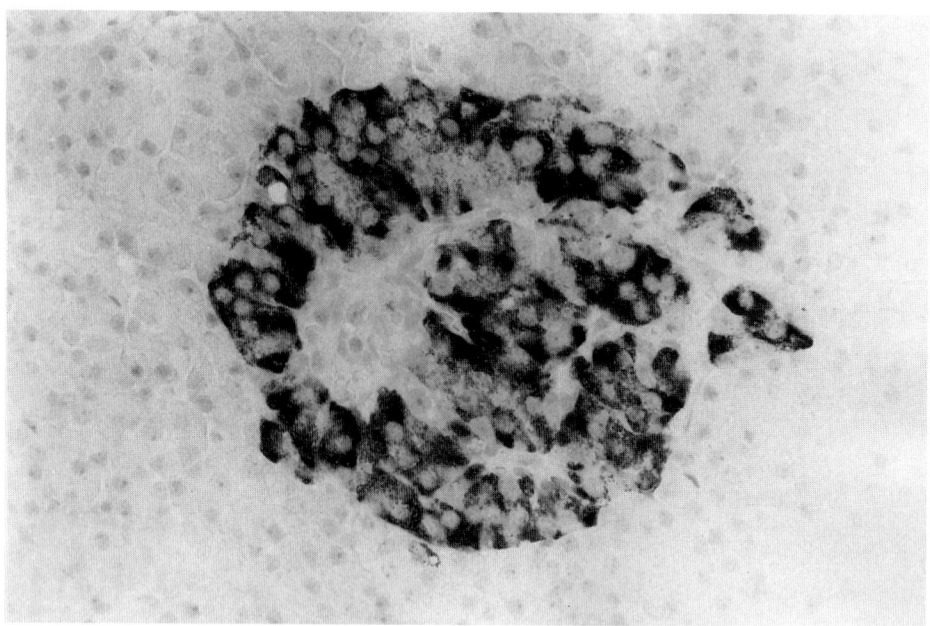

**Fig. 5.** Pancreatic islet staining positively for insulin with anti-insulin antibody. Immunoperoxidase-DAB, original magnification ×300.

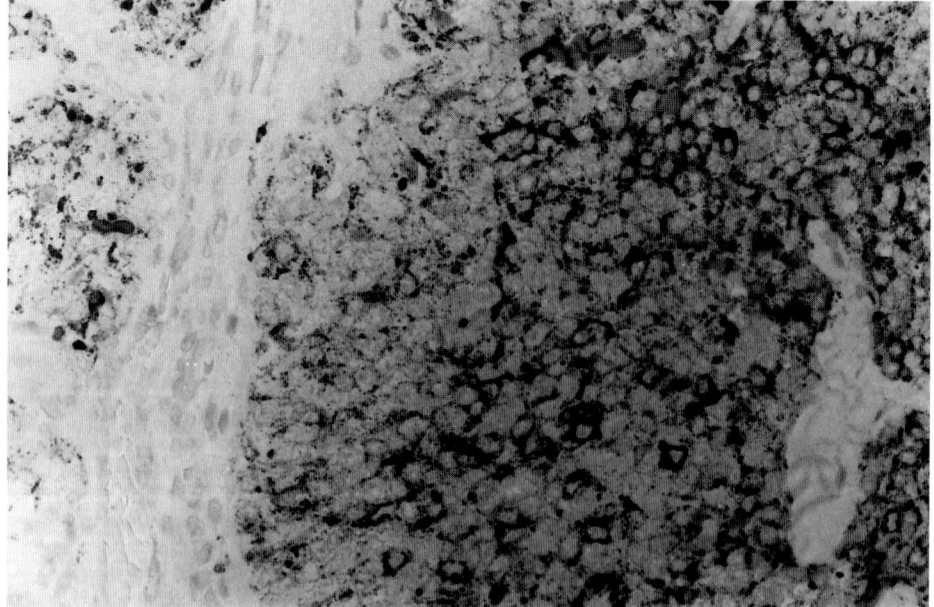

**Fig. 6.** Medullary thyroid carcinoma of the thyroid staining positively for calcitonin. This marker helps to classify this tumor as a thyroid carcinoma derived from the C-cells. Immunoperoxidase-DAB, original magnification ×300.

**Table 3**
**Peptides and Other Markers for Nonneuroendocrine Cells and Tumors**

| Cell type | Neoplasm(s) | Marker |
|---|---|---|
| Thyroid follicular cell | Adenomas and carcinomas | Thyroglobulin, keratin Thyroid transcription factor-1 |
| Adrenal cortex | Adenomas and carcinomas | Inhibin, synaptophysin, keratin, Ad4BP/SF1, vimentin, steroid-metabolizing enzymes |
| Steroid-producing cells of gonads | Benign and malignant neoplasms | Inhibin, vimentin, steroid-metabolizing enzymes |

3. Although the production of a hormone may be associated with specific signs and symptoms, some other substances such as growth factors or related peptides may lead to similar symptoms. For example, although hypoglycemia is commonly associated with excess insulin production, insulin-like growth factor-II (IGF-II) may also be the cause of the hypoglycemia. IGF-II is produced by some neuroendocrine tumors as well as by mesenchymal and other neoplasms *(47,48)*.

4. Failure to detect specific peptides in neuroendocrine tumors does not necessarily mean that the substance is not being produced by the cells. It is possible that the peptide may be rapidly secreted and not stored in the cells. The use of *in situ* hybridization to detect mRNA usually helps to answer this question (*see* Chapter 3).

Additional markers such as S-100 acidic protein are helpful in characterizing normal adrenal medulla and pheochromocytomas, whereas cytokeratin 20 is useful in separating Merkel's cell carcinoma from other neuroendocrine carcinomas.

### Nonneuroendocrine Cells and Tumors

Members of the endocrine system that are not neuroendocrine (i.e., do not contain secretory granules and are not immunoreactive with chromogranin and antibodies) include the adrenal cortical cells, thyroid follicular cells, and steroid-producing cells of the testes and ovaries. A different set of endocrine markers is used to characterize these lesions, as summarized in **Table 3 (Figs. 7–9)**. Thyroglobulin is relatively specific for the thyroid follicular cells **(Fig. 7)**, but other types of tumors may occasionally react with thyroglobulin antibody *(49)*. Antibodies against enzymes in the steroid biosynthesis pathway are also used to characterize adrenal cortical cells *(50–52)*. The adrenal transcription factor Ad4BP/SF1 is a good relatively specific marker for adrenal cortical cells **(Fig. 8)**. Markers such as the inhibins are useful in the diagnosis of adrenal cortical tumors but are not specific for these lesions *(53)* **(Fig. 9)**.

### CONCLUSIONS

Immunohistochemistry has contributed greatly to the characterization and diagnosis of endocrine cells and tumors. Many more advances can be anticipated in the near future with the development of new antibodies for the study of gene expression.

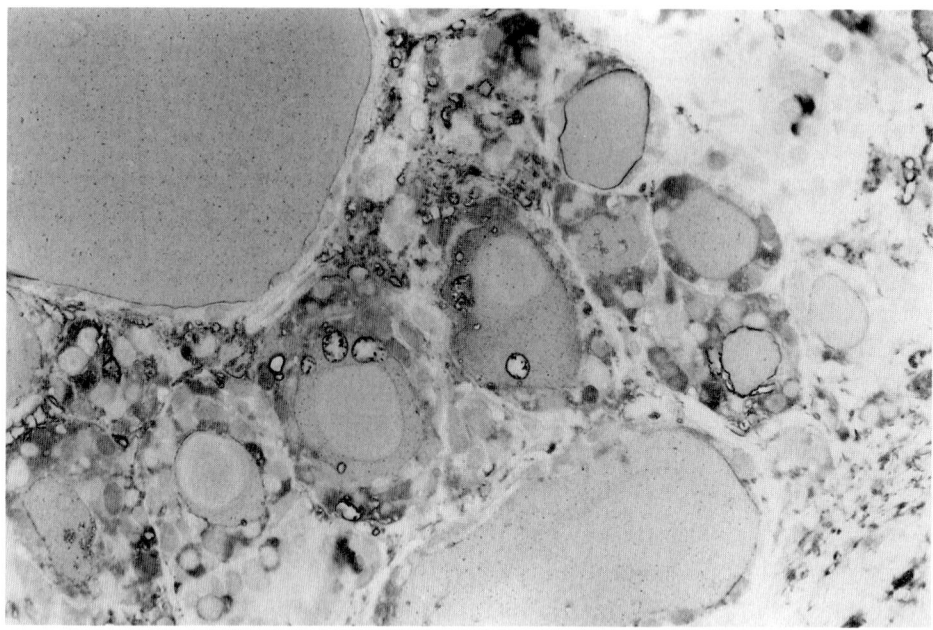

**Fig. 7.** The thyroid follicular cells are nonneuroendocrine cells. These cells express thyroglobulin, which is a relatively specific marker of thyroid follicular cell differentiation. Immunoperoxidase-DAB, original magnification ×300.

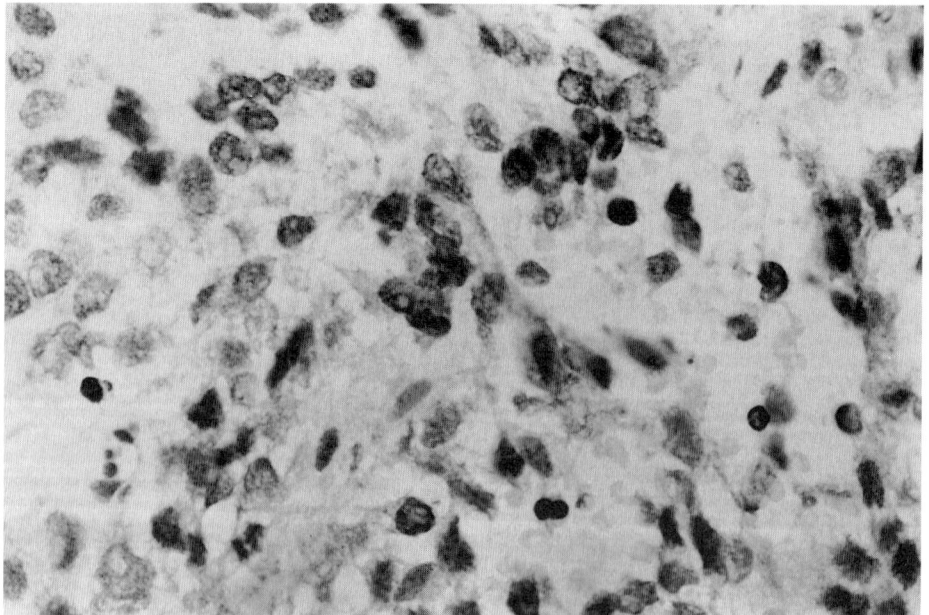

**Fig. 8.** The transcription factor Ad4BP/SF1 is relatively specific for adrenal cortical cells. Positive staining is present in the nucleus of an adrenal cortical carcinoma. Nuclear localization is typically seen with transcription factors. Immunoperoxidase-DAB, original magnification ×200.

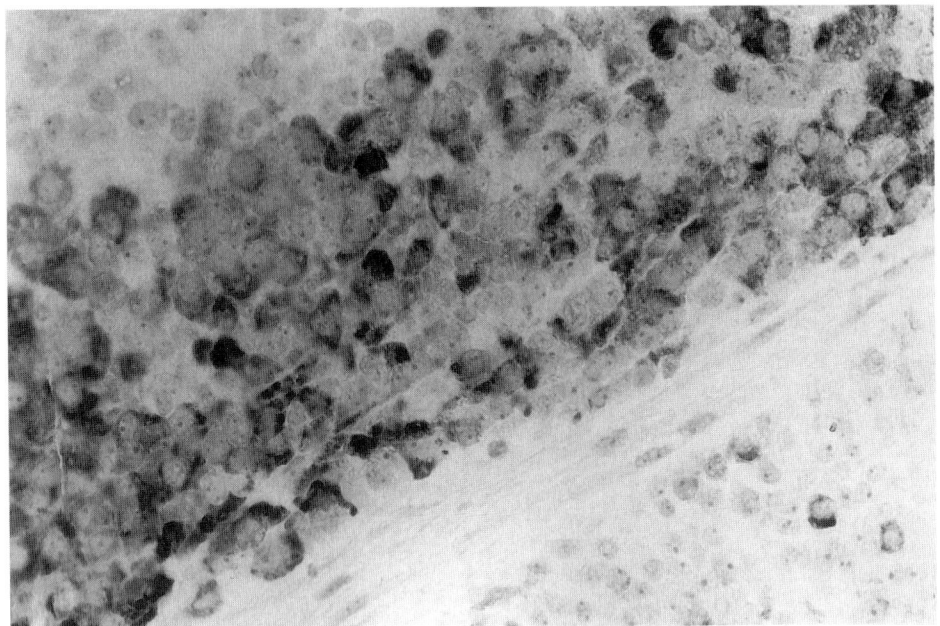

**Fig. 9.** Inhibin A is a polypeptide produced by steroid-producing cells such as those in the adrenal cortex. The cells of this cortical tumor are positive for inhibin A. However, other steroid-producing cells such as those in the ovary and testes also produce inhibin. Immunoperoxidase-DAB, original magnification ×300.

## REFERENCES

1. Pearse, A. G. (1974) The APUD cell concept and its implications in pathology. *Pathol. Annu.* **9**, 27–41.
2. Pearse, A. G. and Takor, T. (1979) Embryology of the diffuse neuroendocrine system and its relationship to the common peptides. *Fed. Proc.* **38**, 2288–2294.
3. DeLellis, R. A. and Wolfe, H. J. (1981) The polypeptide hormone-producing neuroendocrine cells and their tumors: an immunohistochemical analysis. *Methods Achiev. Exp. Pathol.* **10**, 190–220.
4. Lloyd, R. V. (1996) Overview of neuroendocrine cells and tumors. *Endocr. Pathol.* **7**, 323–328.
5. Lloyd, R. V., Mervak, T., Schmidt, K., Warner, T. F., and Wilson, B. S. (1984) Immunohistochemical detection of chromogranin and neuron-specific enolase in pancreatic endocrine neoplasms. *Am. J. Surg. Pathol.* **8**, 607–614.
6. Lloyd, R. V., Sisson, J. C., Shapiro, B., and Verhofstad, A. A. (1986) Immunohistochemical localization of epinephrine, norephinephrine, catecholamine-synthesizing enzymes, and chromogranin neuroendocrine cells and tumors. *Am. J. Pathol.* **125**, 45–54.
7. Wilson, B. S. and Lloyd, R. V. (1984) Detection of chromogranin in neuroendocrine cells with a monoclonal antibody. *Am. J. Pathol.* **115**, 458–468.
8. Wiedenmann, B. and Huttner, W. B. (1989) Synaptophysin and chromogranins/secretogranins—widespread constituents of distinct types of neuroendocrine vesicles and new tools in tumor diagnosis. *Virchows Arch. (B)* **58**, 95–121.
9. Schmid, K. W., Kroll, M., Hittmair, A., et al. (1991) Chromogranin A and B in adenomas of the pituitary. An immunohistochemical study of 42 cases. *Am. J. Surg. Pathol.* **15**, 1072–1077.

10. Al-Khafaji, B., Noffsinger, A. E., Miller, M. A., DeVoe, G., Stemmermann, G. N., and Fenoglio-Preiser, C. (1999) Immunohistologic analysis of gastrointestinal and pulmonary carcinoid tumors. *Hum. Pathol.* **29**, 992–998.

11. Gould, V. E., Lee, I., Wiedemann, B., Moll, R., Chejfec, G., and Franke, W. W. (1986) Synaptophysin: a novel marker for neurons, certain neuroendocrine cells, and their neoplasms. *Hum. Pathol.* **17**, 979–983.

12. Scopsi, L., Gullo, M., Rilke, F., Martin, S., and Steiner, D. F. (1995) Proprotein convertases (PC1/PC3 and PC2) in normal and neoplastic human tissues: their use as markers of neuroendocrine differentiation. *J. Clin. Endocrinol. Metab.* **80**, 294–301.

13. Lloyd, R. V., Jin, L., Qian, X., Scheithauer, B. W., Young, W. F., Jr., and Davis, D. H. (1995) Analysis of the chromogranin A post-translational cleavage product pancreastatin and the prohormone convertases PC2 and PC3 in normal and neoplastic human pituitaries. *Am. J. Pathol.* **146**, 1188–1198.

14. Lloyd, R. V. and Warner, T. F. (1984) Immunohistochemistry of neuron-specific enolase, in *Advances in Immunochemistry* (DeLellis, R. A., ed.), Masson, New York, pp. 127–140.

15. Schmechel, D., Marangos, P. J., and Brightman, M. (1978) Neurone-specific enolase is a molecular marker for peripheral and central neuroendocrine cells. *Nature* **276**, 834–836.

16. Wharton, J., Polak, J. M., Bloom, S. R., et al. (1978) Bombesin-like immunoreactivity in the lung. *Nature* **273**, 769–770.

17. Bostwick, D. G., Roth, K. A., Evans, C. J., Barchas, J. D., and Bensch, K. G. (1984) Gastrin-releasing peptide, a mammalian analog of bombesin, is present in human neuroendocrine lung tumors. *Am. J. Pathol.* **117**, 195–200.

18. Bunn, P. A., Jr., Linnoila, I., Minna, J. D., Carney, D., and Gazdar, A. F. (1985) Small cell lung cancer, endocrine cells of the fetal bronchus, and other neuroendocrine cells express the Leu-7 antigenic determinant present on natural killer-cells. *Blood* **65**, 764–768.

19. Tischler, A. S., Mobtaker, H., Mann, K., et al. (1986) Anti-lymphocyte antibody Leu-7 (HNK-1) recognizes a constituent of neuroendocrine granule matrix. *J. Histochem. Cytochem.* **34**, 1213–1216.

20. Thompson, R. J., Doran, J. F., Jackson, P., Dhillon, A. P., and Rode, J. (1983) PGP 9.5—a new marker for vertebrate neurons and neuroendocrine cells. *Brain Res.* **278**, 224–228.

21. Rode, J., Dhillon, A. P., Doran, J. F., Jackson, P., and Thompson, R. J. (1985) PGP 9.5, a new marker for human neuroendocrine tumours. *Histopathology* **9**, 147–158.

22. Cunningham, B. A., Hemperly, J. J., Murray, B. A., Prediger, E. A., Brackenbury, R., and Edelman, G. M. (1987) Neural cell adhesion molecule: structure, immunoglobulin-like domains, cell surface modulation, and alternative RNA splicing. *Science* **236**, 799–806.

23. Jin, L., Hemperly, J. J., and Lloyd, R. V. (1991) Expression of neural cell adhesion molecule in normal and neoplastic human neuroendocrine tissues. *Am. J. Pathol.* **138**, 961–969.

24. Patel, K., Moore, S. E., Dickson, G., et al. (1989) Neural cell adhesion molecule (NCAM) is the antigen recognized by monoclonal antibodies of similar specificity in small-cell lung carcinoma of neuroblastoma. *Int. J. Cancer* **44**, 573–558.

25. Rygaard, K., Moller, C., Bock, E., and Spang-Thomsen, M. (1992) Expression of cadherin and NCAM in human small cell lung cancer cell liens and xenografts. *Br. J. Cancer* **65**, 573–577.

26. Kibbelaar, R. E., Moolenaar, K. E., Michalides, R. J., et al. (1991) Neural cell adhesion molecule expression, neuroendocrine differentiation and prognosis in lung carcinoma. *Eur. J. Cancer* **27**, 431–435.

27. Eipper, B. A., Stoffers, D. A., and Mains, R. E. (1992) The biosynthesis of neuropeptides: peptide alpha-amidation. *Annu. Rev. Neurosci.* **15**, 57–85.

28. Grimmelikhuijzen, C. J., Leviev, I., and Carstensen, K. (1996) Peptides in the nervous systems of cnidarians: structure, function, and biosynthesis. *Int. Rev. Cytol.* **167**, 37–89.

29. Gether, U., Aakerlund, L., and Schwartz, T. W. (1991) Comparison of peptidyl-glycine

alpha-amidation activity in medullary thyroid carcinoma cells, pheochromocytomas, and serum. *Mol. Cell. Endocrinol.* **79**, 53–63.

30. Lloyd, R. V., D'Amato, C. J., Thiny, M. T., Jin, L., Hicks, S. P., and Chandler, W. F. (1993) Corticotroph (basophil) invasion of the pars nervosa in the human pituitary: localization of proopiomelanocortin peptides galanin and peptidylglycine alpha-amidating monoxygenase-like immunoreactivities. *Endocr. Pathol.* **4**, 86–94.

31. Martinez, A., Montuenga, L. M., Springall, D. R., Treston, A., Cuttitta, F., and Polak, J. M. (1993) Immunocytochemical localization of peptidylglycine alpha-amidating monooxygenase enzymes (PAM) in human endocrine pancreas. *J. Histochem. Cytochem.* **41**, 375–380.

32. Quinn, K. A., Treston, A. M., Scott, F. M., et al. (1991) Alpha-amidation of peptide hormones in lung cancer. *Cancer Cells* **3**, 504–510.

33. Steel, J. H., Martinez, A., Springall, D. R., Treston, A. M., Cuttitta, F., and Polak, J. M. (1994) Peptidylglycine alpha-amidating monooxygenase (PAM) immunoreactivity and messenger RNA in human pituitary and increased expression in pituitary tumours. *Cell. Tissue Res.* **276**, 197–207.

34. Scopsi, L., Lee, R., Gullo, M., Collini, P., Husten, E. J., and Eipper, B. A. (1998) Peptidylglycine α-amidating monooxygenase in neuroendocrine tumors. Its identification, characterization, quantification, and relation to the grade of morphologic differentiation, amidated peptide content, and granin immunocytochemistry. *Appl. Immunohistochem.* **6**, 120–132.

35. Aguado, F., Majó, G., Ruiz-Montasell, B., et al. (1996) Expression of synaptosomal-associated protein SNAP-25 in endocrine anterior pituitary cells. *Eur. J. Cell Biol.* **69**, 351–359.

36. Roth, D., and Burgoyne, R. D. (1994) SNAP-25 is present in a SNARE complex in adrenal chromaffin cells. *FEBS Lett.* **351**, 207–210.

37. Sadoul, K., Lang, J., Montecucco, C., et al. (1995) SNAP-25 is expressed in islets of Langerhans and is involved in insulin release. *J. Cell Biol.* **128**, 1019–1028.

38. Oyler, G. A., Higgins, G. A., Hart, R. A., et al. (1989) The identification of a novel synaptosomal-associated protein, SNAP-25, differentially expressed by neuronal subpopulations. *J. Cell Biol.* **109**, 3039–3052.

39. Johannes, L., Lledo, P. M., Roa, M., Vincent, J. D., Henry, J. P., and Darchen, F. (1994) The GTPase Rab3a negatively controls calcium-dependent exocytosis in neuroendocrine cells. *EMBO J.* **13**, 2029–2037.

40. Fischer von Mollard, G., Mignery, G. A., Baumert, M., et al. (1990) rab3a is a small GTP-binding protein exclusively localized to synaptic vesicles. *Proc. Natl. Acad. Sci. USA* **87**, 1988–1992.

41. Martelli, A. M., Bareggi, R., Baldini, G., Scherer, P. E., Lodish, H. F., and Baldini, G. (1995) Diffuse vesicular distribution of Rab3D in the polarized neuroendocrine cell line AtT-20. *FEBS Lett.* **368**, 271–275.

42. Majo, G., Ferrer, I., Marsal, J., Blasi, J., and Aguado, F. (1997) Immunocytochemical analysis of the synaptic proteins SNAP-25 and Rab3A in human pituitary adenomas. Overexpression of SNAP-25 in the mammosomatotroph lineages. *J. Pathol.* **183**, 440–446.

43. Senden, N. H., Van de Velde, H. J., Broers, J. L., et al. (1994) Cluster-10 lung-cancer antibodies recognize NSPs, novel neuro-endocrine proteins associated with membranes of the endoplasmic reticulum. *Int. J. Cancer Suppl.* **8**, 84–88.

44. Van de Velde, H. J., Senden, N. H., Roskams, T. A., et al. (1994) NSP-encoded reticulons are neuroendocrine markers of a novel category in human lung cancer diagnosis. *Cancer Res.* **54**, 4769–4776.

45. Senden, N. H., Timmer, E. D., Boers, J. E., et al. (1996) Neuroendocrine-specific protein C (NSP-C): subcellular localization and differential expression in relation to NSP-A. *Eur. J. Cell. Biol.* **69**, 197–213.

46. Senden, N. H., Timmer, E. D., deBruine, A., et al. (1997) A comparison of NSP-reticulons

with conventional neuroendocrine markers in immunophenotyping of lung cancers. *J. Pathol.* **182**, 13–21.

47. Daughaday, W. H. (1995) The pathophysiology of IGF-II hypersecretion in non-islet cell tumor hypoglycemia. *Diabetes Rev.* **3**, 62–72.

48. Hoog, A., Sandberg Nordqvist, A. C., Hulting, A. L., and Falkmer, U. G. (1997) High molecular weight IGF-II expression in a haemangiopericytoma associated with hypoglycemia *APMIS* **105**, 469–482.

49. Keen, C. E., Szakacs, S., Okon, E., Rubin, J. S., and Bryant, B. M. (1999) CA125 and thyroglobulin staining in papillary carcinomas of thyroid and ovarian origin is not completely specific for site of origin. *Histopathology* **34**, 113–117.

50. Sasano, H. (1992) New approaches in human adrenocortical pathology: assessment of adrenocortical function in surgical specimen of human adrenal glands. *Endocr. Pathol.* **3**, 4–13.

51. Sasano, H., Miyazaki, S., Sawai, T., et al. (1992) Primary pigmented nodular adrenocortical disease (PPNAD): immunohistochemical and in situ hybridization analysis of steroidogenic enzymes in eight cases. *Mod. Pathol.* **5**, 23–29.

52. Sasano, H., Shizawa, S., Suzuki, T., et al. (1995) Ad4BP in the human adrenal cortex and its disorders. *J. Clin. Endocrinol. Metab.* **80**, 2378–2380.

53. Chivite, A., Matias-Guiu, X., Pons, C., Algaba, F., and Prat, J. (1998) Inhibin A expression in adrenal neoplasms. A new immunohistochemical marker for adrenocortical tumors. *Appl. Immunohistochem.* **6**, 42–49.

# 19
# Ultrastructural Immunohistochemistry

## Sergio Vidal, DVM, PhD, Eva Horvath, PhD, and Kalman Kovacs, MD, PhD

## INTRODUCTION

*He would need practice if he was going to see the things above. First he would most easily see shadows and then the images of men and other things reflected in water, and then those things themselves, and the things in the heavens and the heavens themselves.*
   (Plato 428–348 BC: *The Republic*)

Plato could not imagine the important role that transmission electron microscopy, "the world of shadows," would play in the progress of science. Its crucial role in diagnostic pathology and cell type identification is undisputed. By allowing recognition of subcellular structural features not discernible by light microscopy, electron microscopy has contributed significantly to the better understanding of derivation and function of various cell types in health and disease.

Although transmission electron microscopy has been largely replaced lately by immunocytochemical and nucleic acid hybridization techniques at the light microscopic level, the combination of ultrastructural investigation and immunocytochemistry (immunoelectron microscopy) plays a key role in the accurate localization of immunoreactive substances. This method, by allowing colocalization of two different antigenic epitopes, combines the resolution of ultrastructural features with immunocytochemical specificity. Although immunoelectron microscopy is not the first method of choice in routine study of tissues, it is very useful when ultrastructural and immunocytochemical findings are not pathognomonic and are subject to diverse interpretations.

From their inception in the 1960s, light microscopic immunocytochemical techniques underwent several modifications, and the variants are widely and successfully used. The stringent criterion of optimal preservation of subcellular structures makes electron microscopy much less adaptable and renders immunoelectron microscopy a somewhat tedious technique used chiefly in research.

Preembedding immunoelectron microscopy on vibratome or cryostat sections is a technique not much different from light microscopic immunocytochemistry and can be applied easily. However, the inaccessibility of antigenic sites to antibodies creates serious problems. Postembedding immunolabeling and ultrathin cryosection techniques have solved the problem of inaccessibility. However, these methods are sometimes

From: *Morphology Methods: Cell and Molecular Biology Techniques*
Edited by: R. V. Lloyd © Humana Press, Totowa, NJ

associated with insufficient labeling or other technical difficulties. It should be empha-sized that, in immunoelectron microscopy, virtually every step of the procedure affects the quality of the final result; thus the method of tissue preparation should be determined by the aim of the investigation.

In this chapter we attempt to present an overview of the current methodologies of ultrastructural immunocytochemistry, giving emphasis to tissue preparation methods that provide reliable results at the electron microscopic level.

## IMMUNOELECTRON MICROSCOPIC METHODS

After the introduction of immunocytochemistry at the light microscopic level several approaches have been applied to the processing of tissues for the ultrastructural localiza-tion of specific antigens in intact cells. All these approaches attempted to achieve compromise between structural preservation and reliable visible labeling. Three methods were developed: preembedding methods, postembedding methods, and cryotechniques. These three methods are reviewed in this chapter.

### Preembedding Methods

In preembedding methods, the antibody is applied to relatively thick tissue slices before or after any fixation procedure. Following the immunocytochemical procedure, the tissue is fixed, postosmicated, and routinely processed for ultrastructural examina-tion. Preembedding techniques are difficult and time-consuming procedures requiring immediate tissue handling. For diagnostic purposes, utilization of such techniques in laboratories involved primarily in diagnostic work is extremely cumbersome and unrealistic. The use of preembedding immunocytochemistry remains limited to demon-stration of antigens that cannot withstand exposure to fixatives and/or processing. Alternatively, tissue cryofixation can be applied before postembedding labeling, thereby maintaining antigenic properties.

The major disadvantage of preembedding immunoelectron microscopy is the poor morphologic preservation of ultrastructural details. Another significant problem is poor antibody penetration. In addition, a large amount of tissue is needed, since each antigenic determinant has to be tested on individual tissue slices. The need for fresh unfixed tissue precludes the use of preembeding techniques in retrospective studies utilizing archival material.

These disadvantages notwithstanding, preembedding techniques are of subtantial value, allowing rapid scan of large tissue portions on thick sections (vibratome or cryosections) and selection of specific areas for electron microscopy. This technique is particularly useful to find cells/tissue structures that are rare and difficult to locate. Another notable advantage of the preembedding method is the visualization of mem-brane-bound antigens, such as synaptophysin *(1)*. In postembedding methods, immuno-reactivity is limited to the availability of epitopes exposed on the surface of the ultrathin sections. Membranes usually appear as "lines" in ultrathin sections, and the chance of exposing membrane-bound epitopes is understandably low. In the preembedding proto-col, antibodies can reach the antigenic sites in the tissue before embedding, and the immunocytochemical chromogen product formed can be visualized on the ultrathin sections even if the antigenic sites themselves are not exposed on the surface of the section.

The success of preembedding immunocytochemistry, expressed as efficient labeling as well as good tissue preservation, depends on a number of variables such as fixation and tissue permeability.

*Fixation Procedures*

Accurate identification of antigens is profoundly dependent on tissue preparation. Rapid fixation is essential to render tissue components insoluble, in order to preserve tissue morphology and prevent diffusion or loss of tissue antigens from their original site. Thus the rate of penetration of the fixative will play an important role in the precise localization of soluble antigens *(2,3)*. If the fixative fails to reach and stabilize the antigens before they diffuse from their natural sites, diffused antigens may subsequently be fixed at some other sites, resulting in erroneous localization. This is particularly significant when the antigens are small and/or highly soluble such as some peptide hormones. Thus diffuse immunocytochemical staining was observed in adrenocorticotropic (ACTH) cells of the anterior pituitary and in C-cells of thyroid *(3,4)*.

The ideal fixative should completely preserve all spatial relationships of tissue constituents without affecting the specific binding of antibodies to tissue antigens. However, fixatives that achieve satisfactory preservation of cytologic details typically reduce the number of immunoreactive sites on the antigens, in turn decreasing the number of sites available for antibody binding. This may not prevent the visualization of antigens that are present in high concentration, but antigens present only in trace amounts may be rendered undetectable. The fixation process will be influenced not only by the type of fixative but also by the type of antigen and by its state and/or concentration. For instance, hormones may be more easily denatured while being synthesized than during the process of secretion. Such differences could explain the frequent absence of immunoreactivity within the sacculi of the Golgi complex. However, it is also possible that the lack of immunopositivity is because the antigens have relatively low concentration in the synthetic machinery and are more vulnerable to the effects of fixatives. It has been demonstrated that in pituitary lactotrophs, prolactin is 200 times more concentrated in the secretory granules than within the endoplasmic reticulum *(5)*.

Different strategies have been used to prevent the loss of antigenicity during the immunocytochemical procedures. Excluding fixation would not solve the problem. Antigen immunoreactivity is theoretically maximal in unfixed tissue, but so are the loss of cytologic details and diffusion of antigens. The strategy most successfully used is to match the fixative to the nature of the antigen. If the antigen is a polypeptide, aldehyde fixative should be used. If the antigen is a glycoprotein, a fixative directed toward its carbohydrate moiety would be superior to an aldehyde fixative since the antigenicity of the peptide portion would not be significantly altered. Furthermore, the fixative should preferably react with the antigen molecule at a site distant from the antigenic determinant. For example, if an aldehyde fixative was used to stabilize ACTH, the anti-ACTH should be directed against the C-terminal of the hormone rather than the midportion, since the aldehyde is expected to attach to the lysine residues on the 11th, 15th, 16th, and 21st portions of the molecule *(6)*.

Crosslinking fixatives such as paraformaldehyde and glutaraldehyde have become the most commonly used agents for preembedding electron microscopy. The concentration of fixatives should be sufficient for the preservation of subcellular details but should not significantly affect the immunoreactivity of antigens. If not used in adequate concentration, crosslinking fixatives can induce severe denaturation, especially in large protein antigens, whose reactivity depends not only on primary structure but also on conformational features. This is particularly true for glutaraldehyde, the high concentration of which drastically alters the immunologic properties of proteins and severely limits the penetration of antibodies into tissues. In preembedding electron microscopic immunocytochemistry, there is an inverse relationship between optimal tissue preservation and good antibody penetration, often requiring a compromise between these two variables. To achieve an acceptable middle ground, blockage of excess aldehyde groups after glutaraldehyde fixation was attempted with sodium borohydride *(7,8)*. However, the process is often incomplete and does not prevent nonspecific binding to immunoglobulins, leading to increased background; thus fixatives containing low concentrations of glutaraldehyde are preferable. Fixation of tissues with low concentrations of glutaraldehyde (<1%) in combination with paraformaldehyde not only yielded better tissue fixation than paraformaldehyde alone but considerably reduced nonspecific staining.

It is of note that many tissue antigens cannot withstand even the mildest exposure to glutaraldehyde fixation. In such situations, different fixatives are required to obtain adequate immunocytochemical localization and achieve satisfactory preservation of cellular structure. One option is conventional paraformaldehyde fixation, which does not significantly reduce antigenicity. However, tissue preservation is often not adequate for electron microscopic studies. If glutaraldehyde is to be avoided, a mixture of periodate-lysine-paraformaldehyde (PLP) frequently gives satisfactory results. This mixture was developed by Mclean and Nakane *(9)* to improve fixation by crosslinking carbohydrate moieties that have been oxidized to aldehyde groups by the periodate. Paraformaldehyde mixed with picric acid and/or acetic acid has also been used as a fixative for immunocytochemical staining, although this provide better results for light microscopy than for electron microscopy. These three fixatives are devoid of the harmful effects of glutaraldehyde on protein structure and antigenicity, but ultrastructural preservation is inferior to that obtained with glutaraldehyde.

As has been alluded to, low concentrations of glutaraldehyde are often used to ensure good tissue preservation in preembedding techniques. If tissue blocks are employed, the slow immersion fixation may not be able to prevent substantial antigen loss due to degradation. To achieve rapid and uniform penetration, perfusion of the fixative through vascular channels is the alternative in experimental studies. Using this procedure, even highly diffusible antigens such as serotonin can be accurately preserved and visualized. It is of note that the method of fixation (perfusion vs immersion) may induce structural artifacts, leading to displacement of certain antigens. External pressure during fixation, particularly during perfusion, can introduce major artifactual changes, predominantly in the fine structure of blood vessels and organization of the connective tissue. Therefore perfusion fixation should be limited to situations in which the tissue sample is not easily exposed to the fixative, it is difficult to dissect, or it is inaccessible, as happens in the central nervous system.

*Permeabilization Methods*

Fixed tissue is often treated to enhance antibody penetration. Different methods can be applied for permeabilization, such as cryostat or vibratome sectioning, freezing and thawing, and/or exposure to detergents or ethanol.

The first strategy used to increase antibody permeability in unembedded tissue is to reduce the thickness of sections. Sections of fixed, unembedded tissue may be cut from frozen as well as from unfrozen tissue blocks. Unless special precautions are taken, freezing of tissues will inevitably be accompanied by the appearance of freezing artifacts caused by the formation of ice crystals. Depending on their size, ice crystals can extensively damage tissue morphology and distort localization of immunoreactivity. Unfrozen sections of uniform thickness may be cut on the Vibratome. Free-floating sections of 20–200 μm thickness of high morphologic quality can be collected and processed for immunocytochemistry. The use of Vibratome sections is indicated when high resolution of cytologic details is required.

Even if one is using relatively thin sections, class IgG immunoglobulins (primary and bridge antisera are for the most part IgGs), peroxidase-antiperoxidase (PAP) complexes with molecular weights of approximately 150 and 500 kDa, respectively, and other molecules of this size do not pass readily through extracellular spaces of membranes, especially following fixation. Different strategies have been used to allow antibodies to gain access to intracellular antigens, but the treatments must result in a controlled type and degree of tissue damage.

Nonionic detergents, such as NP-40, Triton X-100, and Tween 20, are commonly used to permeabilize cells after fixation *(10,11)*. These relatively mild detergents are not likely to cause appreciable denaturation of proteins. However, they can interact not only with unfixed lipids to affect permeabilization but also with amphophilic membrane proteins, compromising ultrastructural preservation.

In contrast, use of saponin, a plant glycoside that interacts with cholesterol, allows penetration of immunocytochemical markers into the cells and permits better preservation of membrane ultrastructure than nonionic detergents *(12,13)*. Shortcomings of saponin as a detergent are related to its ability to permeabilize only cholesterol-containing membranes. Saponin proved to be unable to permeabilize mitochondria and the nucleus owing to the low cholesterol content (0 and 5%) of the inner mitochondrial membrane and nuclear membrane, respectively.

Different studies also reported that either freezing after careful cryoprotection or buffered ethanol treatment offers the desired results without major disruption of tissue integrity and loss of antigenicity *(8,14)*. After the tissue has been cryoprotected and frozen, small holes (<400 nm) resembling ice crystal damage can occasionally be seen in the membranes. These holes are secondary to freezing and probably aid the penetration of reagents. The mechanism by which ethanol allows for reagent penetration is not easily discernible under the electron microscope. Ethanol treatment may exert its effect by extracting substances from the membranes or cytoplasm of fixed cells, thereby enhancing the diffusion of antibodies into fixed tissue. However, ethanol washing may also exert its effect by rapidly removing free aldehyde from the tissue. Thus, when the tissue is sufficiently crosslinked after fixation to retain its ultrastructure, treatment with 50% ethanol may wash out free aldehydes. If they are left in the tissue, additional crosslinks could be produced, hindering either antibody penetration or antibody binding. In contrast to freezing procedures, ethanol treatment has the advantage that the severity of tissue damage can easily be reduced by varying the length of exposure and ethanol concentration to adapt to the characteristics of different tissues. Thus membranes appear fragmented if the tissue is washed with too high concentrations of ethanol or with 50% ethanol for too long.

Although it is possible to vary the degree of cryoprotection, freezing procedures are less flexible in terms of controlling the severity of cell damage. However, freezing procedures do have the advantage that the tissue can be sliced at any desired thickness on a sliding microtome while still frozen. In contrast to this advantage, ethanol treatment has been successfully used for locating structures to be examined with the electron microscope in whole-mount preparations. This method has been applied to study peripheral nerves for more than a century and can be adapted to other tissues of small size that can be stained immunocytochemically in whole mounts.

Microwave irradiation of free-floating sections during the incubation with primary antibodies can also be considered for enhancing the penetration of antisera in vibratome and cryostat sections *(15)*. Using microwave technology, incubation time with the primary antibody can be drastically shortened; at the same time the ultrastructural details remain well preserved. This is in contrast to the moderate to severe damage of the ultrastructure observed in tissues fixed by means of microwave irradiation. The question of whether incubation under microwave conditions results in enhanced penetration of the antisera into the sections is still unanswered. The direct and indirect effects of microwave irradiation could account for the acceleration of the immunocytochemical procedure. The direct effect may be the vigorous motion of electrically charged molecules present in the incubating medium and tissue components. The indirect effect, the stimulation of the movement of the bipolar water molecules by the electromagnetic waves, results in an increase of the temperature.

Permeabilization methods permitted increased application of immunocytochemical preembedding procedures. Visualization of antigen-antibody binding *in situ* requires a reporter system. The most frequently applied reporters in preembedding immunocytochemistry for electron microscopy are enzymes, mainly horseradish peroxidase. In this technique, peroxidase is conjugated to a secondary antibody or other immunoreagent (e.g., protein A) that is used to detect the primary antibody directed against the antigen of interest. Variations include labeling the primary antibody directly or labeling a tertiary antibody. The chromogen chosen to visualize the immunocytochemical reaction depends on the requirements of a given procedure or experiment. In this regard, for preembedding immunoelectron the peroxidase-diaminobenzidine (DAB) procedure is favored by many authors, mainly because this technique provides consistent results and allows good visualization of the labeled structures *(16)*. A significant disadvantage of the peroxidase-DAB procedure is that the label is too large for precise antigen localization and that the reaction product tends to diffuse and may bind nonspecifically to other cell structures. Furthermore, it does not allow quantification of the antigen.

In contrast to enzyme-based immunocytochemical methods with DAB as the chromogen, colloidal gold techniques (using gold particles ≥5 nm) provide excellent labeling without diffusion artifacts and a clear recognition of the ultrastructure, allowing visualization of antigenic sites with high resolution. It also permits morphometric evaluation of labeling. Presently, in postembedding techniques the gold particles are usually attached to a secondary antibody, that is associated with poor penetration of gold particles in preembedding techniques. Secondary antibodies labeled with gold particles of about 1 nm diameter (ultrasmall colloidal gold particles prepared in the 1–3-nm size range or gold cluster immunoprobes, Nanogold compounds of 0.8 or 1.4 nm in size) should alleviate many of the penetration problems associated with the use of ≥5 nm colloidal gold and provide the opportunity to localize antigens in permeabilized tissue sections. There is evidence that ultrasmall and Nanogold probes label with greater efficiency than large 5-, 10-, and 15-nm particles. Gold particles of about 1 nm diameter penetrate cells or tissues even under conditions that preclude penetration of 5 nm colloidal gold *(17)*.

The major drawback using ultrasmall or Nanogold probes relates to the difficulty of detection in sections by conventional transmission electron microscopy. This problem can be overcome to a large extent by applying procedures to increase the particle size of the final product. This is usually accomplished by a silver enhancement reaction in which metallic silver is deposited on the surface of the gold particles. One of the first silver enhancement solutions used for electron microscopy was originally designed for light microscopy by Danscher *(18)*. The Danscher solution, although permitting good particle enhancement at the light microscope level, has a very low pH (pH 3.4), and when is used for electron microscopy it causes widespread damage to subcellular structures. The pH of the enhancing solutions can be increased by the use of 10 mM HEPES buffer to pH 6.8 with a noticeable improvement in morphologic preservation. It should be noted that prolonged silver enhancement may result in increased background level. It was also demonstrated that the silver shell may not be stable enough to withstand osmium tetroxide postfixation, which is necessary for adequate preservation and visualization of structures.

To achieve a compromise between structural preservation and visible labeling, gold toning can be successfully applied to stabilize the silver shell around the ultrasmall gold particles. In this technique, silver enhancement with silver acetate, followed by gold toning with chloroauric acid, is used to replace the silver shell with a more stable label, in order to observe immunocytochemical signal after osmium fixation and Epon embedding *(19)*. Gold toning is based on the fact that ultrasmall gold particles can be rendered larger with gold chloride. The method can also be successfully used at the light microscopic level with silver impregnation or with gold colloid enhancement.

Another disadvantage associated with ultrasmall immunogold toning is that the resultant particles are of variable sizes. This fact makes it difficult to combine preembedding labeling with ultrasmall immunogold and postembedding immunogold labeling.

Other experimental variables that markedly influence labeling efficiency in the preembedding protocols (quality of the immunoreagents, temperature, dilutions of reagents, and so on) can be controlled fairly well within one experiment.

This overview of preembedding methods would not be complete without mentioning the different embedding media that can be used. In contrast to postembedding procedures, in preembedding methods the embedding medium does not influence labeling efficiency. However, the use of appropriate media is fundamental to obtain optimal preservation of ultrastructure. The ideal embedding medium permits the necessary dehydration steps, infiltration, and embedding under the mildest possible conditions within a reasonably short time *(20,21)*. The consistency of the embedded material should allow semithin sectioning and possibly ultrathin sectioning for studies to be performed at both the light and electron microscopic levels. It is sometimes also convenient to use water-miscible polymers for conventional enzyme histochemistry. Finally, it is desirable to use polymers of sufficient hardness without britleness for proper sectioning of a variety of tissues. Presently both hydrophobic epoxy (Epon, Araldite) and hydrophilic acrylic (LR white, LR gold) resins are successfully employed for preembedding immunocytochemistry. Because of their physicochemical properties, acrylic resins have an advantage: they can be used following both full and partial dehydration protocols at room temperature or low temperature. Their hydrophilic properties also permit a combination of preembedding immunocytochemistry with postembedding immunostaining or performance of enzyme histochemical methods using semithin sections.

The detailed protocol for preembedding immunocytochemistry given at the end of the chapter provides information on the detection of many different antigens, including some at relatively inaccessible locations.

### Postembedding Methods

Postembedding electron microscopic immunocytochemical techniques require the fixation and embedding appropriate for the cells/tissues under study as well as the preparation of semithin or ultrathin sections thereby exposing the antigens to reagents and antibodies. Some advantages of this approach are immediately obvious. Penetration of antibodies into tissue sections is not necessary. Also, localization of two different antigens can be accomplished by staining adjacent sections from the same blocks. Multiple-labeling immunohistochemistry may be used for demonstration of different antigens in the same cell at the electron microscopic level. Despite the advantages, postembedding electron microscopic immunocytochemistry carries its own limitations in terms of labeling efficiency. Attempts have been made to amplify the end result by using low-temperature tissue embedding in hydrophilic resins, antigen retrieval techniques, introduction of ultrasmall gold markers, and other techniques.

Currently postembedding methods are the most frequently used procedures to study localization of hormones, peptides, and proteins in different cells and tissues at the electron microscopic level. In postembedding immunocytochemistry, the careful planning of a protocol is pivotal to accomplish reliable and reproducible results. The strategic approach should consider the variables that influence labeling efficiency such as fixation, methods available for processing and embedding, or applicability of the preferred immunocytochemical procedure to localize antigenic sites in the tissue. In the last decade there has been an explosion in the number of publications that have employed the various techniques, reviewed the theoretical aspects of their applications, and examined the role of variables. Here we summarize recent advances in these fields and present the current methodology available to yield the highest specificity and signal to noise-ratio.

When selecting a method for investigation, the aspects to consider are tissue type, appropriate fixation, method of dehydration, type of resin, mode of resin polymerization, and immunocytochemical detection method.

*Fixation Strategies*

Optimal fixation is crucial for ultrastructural immunocytochemistry. At present there are no firm rules, and one must explore the options best for the tissues/antigens to be studied. Tissue fixation for postembedding immunocytochemistry can follow one of two paths: the tissue can either be stabilized by chemical crosslinking and/or precipitation, involving the use of chemical fixatives (discussed in the previous section), or it can be physically immobilized by using cryotechniques (discussed in the next section).

Chemical fixation is used most frequently in tissue preparation for postembedding immunocytochemistry. What was previously mentioned for preembedding protocols is also valid for postembedding methods. However, it is important to note that the use of glutaraldehyde is less limited, since inaccessibility of antigenic sites to antibodies is not an issue when using semithin and ultrathin sections. For this reason retrospective immunocytochemical studies of many tissue antigens can be conducted on routinely processed material. Earlier there was little enthusiasm for glutaraldehyde fixation because of its adverse effects on antigen immunoreactivity. At present glutaraldehyde is an integral part of tissue preparation for postembedding immunocytochemistry at the electron microscopic level. However, one has to consider that tissue antigen sensitivity is inversely proportional to the concentration of the glutaraldehyde and the time period for which it is used *(22)*. Minimizing duration of glutaraldehyde fixation without compromising ultrastructural preservation is a constant challenge. Using a high concentration of glutaraldehyde (>1%) for periods of time exceeding 15 min has been shown to be deleterious to antigene preservation. Antigenicity is often lost as a result of conventional postfixation with osmium tetroxide used in diagnostic electron microscopy.

Osmium tetroxide is a very strong fixative reacting with and crosslinking unsaturated fatty acids as well as stabilizing and contrasting cellular membranes. It also cross links protein and polypeptide chains, as well as proteins and unsaturated lipids. However, by reacting with membranes, it slows down the fixation process, impeding its own penetration into the cell. For these reasons it should only be used as a postfixation step. Since omission of osmic acid results in poor membrane contrast and thereby compromises fine structural details, several treatments of the semithin and ultrathin sections have been developed to localize different cellular antigens in tissues postfixed in osmium tetroxide.

Most of these treatments use oxidizing agents such as $H_2O_2$ or sodium metaperiodate, the effects of which on immunocytochemical staining are unclear at present *(23,24)*. Their role is probably related to the removal of osmium by reoxidizing and solubilizing reduced osmium molecules, thereby unmasking antigens. Alternatively, various reagents can be added to the fixative to replace osmium postfixation and prevent its negative effect. These compounds, such as acrolein, en bloc uranyl acetate, dimethylsulfoxide (DMSO), or carboiimides, have been advocated to preserve membrane structure without seriously damaging antigenicity.

Cell suspensions and bacterial and cell cultures are much easier to handle when the appropriate method has been selected. They can be fixed briefly in low concentrations of aldehyde, for example, 0.1% neutral buffered glutaraldehyde for 10 min. Subsequently cell suspensions can be harvested or concentrated by centrifugation or filtration. The cells are then resuspended in agar or in a low-melting-point agarose or gelatin mixture. After cooling, the solid mixture can be cut into small blocks, which can be processed similarly to tissue pieces *(25)*.

*Dehydration and Resin Embedding*

The type of tissue and the method of fixation are important factors in deciding which processing method and resin system should be employed. The application of resin embedding, introduced by Newman and Borysko in 1949 *(26)*, led to the most spectacular advance in electron microscopic studies.

Since embedding media are usually not soluble in water, tissue samples have to be dehydrated in series of solutions that are miscible with the embedding medium. The use of organic solvents can be deleterious to both ultrastructure and antigenicity of tissues. Alcohols, in particular ethanol, are the most popular dehydrating agents, but sometimes an intermediary (to facilitate resin infiltration) is necessary. The process of solvent exchange should be carried out at low temperatures to minimize disruption of protein molecular structure. Processing duration should be as short as possible to minimize extraction of cellular components. To avoid osmotic shock, dehydration has to be carried out in a series of solutions with ascending concentrations of the dehydrating agent. Acetone can also be applied to dehydrate small tissue pieces. It is less used than ethanol since it causes more protein extraction. Acetone may be advantageous in retaining lipid components and mixing with epoxy resins.

Dehydration protocols should be compatible with the embedding procedures. Resins for electron microscopy are subdivided into two major groups, the epoxys and the acrylics. Most of the resins also provide excellent semithin sections for light microscopy. The opposite is not true, since the resins used for light microscopy (methacrylates) lack stability under the electron beam and cannot provide sufficiently thin sections *(20,21)*.

As alternatives to methacrylate, epoxy resins have been successfully employed for ultrastructural investigation for many years. The most frequently used types for electron microscopic immunocytochemistry are Araldite, Epon, and Spurr. The composition of these epoxides is the same; they consist of a mixture of base resin, hardener, and accelerator. Flexibilizers and plasticizers may be added to facilitate sectioning. All these epoxy resins give uniform curing, and when cured they are resistant to the heat and static electricity generated by the electron beam. They are also hydrophobic; thus full tissue dehydration should be carried out before resin infiltration. Alcohols and epoxy resin do not mix; thus an intermediary compound, to assist resin infiltration, is necessary. Propylene oxide is satisfactory for this purpose. It is of note that propylene oxide can bind to the resin during polymerization; if too much of it remains in the tissue, it will change the characteristics of the resin, often making it too soft for sectioning. Although dehydration and infiltration procedures can be carried out at room temperature, the progressive lowering of temperature proposed by Kellenberger, et al. *(27)* can improve preservation of antigenicity. The high viscosity of most epoxy resins would necessitate prolonged room-temperature infiltration; thus the only practical way of curing epoxides to sectionable blocks is heat (60°C).

Although epoxy resins are the embedding medium of choice in conventional electron microscopy, some of their attributes (tough, hydrophobic character, need for heat polymerization, and their stabilizing effect on three-dimensional crosslinking between antigenic determinants and the epoxy groups) constitute major drawbacks in their application in postembedding immunocytochemistry. The resulting loss of sensitivity led to progressive substitution of epoxy with acrylic resins.

The only hydrophilic epoxy resin available presently is Durcupan. Because of its poor sectioning properties, this resin is used only as a dehydration medium, obviating the use of solvents. Unfortunately, this advantage is offset by a fixation effect of the resin, rendering the tissue unsuitable for immunocytochemistry.

Many problems associated with the three-dimensional bonding and hydrophobia of epoxides can be avoided by using acrylic resins. Their favorable properties led to the development of several acrylic resins such as LR white, LR gold, Bioacryl, and the Lowicryls K4M, HM20, K11M, and HM 23. Acrylic resins can be used following both full and partial dehydration protocols, since they are versatile, and some are more miscible with water or 70% organic solvent than others. Also, dehydration can be performed at room temperature, or at low temperature following progressive decrease of temperature or cryoprocedures. LR white is more convenient to handle at room temperature, whereas Lowicryls are more suitable for low-temperature protocols. In addition to their hydrophilic properties and versatility, these resins have several advantages. They are also hydrophilic when polymerized; thus the sections do not require etching to remove the resin before immunolabeling or other staining procedures. Most of them show no nonspecific binding to immunoreagents. Some have very low viscosity, allowing rapid infiltration into tissues; they remain fluid even at −80°C, making them invaluable for low- and very-low-temperature embedding methods.

All acrylics can be polymerized in a variety of ways: heat, chemical catalysis, UV light, or a cool energy source. Acrylic resins preserve antigenicity much better than the conventional epoxy resins. For instance, conventional embedding usually preserves antigenicity only in secretory granules, whereas acrylic resins can demonstrate a variety of substances in various cell organelles such as the rough endoplasmic reticulum, Golgi saccules, and secretory granules and on secretory vesicles.

The most important drawback of acrylic resins is the storage of blocks once polymerization is finished. In contrast to the stability of epoxy resin blocks, acrylic resin blocks can lose immunocytochemical reactivity with time. It seems likely that polymerization continues at a slow rate in blocks if left at ambient temperature, particularly in hot climates. The loss of immunoreactivity can be prevented by storing the blocks at temperatures below −20°C or in a deep freeze. If it is not possible to store blocks at low temperature, storage at room temperature can be conveniently accomplished by the removal of the gelatin capsules. Exposure of the resin blocks to the inhibitory action of atmospheric oxygen prevents further polymerization. Another storage problem is the hygroscopic character of Lowicryl blocks, requiring storage in a dessicator.

*Sectioning*

Sectioning of the tissue blocks should be carried out with care, avoiding any artifact that may interfere with the labeling process. Semithin sections, cut with glass or diamond knives, should be transferred to a droplet of water on a glass slide. It is easier if the slides are coated with chromo-albumin gelatin, which creates round water droplets with a high meniscus. Ultrathin sections should be mounted on nickel grids. The application of nickel or gold grids is strongly recommended for general use and is absolutely required when dealing with tissues that have been postfixed with osmium tetroxide. Indeed, oxidation of the copper grids during various incubations results in the formation and deposition of contaminants. Gold grids have no magnetic attraction for forceps like nickel grids, but they are expensive.

*Immunolabeling*

The precise immunolocalization of proteins requires the preservation of both antigenic and morphologic structures in specimens prepared for electron microscopy. Thus localization of antigen in postembedding immunocytochemical methods may depend on the handling of tissue specimen.

One important limiting factor in postembedding electron microscopy is the nature of the antibody used. Some antibodies are so potent that they can be employed on routinely fixed and embedded specimens. Others are sensitive and will not perform after any type of electron microscopic processing. Antigen concentration and preservation is another factor determining the success or failure of postembedding electron microscopic immunolabeling. Some antigens are concentrated within specific loci in cells, and they can be disclosed with low-affinity purified antibodies. The alterations in antigens during electron microscopic processing explain why some antibodies appear to be inappropriate for postembedding immunocytochemical electron microscopy, whereas they yield excellent results in Western blotting and on paraffin or frozen sections at the light microscopy level.

Indeed, polyclonal antibodies are usually preferred to monoclonal antibodies to perform postembedding immunoelectron microscopy. Polyclonal sera contain a diverse set of antibodies directed to a wide variety of determinants in a single antigen molecule. It is unlikely that all epitopes are modified to the same extent by the tissue processing protocol. On the other hand, monoclonal antibodies are directed toward a single, well-identified epitope in a macromolecule, which makes their use more restrictive. Any positive labeling obtained with a monoclonal antibody, on the other hand, carries a much higher specificity than the one obtained with polyclonal antibodies. However, being restricted to a single epitope, any slight conformational change of the antigen or in its steric hindrance will prevent its recognition by the antibody. The resolution of this problem is the application of a cocktail of various monoclonal antibodies directed toward the same antigen, which greatly improves the chances of obtaining specific labeling.

The procedures required to maximize postembedding labeling on tissue sections can be divided into two groups: those that prepare the sections for staining, collectively called the section pretreatment, and the staining procedures themselves.

RESIN SECTION PRETREATMENT

The requirement for section pretreatment depends on the extent of fixation and the type of resin used. The most commonly applied section pretreatments include resin removal, osmium removal, and, in case of the immunoperoxidase technique, inhibition of endogenous peroxidase. Unmasking of antigen with pepsin, protease, or trypsin is not recommended since they may be ineffective and can also damage sections.

*Resin Removal.* Removal of resin, or etching, is unnecessary if acrylic resin is used. However, it may be essential to improve immunolabeling and staining of epoxy resin semithin sections. Postembedding immunocytochemical staining of semithin sections combines the advantages of high morphologic resolution with the opportunity to stain adjacent sections with different antibodies. An additional application of this procedure is the use of semithin sections for immunocytochemical superimposition techniques *(28)*. This method involves ultrastructural examination of ultrathin sections adjacent to immunostained semithin sections. The bonds linking an epoxide to tissue can be broken, and the resin can be removed by etching. Potential etching agents should be capable of enhancing antigen-antibody interactions and facilitating infiltration of the immunologic reagents by means of partial to complete removal of plastic without damaging tissues and antigenic determinants. Saturated solutions of sodium hydroxide in absolute methanol or ethanol (sodium methoxide or ethoxide) have been widely used for this purpose *(29,30)*.

Other etching agents *(31,32)* providing excellent results for immunocytochemistry on semithin sections have also been described. The extent of etching required is variable from epoxide to epoxide, and even from batch to batch of the same epoxide. The required time for etching for optimal antigen preservation has to be empirically determined. If possible, the resin should only be partially removed by brief (1–5 min) exposure of the section to the etching agent. The etching solution has to be washed away thoroughly with several changes of ethanol/methanol and distilled water.

The procedures described above are not suitable for ultrathin sections. Ultrathin epoxy sections are usually etched with a weak oxidizing agent (hydrogen peroxide, sodium metaperiodate) to remove osmium from osmicated embedded tissue. Although these oxidants are not true etching agents, since they do not remove the plastic, it may well be that weak oxidation breaks surface ester bonds in the resin, thereby inducing superficial hydrophilia.

*Osmium Removal.* As has been alluded to, oxidizing agents are capable of removing osmium from tissue sections, thereby unmasking some antigens. Hydrogen peroxide is used most frequently for semithin sections, whereas saturated aqueous sodium metaperiodate (60 min at room temperature) has been recommended for removing osmium from ultrathin sections. The latter procedure is considered superior for unmasking osmium-sensitive tissue antigens. The use of both hydrogen peroxide and sodium metaperiodate can damage the ultrastructure and contrast of tissue sections; thus they should be used with precaution *(23,24)*. Following osmium removal, sections must be washed thoroughly in several changes of distilled water and then PBS or Tris-buffered saline (TBS). It is important to note that acrylic sections cannot withstand this mode of osmium removal.

*Inhibition of Endogenous Peroxidase.* Blocking of endogenous peroxidase activity is a mandatory step in immunoperoxidase techniques for semithin or ultrathin sections. Since in epoxy-embedded tissues high concentrations of hydrogen peroxide (3–10%) may be harmful to both structure and antigen preservation, a gentler and more specific method employing 0.005% hydrogen peroxide in methanol can be employed to suppress peroxidase activity *(33)*. However, this method can not be used on acrylic sections, which are soluble in methanol. For such sections, gentle methods also suitable for frozen sections or whole-cells preparations can be applied. Several chemicals can be used, such as phenylhydrazine, sodium-azide, or cyclopropanone *(34)*.

RESIN SECTION IMMUNOLABELING

Immunoperoxidase and immunogold postembedding methods are the procedures most frequently used to localize peptides, proteins, and glycoproteins on fixed semithin or ultrathin plastic sections. The successful application of both protocols depends in part on correct specific blocking. The incorrect or gratuitous use of this step can actually contribute to high background, causing serious reduction or even complete abolition of immunolabeling. The problem of background is especially important in immunogold methods since it occurs in every case, but its level has to be kept as low as possible. The nonspecific absorption of the gold probe by the tissue sections could be due to surface charges related to fixation and/or stickiness of the resin or nonspecific adhesion of the antibody molecules.

Specific blockings are of two types. In inhibitive blocking (preblocking), the agent is applied on the section prior to incubation with the primary or secondary antibodies. Competitive blocking takes place when reagents containing the primary or secondary reagents are diluted with the blocking substance. Different blocking solutions have been used such as ovoalbumin, bovine serum albumin (when anti-goat, anti-sheep, or anti-bovine secondary antibodies or gold conjugate are not used), fish cold water gelatin, and skim milk. Because of their wetting capacity, in some cases Triton and Tween-20 may also improve results. Normal serum (goat, rabbit, horse, and so on) is the most common blocking agent in immunoperoxidase protocols. However, it should be avoided in immunogold techniques, particularly when protein A, G-gold complexes are used, since IgG molecules present in normal serum will adhere to the section and be visualized by the probe, causing increased background levels.

Because of their high sensitivity, the immunoperoxidase techniques, the PAP technique, and the ABC technique are widely applied in light microscopic immunocytochemistry using DAB as the chromogen *(35,36)*. These multistep methods are also used for resin section immunolabeling, mainly on semithin sections. These procedures do not significantly differ from those used on paraffin or frozen sections at the light microscopic level and preembedding immunocytochemistry protocols (previously described). According to the principles of the indirect immunocytochemical approach, the labeling protocol in both techniques involves three main incubation steps. The primary antibody is localized using a secondary antibody linked to peroxidase or biotin. Subsequently, in the third step, complexes of antiperoxidase antibody with peroxidase enzyme (PAP) or peroxidase-conjugated avidin (ABC) are used to localize the secondary antibody linked to peroxidase or biotin, respectively. After DAB detection of the reaction product, resin semithin sections are usually treated with osmium tetroxide to enhance the color of the precipitated DAB and obtain better localization. Sodium gold chloride can also be used for the same purpose, providing even more accurate localization. Although both immunoperoxidase techniques have been successfully applied to etched epoxy resin semithin sections, the ABC technique is not recommended for acrylic sections because the very large complex does not penetrate well, leading to less sensitive, uneven surface staining.

Immunoperoxidase protocols have also been used for ultrastructural immunocyto-chemistry, although they carry several limitations in terms of resolution, diffusion artifacts, and quantitative evaluation of labeling. At present, for immunoelectron micros-copy, the enzyme-based immunocytochemical method with DAB as the chromogen has been largely substituted by a more refined protocol using a complex of antibodies or different macromolecules and colloidal particles.

Colloidal gold has a long history as a staining reagent unrelated to microscopy. The first documented application took place in ancient China, and the alchemists of the seventeenth century made use of it as well. Faraday (1791–1867) was the first to perform a scientific study of its properties *(37)*. The first application of colloidal gold as a specific transmission electron microscopic marker for direct preembedding immunolabeling of cell surface antigens, described by Faulk and Taylor, dates back to 1971 *(38)*. The technique was later applied for preembedding indirect labeling of cell surface antigens and for the postembedding labeling of intracellular antigens on tissue sections. Due to its advantages and versatility, the colloidal gold marker found many intriguing applications for transmission electron microscopy, scanning electron microscopy, and light microscopy.

Gold particles can be labeled with a variety of macromolecules (e.g., lectins, glycopro-teins, proteins such as protein A, immunoglobulins), which subsequently maintain their bioactivities *(39)*. The various preparations of monodisperse colloidal gold particles, ranging in size from 5–150 nm, make them eminently suitable for multiple labeling. The gold method is also easily amenable to quantification.

In spite of all these advantages, colloidal gold immunocytochemical techniques carry their own limitations in terms of labeling efficiency. Recently these techniques have been the subject of several modifications to improve resolution and to render the approach more versatile. They have been introduced into various areas of molecular biology and pathology, providing novel information contributing to a better understand-ing of both normal and neoplastic cell functioning. One of the first contributions of immunogold electron microscopy was the demonstration that selective secretion of a hormone can be associated with structural characteristics of the cell type and/or size and shape of its secretory granules. Accordingly, in human and rodent pituitaries, different subtypes of growth hormone (GH) and prolactin (PRL) cells have been described. Immunogold electron microscopy also played a important role in revealing the distribution and exact subcellular localization of different hormones, peptides, and other substances related to the functional activity of the cell. Such findings disclosed important functional implications related to regulation of hormone synthesis and release. The high electron density of gold particles and the employment of gold particles of different diameters made colloidal gold labeling the preferred technique when simultane-ous ultrastructural detection of several antigens is required.

Several double-staining immunocytochemical methods, based on the use of gold particles of different sizes, have been recommended for electron microscopic study. The emergence of the technique for multiple staining is especially valuable in the study of normal and abnormal endocrine organs. For instance, immunoelectron microscopic studies provided conclusive evidence that endocrine cells are frequently plurihormonal, producing more than one hormone; plurihormonal cells may even store more than one different product within the same secretory granules *(40–42)*.

Optimal immunogold labeling depends in part on the choice of colloidal gold probe used. There is no general agreement on whether the direct or indirect immunogold methods give more reliable results. Direct labeling requires large amounts of highly purified specific immunoglobulins, which are usually difficult to obtain, but they provide a high level of resolution. On the other hand, application of the indirect technique with an IgG-gold or protein-gold complex is easier and more versatile, since these probes can be used with a large variety of primary antibodies. The indirect approach also provides greater intensity of labeling.

In this regard, different procedures have been used for amplification of the signal when the labeling is limited to a few gold particles. This may be due to low levels of antigen in the tissue or to loss of antigenicity during tissue processing. Most of the amplification techniques are based on multistep reactions using different proteins or antibody-gold complex. One of the first approaches, introduced by Bendayan and Duhr *(43)*, is based on multistep reactions with an anti-protein A antibody and protein A complexes. Subsequently, other amplification protocols such as the biotin-antibiotin gold technique or the biotin-streptavidin gold technique were successfully introduced. The amplification factor in these techniques varies according to the dilution of the primary antibody, but it is not less than five times the original signal. Recently Mayer and Bendayan *(44)* adapted the catalyzed reporter deposition approach (CARD) for electron microscopy. This technique, originally applied in cell biology for membrane assays, Western blotting, *in situ* hybridization, and light microscopic immunocytochemistry, achieves in the majority of studies greater sensitivity, with reduced background and enhancement of antigen detection of 8–10,000-fold *(45)*. The procedure is based on the ability of horseradish peroxidase to catalyze the deposition of substrate reporter molecules by a short half-life free radical mechanism. It permitted the direct accumulation of colloidal gold or Nanogold particles on target sites by enzymatic activity of horseradish peroxidase, thus improving the efficiency of the immunogold labeling compared with other amplification protocols. Moreover, the introduction of Nanogold particles allowed a combination of the amplification potential of CARD with the ultrasmall gold probe and the silver intensification method. Bendayan and co-workers *(46)* reported that the immunolabeling obtained by the latter amplification strategy appeared specific and highly sensitive, with enhanced intensity.

The size of the gold label has to be carefully selected *(47)*. Smaller sizes provide higher resolution and more intense labeling, but the small gold particles are difficult to visualize at low magnification, and their use requires examination of small areas at very high magnifications. The use of small gold particles also yields higher background labeling because the exposed surface areas of gold particles initiate negative charges, which interact with cationic structures on the tissue section.

A few specific gold complexes have been developed for studies of the extracellular matrix: collagenase-gold complexes, elastase-gold complexes, and amylase–gold complexes have been applied to localize collagen, elastin, and glycogen, respectively.

Immunogold protocols have been widely used for double-labeling immunocytochemistry for the demonstration of different antigens at the electron microscopic level. Several techniques and protocols are available. The immunogold protocol at the end of the chapter was designed to localize mamosomatotrophs (bihormonal pituitary cells containing GH and PRL in their cytoplasm) in a patient with acromegaly and hyperpro-lactinemia.

*Quantitative Evaluation*

Quantitative evaluation became an important part of modern immunocytochemical studies, allowing better assessment of the results as well as comparing levels of labeling in different cellular compartments or between control and experimental groups. In comparative studies, all conditions, such as steps of tissue preparation and labeling protocols, should be identical. It is of note that no correlation can be made between levels of labeling for two different antigens since they may react differently to conditions of tissue processing. In addition, the two different antibodies utilized may display different affinities and different titers.

Compared with other electron microscopic methods, postembedding immunogold protocols have a great advantage in quantitative evaluation of the immunolabeling *(47,49,50)*. When compared with preembedding methods, accessibility of binding sites, diffusion of marker, or thickness of sections are of no concern in postembedding gold techniques. In the latter, only the binding sites at the surface of the section will be revealed, and when gold particles of uniform size are used for labeling, the binding capacity of each particle is assumed to be the same. Although surface labeling may represent a limitation of the technique, it becomes a major advantage for quantitative evaluation. Indeed, regardless of their localization within the tissue, cell, or subcellular compartment, all antigenic sites exposed by sectioning have the same probability to be labeled, because neither the thickness of the section or the diffusion of markers interferes with the labeling.

Since the exact number of molecules per gold particles cannot be determined, absolute amounts of antigen cannot be measured by immunoelectron microscopy. However, information regarding the antigen can still be obtained by relating the quantity of gold particles to structural parameters. Densities of labeling can easily and accurately be related to surface areas of structures and expressed in number of gold particles per unit surface. Thus, to evaluate labeling densities in tissues, cells, or a particular cellular compartment, the surface ($S_a$) of the structure studied should be estimated using an adequate morphometric approach. Upon determination of the area, the number of gold particles ($N_i$) present over this area should be counted and the labeling density ($N_s$) calculated accordingly: $N_s = N_i/S_a$ (number of gold particles per surface unit). Labeling densities can also be evaluated per length of membrane or any linear structure and even in relation to volume using complex stereologic equation.

Quantitative evaluation of labeling densities also has to take into account the nonspecific adherence of gold label, commonly referred to as background. It should be kept as low and even as possible. In the case of high and variable background, controls for the specificity of gold labeling have to be performed before any quantitative study. Another factor to be considered is that gold particles are not simply points, but that they have a certain size (5–30 nm). If only gold particles lying exclusively within the confines of the subject of study were counted, the labeling density could be underestimated if particles were lying over the boundaries of the structure. Increasing the gold size increases the error because large profiles have less chance of lying completely within the compartment studied. To minimize this problem, different unbiased counting rules have been introduced.

## Cryotechniques

Fixatives affect tissues by chemical crosslinking of cellular components, which can reduce or destroy antigenic reactivity as well as by altering membrane permeability and inducing ionic leakage. These changes can led to rearrangements in cellular structure and subcellular organization. Rapid freezing techniques and cryofixation avoid chemical prefixation and offer the opportunity to preserve both structures and antigenic sites that are sensitive even to low concentrations of fixative *(20,51)*.

### Cryofixation

In cryofixation, tissue is not fixed by chemical crosslinking but by physical immunobilization in ice. The advantage of rapid freezing for preserving both structure and antigenic reactivity is offset by the size of the samples that can be frozen successfully. Tissue specimens have to be much smaller and thinner than those required for chemical fixation. This limitation, as well as the difficulties found in cutting ultrathin frozen sections and mounting them on grids, made this technique impractical for use in diagnostic pathology. Currently it is used only for research.

Fernandez-Moran *(52)* was the first to apply cryofixation techniques for electron microscopy. The principle of the technique is rapid freezing in order to immunobilize the cellular structures physically. The temperatures as well as the speed of the procedure are major factors. The objective is to freeze the water present in the specimen in an amorphous state, avoiding crystallization, which would disrupt the structure and introduce major artifacts. The formation of vitreous ice depends on the rapidity of the thermal exchange between the cryogen and the tissue. Several methods developed to perform an optimal cryofixation are listed below.

#### PLUNGING

There are many descriptions of methods for cryofixation by plunging. Liquid nitrogen itself is a very good quenchant liquid, although it is used chiefly to cool and maintain the temperature of more efficient coolant liquids such as the hydrocarbons ethane and propane and the halocarbons, of which freon 22 is the most efficient. Plunging methods are most appropriate for small blocks of tissue (<1 mm thick) in which the cells of interest are not on an accessible surface.

#### SPRAYING

This technique is applicable only to specimens in the form of emulsions, suspensions, or solutions. For these types of specimens the method is advantageous since microdroplets are formed and cooled extremely rapidly when sprayed into propane or ethane.

#### JETTING

This technique is a variant of the plunging technique, although the thermodynamic efficiency of the method can exceed that of plunging. The technique is only applicable to small specimens.

#### HYPERBARIC CRYOFIXATION

This technique depends on the high degree of subcooling that water undergoes when subjected to a pressure of 2000–2100 bar. The advantage of this variant is that relatively large tissue samples can be cryofixed free of visible ice crystals. Its drawback is the extremely high pressure to which tissues are subjected, inducing changes in the structure and composition of tissues.

## METAL MIRROR FREEZING

This is the technique most frequently used for tissue samples. The mirror metal block is usually made of pure copper, which is precooled by liquid nitrogen (–195°C) or liquid helium (–260°C) under vacuum. By breaking the vaccum, the specimen is projected onto the copper block, inducing an immediate freezing of the sample. However, in this procedure only the face of the specimen making contact with the metal block is frozen rapidly enough to permit the vitrification of tissue water. In this 5–40-μm layer of the sample, ultrastructural preservation is good. Deeper in the specimen, ice crystals are formed, increasing progressively with the thickness of the tissue, leading to artifacts such as breakage of membranes and alterations of the nuclear chromatin pattern.

Following cryofixation, the tissue can be processed using various routes as listed below.

### Cryosubstitution

In this process the ice in the tissue is replaced at very low temperature by an organic solvent miscible with the embedding medium. Infiltration and embedding follows at any temperature between 60°C and –80°C. Although Lowicryls are the most frequently used resins, embedding in epoxides is also feasible. Cryosubstitution is favored when dealing with fixation/dehydration sensitive structures.

### Freeze-Drying

In this process the ice sublimates, leaving the tissue completely dry. Subsequently it can be infiltrated with resin; embedding may take place again at any temperature from 60°C to –80°C. This technique may be applied to tissue that is subsequently embedded in epoxide resins. Lowicryls can also be used; however, they are not as favorable for immunocytochemistry as chemical fixation and dehydration in the progressive lowering of temperature technique *(27)*.

Freeze-drying avoids the use of organic solvents, although collapse or aggregation of tissue substances can occur during drying procedures.

Alternatively, tissues following cryofixation can be cryosectioned using cryoultramicrotomy.

### Cryoultramicrotomy

Cryoultramicrotomy is a rapid technique providing blocks for sectioning at 30–60 min following fixation. Cryoultramicrotomy was first introduced for immunocytochemistry by Tokuyasu *(53)*. His method utilizes and relies on prefixed tissue infiltrated with sucrose, frozen, and sectioned at low temperatures. The sections are allowed to thaw out before immunolabeling. The advantage of the technique is that neither dehydration in organic solvents nor plastic embedding is required. Thus both antigens and membranes are optimally preserved. In addition, the epitopes that are otherwise masked by embedding remain accessible to the probes. A disadvantage of the method is the fragility of cryosections, possibly leading to structural distortion and material extraction during the handling process.

Since its introduction in the 1950s *(54)*, ultramicrotomy posed technical challenges *(55)*. At one time the wet sectioning method, similar to that used in conventional plastic ultramicrotomy, was adopted. In this method, the knife trough is filled with a flotation liquid, dimethylsulfoxide, which has a high surface tension yet a low melting point. The fluid facilitates the spreading of the sections before placing them on grids or small plastic rings. The dimethylsulfoxide was replaced by the sucrose drop pick-up method of Tokuyasu *(53)*. The technique is challenging, requiring considerable skills, but it provides high levels of immunolabeling. However, ultrastructure of the specimens is often disappointing, especially when compared with that seen after cryosubstitution. Improvements of this technique are being actively sought to achieve better ultrastructural preservation, while maintaining the level of immunolabeling.

Although it is expensive and a difficult technique to master, cryosectioning yields excellent immunohistochemical results. Some investigators assume that cryosections are permeable to antibody probes, accounting for a high level of immunoreactivity, as opposed to resin sections, on which only surface labeling takes places. Although Stierhof et al. *(56)* provided evidence that antibody binding occurs exclusively on the surface of cryosections as well, we believe that the reasons for the differences in immunolabeling are still debatable.

One of the most exciting recent applications of cryosections is the possibility of correlating light and electron microscopy findings using the fluorescent ultrasmall immunogold probe Fluoro-Nanogold *(57,58)*. Fluoro-Nanogold has the properties of both a fluorescent dye-conjugated antibody for fluorescence microscopy and a gold particle-conjugated antibody for electron microscopy. Therefore, this bifunctional immunoprobe permits correlative fluorescent microscopic and electron microscopic observations of the same cell structures labeled in a single-labeling procedure. Thus, in these techniques there is a one-to-one relationship between fluorescent structures labeled with Fluoro-Nanogold and the organelle profiles labeled with the same silver-enhanced Fluoro-Nanogold in ultrathin cryosections. The methodology carries its own caveats: 1) the quality of ultrathin cryosections for immunocytochemistry is crucial, 2) the use of nickel grids allows for the localization of a given cell/structure by both the fluorescence and electron microscope, 3) anti-photobleaching reagents used as the mounting medium for fluorescence microscopy should be chosen carefully, and 4) the time-course for the silver enhancement reaction is also essential to achieve optimal particle enhancement while minimizing the background signal.

## CRITERIA FOR EVALUATING THE RESULTS

### *Control for Immunogold Labeling*

As in other immunocytochemical techniques, immunogold procedures require controls to establish specificity.

Stirling *(59)* defined the main control methods that should be performed each time a new antibody or new tissue is used to confirm epitope specificity and exclude false-negative results. They can be summarized as follows.

*Omission of the Primary Antibody*

The primary antibody is omitted from the test protocol, and the sections are incubated with the antibody diluent only, followed by the gold probe. This identifies nonspecific binding of the gold probe.

*Substitution of Primary Antibody*

This test should be included to show that the immunocytochemical reaction is genuine. In this control the primary antibody under investigation is substituted by a primary antiserum of the same animal, class, and concentration but that has no expected immunocytochemical affinity with the tissue. This control should replicate the test conditions as closely as possible.

*Absorption of the Primary Antibody with Target Antigen*

The primary antibody is absorbed with purified target antigen, and the resulting solution is used as a substitute for the primary antibody. This negative control is intended to detect contaminating antibodies that might bind to epitopes other than the target, thus indicating specificity of reaction. When the target antigen is not available, the specificity of the immunocytochemical reaction can also be verified by immunobolt analysis. For the immunobolt analysis, protein homogenates of the test tissue and of a positive control undergo gel electrophoresis separately and are transferred to nitrocellulose papers. Subsequently immunobolt must detect bands with similar molecular masses in both samples.

*Removal of Antigen*

In certain situations it is possible to remove the epitope of interest from the tissue by enzymatic digestion or other means. This provides a negative control, which helps to confirm antibody specificity.

*Positive Controls*

A known positive is tested by the same technique to avoid the possibility of a false negative.

## ACKNOWLEDGMENTS

This work was supported in part by the generous donation of Mr. and Mrs. Jarislowski and the Lloyd Carr-Harris Foundation. Dr. Sergio Vidal was supported by a research grant from the Conselleria de Educacion (Xunta de Galicia), Spain. The authors are indebted to Ms. Sandy Briggs, Ms. Sandy Cohen, and Mr. Mark Moreland for technical assistance and the staff of St. Michael's Health Sciences Library for their contribution to the study.

## REFERENCES

1. Tanner, V. A., Ploug, T., and Tao-Cheng, J. H. (1996) Subcellular localization of SV2 and other secretory vesicle components in PC12 cells by an efficient method of preembedding EM immunocytochemistry for cell cultures. *J. Histochem. Cytochem.* **44**, 1481–1488.
2. Hayat, M. A. (ed.) *Principles and Techniques of Electron Microscopy.* Biological applications, 4th ed. Cambridge University Press, Cambridge, United Kingdom, 2000.

3. Moriarty, G. C. (1973) Adenohypophysis: ultrastructural cytochemistry. A review. *J. Histochem. Cytochem.* **21**, 855–894.

4. De Grandi, P. B., Kraehenbuhl, J. P., and Campiche, M. A. (1971) Ultrastructural localization of calcitonin in the parafollicular cells of pig thyroid gland with cytochrome c-labeled antibody fragments. *J. Cell Biol.* **50**, 446–456.

5. Farquhar, M. G., Reid, J. J., and Daniell, L. W. (1978) Intracellular transport and packaging of prolactin: a quantitative electron microscope autoradiographic study of mammotrophs dissociated from rat pituitaries. *Endocrinology.* **102**, 296–311.

6. Nakane, P. K. (1976) Application of enzyme-labeled antibody methods for the ultrastructural localization of hormones, in *Recent Progress in Electron Microscopy of Cells and Tissues* (Yamada, E., Mizuhira, V., Kurosumi, K., and Nagano, T., eds), University Park Press, Baltimore, pp. 189–200.

7. Weber, K., Rathke, P. C., and Osborn, M. (1978) Cytoplasmic microtubular images in glutaraldehyde-fixed tissue culture cells by electron microscopy and by immunofluorescence microscopy. *Proc. Natl. Acad. Sci. USA* **75**, 1820–1824.

8. Llewellyn-Smith, I. J., Costa, M., and Furness, J. B. (1985) Light and electron microscopic immunocytochemistry of the same nerves from whole mount preparations. *J. Histochem. Cytochem.* **33**, 857–866.

9. McLean, I. W. and Nakane, P. K. (1974) Periodate-lysine paraformaldehyde fixative. A new fixation for immunoelectron microscopy. *J. Histochem. Cytochem.* **22**, 1077–1083.

10. Hedman, K. (1980) Intracellular localization of fibronectin using immunoperoxidase cytochemistry in light and electron microscopy. *J. Histochem. Cytochem.* **28**, 1233–1241.

11. Helenius, A. and Simons, K. (1975) Solubilization of membranes by detergents. *Biochim. Biophys. Acta* **415**, 29–79.

12. Willingham, M. C., Yamada, S. S., and Pastan, I. (1978) Ultrastructural antibody localization of alpha2-macroglobulin in membrane-limited vesicles in cultured cells. *Proc. Natl. Acad. Sci. USA.* **75**, 4359–4363.

13. Eldred, W. D., Zucker, C., Karten, H. J., and Yazulla, S. (1983) Comparison of fixation and penetration enhancement techniques for use in ultrastructural immunocytochemistry. *J. Histochem. Cytochem.* **31**, 285–292.

14. Llewellyn-Smith, I. J. and Minson, J. B. (1992) Complete penetration of antibodies into vibratome sections after glutaraldehyde fixation and ethanol treatment: light and electron microscopy for neuropeptides. *J. Histochem. Cytochem.* **40**, 1741–1749.

15. Wouterlood, F. G., Boon, M. E., and Kok, L. P. (1990) Immunocytochemistry on free-floating sections of rat brain using microwave irradiation during the incubation in the primary antiserum: light and electron microscopy. *J. Neurosci. Methods* **35**, 133–145.

16. Priestley, J. V. (1984) Pre-embedding ultrastructural immunocytochemistry: immunoenzyme techniques, in *Immunolabelling for Electron Microscopy* (Polak, J. M. and Varndell, I. M., eds.), Elsevier, Amsterdam, pp. 37–52.

17. Chan, J., Aoki, C., and Pickel, V. M. (1990) Optimization of differential immunogold-silver and peroxidase labeling with maintenance of ultrastructure in brain sections before plastic embedding. *J. Neurosci. Methods* **33**, 113–127.

18. Danscher, G. A. (1983) Silver method for counterstaining plastic embedded tissue. *Stain Tech.* **58**, 365–372.

19. Sawada, H. and Esaki, M. (2000) A practical technique to postfix nanogold-immunolabeled specimens with osmium and to embed them in epon for electron microscopy. *J. Histochem. Cytochem.* **48**, 493–498.

20. Newman, G. R. and Hobot, J. A. (eds.) (1990) *Resin Microscopy and On-Section Immunocytochemistry.* Springer-Verlag, Berlin.

21. Newman, G. R. and Hobot, J. A. (1999) Resins for combined light and electron microscopy: a half century of development. *Histochem. J.* **31**, 495–505.

22. Hayat, M. A. (1986) Glutaraldehyde: role in electron microscopy. *Micron Microsc. Acta* **17**, 115–135.

23. Baskin, D. G., Erlandsen, S. L., and Parsons, J. A. (1979) Influence of hydrogen peroxidase or alcoholic sodium hydroxide on the immunocytochemical detection of growth hormone and prolactin after osmium fixation. *J. Histochem. Cytochem.* **27**, 1290–1292.

24. Bendayan, M. and Zollinger, M. (1983) Ultrastructural localization of antigenic sites on osmium-fixed tissues applying the protein A-gold technique. *J. Histochem. Cytochem.* **31**, 101–109.

25. Vidal, S., Oliveira, M. C., Kovacs, K., Scheithauer, B. W., and Lloyd, R. V. (2000) Immunolocalization of vascular endothelial growth factor (VEGF) in the GH3 cell line. *Cell Tissue Res.* **300**, 83–88.

26. Newman, J. B., Borysko, E., and Swerdlow, M. (1949) New sectioning techniques for light and electron microscopy. *Science* **110**, 66–68.

27. Kellenberger, E., Carlemalm, E., Villiger, W., Roth, J., and Garavito, R. M. (eds.) (1980) *Low Denaturation Embedding for Electron Microscopy of Thin Sections.* Chemische Werke Löwi Gmbh, Waldkraiburg, Germany.

28. Beauvillain, J. C., Tramu, G., and Dubois, M. P. (1975) Characterization by different techniques of adrenocorticotropin and gonadotropin producing cells in lerot pituitary (Eliomys quercinus). A superimposition technique and an immunocytochemical technique. *Cell Tissue Res.* **158**, 301–317.

29. Lane, B. P., and Europa, D. L. (1965) Differential staining of ultrathin sections of epon-embedded tissues for light microscopy. *J. Histochem. Cytochem.* **13**, 579–582.

30. Erlandsen, S. L., Parsons, J. A., and Rodning, C. B. (1979) Technical parameters of immunostaining of osmicated tissue in epoxy sections. *J. Histochem. Cytochem.* **27**, 1286–1289.

31. Romijn, H. J., Janszen, A. W., Pool, C. W., and Buijs, R. M. (1993) An improved immunocytochemical staining method for large semi-thin plastic epon sections: application to GABA in rat cerebral cortex. *J. Histochem. Cytochem.* **41**, 1259–1265.

32. Vidal, S., Lombardero, M., Sanchez, P., Roman, A., and Moya, L. (1995) An easy method for the removal of epon resin from semi-thin sections. Application of the avidin-biotin technique. *Histochem. J.* **27**, 204–209.

33. Hittmair, A. and Schmid, K. W. (1989) Inhibition of endogenous peroxidase for the immunocytochemical demonstration of intermediate filament proteins (IFP). *J. Immunol. Methods* **116**, 199–205.

34. Andrew, S. M. and Jasani, B. (1987) An improved method for the inhibition of endogenous peroxidase non-deleterious to lymphocyte surface markers. Application to immunoperoxidase studies on eosinophil-rich tissue preparations. *Histochem. J.* **19**, 426–430.

35. Sternberger, L. A., Hardy, P. H., Cuculis, J. J., and Meyer, H. G. (1970) The unlabeled antibody method of immunocytochemistry. Preparation and properties of soluble antigen-antibody complex (horseradish peroxidase-anti-horseradish peroxidase) and its use in the identification of spirochetes. *J. Histochem. Cytochem.* **18**, 315–333.

36. Hsu, S. M., Raine, L., and Fanger, H. (1981) Use of avidin-biotin-peroxidase complex (ABC) in immunoperoxidase techniques: a comparison between ABC and unlabeled antibody (PAP) procedures. *J. Histochem. Cytochem.* **29**, 577–580.

37. Roth, J. (1996) The silver anniversary of gold: 25 years of the colloidal gold marker system for immunocytochemistry and histochemistry. *Histochem. Cell Biol.* **106**, 1–8.

38. Faulk, W. P., and Taylor, G. M. (1971) An immunocolloid method for the electron microscope. *Immunochemistry* **8**, 1081–1083.

39. Roth, J. (1983) The colloidal gold marker system for light and electron microscopic cytochemistry, in *Techniques in Immunocytochemistry*, vol. 2 (Bullock, G. R. and Petrusz, P., eds.), Academic, San Diego, pp. 217–284.

40. Vila-Porcile, E. and Corvol, P. (1998) Angiotensinogen, prorenin, and renin are colocalized in the secretory granules of all glandular cells of the rat anterior pituitary: an immunoultrastructural study. *J. Histochem. Cytochem.* **46**, 301–311.
41. Vidal, S., Syro, L., Horvath, E., Uribe, H., and Kovacs, K. (1999) Ultrastructural and immunoelectron microscopic study of three unusual plurihormonal pituitary adenomas. *Ultrastruct. Pathol.* **23**, 141–148.
42. Vidal, S., Cohen, S. M., Horvath, E., et al. (2000) Subcellular localization of leptin in non-tumorous and adenomatous human pituitaries: an immuno-ultrastructural study. *J. Histochem. Cytochem.* **48**, 1147–1152.
43. Bendayan, M. and Duhr, M. A. (1986) Modification of the protein A-gold immunocytochemical technique for the enhancement of its efficiency. *J. Histochem. Cytochem.* **34**, 569–575.
44. Mayer, G. and Bendayan, M. (1999) Immunogold signal amplification: application of the CARD approach to electron microscopy. *J. Histochem. Cytochem.* **47**, 421–430.
45. Bobrow, M. N., Harris, T. D., Shaughnessy, K. J., and Litt, G. J. (1989) Catalyzed reporter deposition, a novel method of signal amplification. Application to immunoassays. *J. Immunol. Methods* **125**, 279–285.
46. Mayer, G., Leone, R. D., Hainfeld, J. F., and Bendayan, M. (2000) Introduction of a novel HRP substrate-Nanogold probe for signal amplification in immunocytochemistry. *J. Histochem. Cytochem.* **48**, 461–470.
47. Bendayan, M. (1995) Colloidal gold post-embedding immunocytochemistry. *Prog. Histochem. Cytochem.* **29**, 1–163.
48. Bendayan, M. (1982) Double immunocytochemical labeling applying the protein A-gold technique. *J. Histochem. Cytochem.* **30**, 81–85.
49. Slot, J. W., Posthuma, G., Chang, L. Y., Crapo, J. D., and Geuze, H. J. (1989) Quantitative aspects of immunogold labeling in embedded and in nonembedded sections. *Am. J. Anat.* **185**, 271–281.
50. Lucocq, J. (1994) Quantitation of gold labelling and antigens in immunolabelled ultrathin sections. *J. Anat.* **184**, 1–13.
51. Elder, H. Y. (1989) Cryofixation, in *Techniques in Immunocytochemistry*, vol. 4 (Bullock, G. R. and Petrusz, P., eds.), Academic, San Diego, pp. 217–284.
52. Fernandez-Moran, H. (1960) Low-temperature preparation techniques for electron microscopy of biological specimens based on rapid freezing with liquid helium II. *Ann. NY Acad. Sci.* **85**, 689–713.
53. Tokuyasu, K. T. (1973) A technique for ultracryotomy of cell suspensions and tissues. *J. Cell Biol.* **57**, 551–565.
54. Fernandez-Moran, H. (1952) Application of the ultrathin freezing-sectioning technique to the study of cell structures with the electron microscope. *Ark. Fys.* **4**, 471–491.
55. Liou, W., Geuze, H. J., and Slot, J. W. (1996) Improving structural integrity of cryosections for immunogold labeling. *Histochem. Cell Biol.* **106**, 41–58.
56. Stierhof, Y. D., Schwarz, H., Durrenberg, M., Villiger, W., and Kellenberger, E. (1991) Yield of immunolabel compared to resin sections and thawed cryosections, in *Colloidal Gold: Principles, Methods and Applications*, vol. 3 (Hayat, M. A., ed.), Academic, San Diego, pp. 87–115.
57. Takizawa, T., and Robinson, J. M. (2000) FluoroNanogold is a bifunctional immunoprobe for correlative fluorescence and electron microscopy. *J. Histochem. Cytochem.* **48**, 481–485.
58. Hainfeld, J. F., and Powell, R. D. (2000) New frontiers in gold labeling. *J. Histochem. Cytochem.* **48**, 471–480.
59. Stirling, J. W. (1993) Controls for immunogold labeling. *J. Histochem. Cytochem.* **41**, 1869–1870.

# PROTOCOLS

## ULTRASTRUCTURAL IMMUNOHISTOCHEMISTRY

### PREEMBEDDING IMMUNOCYTOCHEMISTRY

The following protocol has been extensively used in the central nervous system to localize neurons expressing specific antigens exemplified here by demonstration of tyrosine hydroxylase within dopaminergic neurons in the caudate nucleus of the medulla oblongata of the rat.

### *Tissue Preparation and Fixation*

Tissues are collected from adult female Sprague-Dawley rats of 200–250 g body weight. Under pentobarbital anesthesia, the animals are perfused through the left ventricle of the heart, first with 50 mL of 0.9% saline as a vascular rinse and subsequently with 500 mL of fixative solution containing 4% paraformaldehyde and 0.1% glutaraldehyde in 0.1 $M$ phosphate buffer (PB) at pH 7.4. The brains are removed and postfixed for an additional 2 h in the same fixative solution at 4°C.

### *Penetration Enhancement*

After several washes in 0.1 $M$ PB, the medulla oblongata of the rat brain is serially sliced with a Vibratome, and free-floating sections of 100 μm are brought up to room temperature in PB. Subsequently the sections are washed with 50% ethanol in distilled water for 30 min to improve antibody penetration.

### *Immunocytochemical and Electron Microscopic Processing*

Free-floating sections are processed according to the avidin-biotin peroxidase complex (ABC) procedure, omitting Triton X-100. Processing for immunocytochemistry is carried out as follows.

1. Sections are sequentially incubated in:
   a. 3% bovine serum albumin (BSA) in phosphate-buffered saline (PBS) to minimize nonspecific staining.
   b. Primary antiserum, monoclonal tyrosine hydroxylase diluted 1:1000 in PBS, at room temperature overnight.
   c. Biotinylated anti-mouse IgG diluted 1:100 in PBS, at room temperature for 1 h.
   d. ABC solution for 90 min.
2. Sections are washed at least three times for 10 min in PBS after each incubation.
3. Peroxidase activity is demonstrated with 0.06% DAB in the presence of 0.003% hydrogen peroxide in PBS.
4. Sections are then washed in PBS for 5 min and in PB for a further 5 min.
5. Sections are then postfixed in 1% osmium tetroxide in 0.1 $M$ PB for 1 h.
6. Sections are dehydrated through a graded series of ethanol solutions and propylene oxide.
7. Sections are finally embedded in a mixture of Epon-Araldite, mounted on slides, covered with a plastic cover slip, and cured for 24 h in the oven at 60°C.

8. Neurons positive for tyrosine hydroxylase staining in the caudate nucleus are identified and photographed.
9. Subsequently, a dissecting needle is used to etch a rectangle around the areas of interest to be popped off.
10. To perform the pop-off method, a plastic capsule is filled with liquid resin until the surface is slightly concave. The capsule is inverted, placed over the etched, marked area, and polymerized according to the plastic used.
11. Polymerized capsules are popped off the glass slides by dipping them quickly in and out of a Dewar containing liquid nitrogen.
12. Semithin and ultrathin sections are then cut with a ultramicrotome.
13. Ultrathin sections are mounted on grids stained with lead citrate and examined in a transmission electron microscope.

## IMMUNOGOLD DOUBLE-LABELING IMMUNOCYTOCHEMISTRY

A 41-year-old man with long-standing acromegaly and elevated blood GH and PRL levels had a pituitary tumor, which was removed by the transsphenoidal approach. Part of the specimen was processed for transmission electron microscopy. It was fixed overnight at 4°C in 2.5% glutaraldehyde in Sorensen's phosphate buffer (pH 7.4), osmicated for 1 h at room temperature with 1% $OsO_4$ in Millonig's buffer (pH 7.4), dehydrated in a graded ethanol series, processed through propylene oxide, and embedded in an Epon-Araldite mixture. After staining with uranyl acetate and lead citrate, the sections were examined in a Philips 410LS electron microscope.

Electron microscopy demonstrated that the tumor was bimorphous, consisting of cells with ultrastructural features of either lactotrophs or somatotrophs. To assess the existence of these two different cell types, immunocytochemical staining was performed on semithin plastic sections using anti-human GH (Dako, Carpinteria, CA) and anti-human PRL (donated by Dr. H. Friesen, University of Manitoba, Winnipeg, MAN, Canada) (dilutions: human anti-GH 1:500 and human anti-PRL 1:1000).

### Pretreatment of Sections

1. Semithin sections were etched for 15 min in a saturated solution of potassium hydroxide in ethanol and then washed several times in ethanol and in distilled water.
2. Subsequently, sections were incubated in 4% hydrogen peroxide in methanol for 30 min to remove osmium tetroxide and to block endogenous peroxidase activity.

### Immunocytochemical Labeling

3. Before the immunocytochemical procedure, sections were washed in TBS and preblocked in 3% BSA in TBS for 30 min.
4. Next, they were immunolabeled by the ABC technique *(36)*. First, consecutive semithin plastic sections were incubated with GH or PRL antibodies diluted in TBS with 0.5% BSA at 37°C overnight.
5. Following incubation with the primary antiserum, section were washed in TBS and incubated for 1 h in the biotinylated secondary antibody (dilution 1:100).
6. Subsequently the sections were washed again in TBS and incubated for 1 h in streptavidin-peroxidase complex (dilution 1:100).
7. The final reaction was achieved by incubating the sections for 5 min in a solution of 5 mg DAB and 1% hydrogen peroxide in 100 mL Tris buffer (pH 7.6).

8. Sections were then washed in TBS for 5 min and treated with 1% osmium tetroxide to increase the density of the precipitated DAB and improve localization of DAB deposits.

Study of consecutive sections stained with GH or PRL antibodies showed immuno-labeling for both hormones in the cytoplasm of the same cells, indicating the presence of bihormonal mammosomatotrophs in this pituitary tumor.

To confirm the presence of bihormonal cells and assess subcellular localization of GH and PRL in mammosomatotrophs, immunoelectron microscopy was performed using the double immunogold method of Bendayan *(48)*. In this protocol, it is essential to carry out all incubations by floating. It is important that the reagents should not reach the other face of the grid, to avoid excessive background. The incubations should also be undertaken in a humidified chamber to prevent drying of the immunoreagents. To localize two different antigens on the same ultrathin section by the immunogold method, each reaction must be performed on opposite surfaces of the section. However, when the double staining is carried out using primary antibodies raised in different species, the two reactions to detect different antigens can be done on the same surface of the ultrathin section. The gold-labeled secondary antibody is specific against different species, and crosslinking does not occur. It is also important to consider that in double immunostaining, the gold particles for the second antigen must be smaller than those used for the first antigen.

## Pretreatment of Sections

1. Before labeling, ultrathin sections are pretreated for 1 h in a saturated aqueous solution of sodium metaperiodate.
2. Subsequently sections are washed several times in distilled water for 20 min.

## Immunocytochemical Labeling

3. One side of the nickel grids are preblocked in 0.2 *M* PBS (pH 7.5) admixed with 0.2% cold water fish gelatin.
4. Next they are incubated at 37°C for 12 h with specific antisera directed toward either GH or PRL.
5. Grids are washed at least four times in 0.2 *M* PBS (pH 7.5) admixed with 0.2% cold water fish gelatin.
6. Subsequently, the grids are treated at 37°C for 1 h with gold-labeled, goat anti-rabbit IgG (Biocell Research Laboratories, Cardiff, UK); particle diameters should be either 10 or 20 nm.
7. Grids are washed again in 0.2 *M* PBS (pH 7.5) admixed with 0.2% cold water fish gelatin.

For double immunostaining, the procedure is repeated on the other side of the grid using another specific antibody and a colloidal gold conjugate of a different size. After immunolabeling is completed, sections are stained with uranyl acetate and examined on a Philips 410 LS electron microscope.

Electron microscopy demonstrates colocalization of GH and PRL in the same ade-noma cells. The two different hormones can be localized within the same secretory granules (**Figs. 1** and **2**).

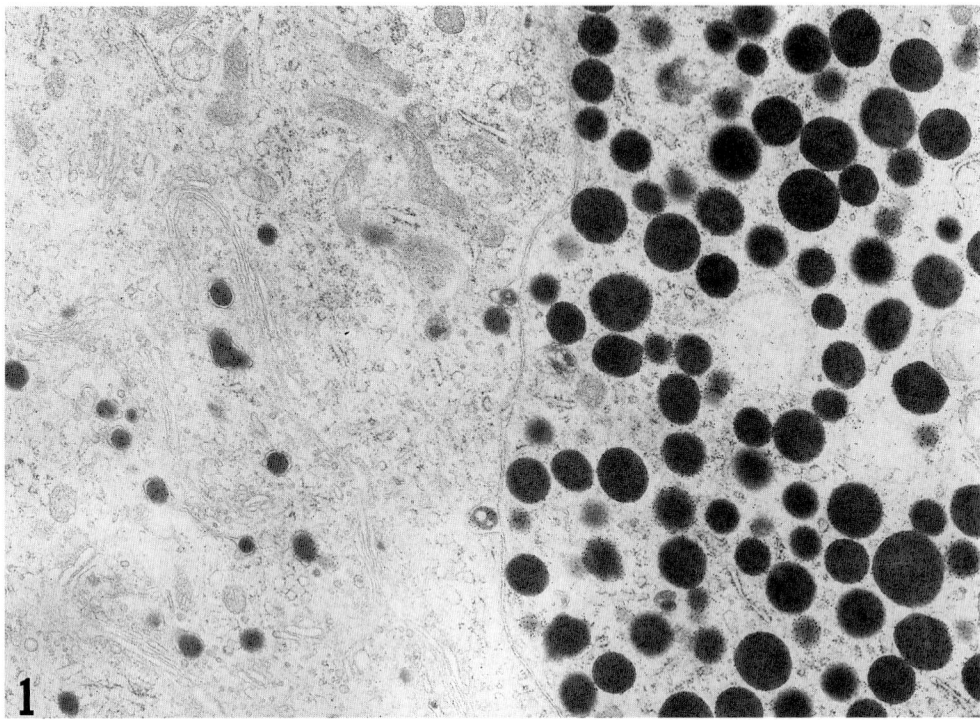

**Fig. 1.** Immunogold labeling in a GH- and PRL-producing human pituitary adenoma composed of mammosomatotrophs demonstrating that adenoma cells with different granule sizes are immunopositive for GH. Gold particles are 10 nm. Original magnification ×19,170.

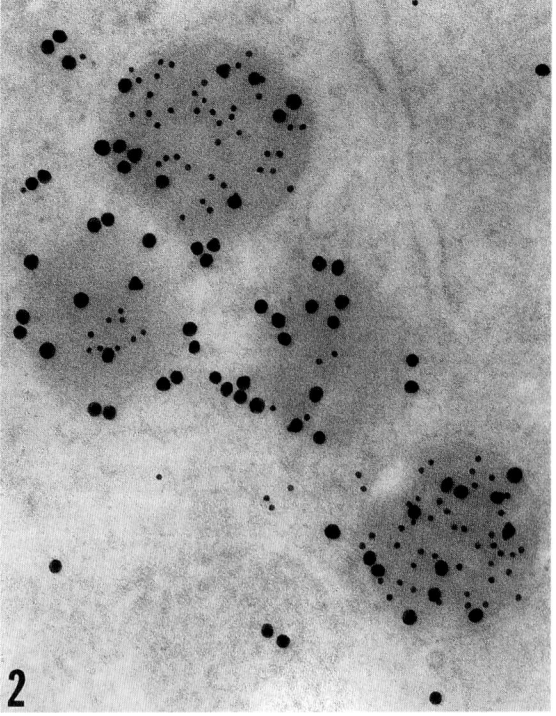

**Fig. 2.** Double immunogold labeling shows mammosomatotrophs in nontumorous human pituitary containing bihormonal granules storing growth hormone and prolactin. Growth hormone 10 nm; prolactin 20 nm. Original magnification ×139,500.

# 20
# Clonality Analysis

Aurel Perren, MD, and Paul Komminoth, MD

## INTRODUCTION

Assessment of clonality is an important factor in differentiating reactive from neoplastic lesions. In tumors with a known genetic defect, clonal expansion can be analyzed by the identification of the genetic changes such as translocations in soft tissue tumors or malignant lymphomas. For examining tumors without known genetic changes, a more general approach is needed.

To compensate for the double amount of X-chromosomal genes in females, either the maternally or paternally derived X chromosome has to be inactivated. This occurs during embryogenesis by random methylation of one X chromosome in each cell *(1,2)*. This process is stable during subsequent cell divisions *(3)*. Normal tissues in adult females are therefore cellular mosaics differing by which of the two X chromosomes is methylated. In contrast, neoplasms derived from a single somatic cell show a uniform pattern of X chromosome inactivation, indicating cellular monoclonality.

The first approach toward analysis of tissue clonality by X chromosome inactivation was based on isoenzymes of glucose-6-phosphate dehydrogenase (G6PD) *(4)*. The low frequency of heterozygosity (2%) for the G6PD isoenzymes and the need for fresh frozen tissue, however, made this approach impractical. All the following approaches take advantage of methylation-sensitive restriction enzymes that selectively cleave nonmethylated DNA, thus permitting discrimination between the active (unmethylated) and inactive (methylated) X chromosome.

First, Southern blotting of the restriction fragment length polymorphisms (RFLPs) in the phosphoglycerate-kinase 1 gene *(PGK-1)* and the hypoxanthine phosphoribosyl transferase *(HPRT)* gene was applied *(5)*. The variable repeat sequence in the Dx255 locus identified by the M27 β probe made clonality analysis more practical because of a high heterozygosity rate of approx 90%, but still all these Southern blotting approaches required high quantities of DNA and hence fresh tissue. The identification of a trinucleotide repeat (CAG$_n$ in exon 1 of the androgen receptor gene *(HUMARA)* *(6)* with nearby methylation-sensitive restriction endonuclease sites for *Hpa*II and *Hha*I *(7)* made a polymerase chain reaction (PCR)-based approach possible. These enzymes are sensitive to methylation of the cytosine residues in their recognition sequence; they only cut the DNA strand if their recognition sequence is not methylated *(8,9)*. Exon

From: *Morphology Methods: Cell and Molecular Biology Techniques*
Edited by: R. V. Lloyd © Humana Press, Totowa, NJ

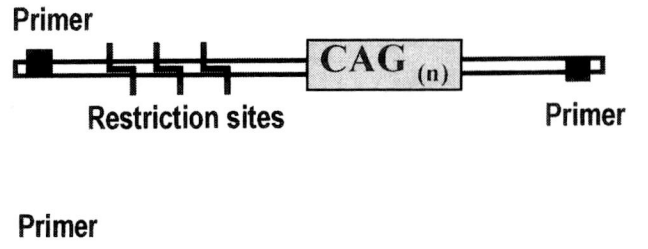

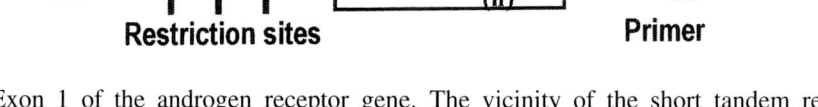

**Fig. 1.** Exon 1 of the androgen receptor gene. The vicinity of the short tandem repeat (STR) to the differentially methylated restriction sites for *Hha*I and *Hpa*II permits a specific amplification of the methylated (inactive) allele by restriction enzyme digestion prior to PCR.

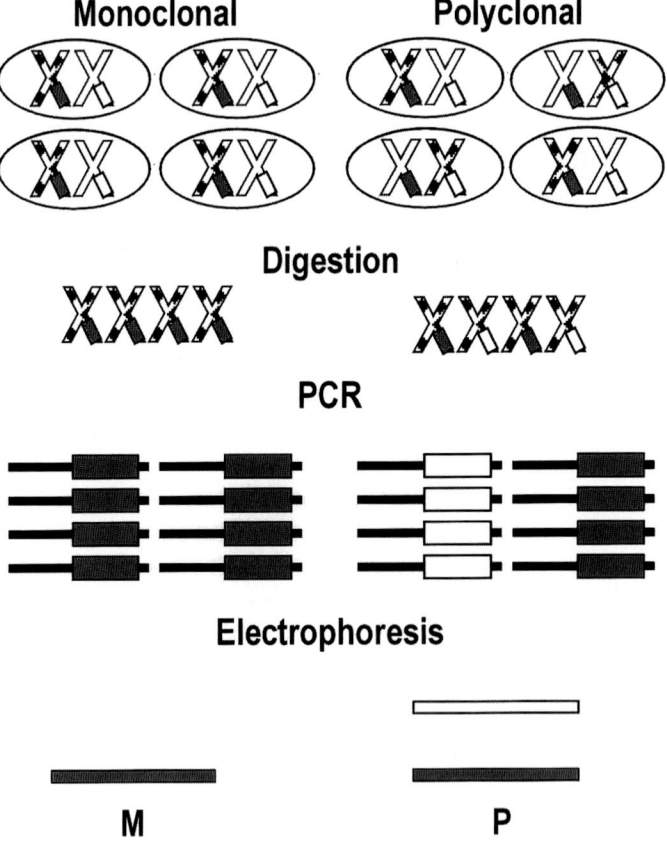

**Fig. 2.** Flow diagram showing expected PCR results in a monoclonal (left) and polyclonal (right) cell population. The active allele (white) of exon 1 of the androgen receptor gene is cleaved because the enzyme restriction sites are not methylated. Thus the methylated inactive allele is selectively amplified (hatched). The different sizes permit discrimination of the two alleles after gel electrophoresis. (Disappearance of one band after restriction enzyme digestion indicates monoclonality).

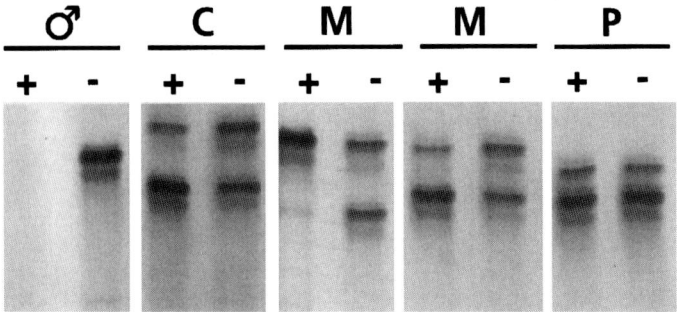

**Fig. 3.** Example of a *HUMARA* assay. +, PCR products of the *Hpa*II-digested samples; −, PCR products of undigested samples; ♂, digestion control using male DNA (PCR after digestion yields a negative result); C, monoclonal control tissue (lymphoma) exhibiting a monoclonal pattern; M, monoclonal samples. The left sample shows a clear monoclonal result, and the right one can only be recognized as monoclonal after direct comparison with the PCR products of nondigested DNA. The remaining upper band is explained by an admixture of "contaminating" normal tissue. P, Polyclonal tissue: note the stronger intensity of the lower band (preferential amplification?).

1 of the androgen receptor gene is shown schematically in **Fig. 1**. Approximately 90% of females are heterozygous with respect to the number of CAG repeats in this region. Restriction enzyme digestion by *Hha*I or *Hpa*II permits a specific amplification of the methylated (inactive) allele. Since the CAG repeats result in PCR products of different sizes from the two alleles, a second restriction enzyme digestion is not necessary. The PCR products can be analyzed directly by electrophoresis. The principle of PCR-based clonality analysis using this CAG repeat is shown in **Fig. 2**.

The critical step of this approach is the restriction enzyme digestion. Incomplete digestion will result in a pseudo-polyclonal pattern of a monoclonal neoplasm. Therefore extra care must be taken in performing this step, and appropriate controls must be used. We recommend a set of different controls: DNA of the same tissue type from a male patient must yield complete absence of PCR products after digestion (**Fig. 3**). Identically fixed tissue with known monoclonality of a female patient (for example, a lymphoma with monoclonal immunoglobulin rearrangement) must yield a monoclonal banding pattern (**Fig. 3**). Another critical issue is the "contamination" of a tumorous lesion with nonneoplastic tissues (connective tissue, blood vessels, inflammatory infiltrate, and so on). This can be solved by careful microdissection of tumor cell groups out of a given tissue. "Contamination" with up to 30–50% of normal tissue still leads to a recognizable monoclonal pattern *(10)*.

For interpretation of the polyacrylamide gels, direct comparison of PCR products of nondigested and digested sample DNA of the same patient is needed. In nondigested DNA, a difference in the intensity of bands can sometimes be observed. Preferential amplification of the smaller allele is the most likely explanation for this phenomenon *(11)*.

Skewing of X-chromosomal inactivation, leading to a monoclonal pattern in nonneoplastic tissues, is known to occur in females carrying X-linked diseases by nonrandom

X inactivation. This phenomenon has also been observed in granulocytes of healthy females, with increasing frequency with aging *(12)*.

Recently a different approach for examining the X-inactivation status has been described. To avoid the critical step of restriction enzyme digestion, a different methylation-sensitive PCR approach has been developed. In short, the DNA is treated with sodium bisulfite prior to PCR, to convert unmethylated, but not methylated cytosines to uracil. Then PCR amplification with primers specifically designed for methylated vs nonmethylated DNA is performed. For details, see refs. *13* and *14*.

## REFERENCES

1. Lyon, M. (1961) Gene action in the X-chromosome of the mouse (*Mus musculus* L.). *Nature* **190**, 372–373.
2. Gartler, S. M. and Riggs, A. D. (1983) Mammalian X-chromosome inactivation. *Annu. Rev. Genet.* **17**, 155–190.
3. Tsukada, M., Wade, Y., Hamade, N, et al. (1991) Stable Lyonization of X-linked pgk-1 gene during aging in normal tissues and tumors of mice carrying Searle's translocation. *J. Gerontol.* **46**, B213–B216.
4. Fialkow, P. J. (1977) Glucose-6-phosphate dehydrogenase (G-6-PD) markers in Burkitt lymphoma and other malignancies. *Haematol. Bluttransfus.* **20**, 297–305.
5. Vogelstein, B., Fearon, E. R., Hamilton, S. R, and Feinberg, A. P. (1985) Use of restriction fragment length polymorphisms to determine the clonal origin of human tumors. *Science* **227**, 642–645.
6. Tilley, W. D., Marcelli, M., Wilson, J. D., and McPhaul, M. J. (1989) Characterization and expression of a cDNA encoding the human androgen receptor. *Proc. Natl. Acad. Sci. USA* **86**, 327–331.
7. Allen, R. C., Zoghbi, H. Y., Moseley, H. B., et al. (1992) Methylation of HpaII and HhaI sites near the polymorphic CAG repeat in the human androgen-receptor gene correlates with X chromosome inactivation. *Am. J. Hum. Genet.* **51**, 1229–1239.
8. Makula, R. A. and Meagher, R. B. (1980) A new restriction endonuclease from the anaerobic bacterium, *Desulfovibrio desulfuricans*, Norway. *Nucleic Acids Res.* **8**, 3125–3131.
9. Garfin, D. E. and Goodman, H. M. (1974) Nucleotide sequences at the cleavage sites of two restriction endonucleases from *Hemophilus parainfluenzae*. *Biochem. Biophys. Res. Commun.* **59**, 108–116.
10. Enomoto, T., Fujita, M., Inove, M., et al. (1994) Analysis of clonality by amplification of short tandem repeats. Carcinomas of the female reproductive tract. *Diagn. Mol. Pathol.* **3**, 292–297.
11. Gale, R. E., Mein, C. A., and Linch, D. C. (1996) Quantification of X-chromosome inactivation patterns in haematological samples using the DNA PCR-based HUMARA assay. *Leukemia* **10**, 362–367.
12. Tonon, L., Bergamaschi, G., Dellavecchia, C., et al. (1998) Unbalanced X-chromosome inactivation in haemopoietic cells from normal women. *Br. J. Haematol.* **102**, 996–1003.
13. Uchida, T., Ohashi, H., Aoki, F., et al. (2000) Clonality analysis by methylation-specific PCR for the human androgen-receptor gene (HUMARA-MSP). *Leukemia* **14**, 207–212.
14. Kubota, T., Nonovarna, S., Tonoki, H., et al. (1999) A new assay for the analysis of X-chromosome inactivation based on methylation-specific PCR. *Hum. Genet.* **104**, 49–55.
15. Volante, M., Papotti, M., Roth, J., et al. (1999) Mixed medullary-follicular carcinoma. (1999) Molecular evidence for a dual origin of tumor components. *Am. J. Pathol.* **155**, 1499–1509.

## MATERIALS

### DNA Extraction of Paraffin Tissue

1. Extraction buffer
   a. 100 m$M$ Tris buffer, pH 8.0 (Trizma-Base, Sigma, St. Louis, MO).
   b. 5 m$M$ EDTA (Triplex III, Merck, Dietikon, Switzerland).
   c. 0.5% NP 40 (Nonidet + P40, BDH, Poole, UK).
   d. 0.5% Tween 20 (Merck).
2. Phenol (Fluka, Buchs, Switzerland).
3. Isoamylalcohol/chloroform 1:24 (Merck).
4. 100% ethanol (Merck).
5. 70% ethanol (Merck).
6. 3 $M$ Na-acetate, pH 5.2 (Fluka).
7. Glycogen, 20 mg/mL (Roche, Rotkreuz, Switzerland).
8. Proteinase K, 50 mg/mL (Roche).
9. Tris buffer, 10 m$M$, pH 8.3 (Sigma).
10. Freezer at −20°C.
11. Freezer at −80°C.
12. Centrifuge.
13. Cold room at 4°C.
14. Spectrophotometer at 260λ and 280λ.

### DNA Extraction of Fresh Frozen Tissue

1. Cell lysis solution (Purgene-Kit, Gentra Systems, Minneapolis, MN).
2. Proteinase K, 50 mg/mL (Roche).
3. RNase, 4 mg/mL (Purgene-Kit).
4. Protein precipitation solution (Purgene-Kit).
5. Isopropanol, 100% (Fluka).
6. Ethanol, 70% (BDH).
7. DNA hydration solution (Purgene-Kit).

### Restriction Enzyme Digestion

1. Restriction enzymes (see *Note 1*).
   a. *Hpa*II (Roche).
   b. *Hha*I/*Cfo*I (Roche).
2. Incubation buffer, 10× (Roche).
3. Ethanol, 100% (BDH).
4. Ethanol, 70% (BDH).
5. Na acetate, 3 $M$ pH 5.2 (Fluka).
6. Glycogen, 20 mg/mL (Roche).
7. Tris-buffer, 10 m$M$, pH 8.3 (Sigma).
8. Freezer −80°C.
9. Incubator 37°C.

### PCR Amplification

1. Oligonucleotide primers, 20–50 pmol/μm (**Table 1**).
2. dNTP, 2 m$M$ (Roche).
3. MgCl$_2$, 25 m$M$ (Perkin-Elmer, Rotkreuz, Switzerland)
4. PCR buffer II, 10× (without MgCl$_2$, Perkin-Elmer).
5. Taq DNA polymerase (Boehringer Mannheim, Mannheim, Germany) (see *Note 2*).

**Table 1**
**Oligonucleotide Primers**

| Primer | Position | Orientation | Sequence |
|--------|----------|-------------|----------|
| AR 1F | 221–240 | Sense | 5′ GAGGAGCTTTCCAGAATCTG 3′ |
| AR 1R | 454–439 | Antisense | 5′ GATGGGCTTGGGGAGA 3′ |

6. 50 μL PCR tubes (Gene Amp thin-walled reaction tubes, Perkin-Elmer).
7. Thermo Cycler (Gene Amp PCR System 9600, Perkin-Elmer).

## Gel Electrophoresis

1. TBE 10×: 0.89 $M$ Trizma base (Sigma), 0.89 $M$ boric acid (Sigma), 0.02 $M$ EDTA Tritriplex III (Merck).
2. Acrylamide, 40% (Bio-Rad, Glattbrugg, Switzerland).
3. Bisacrylamide, 2% (Bio-Rad).
4. Urea (Merck).
5. APS, 10% (Sigma).
6. Tetramethylethylenediamine (TEMED; Sigma).
7. Blue stop loading buffer:
    a. 95% formamide (Sigma).
    b. 20 m$M$ EDTA (Merck).
    c. 0.05% bromphenol-blue (Sigma).
    d. 0.05% xylencyanol (Sigma).

## Silver Staining

1. Glass well.
2. Rain-X anti-Fog (Unelka, Belgium).
3. Saran wrap foil (Dow).
4. Ethanol, 10% (BDH).
5. $HNO_3$, 1% (Merck).
6. $AgNO_3$, 0.2% (Merck) (dissolve 2.02 g $AgNO_3$ in 1 l $dH_2O$ and filter).
7. Filter (Nalgene, Nalge Nuc, Rochester, NY).
8. Developer:
    a. 0.28 $M$ $Na_2CO_3$, 0.02% formaldehyde
    b. 29.6 g $Na_2CO_3$ (Merck) in 800 mL $dH_2O$
    c. 540 μL formaldehyde (Merck).
    d. $dH_2O$ to get 1 l.
9. $CH_3COOH$, 10% (Merck).
10. EDTA, 50 m$M$ pH 8.0 (Merck).
11. $dH_2O$.

## METHODS

### DNA Extraction of Paraffin Tissue

*Tissue Preparation*

Identify tumor and normal tissue on hematoxylin and eosin-stained section

1. Prepare 10-μm serial sections.
2. Use 3–10 sections (according to size of the area) to scrape off the desired area and transfer the tissue to an 1.5-mL Eppendorf tube.

To examine small areas it is possible to use eosin (see *Note 3*) or immunostained deparaffinized sections: cover the tissue with 1 drop of mineral oil (Sigma) and scrape the tissue off under microscopic assistance using an injection needle (24 gauge, Terumo, Leuven, Belgium) (see *Note 4*).

## *Paraffin Removal*

1. Add 500 μL xylene.
2. Dissolve paraffin for 10 min with gentle agitation.
3. Spin down for 10 min at 13,100*g* (14,000 rpm) and then carefully remove xylene.
4. Repeat **steps 5** and **6** twice.
5. Add 500 μL 100% ethanol.
6. Shake gently for 10 min and remove ethanol.
7. Add 500 μL 70% ethanol.
8. Shake gently for 10 min and remove ethanol.
9. Speed Vac for 5 min to dry the tissue.

## *Proteinase K Digestion*

1. Add 500 μL extraction buffer.
2. Add 5 μL proteinase K.
3. Incubate at 54°C overnight in a shaker.
4. Repeat **steps 2** and **3** twice to achieve complete digestion.
5. Heat at 95°C for 10 min to inactivate proteinase K.

## *Phenol/Isoamylalcohol/Chloroform Purification*

1. Add 500 μL phenol and vortex well.
2. Spin for 2 min at 13,100*g* (14,000 rpm).
3. Pipet supernatant in new Eppendorf tube (only take the aqueous phase)
4. Add 250 μL phenol and 250 μL isoamylalcohol/chloroform and vortex.
5. Spin for 2 min at 13,100*g* (14,000 rpm) and pipet supernatant in a new tube.
6. Add 500 μL chloroform and vortex.
7. Spin for 2 min at 13,100*g* (14,000 rpm) and pipet supernatant in a new tube.

## *DNA Precipitation*

1. Add 1000 μL cold (−20°C) ethanol 100%, 50 μL Na-acetate, and 1 μL glycogen.
2. Vortex well and incubate at −20°C overnight.
3. Spin for 10 min at 14,000 rpm in a cold centrifuge (4°C).
4. Gently drain off the ethanol (DNA is visible as small pellet).
5. Add 500 μL cold ethanol 70%; vortex briefly.
6. Spin for 10 min at 13,100*g* (14,000 rpm) in a cold centrifuge (4°C).
7. Gently drain off the ethanol.
8. Air-dry pellet for 15 min; alternatively, Speed Vac for 5 min.
9. Add 20–50 μL Tris buffer (according to expected amounts of DNA).
10. Dissolve DNA at 65°C for 10 min.

## *DNA Concentration*

1. Pipet 5 μL of DNA and 495 μL ddH$_2$O in a new Eppendorf tube.
2. Measure O/D at λ260 nm and λ280 nm.
3. DNA quantity of the stock solution equals λ260 × 5 (in μg/μL).
4. Ratio of λ260:λ280 is a measure of DNA purity and should be between 1.8 and 2.
5. Prepare DNA working solutions by adding ddH$_2$O to get a concentration of 0.1 μg/μL.

## DNA Extraction of Fresh Frozen Tissue

*Tissue Preparation*

1. Put 1 mm$^3$ of fresh frozen tissue in Eppendorf tube.
2. Homogenize tissue.
3. Add 600 µL cell lysis solution.

*Proteinase K Digestion*

1. Add 5 µL of proteinase K.
2. Incubate at 54°C overnight in a shaker.
3. Repeat **steps 1** and **2** twice to achieve complete digestion.

*DNA Purification*

1. Add 3 µL of RNase.
2. Incubate at 37°C for 15 min.
3. Add 200 µL protein precipitation solution and vortex immediately.
4. Spin at 13,000*g* (14,000 rpm) for 3 min.
5. Pipet supernatant in new Eppendorf tube.

*DNA Precipitation*

1. Add 600 µL of isopropanol and immediately mix by gentle inversion until you can see a white precipitate of DNA.
2. Spin down at 13,100*g* (14,000 rpm) for 1 min and discard supernatant.
3. Add 600 µL cold (−20°C) ethanol 70% and vortex.
4. Spin down at 13,100*g* (14,000 rpm) for 1 min and discard supernatant.
5. Air-dry pellet for 15 min; alternatively, Speed Vac for 5 min.
6. Add 100 µL of DNA hydration solution.
7. Dissolve DNA at 37°C for 15 min.

*DNA Concentration*

1. Pipet 5 µL of DNA and 495 µL ddH$_2$O in a new Eppendorf tube.
2. Measure O/D at λ260 nm and λ280 nm.
3. DNA quantity of the stock solution equals λ260 × 5 (in µg/µL).
4. Ratio of λ260:λ280 is a measure of DNA purity and should be between 1.8 and 2.
5. Prepare DNA working solutions by adding ddH$_2$O to get a concentration of 0.2 µg/µL.

## Restriction Enzyme Digestion of the Active Allele

This is the most critical step; therefore *it is very important to use appropriate controls.*

*Restriction Enzyme Digestion*

The following steps are done on ice.

1. Pipet 1 µg of sample DNA (5 µL of DNA working solution) in Eppenorf tube.
2. Add 40 µL of ddH$_2$O.
3. Add 5 µL of incubation buffer 10×.
4. Add 10 U (0.25 µL) of *Hpa*II; alternatively, use *Hha*I (see *note 1*).
5. Vortex and incubate at 37°C overnight. Be careful that no water condensation forms in the cap of the tube.
6. Heat to 94°C for 10 min to inactivate the enzyme.

*DNA Precipitation*

1. Add 125 µL cold ethanol 100% (−20°C).
2. Add 5 µL Na-acetate.

3. Add 1 μL glycogen.
4. Vortex well and precipitate at −80°C for 1 h.
5. Spin for 10 min at 13,100*g* (14,000 rpm) in a cold centrifuge (4°C).
6. Gently drain off the ethanol (DNA is visible as small pellet).
7. Add 55 μL cold ethanol 70%; vortex briefly.
8. Spin for 10 min at 13,100*g* (14,000 rpm) in a cold centrifuge (4°C).
9. Gently drain off the ethanol.
10. Air-dry pellet for 15 min; alternatively, Speed Vac for 5 min.
11. Add 10 μL Tris buffer (giving an approximate DNA concentration of 0.07 μg/μL).
12. Dissolve DNA at 65°C for 10 min.

## PCR Amplification

### PCR Preparation

1. Use 3 μL of digested DNA (approx 0.2 μg).
2. Add 5 μL PCR buffer 10×.
3. Add 3 μL $MgCl_2$.
4. Add 5 μL dNTP.
5. Add 33.8 μL $ddH_2O$.
6. Add 1 μL of each primer.
7. Add 0.2 μL of Taq polymerase.

### Cycling Conditions

1. Initial denaturation and activation of Taq polymerase at 94°C for 300 s.
2. 35 cycles of 3–5 min.
3. Denaturation at 94°C for 45 s.
4. Annealing at 59°C for 45 s.
5. Extension at 72°C for 45 s.
6. Final extension at 72°C for 300 s.

## Gel Electrophoresis

1. For 150 mL of gel use 15 mL of TBE 10×.
2. Add 20 mL acrylamide 40%.
3. Add 20 mL bisacrylamide 2%.
4. Add 63 g urea.
5. Add 150 mL $dH_2O$ and stir till urea is dissolved.
6. Add 1 mL ammonium persulfate (APS) 10%.
7. Add 70 μL tetraethylenediamine (TEMED).
8. Pour gel between glass plates.
9. Let polymerize at room temperature for 1 h.
10. Preheat gel at 80 W for 30 min.
11. Mix 20 μL of PCR product with 2 μL blue stop.
12. Load probes.
13. Run at 80 W for 2 h.

## Silver Staining

1. Wipe glass well with Rain-X to get a hydrophobic film.
2. Pour ethanol 10% in glass well. Use enough solution to have the gel immersed at all times; after incubation, remove the solution using a vacuum pump.
3. Transfer gel to glass well with ethanol using foil; incubate for 5 min with gentle agitation.
4. Wash 2× in $dH_2O$ for 10 min each to remove urea.
5. Incubate in $HNO_3$ for 3 min.

6. Wash 2× in dH$_2$O.
7. Incubate in AgNO$_3$ for 30 min.
8. Wash 2× in dH$_2$O.
9. Rinse with developer; immediately remove precipitate.
10. Add developer again and incubate until the bands are visible.
11. Incubate in CH$_3$COOH 10% for 10 min to stop development.
12. Incubate in EDTA for 5 min.
13. Wash 2× in dH$_2$O.
14. The stained gel can be archived in sealed plastic foil.

## NOTES

1. *Hha*I and *Cfo*I are isoschizomeres. In our settings, the most consistent results were obtained with *Hpa*II alone compared with *Hha*I and a mixture of both enzymes.
2. The use of a Taq polymerase of a different company requires adaptation of annealing temperatures.
3. Use alcoholic eosin; exposure to acid will cause DNA damage.
4. For analysis of very small tissue areas, laser-based microdissection can be used (PALM Laser-Microbeam Systems, Germany) with modified protocols for DNA extraction and restriction enzyme digestion *(15)*.

# Index